BIOLOGICAL INFORMATION

NEW PERSPECTIVES

BIOLOGICAL INFORMATION

NEW PERSPECTIVES

Proceedings of a Symposium held May 31 through June 3, 2011 at Cornell University

Editors

Robert J. Marks II
Baylor University, USA

Michael J. Behe
Lehigh University, USA

William A. Dembski
Discovery Institute, USA

Bruce L. Gordon
Houston Baptist University, USA

John C. Sanford
Cornell University, USA

World Scientific

NEW JERSEY • LONDON • SINGAPORE • BEIJING • SHANGHAI • HONG KONG • TAIPEI • CHENNAI

Published by

World Scientific Publishing Co. Pte. Ltd.
5 Toh Tuck Link, Singapore 596224
USA office: 27 Warren Street, Suite 401-402, Hackensack, NJ 07601
UK office: 57 Shelton Street, Covent Garden, London WC2H 9HE

Library of Congress Cataloging-in-Publication Data
Biological information--new perspectives : proceedings of a symposium held May 31, 2011 through June 3, 2011 at Cornell University / Robert J. Marks II, Baylor University, USA, Michael J. Behe, Lehigh University, USA, William A. Dembski, Discovery Institute, USA, Bruce L. Gordon, Houston Baptist University, USA John C. Sanford Cornell University, USA.
pages cm
Includes bibliographical references and index.
ISBN 978-9814508711 (hardcover : alk. paper)
1. Genomics--Congresses. 2. Molecular genetics--Congresses. 3. Cell interaction--Congresses. 4. Mutation (Biology)--Congresses. 5. Intelligent design (Teleology)--Congresses. I. Marks, Robert J., II (Robert Jackson), 1950–
QH426.B58 2013
572.8'629--dc23

2013016707

British Library Cataloguing-in-Publication Data
A catalogue record for this book is available from the British Library.

Typeset by Stallion Press
Email: enquiries@stallionpress.com

Printed by FuIsland Offset Printing (S) Pte Ltd Singapore

Biological Information — New Perspectives

Proceedings of a Symposium held May 31, 2011 through June 3, 2011 at Cornell University

Edited by

Robert J. Marks II
Baylor University
Department of Electrical & Computer Engineering
Waco, TX, USA

Michael J. Behe
Lehigh University
Department of Biological Sciences
Bethlehem, PA, USA

William A. Dembski
Discovery Institute
Seattle, WA, USA

Bruce L. Gordon
Houston Baptist University
Department of Philosophy
Houston, TX, USA

and

John C. Sanford
Cornell University, NYSAES
Department of Horticultural Science
Geneva, NY, USA

Franzine D. Smith — Technical Editor
Chase W. Nelson — Editorial Assistant

Contents

Acknowledgements

The Editors would like to thank Cornell University's Conference Services and Cornell University's Statler Hotel for their excellent hosting of this symposium.

We thank the various sponsors of this symposium, including: Physicians and Surgeons for Scientific Integrity; Rainbow Technologies, Inc., Aldevron, Inc.; FMS Foundation, and other anonymous donors.

We thank George Montañez and Chris Rupe for the cover artwork.

General Introduction

Biological information, though still in its infancy as a field of study, is widely thought to be well understood in its broad outlines. The conventional, or "old," perspective on biological information is this: biological information, in the first instance, originates through purely chemical processes. These processes produce the first replicators. Once replication is in place, the Darwinian process of natural selection acting on random variation kicks in, continually increasing the information content of these replicators. Eventually, the information generating power of chemical and Darwinian processes results in organisms as complex and sophisticated as human beings. The origin, structure, and dynamics of biological information is thus thought to reduce to a combination of stochastic chemistry and undirected evolutionary forces.

This perspective on biological information is the majority position in the scientific community. Often it fails to be fully articulated because research on chemical evolution (the chemical processes responsible for first life and thus for the first biological information) and biological evolution (the evolutionary mechanisms responsible for the subsequent history of life and thus for the increase of existing biological information) tend to be conducted by different sets of scientists with different areas of expertise. Nonetheless, one occasionally finds this perspective articulated not in pieces but fully. Nobel laureate and origin-of-life researcher Christian de Duve is a case in point. In his book *Vital Dust*, he lays out various "ages" in the history of life: The Age of Chemistry, The Age of Information, The Age of the Protocell, The Age of the Single Cell, etc. Note that chemistry starts the ball rolling and precedes information. De Duve elaborates:

> History is a continuous process that we divide, in retrospect, into ages — the Stone Age, the Bronze Age, the Iron Age — each characterized by a major innovation added to previous accomplishments. This is true also of the history of life. . . . First, there is the Age of Chemistry. It covers the formation of a number of major constituents of life, up to the first nucleic acids, and is ruled entirely by the universal principles that govern the behavior of atoms and molecules. Then comes the Age of Information, thanks to the development of special information-bearing molecules that inaugurated the new processes of Darwinian evolution and natural selection particular to the living world. [1]

The conventional perspective on biological information tends more often to be articulated in pieces. Thus Harvard chemist George Whitesides, focusing on his expertise in chemistry and setting aside the subsequent history of life, speaks to the origin of life and thus to the origin of the first biological information: "This problem [of life's origin] is one of the big ones in science. It begins to place life, and us, in the universe. Most chemists believe, as do I, that life emerged spontaneously from mixtures of molecules in the prebiotic Earth. How? I have no idea." Though short on details, Whitesides is nonetheless confident that his perspective is correct: "I believe that understanding the cell is ultimately a question of chemistry and that chemists are, in principle, best qualified to solve it. The cell is a bag — a bag containing smaller bags and helpfully organizing spaghetti — filled with a Jell-O of reacting chemicals and somehow able to replicate itself." [2]

Once life has originated and biological information is on hand, the subsequent history of life displays massive increases in information content. To explain these increases, the conventional perspective on biological information takes a thoroughly Darwinian line, elevating natural selection as the primary engine for information generation over the course of biological evolution. Richard Dawkins articulates this view as follows:

> In every generation, natural selection removes the less successful genes from the gene pool, so the remaining gene pool is a narrower subset. The narrowing is nonrandom, in the direction of improvement, where improvement is defined, in the Darwinian way, as improvement in fitness to survive and reproduce. Of course the total range of variation is topped up again in every generation by new mutation and other kinds of variation. But it still remains true that natural selection is a narrowing down from an initially wider field of possibilities, including mostly unsuccessful ones, to a narrower field of successful ones. This is analogous to the definition of information with which we began: information is what enables the narrowing down from prior uncertainty (the initial range of possibilities) to later certainty (the "successful" choice among the prior probabilities). According to this analogy, natural selection is by definition a process whereby information is fed into the gene pool of the next generation. [3]

This is the conventional, or old, perspective on the origin and evolution of biological information. All the contributors to this volume question this perspective. In its place, they propose various new perspectives — plural. Some take a clearly teleological approach, advocating intelligent agent causation as the ultimate source of biological information. Others view information as *sui generis*, as a fundamental entity not reducible to purely material factors such as chemical attraction and natural selection. And others still, while accepting a big chunk of the old perspective, think that it needs to be supplemented with self-organizational processes whose information generating powers transcend those of the old perspective. The contributors, rather than presenting a united front, attempt to explore new ground and ask insightful new questions.

But if the old perspective is so well established, why question it? Is it not a sign of recalcitrance to contradict well settled verities of the scientific community? Certainly, this can be a danger. But it is a danger only when those raising the questions are ill-informed and unqualified in the relevant sciences, and have as their main motive to derail rather than foster genuine scientific inquiry. That is the not the case with any of the contributors to this volume. Science progresses not by acceding to consensus but by breaking with it. Moreover, even with well settled scientific theories, it is healthy for science periodically to question whether those theories really hold up.

In any case, there are good reasons, readily accessible to non-experts, for thinking that the old perspective on biological information bears closer scrutiny and may well be false. Take the origin of life, where all biological information begins. Origin-of-life researchers readily admit that they don't know how life began. True, they entertain speculative ideas about life's origin, with RNA-worlds currently heading the pack. But no one in the field claims to have a precisely formulated theory with solid evidential support that explains life's origin.

Thus, Stuart Kauffman, a contributor to this volume, writes, "Anyone who tells you that he or she knows how life started on the earth some 3.45 billion years ago is a fool or a knave. Nobody knows." [4] Origin-of-life researcher Leslie Orgel similarly held that "anyone who thinks they know the solution to this problem is deluded." [5] Or consider science writer Paul Davies: "We are a very long way from comprehending the how [of life's origin]. This gulf in understanding is not merely ignorance about certain technical details, it is a major conceptual lacuna… My personal belief, for what it is worth, is that a fully satisfactory theory of the origin of life demands some radically new ideas." [6]

The origin of life is the most vexing problem facing contemporary science. It has fiercely resisted reductionist approaches to its resolution. All attempts to get life started solely through life's underlying chemistry have come up short. Could it be that although chemistry provides the medium for biological information, the information itself constitutes a message capable of riding free from the underlying medium? Could such information be a real entity — as real as the chemical constituents that embody it, and yet not reducible to them — and, dare we say, have an intelligent cause? Granted, this is itself a speculative possibility, but in a field so rife with speculation, why allow only one set of speculations (those that adhere to the old perspective) and disallow others (those that open up new possibilities)? The contributors to this volume are not offering final answers. Rather, they are raising penetrating questions precisely where the old perspective has failed to offer a promising starting point for understanding the origin of biological information.

Even so, once biological information comes on the scene at the origin of first life, don't we have a well supported theory for the increase of biological information via the Darwinian mechanism of natural selection acting on random variation? In fact, even here the old perspective on biological information comes up short. The problem, as University of Chicago molecular biologist James Shapiro notes in *Evolution: A View from the 21st Century*, is that Darwinism constitutes an oversimplification: "Molecular evidence about genome sequence changes tell us that the simplifying assumptions made in the 19th and early 20th Centuries are plainly wrong. They fail to account for the variety of cellular and genomic events we now know to have occurred." [7] Shapiro continues:

> Living cells and organisms are cognitive (sentient) entities that act and interact purposefully to ensure survival, growth, and proliferation. They possess corresponding sensory, communication, information-processing, and decision-making capabilities. Cells are built to evolve; they have the ability to alter their hereditary characteristics rapidly through well-described natural genetic engineering and epigenetic processes as well as cell mergers. [8]

The picture of life and evolution that Shapiro presents is radically at odds with the old perspective on biological information. Shapiro is not alone. Many biologists are now questioning whether conventional evolutionary theory needs to be rethought from the ground up, notably the "Altenberg 16," who started out as mainstream biologists wedded to the old perspective, but now have jumped ship because the old perspective is no longer working, at least not for them. [9]

So too, notable outsiders are beginning to question whether the old perspective is disintegrating before their very eyes. Thus Robert Laughlin, a Nobel laureate physicist who studies the properties of matter that make life possible, remarks:

> Evolution by natural selection, for instance, which Charles Darwin originally conceived as a great theory, has lately come to function more as an antitheory, called upon to cover up embarrassing experimental shortcomings and legitimize findings that are at best questionable and at worst not even wrong. Your protein defies the laws of mass action? Evolution did it! Your complicated mess of chemical reactions turns into a chicken? Evolution! The human brain works on logical principles no computer can emulate? Evolution is the cause! [10]

Note that Laughlin himself does not disavow evolution. His beef is with ill-considered conceptions of evolution and the facile use of "evolution" as a magic word to conjure away hard scientific problems, when doing so in fact merely cloaks ignorance.

Even Francisco Ayala, an otherwise staunch Neo-Darwinist (himself a protégé of Theodosius Dobzhansky, one of the key architects of the neo-Darwinian synthesis), now questions whether evolutionary theory requires fundamentally new insights: "Unfortunately, there is a lot, lot, lot to be discovered still. To reconstruct evolutionary history, we have to know how the mechanisms operate in detail, and we have only the vaguest idea of how they operate at the genetic level, how genetic change relates to development and to function. I am implying that what would be discovered would be not only details, but some major principles." [11]

In the spring of 2011 a diverse group of scientists gathered at Cornell University with an eye on the major new principles that might be required to unravel the problem of biological information. These scientists included experts in information theory, computer science, numerical simulation, thermodynamics, evolutionary theory, whole organism biology, developmental biology, molecular biology, genetics, physics, biophysics, mathematics, and linguistics. Original scientific research was presented and discussed at this symposium, which was then written up, and constitute most of the twenty-four peer-edited papers in this volume. These papers are presented in four sections: Information Theory and Biology, Biological Information and Genetic Theory, Theoretical Molecular Biology, and Self-Organizational Complexity Theory. Each of these sections begins with an introductory chapter laying out the themes and problems to be discussed there as well providing brief summaries of the papers appearing in that section.

Many of the papers in this volume speak of biological information in the limited context of the multi-dimensional array of information encoded within a cell's genome. Nevertheless, if we define information more broadly as "all that which is communicated," the information within a living cell is much greater than its DNA sequence. All the components of the cell, including all the RNA and protein molecules, are continuously communicating with each other. It is recognized that there are hundreds of thousands of different types of interactions within the cell's "interactome," and most of these interactions in one way or another involve communication. In this sense, the amazing communication network within a cell can very reasonably be compared to the Internet.

If we extend the computer science analogy further, we can consider the genome as stored information (the "hard drive" of the cell), while the RNA, protein, and other structures can be considered the "active information" (the RAM of the cell). While many of the papers given at this symposium deal with the information within the genome, it is very important we do not forget that most biological information in the cell is above and beyond the genome. On a level entirely above and beyond all this communicated information within the cell, information is also being communicated between cells, and between organisms. On a still higher level, we have the little-understood biological information that underlies the human mind, our own intelligence, and human consciousness. All of this is biological information! There exists an unknown number of symbolic languages (the genetic code being just one of many biological codes) underlying this astounding communication labyrinth integrating all levels of biological information.

All this talk about information as a real object of study within the field of biology, however, raises the question, What exactly is information in the first place? Is it a precisely defined measurable entity? Can the study of biological information be turned into an exact science? Does biological information connect meaningfully with information theory as understood in the mathematical and engineering sciences? As University of Texas philosopher of biology Sahotra Sarkar rightly notes, "It is incumbent upon those who think that informational concepts have theoretical value in biology (that is, they explain things rather than being merely metaphors) to produce an appropriate technical concept of information for biological contexts." [12] The first section of this volume is devoted to precisely this concern. Keying off of research on evolutionary search, No Free Lunch theorems, and Conservation of Information, this section attempts to provide the theoretical underpinnings for a full-fledged theory of biological information.

In the last decades, it has become clear that biological information is crucial to our understanding of life. On completion of the Human Genome Project, former Caltech president and Nobel Prize-winning biologist David Baltimore remarked, "Modern biology is a science of information. The sequencing of the genome is a landmark of progress in specifying the information, decoding it into its many coded meanings and learning how it goes wrong in disease. While it is a moment worthy of the attention of every human, we should not mistake progress for a solution. There is yet much hard work to be done..." [13] The contributors to this volume agree and desire that their efforts here will inspire much hard work on the greater project of providing a full-fledged theory of biological information, one that is free of ideological bias and gets at the truth of the matter.

— *The Editors*

References and Notes

1. de Duve C (1995) Vital Dust: Life as a Cosmic Imperative, p. 10. Basic Books, New York.
2. Whitesides GM (2007) Revolutions in Chemistry (Priestly Medalist address). Chem & Eng News 85(13):12–17, available online at http://pubs.acs.org/cen/coverstory/85/8513cover1.html (last accessed November 10, 2011).

3. Dawkins R, "The Information Challenge," published through the Australian Skeptics and available online at http://www.skeptics.com.au/publications/articles/the-information-challenge (last accessed November 10, 2011).
4. Kauffman S (1995) At Home in the Universe: The Search for the Laws of Self-Organization and Complexity, p. 31. Oxford University Press, New York.
5. Quoted in Jason Socrates Bardi, "Life — What We Know, and What We Don't," *TSRI News & Views* (the online weekly of The Scripps Research Institute) 3(30) (October 11, 2004), available online at http://www.scripps.edu/newsandviews/e_20041011/ghadiri.html (last accessed November 11, 2011).
6. Davies P (1999) The Fifth Miracle: The Search for the Origin and Meaning of Life, p. 17. Simon & Schuster, New York.
7. Shapiro JA (2011) Evolution: A View from the 21st Century, p. 128. FT Press Science, Upper Saddle River, NJ.
8. Ibid., 143.
9. Mazur S (2010) The Altenberg 16: An Exposé of the Evolution Industry. North Atlantic Books, Berkeley, CA.
10. Laughlin RB (2005) A Different Universe: Reinventing Physics from the Bottom Down, pp. 168–169. Basic Books, New York.
11. In an interview, reported in Larry Witham, *Where Darwin Meets the Bible: Creationists and Evolutionists in America* (Oxford: Oxford University Press, 2002), 90.
12. Sarkar S (2007) Doubting Darwin? Blackwell, Oxford, p. 119.
13. Baltimore D, "DNA is a Reality beyond Metaphor," *Caltech and the Human Genome Project* (2000): available online at http://marcomm.caltech.edu/events/dna/dnabalt2.html (last accessed November 11, 2011).

Section One — Information Theory & Biology: Introductory Comments

Robert J. Marks II — Section Chairman

All agree there is information in biological structure and function. Although the term *information* is commonly used in science, its precise definition and nature can be illusive, as illustrated by the following questions:

- When a paper document is shredded, is information being destroyed? Does it matter whether the shredded document is a copy of an un-shredded document and can be replaced?
- Likewise, when a digital picture is taken, is digital information being created or merely captured?
- The information on a DVD can be measured in bits. Does the amount of information differ if the DVD contains the movie *Braveheart* or a collection of randomly generated digital noise?
- When a human dies, is experiential information lost? If so, can birth and experience create information?
- If you are shown a document written in Japanese, does the document contain information whether or not you know Japanese? What if instead, the document is written in an alien language unknowable to man?

The answers to these questions vary in accordance to the information model used. However, there are properties of information common to all models. As noted by Norbert Weiner [1, 2], the father of cybernetics:

> "Information is information, neither matter nor energy."

Information can be written on energy. Examples include wireless electromagnetic waves and audio waves that carry the content of conversations. As is the case with books and DVD's, information can also be etched onto matter. But energy and matter serve only as transcription media for information. Information can also reside in structure or phenomena. Varying degrees of information are available in nature. A bacterium obviously contains more information than a grain of sand. Information can be extracted from inspection of information-rich sources. The idea for Velcro came from close examination of burrs stuck to the clothes of a Swiss engineer after a hunting trip [3]. The function of the human eyelid was the inspiration for invention of the intermittent windshield wiper [4]. *The IEEE*

Computational Intelligence Society [5], a professional electrical and computer engineering organization,[1] has as its motto, "Nature inspired problem solving." The implication is that structure in nature, when examined, can be a rich source of information applied to engineering. Unlike mass and energy in physics, a single model or definition of information does not exist. Claude Shannon recognized his theory was not the last word in the mathematical modeling of information [6]:

> "It seems to me that we all define 'information' as we choose; and, depending upon what field we are working in, we will choose different definitions. My own model of information theory... was framed precisely to work with the problem of communication."

Shannon Information

Because of its widespread application and depth of mathematical rigor, the most celebrated information model is Shannon information theory. In an astonishing 1948 paper [7], Claude Shannon single-handedly founded a discipline still celebrated today by professional organizations such as the *IEEE Information Theory Society* who has published The IEEE TRANSACTIONS ON INFORMATION THEORY since the mid-1950's. Shannon's original paper is remarkable. The word *bit*, a contraction of *binary digit*, was first used in this paper.[2] To show that continuous time signals could be represented by discrete time samples, Shannon discussed the sampling theorem[3] that is today a universal staple of undergraduate electrical engineering curricula [9], and dictates how many discrete samples must be captured on DVD's and digital images to faithfully reconstruct continuous time audio signals and images [8, 9]. A relationship between average information and thermodynamic entropy was established by Shannon. In one of the most important applied mathematical results of the twentieth century, Shannon also showed that errorless communication was possible over a noisy channel. Forty five years later, *turbo codes* for the first time came very close to achieving the errorless communication bounds predicted by Shannon [10].

A fundamental contribution of Shannon's paper is a mathematical definition of information. Two axioms are foundational to Shannon information.

---

[1]IEEE, the *Institute of Electrical and Electronic Engineers*, is the world's largest professional society. In 2010, there were 382,400 members.

[2]Shannon credited John W. Tukey, a fellow Bell Labs researcher, with coining the word.

[3]I wrote an entire book dedicated to this topic [8], only one of the amazing contributions of Shannon's paper.

1. As the probability of an event increases, the amount of information decreases. There is little or no information in the statement that the sun will rise tomorrow morning. The probability of the event is nearly one. On the other hand, the event of the sun going supernova tomorrow has a miniscule almost zero probability. Being told the sun is going supernova tomorrow conveys much information.
2. Information of two disjoint events should be additive. That is, if the word "stuttering" conveys information I_1 and "professor" conveys information I_2, then "stuttering professor" should convey information $I_1 + I_2$.

If p denotes the probability of an event, the definition that satisfies both of these axioms is[4]

$$I = -\log_2 p.$$

As required by the first axiom, information increases as probability decreases. If two disjoint (statistically independent) events have probabilities p_1 and p_2, then the probability of both events is p_1p_2 with information $I = -\log_2 p_1p_2 = I_1 + I_2$ where $I_1 = -\log_2 p_1$ and $I_2 = -\log_2 p_2$. The additivity axiom is thus satisfied.

The base of the log in the definition of Shannon information is arbitrary and determines the units of information. If base 2 is used, then the unit of information is a *bit*. If a fair coin is flipped 6 times, we can say there are six bits of information generated since the probability of generating a specific sequence, say HTTHH, is

$$p = \left(\frac{1}{2}\right)^6 = 2^{-I}.$$

The bit can be viewed as probability measured in coin flips. Ten bits, for example, corresponds to successfully forecasting the results of ten coin flips. Pioneering application of Shannon information theory to biology includes the work of Thaxton, Bradley & Olsen [12] and Yockey [13, 14]. There are limitations to Shannon information. Isolated from context, Shannon information measure is divorced from meaning. A *Braveheart* DVD can contain as many bits as a DVD filled with random noise. When applying Shannon information, care must be taken to recognize this property and, if meaning is applicable, to make clear the connection.

[4]Use of the log to measure information dates to 1928 when Ralph Hartley noted that "...our practical measure of information [is] the logarithm of the number of possible symbol sequences." [11] This is equivalent to Shannon information when all symbol sequences are equally probable.

Solomonov-Kolmogorov-Chaitin Information

Shannon information is motivated by communication. *Algorithmic information theory*, also called Solomonov-Kolmogorov-Chaitin information after the three men who independently founded the field[5] [15–22], is a topic in the field of computer science. Whereas Shannon information deals with probability of future or unknown events, algorithmic information deals largely with the complexity of existing structure. To what degree can a thick book, say the KJV Bible, be compressed? The length of the shortest computer program to generate KJV Bible is dubbed the Chaitin-Kolmogorov complexity of the book.[6] A repeated sequence 010101010... for a billion bits has low complexity. The computer program is "Repeat 01 a half billion times." A billion bits generated by repeated flips of a fair coin, on the other hand, is almost certainly incompressible. The shortest program to print the sequence must then contain the sequence, "Print '0110100010....'."

An implication of the word *information*, when used conversationally, is the communication of meaning. Algorithmic information theory's measure of complexity suffers from the same problem as Shannon's model—it does not inherently capture the meaning in the information measured [23]. A digital image of a Caribbean sunset can have the same Chaitin-Kolmogorov complexity as an unfocused image of correlated noise.

The Meaning of Information

Meaning in information is captured by the concept of *specified complexity* popularized by Dembski [24, 25]. The idea can be illustrated using the English alphabet [12]. The phrase

OVER AND OVER AND OVER AND OVER AND OVER AND
OVER AND OVER AND OVER AND OVER AND OVER AND

has specific meaning but has a low Chaitin-Kolmogorov complexity. A program can read "Repeat 'OVER AND' ten times." The phrase

HSUEX SHDF OSJ HDFN SJABXMJ SHBU SZJLK QPRQZ HASKS
FPSCSJSJAA PJKAO DFAJ AJDFHFQWSALA DAFL V AZQEF

[5]Chaitin, born in 1947, was still a teenager when his first groundbreaking work was published in 1966.

[6]The minimum program depends on the computer program used, but the measure from computer to computer varies only by an additive constant.

is complex. The program for this phrase would be "Print HSUEX SHD... ZQEF". This is about the same size of the phrase itself. The phrase however, has no specified meaning. Next, consider the Bob Dylan lyrics[7]

I ASKED FOR SOMETHING TO EAT IM HUNGRY AS HOG SO I
GET BROWN RICE SEAWEED AND A DIRTY HOT DOG.

This sequence of letters, display both a specified meaning and high complexity.

Leslie Orgle notes, regarding the requirement of specified complexity in life:

> "Living organisms are distinguished by their specified complexity. Crystals such as granite fail to qualify as living because they lack complexity; mixtures of random polymers fail to qualify because they lack specificity." [26]

Orgle's statement was independently observed by Yockey and Wickens [12]. Other models of information include universal information [1], functional information [23, 27, 28], pragmatic information [29] and evolutionary informatics [30–32]. Except for functional information, all of these models are addressed in this section.

Papers

The papers in this section on *Information and Biology* fall into three distinct categories.

1. *Information Theory Models*

How can information be modeled to reflect the information residing in biological systems? **Gitt, Compton and Fernandez** [1] define *universal information* as; "A symbolically encoded, abstractly represented message conveying the expected action and the intended purpose." They then show how universal information is resident in biological systems. **Dembski *et al.*** [43] build on the theory of evolutionary informatics [30–32] by developing a generalized search methodology. Using conservation of information ideas popularized by the No Free Lunch theorem [25], evolutionary search is shown to produce no *active information*. The difficulty of the search at hand, measured by *endogenous information*, can be simplified only by access to some source of information. **Oller**'s *pragmatic*

[7] "On the Road Again" by Bob Dylan.

information [29] refers to the content of valid signs — the key that unlocks language acquisition by babies and ultimately leads to human communication through language. Oller shows this same measure is required for "codes" in genetics, embryology, and immunology to work.

2. *Limitations of Evolutionary Models*

A colleague of mine visiting my office noticed my computer buzzing away. When he asked what I was doing, I replied "running a self-organizing evolutionary program." In mocked astonishment, he queried "That's exciting! When will it be able to talk?" The truth in this quip is that evolutionary systems often hit a point after which no further improvement is observed. Behe [37], who coined the phrase *edge of evolution*, documents that biological evolution can also develop to a point where no other improvement is observed. In such case, specified complex information is bounded. **Basener** [38] proves such a ceiling of performance exists in many evolutionary processes. Specifically he finds; "In an evolutionary system driven by increasing fitness, the system will reach a point after which there is no observable increase in fitness." Schneider's *ev* [39] and Avida [40] computer programs that purport to demonstrate biological evolution obey the criteria necessary for Basener's result to apply. No matter how long they run, neither program will ever learn to talk. **Ewert *et al.*** [41] demonstrate that TIERRA, Thomas Ray's attempt to simulate a Cambrian explosion on the computer, also hits Basener's ceiling. Although TIERRA demonstrates fascinating and unexpected behavior, interesting innovations consistently arise only from loss of function. This same phenomenon in biology is reported by Behe [37]. **Montañez *et al.*** [42] assess the probability of information being increased via random mutations within a genome. They show that the probability of improvement drastically diminishes as the number of overlapping codes increases and to the extent that the DNA sequence is already near its optimum.

3. *Thermodynamics, Entropy and Information*

Both information theory and thermodynamics share the concept of entropy referring to maximum disorder and uncertainty. Recognizing that life does not conform to thermodynamics' demand for ever increasing disorder, Erwin Schrödinger coined the term *negentropy* (negative entropy) to apply to life. What is the source of negentropy?

Sewell [35] shows that the decrease of entropy within a non-isolated system is limited not by "compensating" entropy increasing outside the system, but by the type and amount of entropy exported through the boundary. Thus, in open systems, information increases are limited by the information entering through the boundary. In other words, it is not true that *anything* can happen in an open system [36]. **McIntosh** [33] carefully argues that the laws of thermodynamics do not permit the rise of functional devices ('machines') just by the flow of energy into a non-isolated system. Free energy devices available to do useful work are a product of intelligence. If one then considers information itself, one then finds that rather than matter and energy defining the information sitting on the polymers of life (a view held by many today), McIntosh posits that the ***reverse*** is in fact the case. Information has its definition outside the matter and energy on which it sits, and furthermore constrains matter/energy to operate in a highly non-equilibrium thermodynamic environment. He then outlines principles of information interaction with energy and matter in biological systems [34].

A Final Thought

Much work remains on development of a concise mathematical model of information applicable to biological systems. Some physicists have argued that all of the information required for the observable universe, including physical laws and the prescription for life, was created through the Big Bang. The authors of this section appear to unanimously disagree with such an assertion.

References

1. Gitt W, Compton R, Fernandez J (2013) Biological Information — What is It? In: Marks II RJ, Behe MJ, Dembski WA, Gordon BL, Sanford JC (eds) Biological Information — New Perspectives. World Scientific, Singapore, pp. 11–25.
2. Wiener N (1968) Cybernetics: or the control and communication in the animal and the machine. Technology Press, MIT.
3. Freeman A, Golden B (1997) Why Didn't I Think of That: Bizarre Origins of Ingenious Inventions We Couldn't Live Without. John Wiley & Sons, New York. ISBN 0471165115
4. Seabrook J (1993) The Flash of Genius. The New Yorker, 11 January 1993.
5. IEEE Computational Intelligence Society, http://www.ieee-cis.org/

6. Mirowski P (2002) Machine dreams: economics becomes a cyborg science. Cambridge University Press, Cambridge.
7. Shannon CE (1948) A Mathematical Theory of Communication. Bell System Technical Journal 27: 379–423, 623–656.
8. Marks II RJ (1991) Introduction to Shannon Sampling and Interpolation Theory. Springer-Verlag, New York.
9. Marks II RJ (2008) Handbook of Fourier Analysis and Its Applications. Oxford University Press, Oxford.
10. Guizzo E (2004) Closing in on the perfect code. IEEE Spectrum, March 2004.
11. Hartley RVL (1928) Transmission of information. Bell System Technical Journal 7 (3):535–563.
12. Thaxton CB, Bradley WL, Olsen RL (1992) The Mystery of Life's Origin: Reassessing Current Theories, 2nd print edn. Lewis and Stanley. [http://TheMysteryOfLifesOrigin.org/]
13. Yockey HP (1992) Information Theory and Molecular Biology. Cambridge University Press, Cambridge.
14. Yockey HP (2005) Information Theory, Evolution, and The Origin of Life. Cambridge University Press, Cambridge.
15. Chaitin GJ (1966) On the length of programs for computing finite Binary sequences. J. Assoc for Computing Machinery 13 (4):547–569.
16. Chaitin GJ (1966) On the length of programs for computing finite binary sequences by bounded-transfer Turing machines. AMS Notices 13: 133.
17. Chaitin GJ (1966) On the length of programs for computing finite binary sequences by bounded-transfer Turing machines II. AMS Notices 13: 228–229.
18. Chaitin GJ (1966) On the length of programs for computing finite binary sequences. Journal of the ACM 13: 547–569.
19. Kolmogorov AN (1965). Three approaches to the definition of the quantity of information. Problems of Information Transmission 1: 3–11.
20. Kolmogorov AN (1968) Logical basis for information theory and probability theory. IEEE Trans. Inform. Theory, vol. IT-14: 662–664.
21. Solomonoff RJ (1960) A preliminary report on a general theory of inductive inference, Tech. Rept. ZTB-138, Zator Company, Cambridge, MA.
22. Solomonoff RJ (1964) A Formal Theory of Inductive Inference. Information and Control 7 (1):1–22 & 7(2): 224–254.
23. Szostak JW (2003) Functional information: Molecular messages. Nature 423(6941): 689.
24. Dembski WA (2002) Can evolutionary algorithms generate specified complexity? In: Gregersen NH (ed) From Complexity to Life: On The Emergence of Life and Meaning. Oxford University Press, USA.
25. Dembski WA (2007) No Free Lunch: Why Specified Complexity Cannot Be Purchased without Intelligence. Rowman & Littlefield, Lanham, MD.
26. Orgel LE (1973) The Origins of Life. John Wiley and Sons, Inc., New York.

27. Durston KK, Chiu DKY, Abel DL, Trevors JT (2007) Measuring the functional sequence complexity of proteins. Theoretical Biology and Medical Modelling 4:47 doi:10.1186/1742–4682–4-47 [http://www.tbiomed.com/content/4/1/47]
28. McIntosh AC (2006) Functional Information and Entropy in living systems. In: Brebbia CA, Suchrov LJ, Pascolo P (eds) Design and Nature III: Comparing Design in Nature with Science and Engineering. WIT Press, UK.
29. Oller JW (2013) Pragmatic Information. In: Marks II RJ, Behe MJ, Dembski WA, Gordon BL, Sanford JC (eds) Biological Information — New Perspectives. World Scientific, Singapore, pp. 64–86.
30. Dembski WA, Marks II RJ (2009) Conservation of Information in Search: Measuring the Cost of Success. IEEE Transactions on Systems, Man and Cybernetics A, Systems and Humans 39 (5):1051–1061.
31. Dembski WA, Marks II RJ (2010) The Search for a Search: Measuring the Information Cost of Higher Level Search. Journal of Advanced Computational Intelligence and Intelligent Informatics 14(5):475–486.
32. Dembski WA, Marks II RJ (2011) Life's Conservation Law: Why Darwinian Evolution Cannot Create Biological Information. In: Gordon B, Dembski W (eds) The Nature of Nature (Wilmington, Del.: ISI Books, 2011) pp.360–399.
33. McIntosh AC (2013) Information and Thermodynamics in living systems. In: Marks II RJ, Behe MJ, Dembski WA, Gordon BL, Sanford JC (eds) Biological Information — New Perspectives. World Scientific, Singapore, pp. 179–202.
34 McIntosh AC (2009) Information and Entropy — Top-down or bottom-up development in living systems? Int. J. of Design & Nature and Ecodynamics 4(4):351–385.
35. Sewell G (2013) Entropy, Evolution and Open Systems. In: Marks II RJ, Behe MJ, Dembski WA, Gordon BL, Sanford JC (eds) Biological Information — New Perspectives. World Scientific, Singapore, pp. 168–178.
36. Sewell G (2001) Can ANYTHING Happen in an Open System? The Mathematical Intelligencer 23(4): 8–10.
37. Behe MJ (2007) The Edge of Evolution: The Search for the Limits of Darwinism. Free Press, New York.
38. Basener WF (2013) Limits of Chaos and Progress in Evolutionary Dynamics. In: Marks II RJ, Behe MJ, Dembski WA, Gordon BL, Sanford JC (eds) Biological Information — New Perspectives. World Scientific, Singapore, pp. 87–104.
39. Montañez G, Ewert W, Dembski WA, Marks II RJ (2010) Vivisection of the ev Computer Organism: Identifying Sources of Active Information. BIO-Complexity 2010(3):1–6.
40. Ewert W, Dembski WA, Marks II RJ (2009) Evolutionary Synthesis of Nand Logic: Dissecting a Digital Organism. Proceedings of the 2009 IEEE International Conference on Systems, Man, and Cybernetics. San Antonio, TX, USA — October 2009, pp. 3047–3053.

41. Ewert W, Dembski WA, Marks II RJ (2013) Tierra: the character of adaptation. In: Marks II RJ, Behe MJ, Dembski WA, Gordon BL, Sanford JC (eds) Biological Information — New Perspectives. World Scientific, Singapore, pp. 105–138.
42. Montañez G, Marks II RJ, Fernandez J, Sanford JC (2013) Multiple overlapping genetic codes profoundly reduce the probability of beneficial mutation. In: Marks II RJ, Behe MJ, Dembski WA, Gordon BL, Sanford JC (eds) Biological Information — New Perspectives. World Scientific, Singapore, pp. 139–167.
43. Dembski WA, Ewert W, Marks II RJ (2013) A General Theory of Information Cost Incurred by Successful Search. In: Marks II RJ, Behe MJ, Dembski WA, Gordon BL, Sanford JC (eds) Biological Information — New Perspectives. World Scientific, Singapore, pp. 26–63.

Biological Information — What is It?

Werner Gitt[1]*, Robert Compton[2] and Jorge Fernandez[3]

*[1]German Federal Institute of Physics and Technology, Braunschweig, Germany. Former Head of Information Technology, Director and Professor (retired). [2]Department of Veterinary Anatomy and Physiology, College of Veterinary Medicine, Washington State University, Pullman, Washington (retired). [3]President, Mastek Inc., Melbourne, Florida. (*Corresponding author: werner.gitt@t-online.de)*

Abstract

Scientific discoveries, especially over the last six decades, have left no doubt that 'information' plays a central role in biology. Specialists have thus sought to study the information in biological systems using the same definitions of information as have been traditionally used in engineering, computer science, mathematics and in other disciplines. Unfortunately, all of these traditional definitions lack aspects that even non-specialists recognize as being essential attributes of information — qualities such as meaning and purpose. To remedy that deficiency, we define another type of information — Universal Information — that more accurately embodies the full measure of information. We then examine the DNA/RNA protein synthesizing system with this definition of Universal Information and conclude that Universal Information is indeed present and that it is essential for all biological life. Furthermore, other types of information, such as Mental Imaging Information, also play a key role in life. It thus seems inevitable that the biological sciences (and science in general) must consider other-than-the-traditional definitions of information if we are to answer some of the fundamental questions about life.

Key words: information, codes, Universal Information, biological information, scientific laws, laws of nature, transmitter, receiver

Introduction

The title of this symposium is "Biological Information: New Perspectives". But what do we mean by the term "biological information"? We suggest that, at present, it cannot be unambiguously defined. Yet, an unambiguous definition would be extremely helpful because multiple levels of communication systems are being researched: from the DNA-coded information in the genome, to the intra-cellular communication networks involving RNA and proteins, to inter-cellular signaling via entities such as hormones, all the way up to and including the nervous system and the brain. Clearly, identifying all of these communication systems and defining the information that is being transferred will be a challenge. It is clear that there are many subsets (or categories) of biological information, and that many more will be discovered [1]. At some point, perhaps after further research, we will be in a position to more precisely define "biological information". For the interim, we will offer a description of it as a placeholder until we have enough knowledge to define it with scientific rigor.

We propose that biological information includes all manifestations of information in living organisms. This description has the potential to include all categories of information. From recent scientific studies in genetics it is clear that there are many subsets of biological information (codes), and many more wait to be uncovered within the DNA/RNA systems alone [1]. It is reasonable to believe that progress on biological information will be accelerated if each information subset is unambiguously defined. Towards this goal we should begin by defining a definition: *a description or explanation of a word or thing by its attributes, properties or relations that distinguishes it from all other entities* [2]. Even applying this definition carefully is important, because scientifically rigorous results cannot be achieved when using ambiguous terms. A common example of this is the claim that, "Evolution is a fact." The validity of such a claim is certainly going to depend on the precise meaning of the term 'evolution'.

Defining Subsets of Information

This leads us to ask the more general question: *What precisely is information?* Anyone who has studied this field is aware of three working definitions of information:

- **Classical Information Theory**: Shannon (statistical) information [3]; dealing solely with the technical/engineering aspects of communication. This involves analyses including obtaining statistics on the material symbols for data transmission, storage and processing.
- **Algorithmic Information Theory**: Solomonoff/Kolmogorov/Chaitin [4–6]; dealing with the 'complexity' (as this term is defined in the theory) of material symbols in data structures and of objects in general.
- **Complex Specified Information (CSI) Theory**: Dembski [7]; roughly the same as Classical Information Theory but adding the important concept of a 'specification'.

These theories, like modern genetics, focus primarily on the material carriers of the information. On the other hand, American mathematician and National Medal of Science recipient Norbert Wiener in 1968, made his often quoted statement *"Information is information, neither matter nor energy."*[8]

Wiener's statement prompted one of us (W. Gitt) to ask; if information is not matter (mass and energy) then what is it? Gitt therefore started a long quest to define information — at least the information that was most familiar to scientists of that day. As an information scientist, Gitt not only examined the information conveyed within human natural languages, but also the information conveyed within abstract and artificial languages such as machine languages. In his studies he identified five attributes, four of which qualified as *distinguishing* attributes of 'information'. Before we examine these, let us make it clear that these natural and artificial language systems were first studied because at that time they had already been extensively characterized. We used these human information systems as 'known systems', which would most likely be amenable to precise definition.

Distinguishing Attributes of Information

Code plus syntax

At the basic level of information in these languages we find a set of abstract symbols formally known as an alphabet — this set constitutes the code. By abstract we mean that each of these symbols has no resemblance and no inherent physical relationship to the entity that they represent. These symbols have a characteristic two-dimensional configuration that distinguishes them from each other. One way in which this is manifested in the material domain is by inscribing these symbols onto a wide variety of material media and formats.

Examples of this are abundant. For instance, the first five words of Lincoln's Gettysburg Address — "Fourscore and seven years ago" — may be inscribed on paper with ink, or chiseled onto a block of granite, or on a blackboard with chalk, or in the air with smoke signals or with the vibrations of speech, or on a transmission line with electrical dots and dashes as in Morse code, or on a computer's hard drive by properly setting magnetic 'bits', or in many other ways.

With this we see that the actual information is completely independent of the material medium that serves only to 'carry' it. Any one of a multitude of material media and formats may be used to carry exactly the same information. While there is indeed a *correlation* between the material media and format that carries the information, the dictum "correlation does not imply causation" certainly applies here. The material carrier cannot be and is not the *cause* of the information.

Further examination reveals that there is also a set of rules governing what is permissible regarding the arrangements of the symbols — this set of rules constitutes the syntax.

With the combination of abstract code and syntax we are able to generate more complex language structures such as words and sentences. However, at this (formal language) stage *meaning* plays no role. It was at this level *only* that Shannon developed his *Theory of Communication* [3] into the highly useful statistical analyses of the material symbols, solely for the technical purposes of data transmission, storage and processing. Code plus syntax is a necessary distinguishing attribute of all human languages. Let us pause for a moment and reflect on how this all comes about.

In order to develop, learn, or use a code plus syntax system, it requires a high degree of mental effort and intelligence. No one at any time has ever observed this basic attribute of information (i.e., code plus syntax) being established through unguided, purely physicochemical processes. However, we have observed young children learning the alphabet and learning to read, write, and speak words. Also many of us as adults have developed and/or learned machine languages. We may say that people acquired these abilities from their parents and so on down through history. However, this does not in any way explain how the *first* human acquired this ability. If we assume this happened without intelligent guidance, there are only two alternatives: 1) it is an inherent property of matter or, 2) it is possible for these abilities to 'evolve' over time. A person may choose to believe in either of these alternatives but that person would have to also accept that this is a *belief* with no hard science to support it.

Meaning

The next level of the distinguishing attributes of information in human languages is meaning. At this level, words that were formed by short sequences of symbols are assigned to represent 'something' (where that 'something' may be any particular entity or object, event, thought or concept). Additionally, that 'something' must be defined and that definition is also represented by words.[1] Higher levels of meaning and information content are constructed using phrases, sentences and paragraphs

[1] For example, consider the word 'cat'. 'Cat' is an abstract representation of the actual creature. If we then define this creature as 'a four-legged mammal that purrs and meows', we then have other words that are being used to represent both the word 'cat' and the creature. We note that the words being used also be defined with other words — this goes on level after level. For instance, the above definition for 'cat' included the word 'mammal'. That word 'mammal' must be defined (with other words, of course) and then those other words will in turn need to be defined. Thus, a measure of circularity is ultimately unavoidable.

when the meaning from one word is insufficient. In our example of a 'cat' (footnote 1), its definition is a sentence that represents a creature and the word 'cat'.

Meaning is an absolutely essential attribute of information that is conveyed in language and communication. Words, both written and spoken, can be used to represent entities, events and/or concepts — literally anything. The entities need not be present but words, serving as their placeholders, represent and thereby communicate their reality as if they were present. Unguided, purely physicochemical processes have never been observed creating this 'substitutionary' process [10]. We are referring here to natural, unguided, purely physicochemical processes that have no external guiding (control) systems found in information-rich systems. These seem to eliminate all biological information systems as being examples of unguided, purely physicochemical processes.

Expected Action

The examination of sentences or paragraphs in a message reveals an implied request or a command for the receiver of the message to perform some action. These actions start with the receiver reading and understanding the message (this in itself involves very complex actions). From understanding the message, the receiver must decide whether or not he/she will comply fully, partially or not at all with the sender's expected action. If the decision is to fully comply then the receiver performs whatever action was indicated (purposed!) by the sender's message. Here we must distinguish between two types of receivers: 1) an intelligent being that possesses the capability of making free choices and is able to determine the meaning of the message, and 2) a machine that does not have these capabilities. In the former, the intelligent being can respond to the request or command in highly variable ways. With the machine the meaning has been programmed into command signals that 'start' or initiates the action level — the systems control program guides the machine to automatically perform the action. It must be pointed out that in both cases machines are essential for performing the expected actions [10]. In the case of the intelligent being, the machinery of his body may be sufficient or he may need to utilize external machinery (which may be mechanical/electrical machines, animals, other humans, etc.).

Intended Purpose

Prior to issuing an original written or verbal message there must be an internal thought process that motivates the sender to formulate a message. This thought process is necessarily complex and involves need, motivation or intent for something to be achieved. If it is not to be performed by the sender, then the thought process must include selecting a particular receiver and determining whether or not that receiver is capable of performing the expected action. If the whole process is completed successfully, then the sender's original purpose is achieved. Thus we see that information's attribute of intended purpose is essential at the very beginning of a message. The achievement of that purpose is the result of the receiver's performance of the desired action. From this we see that the most important attribute of information is the intended purpose and that it is at both ends of a successfully completed message. Purpose may thus be thought of as the 'bookends attribute'.

The Definition of Universal Information

All four attributes described above are necessary to unambiguously distinguish this subset (category) of information. Due to this, the formal definition of Universal Information (UI) stated below incorporates all four of these distinguishing attributes.

A symbolically encoded, abstractly represented message conveying the expected action and the intended purpose.

Now we can appraise the three previously discussed working definitions of information in light of the attributes of Universal Information.

Shannon's classical information theory concerns itself only with statistical relationships of material symbols found within the code of Universal Information. This was because nothing more was necessary in order to address the technical issues of information transmission and storage. While Shannon stated this point clearly in his landmark paper [9], most modern day evolutionary theorists champion his definition primarily because it allows for the creation of 'information' by randomly assembling symbols. This makes creation of biological information trivial, and separates biological information from biological functionality. The attempt to define biological information in this way is clearly ideologically driven and is obviously not sufficient, since no thinking person would exclude meaning and purpose from biological (functional) information.

Algorithmic Information is a measure of the information content of material systems in terms of the degree of 'complexity' (as algorithmic 'complexity' is defined) of the system. Those material systems displaying greater complexity (more aperiodicity) have higher information content than those material systems displaying less complexity (more periodicity). The four distinguishing attributes of Universal Information are *not* required for algorithmic information.

Complex Specified Information (CSI) exists in all material systems that exhibit a 'specification' and this specification is expressed in terms of functionality or purpose. As a result, CSI requires only UI's distinguishing attribute of purpose. By definition this means that any system exhibiting CSI implies design. Even though all of the distinguishing attributes of UI were necessary during the design and construction phase, these attributes need not be present in the observed complex specified system.

The Nature of Universal Information

Having clearly distinguished Universal Information from other types/definitions of 'information', we now proceed to answer (at least for UI) the question [8]: if 'information' is not mass and energy, what is it? In the following discussion we will use the term 'matter' to include both mass and energy and the term 'nonmaterial entity' to refer to all entities outside the material domain.

There are many significant criteria for distinguishing material entities from nonmaterial entities. Perhaps the most simple, direct and scientific criterion is the fact that all material entities can be measured and thereby 'quantified' using one or a combination of the seven units of measurement established by the System International. These are the meter, kilogram, ampere, kelvin, mole, candela and second. Any entity within the universe that cannot be measured and described with one or a combination of these units is, by definition, a nonmaterial entity. Another criterion is that a nonmaterial entity does not and cannot originate from unguided, purely physicochemical processes [10, 12]. Finally, a nonmaterial entity does not have any direct physicochemical interaction with matter [10].

Universal Information satisfies all of the above criteria for a nonmaterial entity. A material medium is essential for the storage, transmission and processing of UI but, as described earlier, the quantity and type of matter that is used is highly variable and not correlated at all to the value of the Universal Information; i.e., the UI is completely independent of the material medium.

Additionally, the symbols (code level) that are utilized and physically manifested in the material domain display a vast degree of variation. To illustrate this, Figure 1 depicts the words from ten different languages that have the same meaning even though the individual symbols/letters differ markedly from one another. However, regardless of the symbols used the 'content' of the meaning remains essentially the same. Content as used here includes the attributes of meaning, action and purpose.

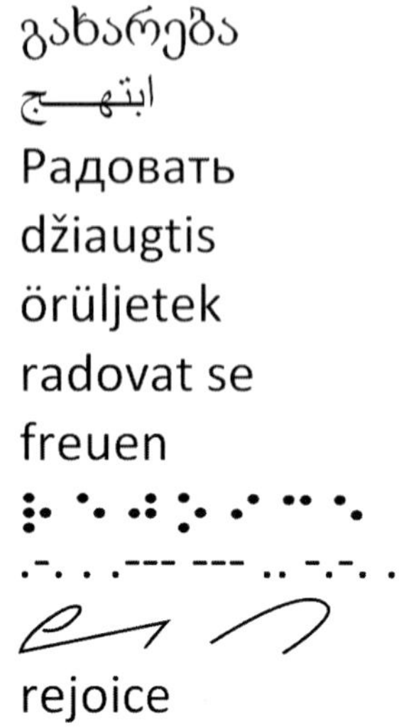

Fig. 1. Different codes expressing the same meaning. The word "rejoice" is represented by means of a selection of different coding systems — from the top down, Georgian, Arabic, Russian, Lithuanian, Hungarian, Czech, German, Braille, Morse code, Shorthand and English.

Does Biological Life Contain Universal Information?

There have been monumental advancements in both information science/theory, and genetics and molecular biology in the last six decades. The processes involved in cellular synthesis of proteins have been explained in great detail. We will examine this DNA/RNA protein synthesizing system to determine if it stores and conveys Universal Information. In order to systematically make this determination we will look for each distinguishing attribute of UI in the cells' protein synthesizing system.

Code plus Syntax

Within DNA/RNA we have a four-letter alphabet — adenine, thymine, cytosine and guanine (A, T, C and G) — in RNA the thymine is replaced by uracil (U). These four letters are arranged into 'words' that are always composed of three letters. These three-letter words are called 'codons'. So we have a Code (a four-letter alphabet) and Syntax (three-letter words). Thus, the first distinguishing attribute of UI is present: code plus syntax.

Abstract Meaning

There are $4^3 = 64$ different three-letter 'words' that may be composed out of the four letters in the Code. Apart from three stop codons, each of the remaining sixty-one three-letter words, or codons, means/represents/denotes one of the twenty amino acids utilized in polypeptide/protein synthesis. The codon for methionine also denotes or represents a start command. Additionally, the specific sequence of codons in messenger RNA (mRNA) represents the specific sequence of amino acids in the polypeptide precursor to the protein. Despite intensive research, no physicochemical bonding relationship has been found between the codons and the amino acids they represent [10, 12]. Hence, the second distinguishing attribute of UI is present: abstract meaning.

The Expected Action

The messenger RNA (mRNA) is transported out of the cell nucleus into the cytosol to a very complex RNA/protein machine — the ribosome. At the ribosome, beginning with a start codon on the mRNA, this specific mRNA codon is joined with an anticodon at one end of the small transfer RNA (tRNA) molecule. At the other end of tRNA is the amino acid specified by the mRNA codon, in this case methionine. The mRNA is then advanced one codon step and another tRNA anticodon is joined to the mRNA codon. At this stage two amino acids have been brought together and the ribosome, utilizing energy, joins the two amino acids together by forming a peptide bond. This process repeats itself until a stop codon is reached on the mRNA. The polypeptide thus formed is then folded by other protein machines into a functional protein with a highly specific three dimensional configuration. This precise synthesis of a unique functional protein by the ribosome (machine) fulfills the third distinguishing attribute of UI: expected action. However, this is only the first level of action — the proteins themselves have higher-level functions, e.g., the ribosome, which is primarily protein. At a macroscopic level the activity of proteins in muscles of higher animals perform useful work. Figure 2 demonstrates that DNA replication during cellular reproduction requires protein nanomachines such as DNA polymerase and usable energy. Next, transcription to mRNA requires a DNA template, several nanomachines (such as RNA polymerase and spliceosome) and usable energy. Finally, synthesis of all protein nanomachines and protein structural elements require mRNA, tRNA and nanomachines such as ribosomes and chaperonins, and usable energy. This essential closed-loop conundrum has stymied researchers for decades as they have attempted to account for the origin of the first living cell through unguided purely physicochemical processes. Their attempt at 'protein first', 'DNA first' or 'RNA first' models have all failed [10, 12]. As demonstrated in Figure 2, all three must be 'first' *simultaneously*.

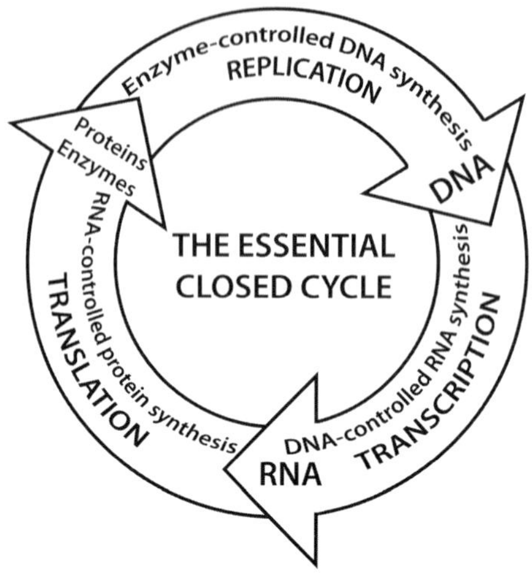

Fig. 2. A simplified representation of a UI-controlled cyclical process in living cells. The translation mechanism (protein synthesis) corresponds to the lowest level of expected action. However, the action of a protein nanomachine (DNA polymerase) is required in the next step of the cycle in DNA replication. The intricate process of mRNA synthesis (transcription) requires the DNA template and nanomachines (RNA polymerase II and spliceosome). Each of these three steps must be present *simultaneously*.

The Intended Purpose

The UI instructions for protein synthesis are stored within the nuclear DNA. The initial purposes of these UI instructions are achieved as the processes of transcription and translation are successfully accomplished. The ultimate physical purpose for the DNA/RNA protein synthesizing system is for the initial creation of organisms, and for their operations, maintenance and reproduction. Undoubtedly the earth's biosphere would not exist if all of the protein components were absent. At the intracellular level, while not identical, the protein requirements are similar in many areas for both plant and animal life. However, in multicellular animals the use of extracellular protein is far more extensive than in multicellular plants. Therefore, the greater diversity of protein in animals than in plants will require more complex amounts of UI stored in the DNA and transcribed into RNA. Further research into this difference as well as comparing the DNA coding for protein with the DNA coding for cellulose synthesis in plants may reveal important features of DNA coding. The multiple purposes achieved by the DNA/RNA protein synthesizing system attests to the fact that the fourth distinguishing attribute of UI (intended purpose) is indeed present.

UI Senders, Transmitters and Receivers

Problems associated with determining the origin and utilization of UI can be somewhat mitigated if we use specific terms to differentiate between the following:

1. An ***original sender*** is an intelligent agent that creates the original UI message. As demonstrated by Gitt *et al.* [10] this intelligent agent must have a nonmaterial component beyond the embedded UI. This is because even UI-guided purely physicochemical processes wholly constrained by natural laws have never been observed to create *de novo* UI despite all scientific efforts to date [10, 12]. Since humans do create *de novo* UI they qualify as original senders. This is strong evidence that humans have a nonmaterial component beyond their embedded UI [10].
2. ***Intermediate transmitters*** receive a UI message and simply copy, transmit, display or broadcast the message. Ideally, an intermediate transmitter will not distort the meaning of the original message in any way [10]. Intermediate transmitters can be intelligent agents or machines that are specifically designed to perform the transmitting processes.
3. ***Machine receivers*** obtain and process the messages and perform the commanded action thereby achieving the purpose intended by the original sender. Machine receivers (either mechanical or biological) do not have the capability to freely interpret the messages. They must be 'preprogrammed' with the capability to receive, then process and then execute the commanded actions *without* requiring that the meaning of the messages be determined. In other words, the programmer must convert the meaning of the messages into a series of preprogrammed executable steps that are initiated by start commands so that the proper actions are performed [10, 11].
4. ***Intelligent receivers*** possess the capability of determining the meaning of the message and also possess the capability of making free choices. This latter capability allows the intelligent receiver to decide whether to perform the expected action fully, partially or not at all.

When the UI in the DNA/RNA protein synthesizing system is expressed in biological life, it guides the transcription/translation processes to produce a specifically controlled amount of a specified protein. This protein will then perform specific functions within the cell or within the organism. This is an example of number 3 above whereby machines are guided by instructions (namely UI) that

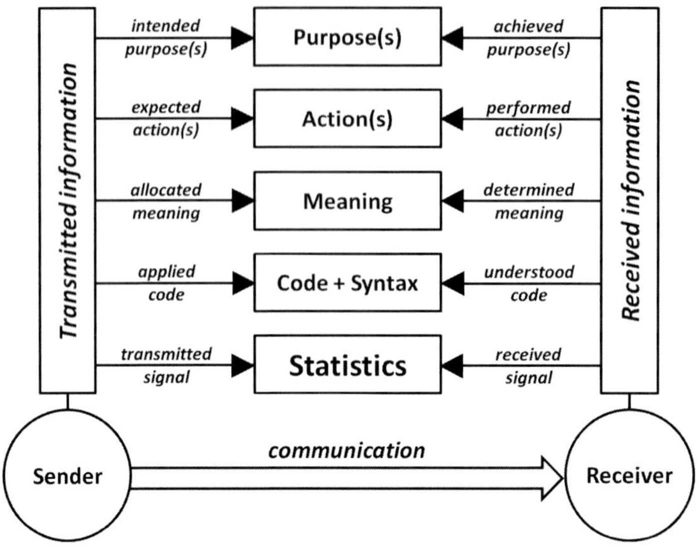

Fig. 3. A comprehensive diagram of the five levels of Universal Information. All five levels are relevant for both the intelligent sender and the intelligent receiver.

was stored in the nuclear DNA of the cell by the original sender of that UI. Figure 3 displays a comprehensive diagram of Universal Information being originated (*de novo*) by an intelligent sender and being received by an intelligent receiver. More complex diagrams that include transmitters and machine receivers can be found in [10].

The Existence, Validity and Significance of Universal Information

While identifying and studying the distinguishing attributes of Universal Information (UI), we discovered and formulated 32 Empirical Statements involving the origin and nature of UI [10]. We have repeatedly verified these Empirical Statements over a 30-year period. Not one of these Empirical Statements has ever been refuted despite wide dissemination of this information and they remain an open challenge to this day.

We then turned our attention to the code discovered in the DNA and the volumes of research describing it. It was easily determined that Universal Information is definitely stored, transmitted and utilized within the DNA/RNA Protein Synthesizing System of all living organisms. In other words, UI is not merely an interesting theoretical concept; UI truly exists. UI is not only a foundational component of human languages and communication, it is also a vital control system found in all biological life on earth.

Undoubtedly the most important activity in science is to utilize factual data and observations to construct reliable and valid conclusions. This goal is achieved via sound, logical arguments that lead to those conclusions. According to Kreeft [13], there are three things that must be in place in order to develop logically sound arguments.

- The significant terms must be unambiguous.
- The premises must be true.
- The conclusion must logically follow from the premises (logically valid).

In order to satisfy these three requirements, we carefully defined all significant terms so that there would be clear, unambiguous formulations of the questions, arguments and conclusions [10]. In addition, the Empirical Statements were continually evaluated by a number of independent individuals in order to ensure clarity of meaning and validity of the statements. With this foundation we then proposed specific Empirical Statements as Scientific Laws and used them, along with verified scientific facts, as premises in our deductions [10]. Therefore, our premises are extracted from the two categories of science — verified facts and scientific laws — that have the highest degree of scientific certainty. Finally, we constructed ten logically-sound deductions that led to ten strong conclusions [10]. By rigorously following this procedure we have minimized investigator bias or interference from our conclusions. This is important for any conclusion in science, but especially so in this case because of the broad significance of these ten conclusions. Also, by minimizing investigator interference these results retain objective validity to the extent that this is possible.

Conclusion

Coming full circle, we return to our original question regarding Biological Information — *What is it?* We have identified an important subset of Biological Information that we call Universal Information that is present in every cell of every living organism.

We the authors of this paper used Universal Information in order to communicate these things to you. This Universal Information was processed through our brains that in turn, controlled our body parts to write the words on this page. These words reach receptors in your visual system that will then send impulses (i.e., messages) to your brain. You then determine the meaning of the words of our message and consider their significance. This too is Universal Information and is also part of Biological Information.

Between our intracellular DNA/RNA systems and our capacity to express thoughts through words there are many levels of highly integrated, organized biological systems which themselves necessarily operate under the control of some type of biological information. At each of these levels there are many structural components and biological machines that perform the required actions. Essentially all of these structures and machines are composed of proteins synthesized by the DNA/RNA protein synthesizing system.

Will we find Biological Information in forms other than Universal Information? We believe that we will. For instance, we are already aware of Mental Image Information (MII). MII is information in which there is meaning, action and purpose but no *abstract* code, syntax or abstract meaning. Recall that two of the distinguishing attributes of Universal Information is an abstract code with syntax and abstract meaning such as that which is manifested in the DNA/RNA protein synthesizing system. We know that MII plays a role in living organisms yet MII does not have an abstract code, syntax or meaning. For example, a 'spoon and fork' on a highway sign *directly* (i.e., not abstractly) represents 'food' or 'eating place' since it resembles the entity that it represents. Another example is the pheromones emitted by certain insects for, say, mating purposes. These pheromones have an inherent physicochemical relationship with the entity they represent. When received, these pheromones convey meaning, expected action and purpose *directly* (i.e., not abstractly) instead of through some intermediate substitute possessing 'abstract meaning' expressed via an abstract code with an associated syntax. In other words, the pheromone molecule is not an abstract substitute for the entity, it is the entity itself.

Just as was the case for Complex Specified Information in Intelligent Design Theory, Universal Information and related topics represent a revolutionary departure from the materialistic approach to information. Since UI and its requisite machines have great explanatory power in biology, a search for machines, even without explicit (embedded) UI, operating at various ranges of scale in the inanimate world may also yield results with great explanatory power [10].

References

1. Montañez G, Marks II RJ, Fernandez J, Sanford JC (2013) Multiple overlapping genetic codes profoundly reduce the probability of beneficial mutation. In: Marks II RJ, Behe MJ, Dembski WA, Gordon B, Sanford JC (eds) Biological Information — New Perspectives. World Scientific, Singapore, pp. 139–167.
2. Funk and Wagnalls Standard Dictionary (1962).
3. Shannon CE, Weaver W (1949) The mathematical theory of communication. University of Illinois Press, Urbana IL.

4. Solomonoff R (1960) A preliminary report on a general theory of inductive inference. Report V-131, Zator Co, Cambridge, MA.
5. Kolmogorov AN (1965) Three approaches to the quantitative definition of information. Problems Inform. Transmission 1 (1): 1–7.
6. Chaitin GJ (1969) On the simplicity and speed of programs for computing infinite sets of natural numbers. Journal of the ACM 16: 407.
7. Dembski WA (1998) The design inference. Cambridge University Press, United Kingdom.
8. Wiener N (1968) Cybernetics: or the control and communication in the animal and the machine. First American Edition (Technology Press, MIT).
9. Shannon CE (1948) A mathematical theory of communication. Bell System Technical Journal Vol.27.
10. Gitt W, Compton B, Fernandez J (2011) Without Excuse. Creation Book Publishers.
11. Seaman J (2013) DNA.EXE: A sequence comparison between the human genome and computer program code. In: Marks II RJ, Behe MJ, Dembski WA, Gordon B, Sanford JC (eds) Biological Information — New Perspectives. World Scientific, Singapore, pp. 385–401.
12. Meyer SC (2009) Signature in the cell. Harper One, New York.
13. Kreeft P (1989) Three philosophies of life. Ignatius Press, San Francisco, CA.

A General Theory of Information Cost Incurred by Successful Search

William A. Dembski[1]*, Winston Ewert[2] and Robert J. Marks II[2]

[1]*Discovery Institute, 208 Columbia Street, Seattle, WA 98104.* [2]*Electrical & Computer Engineering, One Bear Place #97356, Baylor University, Waco, TX 76798–7356.* (**Corresponding author: dempski@discovery.org*)

Abstract

This paper provides a general framework for understanding targeted search. It begins by defining the search matrix, which makes explicit the sources of information that can affect search progress. The search matrix enables a search to be represented as a probability measure on the original search space. This representation facilitates tracking the information cost incurred by successful search (success being defined as finding the target). To categorize such costs, various information and efficiency measures are defined, notably, *active information*. Conservation of information characterizes these costs and is precisely formulated via two theorems, one restricted (proved in previous work of ours), the other general (proved for the first time here). The restricted version assumes a uniform probability search baseline, the general, an arbitrary probability search baseline. When a search with probability q of success displaces a baseline search with probability p of success where $q > p$, conservation of information states that raising the probability of successful search by a factor of $q/p(>1)$ incurs an information cost of at least $\log(q/p)$. Conservation of information shows that information, like money, obeys strict accounting principles.

Key words: Search matrix, targeted search, active information, probabilistic hierarchy, uniform probability, conservation of information

1. The Search Matrix

All but the most trivial searches are needle-in-the-haystack problems. Yet many searches successfully locate needles in haystacks. How is this possible? A successful search locates a target in a manageable number of steps. According to conservation of information, nontrivial searches can be successful only by drawing on existing external information, outputting no more information than was inputted [1]. In previous work, we made assumptions that limited the generality of conservation of information, such as assuming that the baseline against which search performance is evaluated must be a uniform probability distribution or that any query of the search space yields full knowledge of whether the candidate queried is inside or outside the target. In this paper, we remove such constraints and show that conservation of information holds quite generally. We continue to assume that targets are fixed. Search for fuzzy and moveable targets will be the topic of future research by the Evolutionary Informatics Lab.

In generalizing conservation of information, we first generalize what we mean by targeted search. The first three sections of this paper therefore develop a general approach to targeted search. The upshot of this approach is that any search may be represented as a probability distribution on the space being searched. Readers who are prepared to accept that searches may be represented in this way can skip to section 4 and regard the first three sections as stage-setting. Nonetheless, we suggest that readers study these first three sections, if only to appreciate the full generality of the approach to search we are proposing and also to understand why attempts to circumvent conservation of information via certain types of searches fail. Indeed, as we shall see, such attempts to bypass conservation of information look to searches that fall under the general approach outlined here; moreover, conservation of information, as formalized here, applies to all these cases.

In first generalizing targeted search before generalizing conservation of information, we introduce the search matrix. The elements that constitute the search matrix may be illustrated as follows. Imagine a gigantic table that is miles in both length and width. Covering the table are upside-down dixie cups that are tightly packed, such as the following hexagonal packing:

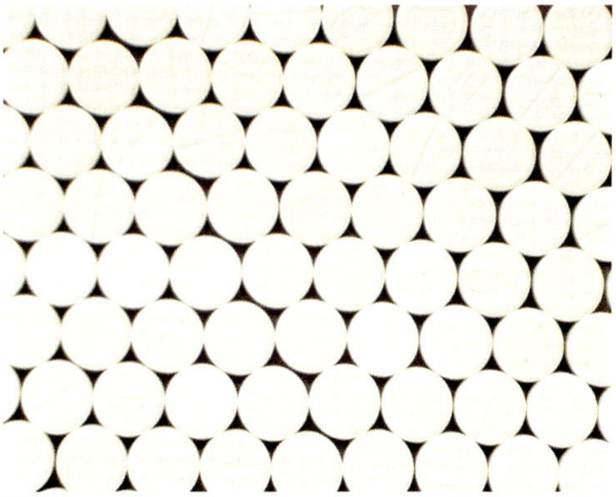

Under each dixie cup resides a single pea. The cups are opaque, so the peas are not visible unless the cup is lifted. The peas come in two varieties, high-yield and low-yield (the difference being that high-yield peas, if planted, produce lots of peas whereas low-yield peas produce only few). The low-yield peas far outnumber the high-yield peas. Our task is to locate a high-yield pea. The high-yield peas therefore form the target. Because the table is so large and the cups are tightly packed, for a human to try to walk around the table and turn over cups is infeasible.

We therefore imagine a remote-controlled toy helicopter flying over the table, hovering over individual cups, and lifting a given cup to examine the pea under it. Each act of lifting a cup to examine the pea under it constitutes a single query.

Because the table is so large and the high-yield peas are so few, this search constitutes a needle-in-a-haystack problem. As with all such problems, the number of queries (i.e., attempts to locate the needle, or, in this case, to locate a high-yield pea) is strictly limited. Moreover, because the needle is so small in relation to the haystack, unless the queries are chosen judiciously, the needle will in all likelihood elude us. Our search therefore is limited to at most m queries, which we call the *sample size*. This, in the case at hand, is the maximal number of dixie cups the search can turn over. We therefore imagine that the remote controlled toy helicopter flies over the gigantic table of dixie cups, hovers over a given cup, turns it over to examine the pea under it, replaces the cup, and then moves on, repeating this action at most m times.

Within the constraints of this scenario, how do we find the target? The helicopter has m queries in which to locate the target (i.e., to find a high-yield pea). At each query, the helicopter does three things:

(1) It identifies a given pea by removing and replacing the dixie cup over it.
(2) It extracts information that bears on the pea's probability of belonging to the target.
(3) It receives information for deciding where to fly next to examine the next pea.

The helicopter's search for the target may therefore be characterized as the following $3 \times m$ matrix, which we call the *search matrix*:

$$\begin{bmatrix} x_1 & x_2 & x_3 & \ldots & x_m \\ \alpha_1 & \alpha_2 & \alpha_3 & \ldots & \alpha_m \\ \beta_1 & \beta_2 & \beta_3 & \ldots & \beta_m \end{bmatrix}$$

Here the first row lists the actual peas sampled, the second row lists the information extracted about the peas that bears on their probability of belonging to the target, and the third row lists the information for deciding where the helicopter is to fly next. Moreover, each column represents a single query, with the columns listed in the order of search. A successful search is then one that on the basis of the search matrix explicitly identifies some x_i in the first row that belongs to the target.

Note that in this general search scenario, given columns may query the same element more than once. Thus, for separate columns that record x_i, x_j, x_k, etc., these first-row elements of the search matrix might all be identical. Such repetitions could occur if each time the helicopter lifts a dixie cup and queries a pea, it can extract only partial information about it, and so the search may need to query the pea again to obtain still more information about it. We could have distinguished between queries that lift a dixie cup to access a pea and queries that subsequently extract further information about a pea once the dixie cup is lifted. But since one may not want to query a pea immediately after having already queried it but rather wait until other peas have been queried (information about other peas might help to elucidate information about the given pea), in the interest of generality it is best to allow for only one type of query, namely, a combined query that lifts a dixie cup and then extracts information about the underlying pea.

The search matrix is not identical with the search. Rather, the search matrix records key information about the progress of the search. Specifically, it records the elements sampled from the search space, including any repetitions (this information appears in the first row); it records information bearing on the probability that an element sampled belongs to the target (this information appears in the second row); and it records information for deciding where to sample next (this information appears in the third row). All this information contained in the search matrix comes to light through the activity of a search algorithm. Success of the search therefore depends on how effectively the algorithm uses as well as fills in the information contained in the search matrix. Does the search algorithm, in sampling x_i, have complete memory of the prior information sampled? Or is it a Markov process with access only to the latest information sampled? Does the algorithm contain additional information about the target so that regardless of the information in rows two and three, the algorithm will, with high probability, output an x_i that is in the target? Or is the target-information available to the algorithm restricted entirely to the search matrix in the sense that its probability of successfully locating the target depends entirely on the information contained in those two rows [2]? The options here are wide and varied.

We consider next several examples of how the search matrix might work in practice.

Example 1.1: Uniform random sampling with perfect knowledge

In this case, each x_i is selected according to a uniform distribution across the dixie cups, each α_i records whether x_i belongs to the target (1 for yes, 0 for no), and each β_i directs the helicopter to take a uniform random sample in locating the next point in the search space (that being x_{i+1}). The reference to "perfect knowledge" here signifies that for each query we know exactly whether the pea sampled (each x_i)

belongs to the target (in which case $\alpha_i = 1$) or not (in which case $\alpha_i = 0$). If any α_i equals 1, we can stop the search right there and produce x_i as an instance of a successful search. Alternatively, we can fill out the search matrix rather than leave it incomplete, and then produce the first x_i for which α_i equals 1 (producing simply x_1 if none of the α_is equals 1). Given that the proportion of high-yield peas (i.e., the target) has probability p, the probability that this search is successful is $1-(1-p)^m$.

Example 1.2: Uniform random sampling with zero knowledge

In this case, as before, each x_i is selected according to a uniform distribution across the dixie cups; moreover, each β_i directs the helicopter to take a uniform random sample in locating the next point in the search space (that being $x_i + 1$). This time, however, examining the peas reveals nothing about whether they belong to the target. This might happen, for instance, if high-yield and low-yield peas are visually indistinguishable and we have no way of otherwise discriminating them (as we might through genetic analysis or actually planting them). The reference to "zero knowledge" therefore signifies that for each query we know nothing about whether the pea sampled (x_i) belongs to the target. In this case, each of the α_is may be treated as equal to 0. Given that the proportion of high-yield peas (i.e., the target) has probability p, the probability that this search successfully identifies a particular x_i in the target is simply p. Accordingly, a sample size of m greater than 1 does nothing here to improve on the probability of locating the target if we have no means of obtaining knowledge about the peas we are sampling.

Note that the probability that some element in the first row of the search matrix belongs to the target is $1-(1-p)^m$. This is the probability of successful search as calculated in the previous example, which presupposed perfect knowledge. Nevertheless, for a search to be successful in the present example, it is not enough for the search matrix merely to have a target element appear in the first row. In addition, it must be possible to explicitly identify one element in the first row taken to be the best candidate for belonging to the target. Moreover, because this is a needle-in-the-haystack problem, successful search requires that the candidate selected must belong to the target with probability considerably larger than p. With zero knowledge about whether elements in the first row of the search matrix belong to the target, however, no candidate selected from that row stands a better chance of belonging to the target than any other. In that case, each such candidate has the very small probability p of belonging to the target.

Example 1.3: Uniform random sampling with partial knowledge

In this example, as in the previous two, each x_i, when first selected, follows a uniform distribution across the dixie cups. Yet, to determine whether a given x_i actually does belong to the target, two agricultural tests may need to be performed on it. The tests work as follows: if both yield a positive result (denoted by a 1), then the candidate x_i belongs to the target; if one or both yield a negative result (denoted by 0), then it does not belong to the target. Moreover, the performance of each of these tests requires a single query. Thus, to determine whether an x_i that is in the target actually does belong to it, the dixie cup over it will have to be removed and replaced twice, meaning that x_i itself will appear twice in the top row of the search matrix, implying that under those appearances the corresponding α_is will both be 1.

On the other hand, if on either of the tests, the first query performed yields a 0, then there's no point in performing the other test, and x_i need appear only once in the top row. Given a query that for the first appearance of x_i yields an α_i equal to 1, x_i will need to be queried again to determine whether it indeed belongs to the target. Once a given x_i's inclusion in or exclusion from the target is determined, the next query is uniformly random across the dixie cups. In this case, the probability p' of hitting the target over m queries will be strictly between the probabilities determined in the last two examples, i.e., $p < p' < 1 - (1 - p)^m$, where p is the zero-knowledge lower bound and $1 - (1 - p)^m$ is the perfect-knowledge upper bound. The exact value of p' will depend on Bayesian considerations relating to how negative results on the two agricultural tests are distributed (in terms of prior probabilities) among the non-target elements.

Example 1.4: Smooth gradient fitness with single peak

In this case, we begin by turning over a randomly chosen dixie cup, examining the pea under it (x_1), and recording its fitness (α_1). We assume that the fitness function over the peas has a smooth gradient (in other words, fitness does not zigzag up and down as we move in a given direction over the large table, which is our search space) and that it has a single peak (in other words, any local maximum is also the [unique] global maximum). In this case, a hill-climbing strategy is appropriate, so each β_i directs the helicopter to search in the neighborhood of the x_i that, so far in the search, has attained the highest value of fitness α_i, looking to increase the fitness still further. There's no reason in this case to repeatedly query a given pea since we assume that fitness can be precisely determined in a given query and that fitness does not vary from one query to another (in other words, the fitness function here is "time independent"). Once m queries in this search have been carried out, we consult the search matrix and choose, as our best candidate for landing in the target, the x_i that attains the highest value of fitness α_i. The probability that such a search is successful will depend on the sample size m, the initialization (i.e., the procedure for deciding where the search begins), the precise characteristics of the fitness function, and how efficiently the search samples points in the neighborhood of an already sampled point to improve fitness (i.e., to "climb the hill").

2. General Targeted Search

A precise mathematical characterization of the general search scenario described in the last section now looks as follows. Let $\Omega = \{\omega_1, \omega_2, ..., \omega_K, \omega_{K+1}, ..., \omega_N\}$ be the search space, which we assume to be finite (this assumption can be relaxed and we have done so in other work, but doing so entails no substantive gain in generality). Let $T = \{\omega_1, \omega_2, ..., \omega_K\}$ be the target and define the probability $p = K/N$. A search of Ω for T then consists of the following 6-tuple: $(\iota, \tau, O_\alpha, O_\beta, \mathcal{A}, \Delta)$. The items in this 6-tuple denote respectively the *initiator*, the *terminator*, the *inspector*, the *navigator*, the *nominator*, and the *discriminator*. Here is what these six items mean:

Initiator. The initiator ι, denoted by the Greek *iota*, starts the ball rolling. It is the procedure by which the search determines where to begin. The initiator ι is responsible for x_1, and possibly additional members of the search space x_2 through x_k, that appear as the first entries in the first row of the search matrix. In many searches the initiator does nothing more than choose a single search space element (i.e., x_1) at random according to some probability distribution.

Terminator. The terminator τ, denoted by the Greek *tau*, provides a stop criterion for ending the search. Because all searches are limited to a maximum number of queries m (i.e., the sample size), the terminator can always simply be identified with the policy to cut off the search after m queries. In practice, however, terminators often end a search before the maximal number of queries have been made because the search is deemed to have achieved success before this maximal number. In that case, the search matrix may be incomplete, with missing entries in the columns to the right of the last column for which a point in the search space was queried. Without loss of generality, we can then fill up the columns that are missing entries by repeating the last complete column. Alternatively, we can just leave the columns empty.

Inspector. The inspector O_α is an oracle that, in querying a search-space entry, extracts information bearing on its probability of belonging to the target T. We assume that O_α is a function mapping into some range of values capable of providing information about the degree to which members of the search space Ω give evidence of belonging to the target T. Quite often, the domain of O_α is merely Ω, and O_α maps into $\{0,1\}$, returning a 1 if an element of Ω is in the target, 0 otherwise. Alternatively, O_α may map into the singleton $\{0\}$, returning the same element regardless of the element of Ω in question, thus providing zero information about target elements. O_α may even assume misleading values, suggesting that search-space entries are in the target when they are not and vice versa. Besides taking on discrete values, O_α may also take on more continuous values, signaling the degree to which a search-space entry is likely to be included in a target, as with a fitness function. The possible forms that O_α can take are wide and varied. Without loss of generality, however, we assume that the range of values that the inspector can take is finite.

As the inspector, O_α's task is to fill the second row of the search matrix and thus provide evidence about the degree to which corresponding elements in the first row may belong to the target. Accordingly, all the α_is in the second row take values in O_α's range. Nevertheless, given that a single query may not provide all the information that the inspector is capable of providing about a given element from the search space, the inspector may perform multiple queries on a given search-space element and may even use information gained from different previously queried elements in answering the present query. Thus, given an element x_i in the search space that's just been selected, its value as assigned by O_α can depend on the entire partial matrix

$$\begin{bmatrix} x_1 & \cdots & x_{i-1} & x_i & ** \\ \alpha_1 & \cdots & \alpha_{i-1} & * & ** \\ \beta_1 & \cdots & \beta_{i-1} & * & ** \end{bmatrix}.$$

Here ellipses denote elements of the search matrix that have been filled in, single asterisks denote individual missing entries, and double asterisks denote possibly multiple missing entries. In the case at hand, O_α uses the partial search matrix given here to determine α_i. If it ignores all entries of the partial search matrix prior to column $i - 1$, then we say that O_α is *Markov*. If it determines α_i solely on the basis of x_i, we say that O_α operates *without memory* (otherwise, *with memory*).

Navigator. Like the inspector O_α, the navigator O_β is an oracle. Given that we are at

$$\begin{bmatrix} x_1 & \cdots & x_{i-1} & x_i & ** \\ \alpha_1 & \cdots & \alpha_{i-1} & \alpha_i & ** \\ \beta_1 & \cdots & \beta_{i-1} & * & ** \end{bmatrix}$$

in the search process, the navigator takes this partial search matrix and returns the value β_i, which directs (navigates) the search as it attempts to locate the next entry in the search space (i.e., x_{i+1}). O_β maps into a fixed range of values, which in the search matrix we denote by βs. As with the inspector, if O_β ignores all entries of the partial search matrix prior to column $i - 1$, then we say that O_β is *Markov*. If it determines β_i solely on the basis of x_i and α_i, we say that O_β operates *without memory* (otherwise, *with memory*).

The type of information that O_β delivers can be quite varied. It can provide distance of search-space elements from the target. It can provide information about the smoothness of fitness. It can provide information about how likely neighbors of a given search-space element are to be in the target. Moreover, it can combine these types of information. Whereas the inspector O_α confines itself to extracting information that bears on the probability of search-space elements residing in the target, the navigator O_β focuses on information that helps guide the search to the target. As with the inspector, we assume that the range of values the navigator may take is finite. For (mathematically) smooth fitness functions, this will entail discretizing the values that the fitness function may assume. Yet, because the degree to which searches can discriminate such information is always strictly limited (in practice, distinct measurements when sufficiently close become empirically indistinguishable), assuming a finite range of values for the navigator entails no loss of generality.

Nominator. The nominator $\mathcal{A}$ is the update rule that, given a search matrix filled through to the ith column and thus incorporating the most current information from the inspector and navigator, explicitly identifies (and thereby "nominates") the next element to be queried, namely x_{i+1}. Thus $\mathcal{A}$ takes us from the search matrix

$$\begin{bmatrix} x_1 & \cdots & x_i & * & ** \\ \alpha_1 & \cdots & \alpha_i & * & ** \\ \beta_1 & \cdots & \beta_i & * & ** \end{bmatrix}$$

to the updated search matrix

$$\begin{bmatrix} x_1 & \cdots & x_i & x_{i+1} & ** \\ \alpha_1 & \cdots & \alpha_i & * & ** \\ \beta_1 & \cdots & \beta_i & * & ** \end{bmatrix}$$

We denote the nominator by $\mathcal{A}$ (for "algorithm") because, in consulting the inspector and navigator to determine the next search-space element to be queried, it acts as the basic underlying algorithm of the search, running through all the target candidates that the search will consider. We say that the nominator is *Markov* if its selection of x_{i+1} depends solely on the ith column of the search matrix. We say that it operates without memory if its selection of x_{i+1} is independent of prior columns of the search matrix.

To say that the nominator nominates an element x_{i+1} based on the partial search matrix

$$\begin{bmatrix} x_1 & \cdots & x_i & * & ** \\ \alpha_1 & \cdots & \alpha_i & * & ** \\ \beta_1 & \cdots & \beta_i & * & ** \end{bmatrix}$$

may seem to entail a loss of generality since in many searches (e.g., genetic algorithms and particle swarms), multiple candidates from the search space tend to be generated in batches. Thus with genetic algorithms, for instance, all candidates of a given reproduction cycle appear at the same time. Accordingly, if, say, 100 offspring are generated at each reproduction cycle, the new partial search matrix is not

$$\begin{bmatrix} x_1 & \cdots & x_i & x_{i+1} & ** \\ \alpha_1 & \cdots & \alpha_i & * & ** \\ \beta_1 & \cdots & \beta_i & * & ** \end{bmatrix}$$

but rather

$$\begin{bmatrix} x_1 & \cdots & x_i & x_{i+1} & \cdots & x_{i+100} & ** \\ \alpha_1 & \cdots & \alpha_i & * & ** & * & ** \\ \beta_1 & \cdots & \beta_i & * & ** & * & ** \end{bmatrix}.$$

Given this last matrix, we can then let the inspector and navigator fill in the columns below x_{i+1} to x_{i+100} one column at a time proceeding left to right. Alternatively, we can simply require the nominator to proceed one column at a time (thus taking a given batch of candidates one by one in sequence), letting the inspector and navigator fill in that column before proceeding to the next. Both cases are mathematically equivalent. For some searches, it makes better intuitive sense for the nominator to nominate a whole batch of search-space elements at a time. But this can always be made equivalent to nominating one element of the batch at a time until the entire batch is exhausted. For simplicity, we tend to adopt this latter approach. Another possibility is to change the search space so that each element consists of multiple elements from the original search space (see example 3.5).

Discriminator. Once a search matrix

$$\begin{bmatrix} x_1 & x_2 & x_3 & \cdots & x_m \\ \alpha_1 & \alpha_2 & \alpha_3 & \cdots & \alpha_m \\ \beta_1 & \beta_2 & \beta_3 & \cdots & \beta_m \end{bmatrix}$$

that's compatible with $\mathcal{A}$ has been formed, it's time to decide which x_i that appears in the first row is most likely to belong to the target T. With a complete search matrix in hand, it's not enough to suspect that some entry somewhere in the first row belongs to T. For the search to be successful, we need to know which of these entries in fact belongs to T or, if definite knowledge of inclusion in T is not possible, then which of these entries is more likely than the rest to belong to T. Choosing from the first row of the search matrix the most likely candidate that belongs to T is the job of the discriminator Δ. As such, the discriminator is a function into the search space Ω from possible search matrices (i.e., from $3 \times m$ matrices whose first row consists of elements from Ω, whose second row consists of elements from the range of the inspector, and whose third row consists of elements from the range of the navigator). For each such search matrix, the discriminator outputs the element x_i in the first row that it regards as most likely to belong to T.

Discriminators can vary in quality. *Self-defeating discriminators* that, whenever possible, select first-row entries belonging outside the target are an option. For a given search matrix, such discriminators minimize the probability of successfully locating the target. Also an option are *independent-knowledge discriminators* that can identify whether a first-row entry belongs to the target with greater certainty than is possible simply on the basis of the information delivered by the inspector and navigator (information found in the second and third rows of the search matrix). Thus, the discriminator might have access to a source of information about target inclusion that is less ambiguous than what is available to the inspector and navigator. Such discriminators would thereby introduce information external to the search matrix to locate those elements in the first row most likely to belong to the target. By contrast, *no-independent-knowledge discriminators* would select x_i from the first row based solely on information contained in the second and third rows of the search matrix. Such variations among discriminators are easily multiplied and formalized. We leave doing so as an exercise to the reader.

Although the discriminator Δ as here described is a function from complete search matrices to the search space Ω, in fact we allow Δ also to be a function from partial search matrices to Ω, in keeping with the terminator's ability to stop a search when success in fewer than m queries has likely been achieved. Recall that partial search matrices can always be filled up with redundant columns and thus turned into complete search matrices. Hence, allowing partial search matrices entails no gain in generality, nor does restricting ourselves to complete search matrices entail a loss of generality.

Each of the six components of a search $S = (\iota, \tau, O_\alpha, O_\beta, \mathcal{A}, \Delta)$ can be stochastic. Thus, the initiator might choose x_1 according to some probability distribution.

Likewise, the terminator may end the search depending on chance factors relevant to success being achieved. The inspector and navigator, at any given stage in forming the search matrix, may draw on a stochastic source to randomize its outputs. So too, the nominator and discriminator may choose their candidates in part randomly. It follows that a search S can be represented as a random search matrix consisting of three discrete stochastic processes X, Y, and Z:

$$S = \begin{pmatrix} X_1 & X_2 & \dots & X_m \\ Y_1 & Y_2 & \dots & Y_m \\ Z_1 & Z_2 & \dots & Z_m \end{pmatrix}.$$

Here X represents the search-space elements delivered by the nominator (or the initiator for X_1), Y the corresponding outputs of the inspector, and Z the corresponding outputs of the navigator. X therefore takes values in Ω, Y in the range of O_α, and Z in the range of O_β.

Alternatively, S can be conceived as a vector-valued stochastic process $\vec{W}$ where each

$$\vec{W}_i = \begin{pmatrix} X_i \\ Y_i \\ Z_i \end{pmatrix},$$

in which case

$$S = \left(\vec{W}_1 \ \vec{W}_2 \ \cdots \ \vec{W}_m\right).$$

Applying the discriminator Δ to this random search matrix thus yields an Ω-valued random variable $\Delta(S)$, which we denote by X_S. As an Ω-valued random variable, X_S therefore induces a probability distribution μ_s on Ω that entirely characterizes the probability of S successfully locating the target T. In this way, an arbitrary search S can be represented as a single probability distribution or measure μ_s on the original search space Ω. This representation will be essential throughout the sequel.

As noted at the start of this paper, this representation of searches as probability measures is central to our formalization of conservation of information. If it were obvious that searches could in general be represented this way, we might just as well have omitted these first three sections. But given that a general characterization of targeted search is itself a point at issue in determining the scope and validity of conservation of information, these preliminary sections were in fact necessary. Logically speaking, however, these sections come up only tangentially in the sequel by guaranteeing that searches can indeed be represented as probability measures.

3. Search Examples

In this section, we consider several further examples of targeted search, expanding on the examples given at the end of section 1.

Example 3.1: Uniform random sampling with perfect knowledge and without replacement

In the last section, we considered a search of m queries in which, at each query, the entire search space was sampled uniformly. This led to independent and identically distributed uniform random variates in the first row of the search matrix, 0s and 1s in the second row depending on whether the corresponding entry in the first row was respectively outside or inside the target, and in the third row a directive simply to continue uniform random sampling. The discriminator in this case simply looked for a first-row entry with a 1 directly below it in the second row. Accordingly, with uniform probability $p = K/N$ of the target T in the search space Ω, we calculated the probability of successful search at $1 - (1 - p)^m$. This probability, however, assumes that the first row of the search matrix was sampled *with replacement* and thus may repeat elements of the search space.

We can, on the other hand, have the navigator direct the search to avoid elements in the search space previously queried (*this implies* a memory of previously queried elements). If all other aspects of the search are kept the same, the search space is then sampled *without replacement* so that each query is uniform with respect to elements of the search space yet to be queried. This sampling procedure yields a hypergeometric distribution. Thus for $\Omega = \{\omega_1, \omega_2, \ldots, \omega_K, \omega_{K+1}, \ldots, \omega_N\}$, $T = \{\omega_1, \omega_2, \ldots, \omega_K\}$, $p = K/N$, and m not exceeding $N - K$, the probability that this search locates the target is then

$$1 - \frac{\binom{N-K}{m}}{\binom{N}{m}}.$$

Moreover, if N is much larger than m, this probability approximately equals the "with replacement probability" $1-(1-p)^m$, underscoring the well-known fact that sampling without replacement only negligibly improves the efficiency of search compared to sampling with replacement unless the sample size m is large [3].

Example 3.2: Easter egg hunt

Imagine a very large two-dimensional grid with Easter eggs hidden under various squares of the grid. You are able to move around the grid by going from one square to an adjacent square. Thus you can move one square vertically, horizontally, or diagonally, like a king in chess:

You start out on a randomly chosen square, which is determined by the initiator. The terminator gives you at most m squares to examine. When you are on a given square, the inspector tells you whether you are over an Easter egg (by saying "yes") or not (by saying "no"). If "yes," uncover the square on which you are standing, locate the egg underneath, and end the search.

Given that you have moved from one square to another with neither being over an Easter egg, the navigator tells you whether the square you are currently on is closer to, the same distance from, or further from the nearest Easter egg (by saying "warmer," "same," or "colder"; distance between squares A and B is calculated as minimum number of steps needed to reach B from A). Notice that the navigator cannot provide such information until the initiator has designated the first square and the nominator has designated the second. Thus, for the very first square chosen by the initiator, the navigator simply puts down "same."

If for your current square the navigator says "warmer," the nominator says to choose that square from your immediate neighborhood that takes you in the same direction as your last move. If for your current square the navigator says "same," the nominator says to choose at random a square that you have not yet visited in the immediate neighborhood of the current square. If for your current square the navigator says "colder," the nominator says to return to the previous square and randomly choose a square in its immediate neighborhood that you have not yet visited. Proviso: the nominator ignores any column with "colder," in subsequent search treating it as though it were not part of the search matrix except for not revisiting its square when sampling nearest neighbors. This proviso prevents the search from getting stuck. Finally, the discriminator returns the first square under which an Easter egg was found if an egg is indeed found; otherwise, it returns the square chosen by the initiator.

The Easter egg hunt so described falls within our general framework for targeted search.

Example 3.3: Competitive search

In competitive search, elements of the search space Ω are conceived as "players" whose skill can be evaluated and ranked according to certain "performance criteria." Evolutionary computing typically employs a single performance criterion given by a fitness function. Fitness thus provides a *single-objective* measure of optimality — one and only one thing needs to be optimized, and when it is optimized we have the undisputed best player. In many circumstances, however, optimality is *multi-objective*, that is, there are several competing things we are trying to optimize simultaneously, where a rise in one leads to a drop in another. Optimization with multiple performance criteria thus requires a balancing or compromise among rival objectives. How these criteria are balanced determines what we regard as the "best players," that is to say, the target.

Just what constitutes the right balance of performance criteria is not written in stone but constitutes a judgment call [4]. Consider a search space consisting of all college men's basketball players in a given year. Professional NBA teams are seeking the best basketball players in this search space — the very best presumably being the player picked first in the first round of the NBA draft. But what determines the best players? Many performance criteria come to mind: speed, height, field-goal percentage, three-point percentage, average number of rebounds per game, average number of blocked shots per game, average number of assists per game, etc. etc. All these performance criteria need to be suitably combined to determine who are the best players and thus what constitutes the target of the search. Some years, this balancing of performance criteria is straightforward, so that one player stands out head and shoulders above the rest. At other times, different teams may have different needs, leading them to emphasize certain performance criteria over others, so that no player is completely dominant and no target is universally agreed upon.

How we combine and balance performance criteria depends on our needs and interests. Suppose, to change examples, you are a college admissions officer. Your search space is all graduating high school students and you are trying to find those who will thrive at your institution. This is your search, that is, to find the "right" students. Prospective students need to take the Scholastic Aptitude Test (SAT). The test provides two main scores, a verbal and a math score (each varying between 200, which is worst, and 800, which is best) [5]. Each of these scores corresponds to a performance criterion and requires a search query. With these two queries performed on each high school student, how do you now select the best students for your school (leaving aside other performance criteria such as GPA and recommendations)? Do you add the scores together, as is commonly done? Or do you weight one more than the other and, if so, how?

If your school focuses mainly on liberal arts, you will want to weight the verbal portion more strongly than the math portion. Thus, even though you may want to see a combined score of 1200 or better, you will favor students who get a 750 verbal/450 math over students who get a 450 verbal/750 math. If, on the other hand, yours is an engineering school, then you will prefer the latter over the former. Some schools don't discriminate the two scores but simply add them to give a combined performance measure for the test. Besides adding scores or weighting them, one can introduce arbitrary cut-offs. Thus, one might require that no student be admitted who performs less than 500 on either test, thereby ensuring that both verbal and math scores exceed a certain threshold. This suggests a maxi-min approach to combining performance measures: take the minimum of the two SAT scores and try to recruit those students whose minima are maximal. The precise formulation of such combined performance measures is straightforward. The trick is to find the right combination that suits one's purposes.

In such examples of competitive search, evaluating how a search space element fares with respect to the various performance criteria is the job of the inspector. To evaluate a search space element's competitiveness, the inspector may need to query it several times. Sometimes, however, a single query is enough.

In basketball, for instance, a player whose free throw percentage is less than 10 percent can be eliminated from consideration for the NBA draft without needing to consult any other performance criteria. Alternatively, a player who scores over 100 points a game on average (a performance achieved just once in the entire history of the NBA, as it is, by Wilt Chamberlain) will rise to the very top of the player pool even if we don't know any of his other stats. Knowing which queries to make conditional on which queries have already been made is essential to constructing an effective competitive search.

Example 3.4: Tournament play

Tournament play is a special case of competitive search in which the players display their competitive abilities by playing against each other. In tournament play, there are as many performance criteria as there are players, each player's competitiveness being gauged by how well one performs in relation to the others. Basketball is an example of tournament play, though in this case the unit of search is not the individual player (as it was in the last example) but the team. In chess, on the other hand, the unit of search tends to be the individual player (though team play is also known, as when local chess clubs play each other).

Tournament play is typically represented by a square anti-symmetric matrix with blanks down the diagonal (players don't play themselves) and opposite outcomes mirrored on either side of the diagonal. For instance, in the St. Petersburg Chess Congress of 1909, the tournament matrix was as follows [6]:

St. Petersburg 1909

		1	2	3	4	5	6	7	8	9	0	1	2	3	4	5	6	7	8	9		
1	Rubinstein	*	1	1	1	½	½	½	1	1	1	½	1	0	1	½	1	1	1	1	14½	875 Rubles
2	Lasker	0	*	½	1	½	1	1	1	½	1	1	1	0	1	1	1	1	1	1	14½	875 Rubles
3	Spielmann	0	½	*	1	0	1	1	½	1	½	½	½	1	0	½	1	½	½	1	11	475 Rubles
4	Duras	0	0	0	*	0	1	½	0	½	1	0	1	1	1	1	1	1	1	1	11	475 Rubles
5	Bernstein	½	½	1	1	*	0	1	0	1	1	1	1	½	0	0	0	½	½	1	10½	190 Rubles
6	Teichmann	½	0	0	0	1	*	0	½	½	½	½	1	1	½	1	½	1	1	½	10	120 Rubles
7	Perlis	½	0	0	½	0	1	*	½	½	1	½	1	1	½	1	½	0	0	1	9½	80 Rubles
8	Cohn	0	0	½	1	1	½	½	*	0	0	1	½	½	0	½	½	½	1	1	9	40 Rubles
9	Schlechter	0	½	0	½	0	½	½	1	*	1	0	0	1	1	½	0	1	½	1	9	40 Rubles
10	Salwe	0	0	½	0	0	½	0	1	0	*	½	1	1	1	½	0	1	1	1	9	40 Rubles
11	Tartakower	½	0	½	1	0	½	½	0	1	½	*	0	0	0	½	1	1	1	½	8½	
12	Mieses	0	0	½	0	0	0	0	½	1	0	1	*	½	1	1	1	0	1	1	8½	
13	Dus Chotimirsky	1	1	0	0	½	0	0	½	0	0	1	½	*	½	½	½	1	0	1	8	
14	Forgács	0	0	1	0	1	½	½	1	0	0	1	0	½	*	½	½	½	0	½	7½	
15	Burn	½	0	½	0	1	0	0	½	½	½	½	0	½	½	*	1	½	½	0	7	
16	Vidmar	0	0	0	0	1	½	½	½	1	1	0	0	½	½	0	*	½	1	0	7	
17	Speyer	0	0	½	0	½	0	1	½	0	0	0	1	0	½	½	½	*	½	½	6	
18	Von Freyman	0	0	½	0	½	0	1	0	½	0	0	0	1	1	½	0	½	*	0	5½	
19	Znosko Borovsky	0	0	0	0	0	½	0	0	0	0	½	0	0	½	1	1	½	1	*	5	

Emanuel Lasker, who tied with Akiba Rubinstein for first place, was at the time the world champion. Even so, Rubinstein, though he never played Lasker in a title match (back then challengers had to raise sufficient funds before they could arrange such a match with the world champion), was in the five years preceding World War I regarded as the strongest player in the world (in 1912 he won five international tournaments in a row, a feat unparalleled). Yet both Rubinstein and Lasker were defeated by Dus Chotmirsky, a chess player who would be lost to history except for this feat.

In chess tournaments, winners are decided by summing performance across all games (assigning 1 to a win, ½ to a draw, and 0 to a loss) and then selecting the player(s) with the highest total. This is standard practice, but one can imagine variations of it. We might want simply to focus on victories and count them. In that case, Lasker would have won the tournament outright (with 13 victories), Rubinstein would have come in second (with 12 victories), and Duras would have come in third (with 10 victories). On the other hand, we might want to choose the victor based on fewest losses. In that case Rubinstein would have been the outright victor (with a single loss), Lasker would have taken second (with 2 losses), and Spielmann would have taken third (with 3 losses). Other options for balancing performance criteria in tournament play are possible as well. For instance, Chotmirsky, for having defeated the two top performing players as determined by conventional tournament standards, might have been rewarded extra points for doing so.

In tournament play, exhaustive search means each player playing all the other players and recording the outcomes. Most chess tournaments, however, have so many players that an exhaustive search is not possible. The St. Petersburg tournament was a select invitational meet. Most tournaments are open to the chess community. In such tournaments, players are initially matched in line with their official chess ratings (1600 for amateur, 2000 for expert, 2200 for master, 2500 for grandmaster), with weaker players initially playing stronger players so that the best players don't cancel each other out early. Then, as the rounds proceed (typically between six to eight rounds total), players with the same tournament record, or as close a record as available, are matched. Note that weaker players, as gauged by their rating coming into the tournament, tend at each round to be matched with stronger players. At the end of the tournament, one's wins and draws are summed (1 for a win, ½ for a draw). The tournament winner(s) are then decided in terms of this raw total as well as an algorithm that, in the case of ties, takes into account the strength, as determined by chess rating, of one's opponents.

In terms of the search matrix, especially when the pool of tournament players is quite large, each player can play only a few other players (each game played constituting a single query, the outcomes of these games being noted by the inspector). Given that one wants to discover the strongest players (for some specified method of balancing performance criteria, these players, taken jointly, constitute the target), the search needs to be judicious in the choice of players it uses to query individual players. Is it more effective, given at most m queries, to query as many players as possible by playing them against only one or a few other players? Or is it better to hone in on a few players and put them through their paces by having them play a wide cross-section of other players? It all depends. It depends on whether player strength tends to function transitively (if A is able to defeat B and B is able to defeat C, is A able to defeat C?). It depends on whether, in the case of a single player testing other players, this player is strong or weak.

The discriminating power of strong players is important in tournament play. A strong player, by definition, loses only to a few players (i.e., to other strong players), and thus will clearly discriminate strong from weak players. In contrast, a weak player, by losing to most players, will fail to discriminate all but the weakest players. To change the game from chess to baseball, if the test of a team is whether it performs well against the New York Yankees, any team that does reasonably well against them could rightly be considered excellent. But if the "test team" is drawn from the local little league, then it would not provide a useful way of determining teams of national excellence. But notice, using the New York Yankees as a test team may not be very useful either — a team that beats or keeps it close with the Yankees is surely top notch, but if all the teams tested fail miserably against the Yankees, we may learn nothing about their relative strength. Players that are overly strong or overly weak are poor discriminators of play excellence.

In sum, tournament play is a special case of competitive search that fits within our general search framework but in which performance is assessed by the players (i.e., the search space elements) playing each other directly and then by noting the winner and, if applicable, the margin of victory [7].

Example 3.5: Population search

For some searches, the concern is not simply with individuals exhibiting certain characteristics but with whole populations exhibiting those characteristics. Thus, in population genetics, the emergence of a trait in a lone individual is not enough. Traits can come and go within a population. The point of interest is when a trait gets fixed among enough members of a population to take hold and perpetuate.

To represent such a scenario, we might imagine all the members of a population as drawn from a set of individuals Ω'. Moreover, Ω' may contain a target T'

consisting of all members exhibiting the trait(s) in question. Yet the actual search is not for T' in Ω' but for sets, whether ordered or unordered, of members from Ω'; what's more, the target will consist of such sets that have a sufficient proportion of members from T'. Thus, in a standard evolutionary computing scenario, the actual search space Ω might consist of all 100-tuples of elements drawn from Ω' (each element of the first row of the search matrix would belong to this Ω). Moreover, the actual target T might, in this case, consist of 100-tuples drawn from Ω' for which 75 or more of their elements belong to T'. In this case, successful search would require 75 percent of the population to have acquired the given trait(s).

Many ways of transitioning from Ω' to Ω and T' to T are possible here depending on the population size (is it fixed or variable?), on whether the order of elements in the population is important, and on the threshold that determines whether individually determined characteristics are widespread enough for the population to have properly acquired them. Even though the natural search space may seem to be one we have called Ω', representing the search within the general framework outlined in this paper may require identifying another space, which we called Ω. The actual target we are trying to locate would thus belong not to Ω' but to Ω. Note that such an Ω will invariably have more structure than Ω', even supplying a metric of comparison in terms of how many members of Ω' the members of Ω share.

4. Information and Efficiency Measures

In a general theory of search that avoids arbitrary assumptions about underlying probability distributions, uniform probabilities nonetheless play a salient role. Consider our general set-up, a search space $\Omega = \{\omega_1, \omega_2, ..., \omega_K, \omega_{K+1}, ..., \omega_N\}$ with target $T = \{\omega_1, \omega_2, ..., \omega_K\}$, where the target has uniform probability $\mathbf{U}(T) = p = K/N = |T|/|\Omega|$, where $|*|$ is the cardinality of $*$. In any such scenario, we can always do at least as good as take a single uniform random sample and thereby attain a target element with probability p. We might conduct our search to improve this probability or we might conduct it to diminish this probability. The natural probability distribution attaching to Ω, given the idiosyncrasies of this search space, may be very different from uniform. But it is always, in principle, possible to enumerate the elements of a finite space Ω and choose one of them randomly so that no element is privileged over any other. Uniformity, even if destined to miss the target in any nontrivial search, is always an option.

To take a single uniform random variate from the search space Ω will be called the *null search*. This search becomes the baseline against which we compare all other searches. Any search different from the null search will be called an *alternative search* [8]. The null search induces the uniform probability distribution **U** on Ω (see section 2). This is the probability measure we get by setting our sample size at $m = 1$ and letting the discriminator act on the corresponding 3×1 search matrix whose first row element is simply a uniformly chosen element of the search space. In practice, p is so small that a null search over Ω for T is extremely unlikely to succeed. Success therefore demands that in place of the null search, an alternative search S be implemented that succeeds with a probability q that is considerably larger than p. The search S thus induces, in the notation of section 2, a probability distribution μ_s on Ω that entirely characterizes the probability of S successfully locating the target T. For simplicity, we denote μ_s simply by μ. In this way, an alternative search S reduces to a single probability distribution μ on the original search space Ω where the probability of the target is $\mu(T) = q$.

In comparing null and alternative searches, it is convenient to convert probabilities to information measures (note that all logarithms in the sequel are to the base 2). We therefore define the *endogenous information* I_Ω as $-\log(p)$, which measures the inherent difficulty of a blind or null search in exploring the underlying search space Ω to locate the target T. We then define the *exogenous information IS* as $-\log(q)$, which measures the difficulty of the alternative search S in locating the target T. And finally, we define the *active information* I_+ as the difference between the endogenous and exogenous information: $I_+ = I_\Omega - I_S = \log(q/p)$. Active information therefore measures the information that must be *added* (hence the plus sign in I_+) on top of a null search to raise an alternative search's probability of success by a factor of q/p.

In the null search, the sample size is fixed at 1 (a single uniform random variate is taken) whereas in the alternative search the sample size is m (m queries are made). If we make m explicit, then we can define q_m as the probability that the alternative search locates the target in m queries, and write $I_s^m = -\log(q_m)$ and $I_+^m = I_\Omega - I_s^m$. The behavior of I_+^m as a function of m now provides a measure of the efficiency of search. Suppose, for instance, that S conducts its search by taking independent and identically distributed random variates. In that case, assuming m to be much less than $1/p$, $q_m = 1 - (1 - p)^m$ is approximately equal to mp, and I_+^m is approximately $\log(m)$. If, instead, S conducts its search by, at each query, cutting in half the search space ("interval halving"), then the probability of finding the target increases by a factor of 2 for every query, and I_+^m is approximately m (i.e., the active information is linear rather than logarithmic in m). Interval halving is therefore a much more efficient search strategy (if it can be implemented) than uniform random sampling. I_+^m, as a function of m, therefore measures the efficiency of the search.

By comparing the performance of a search S against the endogenous information baseline I_Ω, I_+^m provides an absolute measure of efficiency of the search. Indeed, in specifying S, we define $I_+^m(S) = I_+^m$, conceived as a function of m, as the *absolute efficiency* of S. Given two searches, S and S', we define $(I_+^m(S' \mid S) = I_+^m(S) - I_+^m(S')$, again conceived as a function of m, as the *relative efficiency* of S' given S. Thus, if S represents uniform random sampling and S' represents interval halving, the relative efficiency of S' given S, $I_+^m(S' \mid S)$ is $m - \log(m)$. In general, for a given m, if S' induces a probability of r_m on T and if S induces a probability of q_m on T, then $I_+^m(S' \mid S) = \log(r_m/q_m)$. Absolute and relative efficiency can also be negative: for a given m, S does worse in locating the target than a single uniform random sample if and only if $I_+^m(S) < 0$; likewise, for a given m, S' does worse in locating the target than S if and only if $I_+^m(S' \mid S) < 0$. Note that if S represents a single uniform random sample, so that the search matrix has only a single column and is incomplete for all remaining $m - 1$ columns (the first entry in the first row is therefore a uniform random variate), then $I_+^m(S' \mid S) = I_+^m(S')$.

5. Liftings and Lowerings

Conservation of information tracks the information that goes into constructing a search, showing that the amount of information exhibited by the search in locating a target can never exceed the amount of information inputted in its construction. Accordingly, conservation of information addresses not just the search for a given target in the original search space, but also a search for the information that goes into rendering such a search successful. Conservation of information therefore is not about search per se but about the search for a search. In other words, it is about a higher-level search for the information required to render a lower-level search successful. We abbreviate "the search for a search" by S4S.

In section 2 we represented an arbitrary search (i.e., S) as a probability measure on a search space (i.e., μ_s). Given that the search for a search (S4S) is itself a search, it must likewise be representable as a probability measure. Such an S4S probability measure assigns probabilities to a higher order search space consisting of probability measures on the original search space. Formulating conservation of information requires the ability to project probability measures up and down a probabilistic hierarchy of search spaces. We show how this is done in this section. This section thus provides the formal background for the conservation of information theorems proved in the next section.

We consider again our general set-up, a search space $\Omega = \{\omega_1, \omega_2, ..., \omega_K, \omega_{K+1}, ..., \omega_N\}$ with target $T = \{\omega_1, \omega_2, ..., \omega_K\}$, where the target has uniform probability $\mathbf{U}(T) = p = K/N = |T|/|\Omega|$. We assume that $1 \le K < N$. Define $\mathbf{M}(*)$ as the set of all Borel probability measures on * where * is any compact metric space. Ω, as a finite set, is compact in the discrete topology, which is given by any metric on it. Any probability measure m on Ω therefore has the form

$$\sum_{i=1}^{N} a_i \delta_{x_i},$$

where each a_i is a nonnegative real number, the a_is together sum to 1, and each δ is a point mass (assigning probability 1 to the corresponding x_i). It follows that $\mathbf{M}(\Omega)$ is the set of all these convex linear combinations of point masses. Note that when each a_i equals $1/N$, this sum of point masses is the uniform probability $\mathbf{U}$ in $\mathbf{M}(\Omega)$.

We can think of the point masses δ_{x_i} (for $1 \leq i \leq N$) as N independent vectors in an N-dimensional vector space. Because these vectors are all added as convex linear combinations to form $\mathbf{M}(\Omega)$, $\mathbf{M}(\Omega)$ in fact sits in an $(N-1)$-dimensional subspace, forming an N-simplex with Euclidean metric. Moreover, as a closed, bounded subset of Euclidean space, $\mathbf{M}(\Omega)$ is compact. It follows that the uniform probability on $\mathbf{M}(\Omega)$ is ordinary Lebesgue measure (suitably normalized). We denote this uniform probability over $\mathbf{M}(\Omega)$ as $\bar{\mathbf{U}}$. $\bar{\mathbf{U}}$ resides in $\mathbf{M}(\mathbf{M}(\Omega))$. For convenience, we therefore define $\mathbf{M}^0(\Omega) = def\, \Omega$, $\mathbf{M}^1(\Omega) = def\, \mathbf{M}(\Omega)$, $\mathbf{M}^2(\Omega) = def\, \mathbf{M}(\mathbf{M}(\Omega))$, and in general $\mathbf{M}^{j+1}(\Omega) = def\, \mathbf{M}(\mathbf{M}^j(\Omega))$.

Thus, to recap, the uniform probability $\mathbf{U}$ over Ω resides in $\mathbf{M}(\Omega)$ and is defined as

$$\mathbf{U} = \frac{1}{N}\sum_{i=1}^{N} \delta_{x_i};$$

moreover, the uniform probability $\bar{\mathbf{U}}$ over $\mathbf{M}(\Omega)$ resides in $\mathbf{M}^2(\Omega)$ and is isomorphic to normalized Lebesgue measure on the N-simplex $\{(a_1, \ldots, a_N) \in \mathbf{R}^N \mid a_i \geq 0, \Sigma_{1 \leq i \leq N} a_i = 1\}$. We call the various $\mathbf{M}^j(\Omega)$, taken together, the *probabilistic hierarchy* over the search space Ω. Note that we give each of these spaces in the probabilistic hierarchy the weak topology. It then follows by Prohorov's theorem that each of these spaces is compact (indeed, they form compact metric spaces in the Kantorovich-Wasserstein metric, which induces the weak topology on these spaces) [9][i].

The probabilistic hierarchy allows for considerable interaction among its measure spaces, so that structures associated with $\mathbf{M}^j(\Omega)$ have corresponding structures both up and down the hierarchy at $\mathbf{M}^{j+1}(\Omega)$ and $\mathbf{M}^{j-1}(\Omega)$. We speak of a structure at $\mathbf{M}^j(\Omega)$ projected up to $\mathbf{M}^{j+1}(\Omega)$ as a *lifting* and a structure at $\mathbf{M}^{j+1}(\Omega)$ projected down to $\mathbf{M}^j(\Omega)$ as a *lowering*. To see how this works, we take the higher-order space $\mathbf{M}(\Omega)$ and the lower-order space Ω and examine how structures associated with these spaces can be projected to the other. Our discussion here will focus on the base of the probabilistic hierarchy (i.e., Ω and $\mathbf{M}(\Omega)$), but our observations readily generalize up the probabilistic hierarchy. Accordingly, structures associated with Ω may be *lifted* to structures associated with $\mathbf{M}(\Omega)$ and structures associated with $\mathbf{M}(\Omega)$ may correspondingly be *lowered* to structures associated with Ω.

To start, consider a real-valued function f on Ω (note that because Ω is finite and has a discrete topology, f is bounded, measurable, and even topologically continuous). The function f now lifts to a real-valued (continuous) function $\overline{f}$ on $\mathbf{M}(\Omega)$ that takes any probability measure θ in $\mathbf{M}(\Omega)$ and assigns its integral with f, i.e., $\overline{f}$ is the mapping from $\mathbf{M}(\Omega)$ to $\mathbf{R}$ such that

$$\theta \mapsto \int_{\Omega} f(x) d\theta(x).$$

Note that for $\theta = \delta_x$ (i.e., the point mass at x), $\overline{f}(\theta) = \overline{f}(\delta_x) = f(x)$. Call $\overline{f}$ the *lifting* of f from Ω to $\mathbf{M}(\Omega)$. Likewise, for F a real-valued function on $\mathbf{M}(\Omega)$, define $\tilde{F}$ on Ω as $x \mapsto F(\delta_x)$. Call $\tilde{F}$ the *lowering* of F from $\mathbf{M}(\Omega)$ to Ω. It then follows that $\tilde{\overline{f}} = f$, but it need not be the case that $\overline{\tilde{F}} = F$ (lowerings can lose information whereas liftings do not). In general, under the weak topology, liftings and lowerings of functions preserve measurability and continuity.

Next, consider a probability measure μ on Ω (μ is therefore in $\mathbf{M}(\Omega)$). Because Ω is finite, all probability measures in $\mathbf{M}(\Omega)$ are absolutely continuous with respect to the uniform probability $\mathbf{U}$. Absolute continuity of μ with respect to $\mathbf{U}$ means that every set of nonzero probability under μ also has nonzero probability under $\mathbf{U}$. By the Radon-Nikodym theorem, it follows that μ can be rewritten as the product of a density, denoted by $\frac{d\mu}{d\mathbf{U}}$, times the measure $\mathbf{U}$. This means that for $f = \frac{d\mu}{d\mathbf{U}}$, μ can be written as $f \cdot d\mathbf{U}$. In other words, for a set A contained in Ω,

$$\mu(A) = \int_A f(x) d\mathbf{U}(x)$$

In particular, if $\mu = \sum_{i=1}^{N} a_i \delta_{x_i}$, $f(x_i) = \frac{d\mu}{d\mathbf{U}}(x_i) = a_i \cdot N$.

It follows that by lifting f from a function on Ω to a function $\overline{f}$ on $\mathbf{M}(\Omega)$, we can now lift μ from a probability measure in $\mathbf{M}(\Omega)$ to a probability measure $\overline{\mu}$ in $\mathbf{M}^2(\Omega)$. Specifically, for B a measurable subset of $\mathbf{M}(\Omega)$, we define

$$\overline{\mu}(B) = \int_B \overline{f}(\theta) d\overline{\mathbf{U}}(\theta)$$

where

$$\bar{f}(\theta) = \int_{\Omega} f(x)d\theta(x).$$

To see that $\bar{\mu}$ is indeed a probability measure over $\mathbf{M}(\Omega)$, we need the following result.

Proposition 5.1 (Consistency of Uniformity)

$$\mathbf{U} = \int_{\mathbf{M}(\Omega)} \theta \, d\bar{\mathbf{U}}(\theta).$$

Remarks. The integral on the right side of this equation is vector-valued [10]. Such integrals exist provided that in applying continuous linear functionals to them (which, in this case, amounts to integrating with respect to all bounded continuous real-valued functions on Ω), one gets the same result as integrating over the continuous linear functions applied inside the integral. Linear functionals thereby reduce vector-valued integration to ordinary integration. Thus, the equality in the statement of this theorem means that for all continuous real-valued h on Ω,

$$\int_{\Omega} h(x)d\mathbf{U}(x) = \int_{\mathbf{M}(\Omega)} \left[\int_{\Omega} h(x)d\theta(x) \right] d\bar{\mathbf{U}}(\theta).$$

Because Ω is finite, all real-valued functions on Ω are continuous, so this equality needs to hold for all real-valued h. As we move up the probabilistic hierarchy, subsequent $\mathbf{M}^j(\Omega)$ are compact metric spaces, so continuity actually does place a restriction on the continuous linear functionals used in calculating vector-valued integrals. Because these are all compact metric spaces, existence and uniqueness of such vector-valued integrals is not a problem [11]. For equality to hold in $\mathbf{U} = \int_{\mathbf{M}(\Omega)} \theta \, d\bar{\mathbf{U}}(\theta)$ means that averaging all probability measures on $\mathbf{M}(\Omega)$ with respect to the uniform probability $\bar{\mathbf{U}}$ is just the uniform probability $\mathbf{U}$ on Ω. This establishes measure-theoretic consistency in lifting the uniform probability $\mathbf{U}$ on Ω to the uniform probability $\bar{\mathbf{U}}$ on $\mathbf{M}(\Omega)$.

Proof. This result follows from exchangeability — $\mathbf{U}$ is the only probability measure invariant under permutation of the elements of Ω. The vector-valued integral in question can immediately be seen to have this same property — its value does not depend on any point in Ω to which it is applied. A detailed proof is available elsewhere [12].

Suppose now that μ is a probability measure on Ω that is absolutely continuous with respect to $\mathbf{U}$ (in fact, because Ω is finite, this assumption holds for all probability measures on Ω). Let $\frac{d\mu}{d\mathbf{U}}$ denote the Radon-Nikodym derivative of μ with respect to $\mathbf{U}$ and let $\overline{\frac{d\mu}{d\mathbf{U}}}$ denote its lifting. If we now define the lifting of μ as $\bar{\mu} = \overline{\frac{d\mu}{d\mathbf{U}}} d\bar{\mathbf{U}}$, then $\bar{\mu}$ is a probability measure on $\mathbf{M}(\Omega)$. Moreover, since $\mathbf{U}$ is absolutely continuous with itself such that $\frac{d\mathbf{U}}{d\mathbf{U}}$ is identically equal to 1 on Ω, it follows that the lifting of $\frac{d\mathbf{U}}{d\mathbf{U}}$, i.e., $\overline{\frac{d\mathbf{U}}{d\mathbf{U}}}$, is identically equal to 1 on $\mathbf{M}(\Omega)$, and thus the lifting of $\mathbf{U}$, as so defined, is in fact the uniform probability on $\mathbf{M}(\Omega)$. Thus, whether we interpret $\bar{\mathbf{U}}$ as the uniform probability on $\mathbf{M}(\Omega)$ as ordinary Lebesgue measure (suitably normalized) on an N-simplex (which is isomorphic to $\mathbf{M}(\Omega)$), or as the lifting of the uniform probability $\mathbf{U}$ on Ω, both signify the same probability measure on $\mathbf{M}(\Omega)$.

To see that all the claims in the previous paragraph hold, it is enough to see that $\bar{\mu}$ is indeed a probability measure on $\mathbf{M}(\Omega)$, and for this it is enough to see that

$$\begin{aligned}
\int_{\mathbf{M}(\Omega)} \overline{\frac{d\mu}{d\mathbf{U}}}(\theta) d\bar{\mathbf{U}}(\theta) &= \int_{\mathbf{M}(\Omega)} \left[\int_{\Omega} \frac{d\mu}{d\mathbf{U}}(x) d\theta(x) \right] d\bar{\mathbf{U}}(\theta) \\
&= \int_{\Omega} \frac{d\mu}{d\mathbf{U}}(x) d\left[\int_{\mathbf{M}(\Omega)} \theta \, d\bar{\mathbf{U}}(\theta) \right](x) \\
&= \int_{\Omega} \frac{d\mu}{d\mathbf{U}}(x) d\mathbf{U}(x) \text{ [by Cons. of Unif.]} \\
&= \int_{\Omega} d\mu \\
&= 1.
\end{aligned}$$

Lastly, we need to be able to lift targets from Ω to $\mathbf{M}(\Omega)$. Thus, given the target T in Ω, we define a corresponding higher-order target $\bar{T}_q$ in $\mathbf{M}(\Omega)$, indexed by q in the unit interval ($0 \le q \le 1$), namely,

$$\bar{T}_q = \{\theta \in \mathbf{M}(\Omega) \mid \theta(T) \ge q\}.$$

$\bar{T}_q$ equals $\mathbf{M}(\Omega)$ when q is 0 and grows smaller as q increases. Elsewhere [13] we have shown that for the search space $\Omega = \{\omega_1, \omega_2, ..., \omega_K, \omega_{K+1}, ..., \omega_N\}$ with target $T = \{\omega_1, \omega_2, ..., \omega_K\}$, where the target has uniform probability $\mathbf{U}(T) = p = K/N = |T|/|\Omega|$, the (higher-order) uniform probability of $\bar{T}_q$ is given by

$$\bar{\mathbf{U}}(\bar{T}_q) = \frac{\Gamma(N)}{\Gamma(N(1-p))\Gamma(Np)} \int_0^{1-q} t^{N(1-p)-1}(1-t)^{Np-1} dt.$$

Note that this last expression describes a cumulative beta distribution with first parameter $r = N(1 - p)$ and second parameter $s = Np$ [14].

6. Conservation of Information — The Uniform Case

We are now in a position to prove two conservation of information theorems: the special case for uniform probabilities, which we have proved elsewhere and recap here in this section; and the general case for arbitrary probabilities, which we prove for the first time in the next section [15]. We begin with the special case.

Theorem 6.1 (Conservation of Information — Uniform Case)

Let T be a target in Ω. Assume Ω is finite and T is nonempty. Let $\mathbf{U}$ denote the uniform probability distribution on Ω and let $p = |T|/|\Omega| = \mathbf{U}(T)$ (which we take to be extremely small). Next, let μ be a probability distribution on Ω such that $q = \mu(T)$ (which we take to be considerably larger than p). Suppose that μ characterizes the probabilistic behavior of an alternative search S, so that the endogenous information is $I_\Omega = -\log(p)$ and the exogenous information is $I_S = -\log(q)$. Then the (higher-order) uniform probability of $\overline{T}_q$ in $\mathbf{M}(\Omega)$, denoted by $\overline{\mathbf{U}}(\overline{T}_q)$, is less than or equal to p/q. Equivalently, the (higher-order) endogenous information associated with finding the (higher-order) target $\overline{T}_q$ in $\mathbf{M}(\Omega)$, i.e., $-\log(\overline{\mathbf{U}}(\overline{T}_q))$, is bounded below by the (lower-order) active information $I_+ = -\log(\mathbf{U}(T)) + \log(\mu(T)) = \log(q/p)$.

Proof. Let $\Omega = \{x_1, x_2, ..., x_K, x_{K+1}, ..., x_N\}$ and $T = \{x_1, x_2, ..., x_K\}$ so that $p = K/N$. As we saw in the last section, it then follows that

$$\overline{\mathbf{U}}(\overline{T}_q) = \frac{\Gamma(N)}{\Gamma(N(1-p))\Gamma(Np)} \int_0^{1-q} t^{N(1-p)-1}(1-t)^{Np-1}\,dt,$$

which is a cumulative beta distribution with first parameter $r = N(1 - p)$ and second parameter $s = Np$.

Integration by substitution shows that this expression can be rewritten as

$$\frac{\Gamma(N)}{\Gamma(Np)\Gamma(N(1-p))} \int_q^1 t^{Np-1}(1-t)^{N(1-p)-1}\,dt,$$

which describes a cumulative beta distribution with first parameter $r = Np$ and second parameter $s = N(1-p)$. It is well known that the mean for this distribution is $r/(r+s)$ [16]. In consequence,

$$\begin{aligned}
\frac{\Gamma(N)}{\Gamma(Np)\Gamma(N(1-p))}\int_q^1 t^{Np-1}(1-t)^{N(1-p)-1}\,dt &= \frac{\Gamma(N)}{\Gamma(Np)\Gamma(N(1-p))}\int_q^1 \frac{q}{q}\cdot t^{Np-1}(1-t)^{N(1-p)-1}\,dt \\
&= \frac{1}{q}\cdot\frac{\Gamma(N)}{\Gamma(Np)\Gamma(N(1-p))}\int_q^1 q\cdot t^{Np-1}(1-t)^{N(1-p)-1}\,dt \\
&\le \frac{1}{q}\cdot\frac{\Gamma(N)}{\Gamma(Np)\Gamma(N(1-p))}\int_0^1 t\cdot t^{Np-1}(1-t)^{N(1-p)-1}\,dt \\
&= \frac{1}{q}\cdot\frac{Np}{Np+N(1-p)} \\
&= \frac{p}{q}.
\end{aligned}$$

It follows that –log $(\bar{\mathbf{U}}(\bar{T}_q))$, is bounded below by the active information $I_+ = \log(q/p)$. This proves the theorem.

This theorem characterizes the probability costs incurred by a search for a search. Given a vast search space Ω and a tiny target T, the probability of finding the target via the null search is effectively nil ($p = |T|/|\Omega|$). To find the target, we thus need an alternative search S that is able to find it with a probability q that is much larger than p. But where did S come from? Because the complexities and idiosyncrasies associated with the construction of searches in general, the first three sections of this paper focused on simplifying our representation of searches, first by representing them as search matrices and then by representing them as probability measures μ on the original search space Ω such that $\mu(T) = q$.

So the question now is, Where did μ come from? In statistics, whenever confronted with a given outcome, the statistician attempts to situate it among a collection of possible outcomes that are *at least as extreme* as the one in question and then inquires into the improbability of that collection. For instance, given a thousand coin tosses and six-hundred heads, the statistician's first impulse will be to ask how likely it is that a fair coin (the null hypothesis) could have led to six-hundred *or more* heads. In this case, the statistician wants the probability of the tail of a binomial distribution. Leaving aside Bayesian considerations, which can always be incorporated later, if the probability of this tail is extremely small, the statistician will be inclined to question whether the coin responsible for six-hundred heads was fair, thereby implicating an alternative hypothesis. As it is, six-hundred or more heads in a thousand coin tosses represents a departure from expectation by more than six standard deviations. Such a result with a fair coin would be very improbable indeed.

Returning now to our search space Ω and target T, the outcome that confronts us is not a sequence of coin tosses but a search S represented by the probability measure μ. If we set aside that the search is the product of intelligent design, then μ presumably results from some statistical process. Moreover, the collection of outcomes as extreme as μ is then

$$\overline{T}_q = \{\theta \in \mathbf{M}(\Omega) \mid \theta(T) \geq q\}.$$

In our analogy with statistical practice, $\overline{T}_q$ may then be conceived as the "tail" associated with the "outcome" μ. It would follow that the improbability of this tail is crucial to deciding whether μ is the outcome of a (higher-level) null search.

The parallel here between coin tossing and the search for a search, though far from exact, is suggestive and illuminating. Each coin toss, under the null hypothesis, is a Bernoulli trial, with probability of ½ for heads and ½ for tails. These trials are probabilistically independent, and thus in one-thousand trials should conform to a null hypothesis characterized by a binomial distribution with parameters $N = 1{,}000$ and $p = ½$. The lower-order Bernoulli trials, as it were, "lift" to a higher-order binomial distribution. Similarly, the null search of Ω for T, characterized by the uniform probability $\mathbf{U}$ on Ω, lifts to a null search of $\mathbf{M}(\Omega)$ for $\overline{T}_q$, characterized by the (higher-order) uniform probability $\overline{\mathbf{U}}$. Conservation of information then shows that the uniform probability of this higher-order target is bounded above by p/q.

Conservation of information is essentially an accounting rule for probabilities associated with search. Here is how it works: finding the original target T within Ω had the very low probability of p under the null search. Fortunately, an alternative search S was available to raise this probability to q. But the probability cost of locating this alternative search, represented by μ, was less than or equal to p/q. Thus, when the cost of locating the alternative search is factored in, nothing is gained over the original null search. The original search, as it were, purchased the target at the "high" probability cost p. The alternative search, correspondingly, purchased the target at the "cheaper" probability cost q, but then itself incurred a probability cost of at least p/q in a higher-order search space since the alternative search itself had to be accounted for. Thus, when the full probability costs incurred by the alternative search are factored in, the total cost is the same as or even worse than the probability cost associated with the original null search.

In fact, the cost tends to be much worse. Conservation of information in the uniform case states that $\bar{\mathbf{U}}(\bar{T}_q) \leq p/q$. Nevertheless, we have shown elsewhere [17] that for $\Omega = \{x_1, x_2, ..., x_K, x_{K+1}, ..., x_N\}$ and $T = \{x_1, x_2, ..., x_K\}$, provided that $p = K/N$ and $N \geq (2q-1)/(q-p)$,

$$\bar{\mathbf{U}}(\bar{T}_q) < \sqrt{N} \cdot \frac{\sqrt{p}}{q} \cdot \left[1-(q-p)^2\right]^N.$$

This (strict) inequality shows that the (higher-order) uniform probability of the lifted target $\bar{T}_q$ decreases exponentially with the absolute size N of the search space Ω. As an upper bound on $\bar{\mathbf{U}}(\bar{T}_q)$, p/q is therefore very conservative.

To see how the probability costs associated with null and alternative searches relate, it is instructive to consider the following two quasi-Bayesian ways of reckoning these costs:

$$\begin{aligned}
\mathbf{P}(\text{locating } T \text{ via null search}) &= \mathbf{P}(\text{null search locates } T \text{ \& null search is available})\\
&= \mathbf{P}(\text{null search locates } T|\text{null search is avail.})\\
&\quad \times \mathbf{P}(\text{null search is avail.})\\
&= \mathbf{U}(T) \times 1 \text{ [because the availability of null search is}\\
&\quad \text{taken for granted]}\\
&= p.
\end{aligned}$$

$$\begin{aligned}
\mathbf{P}(\text{locating } T \text{ via alt. search}) &= \mathbf{P}(\text{alt. search locates } T \text{ \& alt. search is available})\\
&= \mathbf{P}(\text{alt. search locates } T \mid \text{alt. search is avail.})\\
&\quad \times \mathbf{P}(\text{alt. search is avail.})\\
&= \mu(T) \times \bar{\mathbf{U}}(\bar{T}_q)\\
&\leq q \times p/q\\
&= p.
\end{aligned}$$

It follows that $\mathbf{U}(T) \geq \mu(T) \times \bar{\mathbf{U}}(\bar{\mathrm{T}}_q)$ and therefore, by taking negative logarithms, that $I_\Omega \leq I_S - \log(\bar{\mathbf{U}}(\bar{T}_q))$, or equivalently that $-\log(\bar{\mathbf{U}}(\bar{T}_q)) \geq I_+ = \log(q/p)$, inasmuch as $I_+ = I_\Omega - I_S$, $I_\Omega = -\log(\mathbf{U}(T)) = -\log(p)$, and $I_S = -\log(\mu(T)) = -\log(q)$. According to conservation of information, the higher-order endogenous information $-\log(\bar{\mathbf{U}}(\bar{T}_q))$, required to find a search qua probability measure that has probability q or better of locating T, is always at least that of the lower-order active information I_+. To say that information is conserved is thus really to say that in the search for a search, information leading to success of the original search is at best conserved when moving to a higher-order search space and may in fact grow considerably higher (in some circumstances, exponentially higher). This rise in the information/probability cost associated with higher-level search should not be surprising given that spaces comprising searches tend to be bigger and structurally richer than the spaces they are searching [18].

7. Conservation of Information — The General Case

We turn now to a generalization of the previous conservation of information theorem. The previous theorem was formulated in terms of a uniform probability baseline. We now lift this restriction. Processes that exhibit stochastic behavior arise from what may be called a *natural probability*. The natural probability characterizes the ordinary stochastic behavior of the process in question. Often the natural probability is the uniform probability. Thus, for a perfect cube with distinguishable sides composed of a rigid homogenous material (i.e., an ordinary die), the probability of any one of its six sides landing on a given toss is 1/6. Yet, for a loaded die, those probabilities will be skewed, with one side consuming the lion's share of probability. For the loaded die, the natural probability is not uniform. Now, if the natural probability for all search spaces Ω were the uniform probability $\mathbf{U}$, we'd be done — the conservation of information theorem proved in the last section would suffice. Yet despite Bernoulli's principle of insufficient reason, which we have argued elsewhere rightly makes the uniform probability the default in many searches [19], the natural probability associated with some searches need not be uniform.

Given structural and external factors influencing search, the natural probability need not be $\mathbf{U}$ but some probability measure μ that assigns probability q to the target T. It's thus convenient to extend the notion of a null search to include not just uniform or blind searches but any searches that accord with such a natural probability. Accordingly, we may then say that μ characterizes the null search of Ω for T. Moreover, the alternative search will then be characterized by a probability measure ν that assigns probability r to T. As the natural probability on Ω, μ is not confined simply to Ω but lifts to $\mathbf{M}(\Omega)$, so that its lifting, namely $\overline{\mu}$, becomes the natural probability on $\mathbf{M}(\Omega)$ (this parallels how the uniform probability $\mathbf{U}$, when it is the natural probability on Ω, lifts to the uniform probability $\overline{\mathbf{U}}$ on $\mathbf{M}(\Omega)$, which then becomes the natural probability for this higher-order search space). When μ is the natural probability associated with a search space, treating it as the null search and ν as the alternative search now leads to a general conservation of information theorem, one that point for point parallels the previous formulation for uniform probabilities.

Theorem 7.1 (Conservation of Information — General Case)

Let T be a target in Ω. Assume Ω is finite and T is nonempty. Let $\mathbf{U}$ denote the uniform probability distribution on Ω and let $p = |T|/|\Omega| = \mathbf{U}(T)$ (which we take to be extremely small). Next, let μ and ν be probability measures on Ω such that $q = \mu(T)$ and $r = \nu(T)$. We assume that $p \leq q < \mathrm{r}$ (the rationale for assuming that q is no less than p is discussed at the end of this section). Suppose that μ characterizes the probabilistic behavior of a search S and that ν characterizes the probabilistic behavior of a search S'. We treat μ as the null search and ν as the alternative search, thus making μ the natural probability associated with Ω. Accordingly, $I_S = -\log(q)$ becomes the endogenous information and $I_S' = -\log(r)$ the exogenous information. It then follows that the (higher-order) natural probability of $\overline{T}_r$ in $\mathbf{M}(\Omega)$, i.e., $\overline{\mu}(\overline{T}_r)$, is less than or equal to q/r. Equivalently, the (higher-order) endogenous information associated with finding the (higher-order) target $\overline{T}_r$ in $\mathbf{M}(\Omega)$, i.e., $-\log(\overline{\mu}(\overline{T}_r))$, is bounded below by the (lower-order) active information $I_+ = -\log(\mu(T)) + \log(\nu(T)) = \log(r/q)$.

Remark 1. The probabilities r and q in this theorem correspond respectively to q and p in Theorem 6.1. We changed notation because it seemed best to let p continue to denote the uniform probability of the target. Outside the notation of this theorem, however, we shall typically refer to a null search as setting a baseline probability p and an alternative search as giving an improved probability of success q. Thus, outside the notation of this theorem, we shall generally refer to the active information cost of search in terms of $\log(q/p)$ rather than $\log(r/q)$.

Remark 2. Regressing up the probabilistic hierarchy (i.e., Ω, $\mathbf{M}(\Omega)$, $\mathbf{M}^2(\Omega)$, $\mathbf{M}^3(\Omega)$, etc.) does nothing to mitigate the information cost of successful search. In fact, it intensifies the cost. Searching for a target T in the original search space Ω against a baseline natural probability μ in $\mathbf{M}(\Omega)$, we find that the difficulty of the search is only exacerbated by searching for the higher-order target $\overline{T}_r$ with respect to the higher-order natural probability $\overline{\mu}$ in $\mathbf{M}^2(\Omega)$. The proof below can now be applied again up the probabilistic hierarchy, showing that the search for a still higher-order target aimed at resolving the original search requires the still higher-order natural probability $\overline{\overline{\mu}}$ in $\mathbf{M}^3(\Omega)$, and that this move again only intensifies the difficulty of the search. And so on, up the probabilistic hierarchy.

From the vantage of conservation of information, *searches are no less real than the objects being searched.* Just as the existence and structure of objects require explanation, so too the existence and structure of the searches that locate those objects require explanation. It follows that searches, by residing in a space of searches, are themselves objects to be searched. This implies a hierarchy of searches: the original search, the search for that search, the search for the search for that search, etc. Conservation of information entails that as we regress up this search hierarchy, the search problem never becomes easier and may in fact become more difficult.

Proof. Let $\Omega = \{x_1, x_2, ..., x_K, x_{K+1}, ..., x_N\}$ and $T = \{x_1, x_2, ..., x_K\}$ so that $p = K/N$. Since Ω is finite, the probability measures μ and ν are absolutely continuous with respect to $\mathbf{U}$, and so there exist non-negative real-valued functions f and g such that $\mu = f \cdot d\mathbf{U}$ and $\nu = g \cdot d\mathbf{U}$. As we saw in section 5, the lifting of μ is now defined as $\overline{\mu} = \overline{f} \cdot d\overline{\mathbf{U}}$, where $\overline{\mathbf{U}}$ is the uniform probability on $\mathbf{M}(\Omega)$. Thus, for B a measurable subset of $\mathbf{M}(\Omega)$,

$$\overline{\mu}(B) = \int_B \overline{f}(\theta) d\overline{\mathbf{U}}(\theta),$$

where for θ in $\mathbf{M}(\Omega)$,

$$\overline{f}(\theta) = \int_\Omega f(x) d\theta(x).$$

In section 5 we showed that $\overline{\mu}$ is indeed a probability measure over $\mathbf{M}(\Omega)$. Because μ is the natural probability on Ω, $\overline{\mu}$ is the natural probability on $\mathbf{M}(\Omega)$.

Next, for the lifted target $\overline{T}_r = \{\theta \in \mathbf{M}(\Omega) \mid \theta(T) \geq r\}$ define the following measure on Ω resulting from vector-valued integration:

$$\mathbf{V}_r = def \int_{\overline{T}_r} \theta \, d\overline{\mathbf{U}}(\theta)$$

This definition holds for any r in the unit interval. Note that when $r = 0$ (an equality that does not in fact holds since we assume that r is no smaller than p), then $\overline{T}_r$ is all of $\mathbf{M}(\Omega)$; on the other hand, when $r > 0$, then $\overline{T}_r$ is a proper subset of $\mathbf{M}(\Omega)$. It follows that $\mathbf{V}_r$ is a probability measure on Ω only if $r = 0$ and is a sub-probability measure otherwise (i.e., it assigns measure less than 1 to Ω if $r > 0$).

What value less than 1 does $\mathbf{V}_r$ assign to all of Ω? The answer can be seen from the following equation:

$$\mathbf{V}_r(\Omega) = \left[\int_{\overline{T}_r} \theta \, d\overline{\mathbf{U}}(\theta) \right](\Omega) = \int_{\overline{T}_r} \theta(\Omega) d\overline{\mathbf{U}}(\theta) = \int_{\overline{T}_r} 1 \cdot d\overline{\mathbf{U}}(\theta) = \overline{\mathbf{U}}(\overline{T}_r),$$

which, by conservation of information in the uniform case (Theorem 6.1), we know to be bounded above by p/r.

By consistency of uniformity (Proposition 5.1), we know that $\mathbf{V}_0$ is just the uniform probability $\mathbf{U}$. For $r > 0$, it is easily seen that $\mathbf{V}_r$ is exchangeable on T and on its complement T^c. In other words, $\mathbf{V}_r$ is invariant under permutations of T and of T^c. Since the only such exchangeable measures are those proportional to uniform probabilities, this means that there exist positive constants a_r and b_r such that

$$\mathbf{V}_r = a_r \mathbf{U}(\cdot \mid T) + b_r \mathbf{U}(\cdot \mid T^c)$$

Here $\mathbf{U}(\cdot|T)$ is the uniform probability on T and $\mathbf{U}(\cdot|T^c)$ is the uniform probability on T^c. Note that because $\mathbf{V}_0$ is the uniform probability on Ω, $a_0 = p$ and $b_0 = 1 - p$.

Now, because $\mu = f \cdot d\mathbf{U}$ with $\mu(T) = q$ and $\mu(T^c) = 1 - q$, integrating f with respect to $\mathbf{V}_r$, which is proportional to a uniform probability measure on T and on T^c, is the same as integrating the function $\frac{q}{p}1_T + \frac{1-q}{1-p}1_{T^c}$ with respect to $\mathbf{V}_r$, where 1_T and 1_{T^c} are the indicator functions for T and T^c respectively. This is because integrating a real-valued function with respect to a uniform probability measure is the same as integrating its average value with respect to a uniform probability measure (the average of f on T being $\frac{q}{p}$ and the average of f on T^c being $\frac{1-q}{1-p}$).

It follows that

$$
\begin{aligned}
\bar{\mu}(\bar{T}_r) &= \int_{\bar{T}_r} \bar{f}(\theta)d\bar{\mathbf{U}}(\theta) \text{ [unpacking definitions]} \\
&= \int_{\bar{T}_r}\left[\int_{\Omega} f(x)d\theta(x)\right]d\bar{\mathbf{U}}(\theta) \text{ [unpacking definitions]} \\
&= \int_{\Omega} f(x)d\left[\int_{\bar{T}_r}\theta\, d\bar{\mathbf{U}}(\theta)\right](x) \text{ [by vector-valued integration]} \\
&= \int_{\Omega} f(x)d\mathbf{V}_r(x) \text{ [by definition]} \\
&= \int_{\Omega}\left[\frac{q}{p}1_T + \frac{1-q}{1-p}1_{T^c}\right]d\mathbf{V}_r(x) \text{ [as noted above]} \\
&\leq \frac{q}{p}\cdot\int_{\Omega} d\mathbf{V}_r(x) \text{ [because } q \geq p\text{]} \\
&= \frac{q}{p}\cdot\bar{\mathbf{U}}(\bar{T}_r) \text{ [because } \mathbf{V}_r(\Omega) = \bar{\mathbf{U}}(\bar{T}_r)\text{]} \\
&\leq \frac{q}{p}\cdot\frac{p}{r} \text{ [by cons. of info., unif. case]} \\
&= \frac{q}{r}.
\end{aligned}
$$

This proves the theorem.

The theorem just proved assumes that the null search assigns a probability q to T that is at least as large as the uniform probability p. But what if the "natural" probability on the search space entails a null search that is worse at locating the target than uniform random sampling, so that q is strictly less than p? We put "natural" in scare quotes here because, we submit, natural probabilities need never do worse than uniformity. To see this, consider a deck of cards and imagine we are searching for the ace of hearts. Presented with a deck, face down, we are told to draw the first card on top. What is the probability that it will be the ace of hearts? If we knew that the deck had just been thoroughly shuffled, then we would be justified in assigning the uniform probability of 1/52 to the top card being the ace of hearts. But if we knew absolutely nothing about how the deck came to assume its order, the uniformity assumption becomes questionable, requiring for its justification Bernoulli's disputed principle of indifference [20].

Now imagine we learn that that the deck gets thoroughly shuffled, but that whenever the ace of hearts appears on top, a coin is flipped so that heads leaves it there but tails moves it to the bottom of the deck. Given this way of randomly arranging the deck, the probability of the top card being the ace of hearts is not 1/52 but $1/52 \times 1/2 = 1/104$. In this case, the probability of drawing the ace of hearts is strictly less than its uniform probability. Considerations such as this suggest that the uniformity assumption, though appropriate in many circumstances, doesn't hold universally for search. But this example additionally suggests that we don't need to stay with a sub-uniform probability when conducting a search. Precisely because we are searching for the ace of hearts, we don't have to sample the first card at the top of the deck. Search implies we have freedom to move about the search space and thus, in the present example, to sample any card in the deck. Hence, by suitably randomizing the selection, we can ensure that the card picked had the uniform probability 1/52 of being the ace of hearts. In general, then, when conducting a search, we are in our rights to assume that we can always do at least as well as uniformity. Doing worse, at least for search, is *unnatural*. Thus, in the context of search, any natural probability that replaces the uniform probability may be taken to assign a higher probability of success in locating the target than the uniform probability.

8. Regulating the Information Industry

Conservation of information supplies the information industry with a balance sheet, ensuring that the information output on one side of the ledger does not exceed the information input on the other. Specifically, conservation of information guarantees that any search that proportionately raises the probability of finding a target by q/p requires, in its construction, an amount of information not less than the active information $I_+ = \log(q/p)$. Simply put, raise the probability of successful search by a factor of q/p, incur an information cost of $\log(q/p)$.

At the time of this writing, the United States government is much exercised about regulating the financial industry. Essential to any such regulation is accurate accounting of money — where it originates, how it flows, and where it ends up. Conservation of information shows that active information, like money, obeys strict accounting principles. Just as banks need money to power their financial instruments, so searches need active information to power their success in locating targets. Moreover, just as banks must balance their books, so searches, in successfully locating targets, must balance their books — they cannot output more information than was inputted.

Regulation of the financial industry is necessary because it is too easy to mask liabilities as assets and thereby attempt to escape one's obligations. Likewise, regulation of the information industry is necessary because it is too easy to focus on the success of a search and forget the information that paid for that success. The temptation is to inflate the creative power of search programs by conveniently forgetting the creative power of the programmers who impart the information that makes those programs successful. In short, the temptation is to ignore conservation of information in the hopes of a free lunch.

Conservation of information keeps the search practitioner honest.

Acknowledgment

The authors thank Peter Olofsson and Dietmar Eben for helpful feedback on previous work of the Evolutionary Informatics Lab, feedback that has found its way into this paper.

References and Notes

1. See, for instance, William A. Dembski and Robert J. Marks II, "Conservation of Information in Search: Measuring the Cost of Success," *IEEE Transactions on Systems, Man and Cybernetics A, Systems & Humans* 5(5) (September 2009): 1051–1061 and also William A. Dembski and Robert J. Marks II, "Life's Conservation Law: Why Darwinian Evolution Cannot Create Biological Information," in Bruce Gordon and William Dembski, eds., *The Nature of Nature* (Wilmington, Del.: ISI Books, 2010).
2. Compare Joseph Culberson's remarks about evolutionary and adaptive algorithms: "Evolutionary algorithms (EAs) are often touted as 'no prior knowledge' algorithms. This means that we expect EAs to perform without special information from the environment. Similar claims are often made for other adaptive algorithms." From

Joseph C. Culberson, "On the Futility of Blind Search: An Algorithmic View of 'No Free Lunch'," *Evolutionary Computation* 6(2) (1998): 109–127.

3. For the hypergeometric distribution and its limiting behavior, see G. R. Grimmett and D. R. Stirzaker, *Probability and Random Processes* (Oxford: Oxford University Press, 1982), 46–47.
4. "In contrast to single-objective optimization, a solution to a multi-objective problem is more of a concept than a definition. Typically, there is no single global solution, and it is often necessary to determine a set of points that all fit a predetermined definition for an optimum. The predominant concept in defining an optimal point is that of Pareto optimality." Quoted from R. T. Marler and J. S. Arora, "Survey of Multi-Objective Optimization Methods for Engineering," *Structural and Multidisciplinary Optimization* 26 (2004): 371.
5. In 2005, the SAT introduced a writing component and thus another score from 200 to 800. For simplicity, we ignore this change in the test.
6. Taken from http://www.endgame.nl/stpeter.htm (last accessed April 7, 2010).
7. Because our framework for search includes these last two examples, all the results in this paper apply to coevolutionary scenarios of the sort described in David Wolpert and William Macready, "Coevolutionary Free Lunches," *IEEE Transactions on Evolutionary Computation* 9(6) (December 2005): 721–735.
8. In other work of the Evolutionary Informatics Lab (www.evoinfo.org), we have referred to the null search as an "unassisted search" or a "blind search" and the alternative search as an "assisted search" — see for example Dembski and Marks, "Conservation of Information in Search." The language of "null" and "alternative" searches is in analogy to statistics.
9. See William A. Dembski and Robert J. Marks II, "The Search for a Search: Measuring the Information Cost of Higher Level Search," *Journal of Advanced Computational Intelligence and Intelligent Informatics* 14(5) (2010): 475–486.
10. For vector-valued integration, see Nicolae Dinculeanu, *Vector Integration and Stochastic Integration in Banach Spaces* (New York: Wiley, 2000).
11. See, for instance, Donald L. Cohn, *Measure Theory* (Boston: Birkhäuser, 1997), 220, theorem 7.3.5.
12. See Dembski and Marks, "The Search for a Search."
13. This result is somewhat buried in Dembski and Marks, "The Search for a Search." It is proven explicitly in William A. Dembski, "Searching Large Spaces," typescript, available at www.designinference.com (last accessed April 27, 2010). Note that the result, as stated in this last paper, is for $q > p$. In fact, nothing in the proof requires this restriction. Indeed, q is free to vary across the entire unit interval.
14. Robert J. Marks II, *Handbook of Fourier Analysis and Its Applications* (Oxford: Oxford University Press, 2009), 165.

15. We proved the special case in Dembski and Marks, "Life's Conservation Law."
16. See Marks, Handbook of Fourier Analysis, 165.
17. Dembski, "Searching Large Spaces."
18. Search spaces whose elements are themselves searches may be thought of as consisting of probability measures or fitness landscapes or other functions on the original search space. Spaces of functions on an original space are always richer than the original space and, in case the original space is finite, grow exponentially or even super-exponentially in cardinality. Spaces of probability measures, for instance, are always infinite. Cf. William A. Dembski, *No Free Lunch: Why Specified Complexity Cannot be Purchased without Intelligence* (Lanham, Md.: Rowman and Littlefield, 2002), ch. 3.
19. William A. Dembski and Robert J. Marks II, "Bernoulli's Principle of Insufficient Reason and Conservation of Information in Computer Search," *Proceedings of the 2009 IEEE International Conference on Systems, Man, and Cybernetics*, San Antonio, Texas (October 2009): 2647–2652.
20. Though compare the reference in the previous note, which argues that uniformity, even if not holding precisely, is, in the absence of definite knowledge about the underlying probabilities, the probability distribution most reasonably assumed.

Pragmatic Information

John W. Oller, Jr.

Doris B. Hawthorne Board of Regents Support Fund Professor IV,
Department of Communicative Disorders, University of Louisiana at Lafayette,
Lafayette, LA 70504, USA. joller@louisiana.edu

Abstract

The goal of this paper is to define *pragmatic information* with a view toward measuring it. Here, *pragmatic information* means the content of valid signs — the key that unlocks language acquisition by babies and to human communication through language — also the content that enables biological "codes" in genetics, embryology, and immunology to work. In such systems, the inter-related layers appear to be ranked as in a hierarchy. Sounds are outranked by syllables, in turn outranked by words, and so on. In DNA, nucleotide pairs are outranked by codons, which are outranked by genes, and so on. As signs of lower rank combine to form signs of any higher rank, combinatorial "explosions" occur. With each increase in rank, the number of possible combinations grows exponentially, but the constraints on valid strings and, thus, their pragmatic value, sharpens their focus. As a result with each explosive increase in the number of possible combinations the relative proportion of meaningful ones diminishes. Consequently, random processes of forming strings or changing them must tend increasingly toward meaninglessness (invalid and nonviable) strings. The consequent outcome of random mutations is mortality of individuals and in deep time an increasing number of disorders, diseases, and the eventual extinction of populations.

Key words: communication disorders, combinatorial explosion, pragmatic information, child language acquisition, biomolecular cryptology, pragmatic mapping, true narrative representations

Introduction

To show that sign systems are ranked and layered, consider that this is obviously true of the highest cortical functions of human beings. Layering and ranking can be demonstrated easily for our brains and are also found in biological systems. Combinatorial explosions occur as signs of lower rank are combined to form signs and strings of the next higher level up. As the complexity and number of possible strings increases along with the constraints on valid sequences at each higher level, the likelihood of generating them by random processes diminishes toward a vanishing point. As a result, random mutations (or injuries) in sign systems tend to produce disorders, genetic diseases, death, and, eventually, the extinction of populations. In this paper, I limit myself to explaining what pragmatic information is and how it increases with each combinatorial explosion in child language development and in genetic systems. The larger goal is to work toward an empirical measure of pragmatic information in the future.

Ranking in Sign Systems

At the Cornell symposium, since my starting time was an hour after lunch, to get the blood flowing and to give folks a chance to make it to our next coffee break, I asked the audience please to stand. I asked them to perform a few simple movements: a right handed thumbs up; then, a left; then, with both hands. I demonstrated and the audience followed along. Next, we wrote our names in bold strokes in the air with the dominant hand. I demonstrated writing "John" with my right hand. Then, we tried it with both hands. First, we allowed the subordinate hand, the left for most of us, to write the mirror image; then, using both hands in parallel, we wrote our respective names simultaneously with both hands. The reader may easily repeat the experiment and show that it is possible to do something with the subordinate hand that hardly anyone, apart from this sort of experiment, can do with the subordinate hand. For instance, I cannot fluently write the mirror image of my name with my left hand. However, when the subordinate hemisphere of the brain is slaved to the dominant linguistic hemisphere, the subordinate hand can easily do something it has never practiced — fluently writing the mirror image of a sequence of letters. How is this possible?

The actions just described provide a pragmatic (active and dynamic, real) demonstration of the ranking and layering of biocontrol systems at the highest cortical level in human beings. The ranking is shown in the exercises just described in three ways: For one, each compliant member of the audience subordinated himself or herself, to the whole group as led by the speaker. They subordinated their actions to my words. For another evidence of ranking, the speaker, in turn, subordinated himself to the organizers of the conference. The object of all this subordination was to make the ranking of biocontrol systems, combinatory explosions, and their consequences for pragmatic information, as intelligible, relevant, and memorable as possible to the participants at the symposium. For yet another, the slaving of the subordinate hand and the subordinate "mute" hemisphere of the brain to the dominant "talking" hemisphere of the brain — in the parallel and mirror-image writing by the subordinate hemisphere — also shows that linguistic signs at the highest cortical level are dominant.

Every person who performed the requested actions demonstrated the ranking summed up in Figure 1. In that diagram, let S represent the conventional signs (the words) of any natural language; let π represent acts of mapping those signs onto

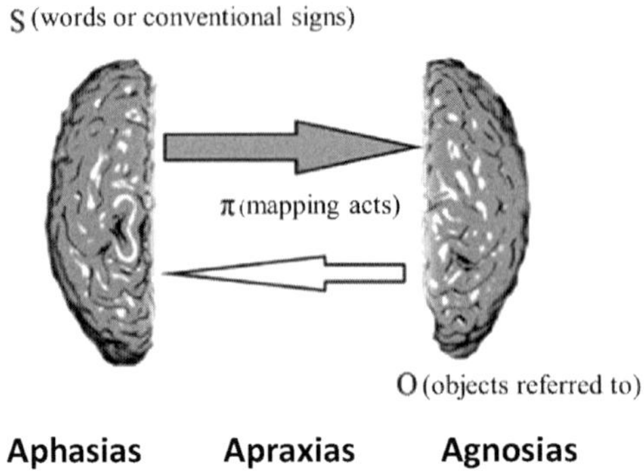

Fig. 1. Pragmatic mapping in the brain.

whatever they are about; and let O represent the logical object(s) referred to. The O may consist of an event or sequence — say, writing the mirror image of a name, or, attending the Cornell symposium, or the exchanges and acts leading to this paper, or the book in which it appears, or the whole network of connections through its cited references.

Keeping in mind that the motor and sensory functions of each side of the body are mapped to the opposite hemisphere in the brain, the physical acts of the exercises, show that the symbolic (word producing and arranging) hemisphere is dominant. It can take nearly complete control of the subordinate hemisphere. The dominant system can "slave" the subordinate one. In between the hemispheres is the corpus callosum (190 million fibers connecting them) — not shown in Figure 1, but implied in the arrows between the hemispheres. Interestingly, random mutations (by disease or accident) or selective ones (by surgery) of the brain often result in disorders. If they impact the dominant hemisphere they commonly produce disorders of language, *aphasias*; damage to the subordinate hemisphere generally results in disorders of recognition, holistic knowledge, and feelings about things, persons, and events, *agnosias*; and damage to the corpus callosum disrupts knowledge and control of action sequences which yields *apraxias*.

The simplest of the valid representations produced when all of our faculties are working well and when we merely report faithfully on actual experience are true narrative representations (TNRs). For instance, if I say truthfully, "I had lunch with Berkley Gryder, Robert Carter, and John K. Park on the second day of the Cornell symposium," I illustrate the sort of valid pragmatic mapping that is required in order to explain pragmatic information. A simpler instance of such a valid mapping can be found in a proper name applied correctly to the person who goes by that name. Analogous to the macro-cortical level seen in Figure 1, in Figure 2 — at a much more focused level — the name can be construed as a

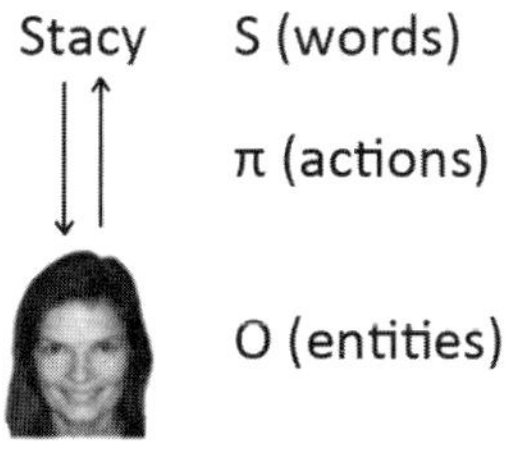

Fig. 2. Naming as a pragmatic mapping of S through π onto O.

symbol, S; the person named as its logical object, O; and the mapping of the name to the person named as an action, π. If the name is applied to the person who actually goes by that name, this sort of mapping captures the essence of all TNRs. It expresses their common form in a simplest instance. The action in validly referring to some logical object as shown in Figure 2, sums up the sort of things we do in giving any valid report. Consider my statement mentioning the persons with whom I had lunch on the second day of the symposium. Biological examples of such valid acts of reference would include complex mappings such as the recognition (or production) of the major histocompatibility complex (MHC) on the surface of a bodily cell enabling the body's own immune systems to identify the marked cell as one of its own — that is, as "self" as contrasted with a "non-self" cell or some foreign entity to be attacked, killed, and dismantled, or merely to be transported to a detention center for interrogation and further identification before it is taken apart piece by piece [1]. The SπO relation of Figure 2 would also include something as simple as the correct rendering of a UUU codon in a gene into the amino acid phenylalanine in a corresponding protein sequence. These and countless other examples, are special instances of the general SπO relation.

In natural human languages (as suggested by Figure 2), the mapping from S to O shows linguistic comprehension while mapping from O to S, shows linguistic production. What may not be so obvious, but must be taken into account, is that the name, or any referring sequence of symbols, S, is inherently abstract and general with respect to its generalized *semantic meaning*, but it is both arbitrary and conventional with respect to its surface form (its sounds, syllables, and syntax — that is the spatio-temporal arrangement of its components). By contrast the logical object O of the S in an ordinary naming relation, for instance, is concrete, particular, and actual. We may say that the *pragmatic meaning* of the S is materially instantiated in its particular logical object, O. In the case of ordinary proper names, we may say that the O involves a unique identity — as it also evidently does in the case of any MHC in the cells of a given individual. Abstracting from all of this, by the term *pragmatic information* I mean the useful content of TNRs, that is, reports or narrative-like representations that involve valid SπO relations.

The pragmatic mapping process, illustrated in naming, is considerably more complex than it might seem on the surface, and, as argued in a series of papers and books elsewhere [2–6], it forms the foundation for valid referring relations — which are invariably embedded in true narrative representations (TNRs). Valid referring relations, $S\pi O$, and all TNRs are true in the ordinary sense of "truth" because they conform to the normal conventional applications of their signs, S; they are narratives in all cases because it is impossible to refer to any particular material entity whatsoever apart from some context of experience that involves events unfolding over time; and they are representations because the S in each case invariably stands for something other than itself. It has been argued that TNRs are crucial to the discovery of pragmatic information in sign systems in general [7–9]. Because our world is so pervaded by valid $S\pi O$ representations from the highest cortical processes downward, their very familiarity makes the pragmatic mapping of a name onto a certain person seem much simpler than it is. Also, many philosophers have been lured into the false notion that names (or referring terms) are non-essential elements on account of the ubiquitous fact that not all signs are names; added to the fact that fictional, erroneous, and deliberately deceptive uses along with nonsensical ones are also possible. A few lines from Shakespeare serve to remind us of the tendency to regard some exceedingly complex relations as simple:

> But man, proud man,
> Drest in a little brief authority,
> Most ignorant of what he's most assured,
> His glassy essence, like an angry ape,
> Plays such fantastic tricks before high heaven
> As make the angels weep [10].

Tampering with the Sign Architecture

Among the "tricks" done on human beings that have certainly made some humans weep are "split-brain" surgeries where the corpus callosum — the bundle of about 190 million fibers [11] enabling the left hemisphere to communicate with the right and vice versa — was cut on the theory that doing so would prevent the spreading of an epileptic event between the hemispheres. The justification has been the claim that in a substantial majority of surviving patients the surgery would prevent full blown life-threatening seizures. Such surgeries and other sources of disease and injury to the brain demonstrate the foundational division of labor, and the ranking of major classes of signs, in the highest cortical functions of human beings as summed up in Figure 1 above. In fact, at Glenn Fulcher's web site on language testing, I have explained the pragmatic mapping process and there I illustrate it with video clips some of which were also presented at the Cornell symposium [12].

At the language testing site, thanks to Fulcher and the BBC in sharing materials from the educational series entitled "The Brain: A Secret History" [12], it is possible to see an extreme instance of what is known as *alien hand syndrome* in which the normal controlling role of the dominant hemisphere is disrupted by severing of the corpus callosum. The alien hand result offers straightforward evidence both of the normal ranking of sign systems in the human brain (as described above in Figures 1 and 2) and also of the fact that things can go very wrong when the normal ranking is disrupted by surgery, disease, or mutation.

After her surgery, Karen Burns discovered to everyone's dismay that her left hand (under the control of her subordinate, right hemisphere) suddenly had a mind of its own, producing a strange conflict with her right hand (under the control of her dominant, left hemisphere). After the surgery, her left hand would disconnect the phone by depressing the "clicker" just after she answered a call with her right hand. Her left hand would put out the cigarette she had just lit with her right hand. Her left hand would unbutton her blouse while her right was trying to button it again. After her surgery, when Karen began to regain consciousness, the attending personnel in recovery, immediately called for the doctor. The neurosurgeon arrived minutes later and found Karen's left hand beating her face black and blue. He asked her to give him a thumbs up. She did so with her right hand but her left hand was unresponsive to the linguistic request. Karen's difficulty was focused specifically in the inability of the dominant hemisphere to take charge of the subordinate hemisphere through the corpus callosum. Karen would have been unable to slave her subordinate hemisphere to perform the mirror writing that the audience at Cornell was able to do easily as described earlier in this paper.

At the symposium, I also gave an example of aphasia owing to damage to the left hemisphere of trilingual Julia Sedera. The relevant video clip can also be found in my feature presentation on the Fulcher site [12]. Julia's injury was owed to a stroke leaving her with a surprising inability to name an object, such as a "pineapple," for instance, though she knew well what the object was (via her relatively intact right hemisphere). Even when the neurologist modeled the first syllable of the word "pineapple" Julia was still unable to say the word.

Looking to the subordinate hemisphere that specializes in handling whole scenes, entities, faces, and in generating the feelings that are ordinarily associated with a sequence of events — the famed psychiatrist and author, Oliver Sacks, describes his special agnosia. He has prosopagnosia — difficulty recognizing faces and places — even his own face or the house where he lives. In the video clip of Sacks [12], he describes how he is apt to mistake an image of himself in a mirror or plate glass window for someone else. Or, when seeing a large bearded man on the opposite side of a window, the reverse has also occurred, where he finds himself preening what he takes for his own reflection only to discover that the bearded man on the other side of the glass is not preening his beard, but is looking rather strangely at Dr. Sacks.

In studies of split-brain patients that won him a Nobel Prize in 1981, neurologist Roger Sperry wrote: "The [dominant] speaking hemisphere in these patients could tell us directly in its own words that it knew nothing of the inner experience involved in test performances correctly carried out by the [subordinate] mute partner hemisphere" [13]. Again, there is video footage from Sperry's studies of such split-brain patients [12]. The relevant video clip reveals that split-brain patients can produce and comprehend language with the dominant hemisphere but are unable to do so with the subordinate hemisphere. Similarly, the subordinate hemisphere can reconstruct a pattern with blocks while the dominant hemisphere makes a hash of the same task.

Not only does the subordinate hemisphere excel at handling holistic scenes, patterns, and images, but it is also evidently in charge of producing feelings about whole patterns and sequences of events. In the BBC footage, a man named Dave, who lost a significant portion of the frontal lobe of his right hemisphere when a tumor was removed, also lost the ability to generate feelings toward the persons and events of his own experience. His wife commented that after the surgery he was not the same. Beforehand he used to do "nice things" to make her feel more comfortable, but afterward, he was no longer able to have normal feelings. They were divorced but she still takes him to his neurological appointments. Dave himself describes how he can remember feelings but no longer generates them. At the end of his post-surgery narrative he says in a near monotone, "The longer I go basing what I should feel on memory, I'm kinda nervous that eventually the memory will fade and then trying to remember what the actual emotion felt like will be more mysterious. At least now I have the memory so I can at least go through life with that understanding. . . if I didn't have that memory, I . . . I guess it would be a lonely . . . lonely existence" [12].

In another segment, Dr. Michael Mosely, who narrates the BBC series [14], talks through his own experience in confronting his fear of being closed in. He does so by going down into a very dark and small cave. Before starting out he is equipped with gloves to stop him from "ripping his fingernails off" if and when he gets stuck and panics. On seeing the entrance to the cave he sighs, "Gosh, well, that's small, isn't it. I was imagining something large," and then he sighs loudly, "Haaaahhhh!" Later, in the video clip [12], he gets stuck in a passageway with one arm pinned beneath him in a prone position. He is barely able to move enough to breathe and the fear momentarily takes over.

Undoubtedly, it is Mosely's right hemisphere (and that of anyone who empathizes with him) that generates the feeling associated with the whole sequence of events leading up to and including Mosely's predicament in the cave. The feeling remains intense for him (and for me as a viewer) even after he is extricated by somehow wriggling out or being helped out of the tight spot by the BBC camera crew. The video does not show how he gets out, only him gasping head in hands afterward, still in the cave saying, "That was bloody awful." Presumably, he would scarcely have put himself in such a situation if it were not for a linguistically guided decision — a dominant hemisphere commitment — to enter the cave despite his fear. Clearly the dominant hemisphere can over-rule the protesting subordinate hemisphere. Would he experience the same sort of fear if he had the sort of brain injury that Dave experienced to his subordinate hemisphere? Probably not. Could Mosely have the same fear if he were anesthetized and then placed in exactly the same posture in the narrow passageway? Again, probably not, as the pragmatic information about the sequence of events would be unavailable to him. But the point is, in ordinary conscious experience, there is a division of labor involving a ranking of the highest sign systems of human cortical functions. Even something as overwhelming as near complete terror (a subordinate hemisphere function) can be dominated by the rational power of the linguistic, speaking hemisphere.

Next, it is useful to note that the ranking of distinct layers of sign systems just demonstrated for the highest cortical functions can also be found in biocontrol systems right down to the molecular levels of DNA, RNA, and proteins. Figure 3

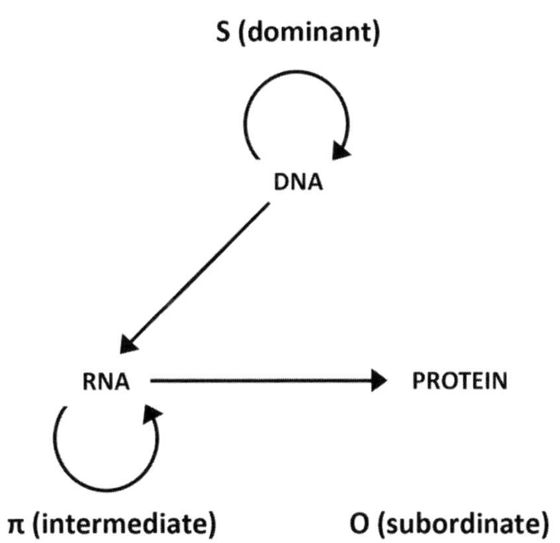

Fig. 3. Crick's dogma and the ranking of biological signs.

shows how Francis Crick's famous dogma [15,16] — though we now know it needs modification to take account of epigenetic interactions between RNAs and DNA (and no doubt other interactions beyond these) — reflected the same sort of ranking of genetic sign systems as we have seen in the highest human cortical functions (Figure 1) and in the linguistic process of pragmatic mapping as summed up in the naming relation (Figure 2). Although Crick's dogma is still defended as standard doctrine in many current biology texts, the interactions between the named systems are more complex, more constrained, and more deeply layered than the dogma suggested. Nevertheless, the point here is merely that the valid ranking proposed in Crick's dogma is consistent with that in the human neuro-architecture and in pragmatic mapping in general.

Pragmatic Mapping

Weinberger (2002) defined pragmatic information as the likelihood that a given message will change another person's conduct [17]. While the measure proposed by Weinberger may be relevant and suggestive, I am aiming for a more general definition of pragmatic information on which all meaningful sign systems depend for their representational power as displayed in the process known as *pragmatic mapping* [2]. Such an approach suggests the question of how pragmatic information enriches the capacity for representation in general — that is, in any representational system. I want to characterize the sort of pragmatic information that seems to be crucial not only to language acquisition, ordinary linguistic communication, and valid reasoning, but also to the biocontrol systems involved in genetics, metabolism, embryological development, immune defenses, and so on. A more recent paper (Gatherer, 2007) reported on the ongoing search for an algorithm to discover what he and others believe will turn out to be the discrete words and phrases, the meaningful/functional strings, in protein texts [18].

Gatherer points out that molecular biologists have commonly compared "genomes . . . to libraries of genetic information, with each chromosome as a book, genes as chapters, and DNA bases as the letters in which the text is written" (p. 101). With this linguistic metaphor in mind, Gatherer and others have suggested that discovering meaningful sequences in biological texts is like cryptology — with geneticists working as "biomolecular cryptologists" [19, 20] — like Jean-François Champollion seeking out the sounds, words, and meanings of Egyptian hieroglyphics [21]. In biology the units would be "nucleotides, codons, motifs, domains, exons, genes, genomes, etc... up to cells and organisms" (John Sanford, personal communication). The purpose of genetic cryptology, according to Gatherer's approach, is to devise an algorithmic discovery procedure to find the meaningful strings embedded, presumably, in the protein languages of various organisms. To test several options and combinations of rules, Gatherer tried them out not only on the deciphered proteins — the "proteomes" of various organisms — but also on various linguistic texts of which the shortest was *Alice in Wonderland*. In that text, his system found 85% of the 2,593 distinct words in the 26,587 word text.

With the cryptology metaphor in mind — as well as Shakespeare's lines about "proud man" and our "glassy essence" — a different metaphor for the difficult problem of deciphering biological language systems can be suggested. Perhaps molecular biologists could learn from normal babies acquiring any one, two, or even three at a time [22] of the 6,909 languages of the world [23]. Babies can solve them all, and as is becoming increasingly evident, biologists also, evidently, have a lot of distinct layers of language systems to decipher. In addition to the DNA codons corresponding to the amino acids of proteins, there are, of course, the RNA intermediaries and there is the protein language itself. In addition there are the partially understood "12 Trifonov codes" [24] and the codes for nucleosome building sites, cohesin protein binding, RNA transcription, splicing, RNA binding/folding, pyknons, isochores, and three dimensional nuclear architecture. According to remarks by Sanford on the paper by Montañez *et al.* (this volume [25]) there may also be codes involving triplex and quadruplex strands of DNA as well as electromagnetic coding, tandem repeat codes, and perhaps even vibrational codes as discussed by Dent (this volume [26]). Also relevant here is the paper by Dembski *et al.* (this volume [27]).

Building on the cryptology metaphor, I would like to propose that the manner in which babies solve natural language systems of the world may be relevant. If normal human babies can solve for the meanings of any unknown natural language, perhaps intelligent adults can figure out how they do it so that linguists, geneticists, and "biomolecular cryptologists" can learn why some discovery procedures for deciphering unknown languages can work where others will not. A clue concerning what advances in child language studies are teaching us about how infants decipher an unknown language can be found in Gatherer's results in trying to identify algorithmically all the meaningful words in *Alice in Wonderland*. Keeping in mind the deceptively simple SπO relation — one exemplified in every valid use of a name or referring term — the clue I have in mind is suggested in these questions: (1) What is the most important entity referred to in the *Alice in Wonderland* text? (2) What referring terms (meaningful words) in the text refer specifically to that entity? (3) Of the 2,953 different words in the 26,587 word text, what consistent referring term occurs most frequently? What term is critical to making the story hang together? What gives the fiction its sense of continuity? Or, to connect back to the cryptology problem of Champollion, what word was crucial to his solving of the hieroglyphics in the Rosetta stone? Similarly, bearing in mind

increasing from word to phrase to sentence to paragraph, chapter, book, series, and so on, the constraints restricting the range of valid constructions (or meaningful continuations) in a given string are converging toward a theoretical limit of unity. Practically speaking, it is the sort of unity exemplified when folks at the Cornell symposium, for example, understood and followed the directions in the opening exercises.

To illustrate combinatorial explosions we may apply the simplified equation, $N = v^l$, to the sounds of English estimated at approximately 35 for General American English (24 distinct consonants and 11 vowels), and setting a limit of syllable length at that of the monosyllabic word "strengths" consisting of the 8 segments transcribed in the International Phonetic Alphabet as [stɹɛŋkθs] gives a possible number of 35^8 or approximately 2.25 trillion combinations (2.25×10^{12} or 2,251,875,390,625 to be exact). Of those strings, only a few thousand (estimated at about 3,000 to 4,000) are syllables actually allowed by English phonology. As a result, even if we suppose 10,000 of the possible combinations are valid syllables in English, this would mean that fewer than 1 string per $10^7$of the possible strings would be a valid syllable in English. Jumping over the levels of words and phrases and advancing to sentences, given that the *Oxford English Dictionary* lists approximately 600,000 words, even if we restrict the number of words in a sentence to 12, the number of strings of that length would be $600{,}000^{12}$ or 2.177×10^{69}. However, only a relatively small proportion of that number would form meaningful sentences of 12 words in length. Because of grammatical constraints only a tiny fraction of the strings in such a vast list would be meaningful, and if we restricted the list to just TNRs, the ratio would become vanishingly small.

George A. Miller estimated on the basis of empirical studies of English texts that about 10 words on the average can form an appropriate continuation at any given point in any meaningful English text [31]. Using his estimate, the number of meaningful 12 word sentences, would be about on the order of 10^{12} enabling us to estimate that the ratio of meaningful 12 word sentences in English to all the strings that could be formed from all the words in the *OED*: it comes out to be about $4.59\text{x}10^{-58}$. Finding the few meaningful strings by chance in a heap of such nonsense would be a little like trying to find some very tiny needles in a really huge haystack (a serious problem as pointed out by Dembski *et al.* this volume [27]). Consider next that if we move the combinatorial explosions up several notches to the length of a short novel, say, 30,000 words (rounding up from the length of *Alice in Wonderland*), the number of possible strings explodes to $600{,}000^{30{,}000}$ as contrasted with — again, using Miller's method of estimating the number of meaningful texts of that length — about $10^{30{,}000}$. At the level of a short novel, the ratio of meaningful strings to possible ones has diminished to a complete vanishing point for all practical purposes. Not only is there no random process that could generate one of the desired strings, neither is there any possible way to list them, much less to search through the list. The difficulty is that if each possible text could be written on something as small as an electron, the writer would run out of places to write before a measurable fraction of the task could be completed.

The Logical Sequence for Discovering Meaning

So, the question remains, how do all normal human infants routinely solve problems of such great magnitude? Normal child language development follows a strict sequence of logical steps [32]. From birth forward babies are solving for the O of SπO relations. Perhaps the most primitive solution of that type is the newborn's mapping of mom's familiar voice to her moving face as she talks. From prior experience in the womb mom's voice is a familiar vocal sign, S, and the O that moves when mom talks to the baby is marked by just that particular voice which is π-mapped onto the moving face, O. In fact, the auditory movements in the normal baby's ear are quite perfectly coordinated with the modulation of mom's voice just as movements right down to the molecular level in the baby's eyes are coordinated with movements in mom's face. These near perfect correlations converge in the understanding that the voice is coming from mom [33].

The normal baby, while paying special attention to entities that talk, also works diligently in finding the boundaries of many objects of experience. By about three months, the baby will be seen to extend the index finger as if having already understood that such a gesture is used to single out things for attention [34]. After solving a substantial repertoire of Os, the baby begins to solve π-mappings that involve significant bodily movements that accompany speech. By about 4.5 months the baby typically demonstrates interest in an often repeated S which is distinct from others — such as the baby's own name, for instance — by looking toward the adult who says it [30]. A month or two later, the baby typically begins to produce repetitive babbling, /bababa/ or /mamama/ and so on, followed by differentiated syllables, /aǰadaǰaba/ and the like [35]. By about month 6 or 7, the baby will typically display comprehension of distinct SπO mappings by looking toward or handing over an object asked for by an adult. However, it will usually take 5 or 6 more months for the child to achieve sufficient motor control of the articulators to be able to produce his or her own "first word."

If the child is learning English, for instance, adults who already know the language will be able to understand that "first word" according to the conventions of the language in use. For instance, if the child's first word is the name of the household pet (as it was for my son Stephen D. Oller), say a dog that answers to the name of "Chester," consider the constraints that must be met in order for adults to share a common understanding with the child. If the phonological target is "Chester" — phonemically /čɛstɹ/ — the utterance of it must be close enough to be recognized as that word and no other. The standard of comprehension is a convergence to the limit of unity — approximating the extreme limit of "absolute certainty" suggested by Weinberger [17]. That is, all the parties concerned think they understand and know what the child is talking about. They are so sure of this that they would consider it odd to question their belief. But the convergence and the agreement achieved is remarkable.

Considering how large the possible set of strings of that length must be — estimated at 35^5 — the target in question occupies a tiny position in a large field. It is a particular string among 52,521,875 possible strings of the 35 phonemes of English. Assisting the adult interpreter(s) in correctly understanding the S and its O is, in many instances, the bodily dog that answers to the name "Chester," the logical object itself. That is, the syntactic tree (in the shape of Figure 2 above) that π-maps the name, S, to the entity, O, assists interlocutors to achieve common understanding. They look where the child is looking, pointing, and so on. Nevertheless, considering the number of potential objects, O, that might be referred to on any given occasion, or the number of babbled strings that might be uttered naming nothing in particular, the discovery of an intended referent, a dead center hit, is much more difficult to explain than a miss.

But the correct result will subsequently be confirmed again and again as the same unity is attained repeatedly not only with the word "Chester" but as the vocabulary explosion kicks in, it will be confirmed thousands of times over with a growing repertoire of more than 50 meaningful one-word utterances. After that a series of much greater combinatorial explosions will occur as the child — now about 2 years old — progresses through the two-word stage and beyond. The key to the combinatorial progress as one of my former PhD students, Ibrahim Al-Fallay, referred to it obliquely, is the child' ability to "climb the syntactic tree." He explained why another student dropped out, "Because," Ibrahim said in his Arabic accent, "He couldn't climb the syntactic tree." So, how is it that normal 2 year olds are able to do it? The answer reveals a severe (absolute) pragmatic constraint on the syntax of abstract predicates. There must be a syntactic tree to climb. Valid signs require objects.

Plainly a name, number, or referring term, that might apply to everything, anything, or nothing at all, has no power to inform us of anything other than itself. It may be a babbled sequence of sounds or syllables, or a random cipher pulled out of the air — an S without any determinate mapping to any O. Even less informative would be something without any consistently noticeable surface form at all. It cannot qualify as an S, or any particular form of nonsense, because it has no formal resemblance to any S. If we cannot recognize the sign itself as distinct from other signs and as a particular form on different occasions, how will we be able to associate it with any language, much less with any content? Although some philosophers have claimed that predicates grounded in referring terms cannot possibly account for abstractions such as love, justice, prime numbers, matrix algebra, etc., all such arguments fail when we see how infants easily climb the syntactic tree to solve abstractions. Invariably they start with referential entities that are well-grounded in valid SπO relations.

Consider the fact that discovering the meaning of a verb such as "bark" in the sentence, "Chester is barking," is materially assisted by the barking of the dog. The action contrasts with the state of affairs when the dog is not barking, or is jumping, running, chasing his tail, or the cat, crossing the road, dreaming about chasing the cat, etc. In his "Logic of relatives" — actually the "logic of relations" generalizing the Boolean algebra from binary to all possible relations — C. S. Peirce claimed as one of his first results that there cannot be any predicates so abstract that they cannot be grounded in relations between material entities in the world of experience [7]. Peirce's proofs in that treatise and many others have stood scrutiny for more than a century. The gist of the argument is suggested by noting how difficult it would be to discover the meaning of a verb such as "dance" without a dancer, or a relation such as "greater than" or "equal to" if it were impossible to find any instantiations to illustrate their meanings. It follows that there are no pragmatically unconstrained predicates no matter how abstract they might be. With pragmatic constraints come syntactic ones and semantic ones: "Pilot the bit dog the," is syntactically disallowed, while "The pilot bit the dog" is okay syntactically (in its spatio-temporal arrangement) and semantically also in terms of its abstract meaning. However, because our pragmatic experience makes it unlikely that a pilot would bite a dog, we might infer that an error has been made, and that "The dog bit the pilot" is what was intended. Children will often correct an odd form, e.g., "Can the blindfolded dolly be seen by you?" and will answer a more sensible one, "No," the child is apt to say, "the dolly can't see me." The researcher asking the question may suppose the child has answered incorrectly, not understanding the passive voice, when, in fact, the child adjusted the question to one that makes sense. The child thinks something like: It's the dolly that is blindfolded, not me. She must mean, "Can the dolly see you?" And so forth [36].

So, again, how do normal children progress to such knowledge and what are the implications for molecular cryptologists in trying to generate viable strings in biological systems? To show how and sum up the sequence, followed by normal children, we require some additional markings on the basic SπO relation. Let $S\pi\underline{O}$ represent the generalized form of a hypothetical, fiction, or fantasy. At the symposium I suggested that participants imagine an elephant standing next to me on the stage. To do so, they would have to conjure the elephant, because there was none on the stage — hence, the underlined $\underline{O}$ to suggest π-mapping the conjured elephant into the blank space. Babies typically solve valid SπO mappings by about 12 months of age but require another year to distinguish a true report from a fiction, by about age 2. Just to understand the example fiction, for instance, the person doing the imagining of an elephant not present must know the meaning of the word "elephant." Thus, an SπO mapping showing what the S means must come first. Errors are more complex. Suppose someone says, "Good morning, Mimi," when Ruthie is present, not Mimi. To correct the error, my grandson not yet 3 years old, had to take the $\underline{\not{S}}\pi\underline{\Theta}$ form and replace it with an SπO. The fictionalized and mistaken $\underline{\Theta}$ which is supposed to be Mimi (his grandmother), but which is in fact Ruthie (his adult aunt), must be replaced with Ruthie, and the fictionalized and erroneous name $\underline{\not{S}}$ must be changed from "Mimi" to "Ruthie." Children typically can correct an error, in this way, by about age 3. Distinguishing a deliberate lie, $\underline{\underline{\not{S}\not{\pi}\Theta}}$, from an unintentional error, $\underline{\not{S}}\pi\underline{\Theta}$, takes 2 or 3 more years of development [37]. Normal children are able to do so by about age 6. In a lie all three of the underlying elements of the SπO relation are erroneous, fictionalized, and intended to cause the lie to be mistaken for a true representation. For instance, if a certain former U.S. President (notably Bill Clinton) said he didn't "have sex with that woman," but it turns out that he was lying, all the elements must be changed to truly represent the relevant facts.

Next, consider how much more degenerate the italicized string *i io mN"o "Dgmon mrgi* is than a fiction, error, or even a lie. It has the same letters, punctuation marks and spaces, as one of the degenerate representations in the preceding paragraph. Is it easy to see which one? It is a nonsensical variant, a jumble, that started as an $\underline{\not{S}}\pi\underline{\Theta}$ (to narrow the field if the reader aims to solve the puzzle), but it is less coherent than any ordinary fiction, error, or lie. The fact is that in languages — and it seems in biological systems as well — fictions, errors, and outright lies are more coherent than scrambled versions of any of even these degenerate forms tend to be. In biology, I suppose a suppressed gene would be an example of a fictional representation; a genetic flaw resulting in, say, sickle cell anemia, or a viable cell mistaken for an invading foreign disease agent by the immune system would be examples of errors; and polyoma viruses, bacteria, or cancer cells impersonating the body's own RNA, DNA, or self cells, respectively, would be examples of biological lies. The fragments of a foreign peptide, or of a cell undergoing apoptosis, would probably qualify as some grade of biological nonsense, say, in the protein language of a given organism.

Typically, evolutionary biologists have sought to imagine ways to generate strings of meaningful signs from the bottom up. Theoreticians have often noted, as Gatherer does, that from letters to words, to phrases, to sentences, and so on (relying on the linguistic metaphor) the number of possible strings repeatedly explodes with a growing vocabulary of signs and an increasing string length at each higher rank. However, if we think from the top downward, we find that the constraints on coherence are greatest (all else being equal) at the highest rank. For instance, if we take historical biographies as an approximation to true narratives rich in pragmatic information, setting them as a kind of "gold" standard (flawed though it may be), it is possible to degenerate one or many such texts by degrees. Holding constant, say, the vocabulary of elements used to create the coherent string and the length of the string, the whole of it or some part can be chopped and scrambled stepwise at distinct ranks. Opposite the level of pragmatic information exemplified in the whole of a true biography, or in several volumes aiming to tell the history of the same person, a zero order of coherence can be found empirically at the place where the entire text is obliterated by reducing all its elements to blank spaces or mere random pixels. Between those limits it should be possible, even easy, with current technologies, to systematically sample and measure empirically the changes in coherence at distinct ranks. Empirical studies of discourse processing in natural languages show that scrambling at any rank or length of string reduces coherence and conversely that access to longer segments of a coherent text enhances comprehensibility, recall, and ability to replace missing elements (letters, words, phrases, and so on). All else being held equal, longer intact strings are increasingly constrained and therefore easier to process (comprehend, recall, and so forth) than the same elements in a cut and scrambled order [38].

Conclusions

Because of the series of combinatorial explosions that occur in progressing up the ranks in any layered hierarchy of representational systems, to find or generate any string that will qualify as a valid representation of any actual sequence of events in ordinary experience, or as a viable representation of any organism or any actual part of one, diminishes rapidly toward a vanishing point. Meanwhile, as the number of strings that are possible are exploding, the ratio of meaningful to meaningless strings at every level diminishes with each increase in the rank of signs and/or the length of allowable strings. As a consequence, the problem of finding (or generating) any valid (viable) biological strings by random processes is like the needle in a haystack problem magnified many times over. As Dembski, Ewert, and Marks [27] showed (this volume), the search for a needle presupposes a searcher. But the problem of randomly generating the searcher is vastly more difficult than any of the seemingly impossible searches we might hope for that person? robot? algorithm? to conduct.

But the difficulty does not end there. Linguistic analysis of natural language systems shows another profound problem, as was illustrated by Montañez, Marks, Fernandez and Sanford (this volume [25]). As valid (meaningful and viable) strings increase, the difficulty of generating them by stochastic processes rapidly increases. Also, as I have argued here, with each combinatorial explosion as we progress upward through sign ranks to their highest level, the ratio of valid strings to all that are possible diminishes toward a vanishing point with a numerator of unity and a divisor representing an uncountable multitude of multitudes.

In 1948, Claude Shannon proposed to measure information as the improbability of any particular message "*selected from a set* of possible [equally likely] messages" [39]. He noted that "the messages" frequently "refer to or are correlated according to some system with certain physical or conceptual entities" which he referred to as "semantic [*sic*] aspects of communication" (p. 379) and which he set aside. In doing so, he conflated the abstract and general sort of meaning properly termed "semantic" (associated with generalized conventional Ss) and also the particular and concrete "pragmatic" content (associated with particular concrete Os — the actual persons, places, events, and the "syntactic" relations between them in space and time (the π-mappings). I suppose that the crucial meaning that Shannon set aside is precisely the kind connecting intelligible signs to the facts of ordinary experience — *pragmatic information*. I agree with what I understood Baumgardner to say in one of the early discussion sessions at the Cornell symposium: When talking about information we need to work with the sort of meaning that is distinctly "linguistic in nature" (also see Baumgardner 2009 [40]). I believe that we need to consider the dynamic character of *pragmatic information* as I have described it here. It seems to be as essential in biology as it is in linguistics.

One of the reasons, I think, that we tend to over-estimate our understanding of "our glassy essence" — and to underestimate the richness of the simplest signs — is that we tend to look right through the π-mapping of any valid S to its O. As the sign systems of a child come to maturity, the generality of the S reaches out very easily to signify all possible instances of the O greatly exceeding the relatively few actual instances that have been or will ever be encountered in a life-time of experience. The agreement attained between the valid π-mapping of any S to its O in a TNR thus achieves what Peirce referred to as the "unity of coherence" [41] — like a glove perfectly fitting a hand, or the bite when the upper and lower teeth fit together. The completed, well-formed-system, is a unified triad of the SπO kind. It enables the closest we can reasonably get, I suppose, to anything like "complete certainty" in the material world. Thus every TNR, though triadic in its internal elements, as a signifying unity singles out a stream of particular facts that are both distinct from all the rest and yet, by virtue of being a part of the whole material world, are connected with the rest of it and with all the other TNRs. As a consequence, they enable, as shown in the earlier analysis of child language development, valid generalizations beyond what is experienced.

I agree with Edward T. Weinberger's comment [17] that "a theory [Shannon's] that totally ignored semantics was, in some sense, incomplete" (p. 105). Weinberger went on to urge a definition of "pragmatic information" in terms of "usefulness in making an informed decision" (p. 106). I would only want to generalize his approach to account for all intelligent judgments of any kind about the facts of experience. To me Weinberger's most intriguing claim is that "the maximal amount of pragmatic information accrues to messages that engender complete certainty" (p. 109). In my linguistic approach to pragmatic information, a maximally informative representation would be the sort found in a name mapped onto a particular identity appearing throughout a faithfully reported true narrative. With a view toward measuring pragmatic information, we can say that it varies from a limit of meaninglessness at one extreme, near 0, to a limit of what seems to be the gold standard where the unity of coherence, near 1, is commonly achieved. Simple SπO mappings, at the foundation of valid representations such as we find in ordinary TNRs and in viable biological codes, exemplify the sorts that can be used to calibrate the high end of a scale of pragmatic information, and as I suggested, we can step down from there toward the lower end by degrees.

Addendum

Due to a delay in publication of these proceedings, I wish to add the following publications which have appeared in the interim. Pertinent to the strict sequence of steps followed by infant language learning per reference [32], see Oller, J.W., Oller, S.D., Oller, S.N.: Milestones: Normal speech and language development across the life span. 2nd edition. Plural Publishing, Inc., San Diego (2014); and in addition to references [25, 26] suggesting various biocontrol systems yet to be discovered, the following entries should be added: Davidson, R.M., Seneff, S.: The initial common pathway of inflammation, disease, and sudden death. Entropy 14(8), 1399–1442 (2012); Dietert, R., Dietert, J.: The Completed self: an immunological view of the human-microbiome superorganism and risk of chronic diseases. Entropy 14(11), 2036–2065 (2012); Seneff, S., Davidson, R.M., Liu, J.J.: Is cholesterol sulfate deficiency a common factor in preeclampsia, autism, and pernicious anemia? Entropy 2012, 14(11), 2265–2290; and Gryder, B.E., Nelson, C.W., Shepard, S.S.: Biosemiotic entropy of the genome: Mutations and epigenetic imbalances resulting in cancer. Entropy 15, (2013).

Acknowledgments

Much is owed to John C. Sanford, Robert J. Marks II, and especially to John Baumgardner all of whom offered comments on one or more drafts of this paper. I also thank Berkley Gryder, Robert Carter, and John K. Park for their comments and encouragement. Any remaining errors or omissions are my sole responsibility.

References

1. Mayer G, Nyland J (2010) Major histocompatibility complex (MHC) and T-cell receptors — role in immune responses. In: Microbiology and immunology on-line. University of South Carolina School of Medicine, Columbia. http://pathmicro.med.sc.edu/bowers/mhc.htm.
2. Oller JW (1975) Pragmatic mappings. Lingua *35*:333–344.
3. Narremore RC (1985) Explorations of language use — pragmatic mapping in L1 and L2. Topics in Language Disorders *5*:66–79.
4. Macnamara TJ (2000). Language testing. Oxford University Press, New York.
5. Ward G (2004) Equatives and deferred reference. Language 80:262–289.
6. Goldberg A, Suttle L (2010) Construction grammar. Interdisciplinary Reviews: Cognitive Science 1:468–477.
7. Peirce CS (1897) The logic of relatives. The Monist 7:161–217.
8. Tarski A (1949) The semantic conception of truth. In: Feigl H, Sellars W (eds) Readings in philosophical analysis, pp. 341–374. Appleton, New York.
9. Oller JW (1996) How grammatical relations are determined. In: Hoffer B (ed) The 22nd Linguistic Association of Canada and the United States (LACUS) forum, pp. 37–88. LACUS, Chapel Hill, NC.
10. Shakespeare W (1623) *Measure for measure*, Act 2, Scene 2. http://www.william-shakespeare.info/shakespeare-play-measure-for-measure.htm.
11. Hofer S, Frahm J (2006) Topography of the human corpus callosum revisited — Comprehensive fiber tractography using diffusion tensor magnetic resonance imaging. Neuroimage 32:989-994.
12. Oller JW (2011) Language assessment for communication disorders. http://languagetesting.info/features/communication/disorders.php.
13. Sperry R (1981) Some effects of disconnecting the cerebral hemispheres. http://nobelprize.org/nobel_prizes/medicine/laureates/1981/sperry-lecture.html.
14. BBC (March 28, 2011) The brain: A secret history. Narrated by Dr. Michael Mosely. http://www.bbc.co.uk/programmes/b00xhgkd.

15. Crick FHC (1958) On protein synthesis. Symp Soc Exp Biol 12:139–163.
16. Crick FHC (1970) Central dogma of molecular biology. Nature 227:561–563.
17. Weinberger ET (2002) A theory of pragmatic information and its application to the quasi-species model of biological evolution. Biosystems 66:105–119.
18. Gatherer D (2007) Peptide vocabulary analysis reveals ultra conservation and homonymity in protein sequences. Bioinformatics and Biology Insights 1:101–126.
19. Konopka AK (1994) Sequences and codes: Fundamentals of biomolecular cryptology. In: Smith DW (ed) Biocomputing Informatics and Genome Projects, pp. 119–174. Academic Press, San Diego.
20. Spencer M, Davidson EA, Barbrook AC, Howe CJ (2004) Phylogenetics of artificial manuscripts. J Theor Biol 227:503–511.
21. Champollion JF (1828) *Précis du système hiéroglyphique des anciens Égyptiens* [A synopsis of the hieroglyphic system of the ancient Egyptians.] Paris: Reissued 2006 in the Elibron Classics Series, Adamant Media Corporation.
22. Swain MK (1972) Bilingualism as a first language. Unpublished PhD Dissertation, University of California at Irvine. Ann Arbor: University Microfilms.
23. Lewis MP, ed. (2009) Ethnologue: Languages of the world, 16th edn. SIL International, Dallas. http://www.ethnologue.com/.
24. Trifonov EN (1997) Genetic sequences as product of compression by inclusive superposition of many codes. Mol Biol 31:647–654.
25. Montañez JF, Marks II RJ, Fernandez J, Sanford JC (2013) Multiple overlapping genetic codes profoundly reduce the probability of beneficial mutation. In: Marks II RJ, Behe MJ, Dembski WA, Gordon B, Sanford JC (eds) Biological Information — New Perspectives. World Scientific, Singapore, pp. 139–167.
26. Dent L (2013) A new model of intracellular communication based on coherent, high-frequency vibrations in biomolecules. In: Marks II RJ, Behe MJ, Dembski WA, Gordon B, Sanford JC (eds) Biological Information — New Perspectives. World Scientific, Singapore, pp. 435–449.
27. Dembski WA, Ewert W, Marks II RJ (2013) A general theory of information cost incurred by successful search. In: Marks II RJ, Behe MJ, Dembski WA, Gordon B, Sanford JC (eds) Biological Information — New Perspectives. World Scientific, Singapore, pp. 26–63.
28. Augustine, St (1952) In: Hutchins RM (ed) Great books of the western world: The confessions of St. Augustine (Vol 18). Chicago: Encyclopedia Britannica. (Original work published ca. 401 AD).
29. Werker JF, Pegg JE, Mcleod PJ (1994) A cross-language investigation of infant preference for infant-directed communication. Infant Behavior & Development 17:323–333.
30. Werker JF, Tees RC (1999) Influences on infant speech processing: Toward a new synthesis. Ann Rev Psych 50:509–535.
31. Margolis E, Lawrence S (2007) The ontology of concepts — abstract objects or mental representations? Noûs 41:561–593.

32. Oller JW, Oller SD, Badon LC (2006) Milestones: Normal speech and language development across the life span. Plural Publishing, Inc, San Diego.
33. Sai FZ (2005) The role of the mother's voice in developing mother's face preference: Evidence for intermodal perception at birth. Infant & Child Development 14:1–29.
34. Masataka N (1995) The relation between index-finger extension and the acoustic quality of cooing in 3-month-old infants. Journal of Child Language 22:247–257.
35. Oller DK (1980) The emergence of the sounds of speech in infancy. In: Yeni-Komshian G, Kavanaugh J, Ferguson C (eds) Child phonology, pp. 93–112. Academic Press, New York.
36. Chomsky CS (1971) The acquisition of syntax in children from 5 to 10. MIT Press, Cambridge, MA.
37. Berthoud-Papandropoulou I, Kilcher H (2003) Is a false belief statement a lie or a truthful statement? Judgments and explanations of children aged 3 to 8. Developmental Science 6:173–177.
38. Oller JW, Chen L, Oller SD, Pan N (2005) Empirical predictions from a general theory of signs. Discourse Processes *40*:115–144.
39. Shannon CE (1948) A mathematical theory of communication. Bell Sys Tech J 27: 379–423, 623–656.
40. Baumgardner J (2009) Language, complexity, and design. In: Seckback J, Gordon R (eds) Divine action and natural selection: science, faith and evolution, pp. 939–980. World Scientific, Singapore.
41. Peirce CS (1866) *[On a method of searching for the categories]*. MS 133. In: Fisch M, Klosel CJW, Moore EC, Roberts DD, Ziegler LA, Atkinson NP (eds) (1982) Writings of Charles S. Peirce: A chronological edition, Vol 1, pp. 515–528. Indiana University Press, Indianapolis.

Limits of Chaos and Progress in Evolutionary Dynamics

William F. Basener

Rochester Institute of Technology School of Mathematical Sciences Rochester, NY, 14623, USA, william.basener@rit.edu

Abstract

There are a number of standard models for the evolutionary process of mutation and selection as a mathematical dynamical system on a fitness space. We apply basic topology and dynamical systems results to prove that every such evolutionary dynamical system with a finite spatial domain is asymptotic to a recurrent orbit; to an observer the system will appear to repeat a known state infinitely often. In a mathematical evolutionary dynamical system driven by increasing fitness, the system will reach a point after which there is not observable increase in fitness.

Key words: population dynamics, evolutionary dynamics, evolutionary network, evolutionary equilibrium, fitness space, fitness network

1. Introduction

1.1 *Goals and Perspective.*

The goal of this paper is to apply standard mathematical theorems from topology and dynamical systems to mathematical models of evolution. Mathematical topology is the logical study of the shape of objects without using specific measurements such as angles and curvature — for example an oval, a square and a circle are all topological the same — and mathematical dynamical systems involves the application of topology to processes that change over time, often without precise formulation of the process.

Like most cross-disciplinary research, this paper requires the difficult task of attempting to speak across the language and style of two disparate technical fields. To those trained in one field or another, exposition in their own field will appear trivial and work presented from the other takes time to digest. The author is an applied mathematician and the tools of this paper are mathematical, and so despite the author's best efforts the style will inevitably tend toward that field, especially when dealing with theorems and proofs.

Historically, when the apparent chasm between fields is breeched results can be profound. Mathematics offers tools — rigorous ways to understand things that can be rigorously described — and science offers objects and processes to be understood. Over the past century, applications of topology have been central to progress in several scientific fields, and to understand the work in this paper it will be helpful to review how topology has been applied in the past.

We should make an important distinction regarding terminology. A dynamical system most generally is anything that changes over time governed by a set of rules. A *mathematical dynamical system* is one that is defined in mathematical logic — it consists of a state space X and a function which, for a given initial state, will determine the state of the system at future times. (This definition will be made in more detail and more broadly involving the possibility of randomness later in this paper.) A *biological dynamical system* consists of organisms that reproduce and grow. We will use the terms mathematical dynamical systems and biological dynamical system to distinguish between the two when not clear from the context. We will also use the term *model* to refer to a mathematical dynamical system that is designed to model a biological one.

Accordingly, we can prove theorems about mathematical dynamical systems and these theorems would only be applicable to biological dynamical systems to the extent to which the mathematics accurately models the biology. In physics, where dynamical systems originated, the distinction is not usually made because the process involved are the result of physical laws such as Newtonian or Relativistic mechanics; conclusions proven about mathematical systems are taken as automatically pertinent to physical ones. Biological organisms are not subject to the same types of laws; individuals are assumed to have probabilities regarding specific behavior and the ability to model the behavior of system as a whole results from averaging the probabilities across a large number of organisms, for example as with the quasispecies equation (See [1]). This is analogous to statistical mechanics and thermodynamics, where the predictability of the collective whole is assumed by averaging out over many individual components.

It is broadly accepted that the process of evolution can be effectively modeled using mathematics. The study of mathematical dynamical systems modeling evolution is called *evolutionary dynamics* and the interested reader is referred to Novak's excellent introduction *Evolutionary Dynamics, Exploring the Equations of Life* [1]. Mathematical dynamical systems modeling evolution are the topic of study in this paper and using the proper tools we prove restrictive behavior about very broad classes of such models. Determining which models are accurate or appropriate for evolution is beyond the scope of this paper.

Mathematical models are developed by formulating some assumed governing scientific principles into mathematics and the resulting behavior of the model is taken to be the logical consequence of the assumed principles. Models can be used in a predictive manner (ie what will happen to a certain species if the harvesting rate is increased) or in an explanatory manner (ie why did the codfish population decrease) and we are concerned with the latter in this paper.

Using mathematical models to explore underlying causes requires a proper understanding of what the models can and cannot tell. In models where the governing principles are derived from laws (ie physics and chemistry), the behavior of the model is taken as the behavior of the physical system in the ideal case. In models where the governing principles are not derived directly from scientific laws (ie economics and ecology), the behavior of the system is only understood to match the behavior of the physical system *if* the assumed governing principles where the most important factors in the process. Thus, it is impossible to prove that certain principles result (or resulted) in observed behavior, but it is possible to prove that certain behavior is impossible as a consequence of certain governing principles. In short, mathematical models cannot demonstrate what is *true* about a physical system, but they can demonstrate what is *false* by way of a hypothesis test; if the behavior in a mathematical model does not match observed phenomena, then original assumed principles cannot be the cause of the observation.

The main results of this paper are for a mathematical dynamical system modeling evolution: 1) If the state space is compact (ie the physical system exists in a finite area) and the genotype has a bounded finite length then the change in phenotype with either stop or appear to repeat some state and the amount of increase in fitness is bounded, stated formally in Theorem 3; and 2) If the system is chaotic (and the fitness is a continuous function that is nondecreasing on orbits) then there is no increase in fitness, stated formally in Theorem 4. The first might not be surprising, although by way of this result we suggest a focus on the bounds of evolution in mathematical models, for example using information theory to quantify the bounds. The second result seems contrary to the prevailing understanding of evolutionary dynamics.

1.2 *History and Application of Topology and Dynamical Systems*

To bridge the gap between mathematical definitions and theorems of topology and their role of in science, we discuss the history of applied topology over the past century. Topology began as a theory in the late 1800s out of attempts to answer two seemingly separate questions — one abstract mathematical question and one applied scientific.

In the late 1800s, German mathematician Gregor Cantor was attempting to define dimension as part of his quest to develop a rigorous theory of points and sets, things that had been taken for granted since the investigations of Greek mathematicians (See [2]). A by-product of this re-development of the foundations of mathematics was the discovery that some sets have a dimension greater than a line but less than a plane — that is some sets have a fractional dimension — the most famous of these sets being the Cantor Set. These sets are what we now call *fractals* (a term coined by Benoît Mandelbrot in 1975). The tools required to study them is not the lines, angles and curves of geometry and calculus, but a more general class of definitions and theorems that make up topology.

Also in the late 1800s, French mathematician Henri Poincaré was studying planetary motion using calculus and differential equations. In his attempt to solve the equations of motion for multiple heavenly bodies, Poincaré wrote his *Les méthods nouvelles de la mécanique Célest*; New Methods in Celestial Mechanics (See [3]). The first printed version of this manuscript contained an error, and in correcting the error Poincaré discovered that equations for planetary motion have solutions that are too complex to be explicitly written in the usual formulas from calculus. Having shown that the solutions are too complex to be solved via calculus, Poincaré developed a new set of tools which we now call topology. Having discovered that the solution to some problems lies not in the formulas but in the general shape and behavior, Poincaré developed a new approach to understanding motion without reference to exact formula, which we now call *dynamical systems*. The type of behavior that Poincaré encountered in his solutions is what we now call *chaos*, a term coined by Jim York in 1975 [4]. The tools of topology have been applied to dynamical systems continually since the time of Poincaré. (See Strogatz [5] for an excellent applied introduction.)

The utility of applied topology comes from the ability to prove mathematical properties of very general classes of objects and phenomena. Since Poincaré's pioneering work, this has been exploited in a number of disparate fields.

In 1950-51, John Nash used topology (in particular the Brower Fixed Point Theorem) to demonstrate the existence of Nash Equilibrium in a very broad class of non-cooperative games. (See [6] and [7, Chapter 4.7]). This result revolutionized game theory with applications in economics, politics and biology. Topology enables the proof of existence of Nash Equilibria in mathematical games even when the exact formulation of the player's strategies are not known, and has application to human conflicts where no precisely defined game or strategy exists. Because of the applicability of topology to a very broad class of games, this result is assumed to apply even to real games where the strategies are not mathematical but are derived from the psychology of the players.

In condensed matter physics, states of matter other than solids liquids and gases can occur as the result of collective behavior of interactions between molecules. Symmetries of forces result in behavior more structured then that of a liquid but less rigid than a crystal or solid. A familiar example is the liquid crystals in a computer display. Pressing on the display creates outward swirls of rotation resulting from the local pressure. The patterns are studied with topology; the twists and singularities, or defects, exist to maintain a consistent global topology even when the exact local positions are not known. This has proven important for understanding states and collective behavior of matter such as superconductors. (See [7-10]).

One of the grand questions in cosmology has been the shape of the universe. Since Aristotle conjectured that the universe is a great sphere, cosmologists have been attempting to infer the structure from observations. Inferring this topological and geometric structure has been one of the main purposes of the NASA WMAP (Wilkenson Microwave Anisotropy Probe) — patterns in the anisotropic cosmic microwave background radiation could be used to determine the topology of the universe. The role of topology is beyond the scope of this paper, but the interested reader is referred to Weeks [11] for an excellent exposition or to Basener [7].

The goal of this paper is to apply some basic theory from the mathematical field of dynamical systems to mathematical models of evolution. The reason we employ the mathematical theory from topology is twofold. First, as with the examples cited in this section, we are then able to prove theorems for broad classes of models; the machinery of topology and dynamical systems allows us to prove theorems about mathematical models of evolution without an exact formulation of the models. Second, in addressing chaotic dynamical systems we are required to use topology (or some equivalent machinery, for example geometry if we assume a suitable state space) as even the definition of chaos requires some level of topology.

The mathematics is basic topology and the theorems we prove are quite simple; they could be basic homework exercises in an upper level undergraduate course in dynamical systems. However, the insights resulting from the application do not seem to be generally known or understood in the study of evolutionary dynamics, either in theory or application. The remainder of this paper consists of a series of expository examples of evolutionary dynamics with application of dynamical systems theory, building up to the main results in Theorems 3 and 4.

1.3 *General Questions in Evolutionary Models*

Every living organism has a genotype, its genetic sequence, and phenotype, the phenomenological manifestation of the genotype. The standard model of evolution is that the genotype determines the phenotype, and combined with other factors this determines the fitness level of the organism in its environment, and this fitness level determines the probability of survival of the organism in competition with other organisms. Reproduction and random mutations create organisms with new genotypes, and the fitness of the new genotypes determines their subsequent survival rates. Consequently, the genotypes of organisms dynamically migrate to those with generally higher fitness levels.

A sort of evolution can be observed experimentally [12] using a series of test tubes each of which contains the four nucleotides ATP, GTP, UTP, and CTP as well as the enzyme $Q\beta$ replicase. An RNA template is added to the first test tube, left for 30 minutes, then a fraction of the solution from the first is added to the second, and the process is repeated. The $Q\beta$ replicase creates almost prefect copies of the RNA molecules in each test tube, and after a series of transfers the RNA will consist of a modified variant that is replicated more quickly than the original. While this biological process is not actual evolution of living organisms, the 'genotype' in this experiment corresponds to the RNA sequence and the 'phenotype' is the resulting replicating performance. The resulting rate of replication by $Q\beta$ replicase determines the 'fitness' of the RNA molecules. The type of RNA sequence in the final equilibrium state is determined by the environment of the solution.

Observe that the dynamic behavior of the $Q\beta$ RNA system is very simple; the RNA 'genotype' goes to an equilibrium which is determined by the parameters of the system. This is the typical behavior of evolutionary dynamical systems based on evolutionary genetics. This raises the question of whether the genetic processes are sufficient to account for macroevolution; quoting John Maynard Smith [12, p.273]: "This book has been concerned with processes that can be studied in contemporary populations over short periods of time. Our picture of evolution on a larger scale — macroevolution — comes from comparative anatomy and embryology, from taxonomy and geographical distribution. The question naturally arises whether the processes of population genetics are sufficient to account for macroevolution. Very different views can be held on this..."

The goal of this paper is to apply basic structure theorems from topological dynamics to answer, at least in part, Maynard's question. We investigate conditions on evolutionary models that guarantee behavior observed in the $Q\beta$ RNA system — evolution progressing for period of time and then ceasing. We show in a very general class of evolutionary models, which includes the standard continuous (differential equations), discrete (iterations of maps), deterministic, stochastic, and spatial evolutionary genetics — based models, this is the only possible behavior.

This is really not surprising. In evolutionary progression that can be studied in contemporary populations over short periods of time, we observe a process that does a finite amount of increase in fitness and then ceases; we do not directly observe evolutionary progress of a species through continually higher, more complex, more fit, genotypes-phenotypes. It is also the behavior observed in standard dynamic models for evolution.

Perhaps the only potentially surprising result is that no evolution takes place within chaotic dynamics, Theorem 4. Chaotic behavior is sometimes offered as an explanation of how complex systems might come from simple governing laws. For example, Novak [1, p.6] writes "Chapter 9 gives an account of evolutionary dynamics on spatial grids. ... We will observe evolutionary kaleidoscopes, dynamic fractals, and spatial chaos. There is all the complexity one could wish for — making it unnecessary for God to play dice." The suggestion seems to be that complex features of nature, implicitly complex organisms resulting from evolution, can result from chaotic dynamics. Theorem 4 shows that, to the contrary, no sustained increase in complexity or fitness is possible within a chaotic dynamical system. Specifically, to within any small amount of observational error, a chaotic system repeats each given state infinitely often. Subsequently, an evolution trajectory that is asymptotic to a chaotic set receives no more increase in fitness than one that is asymptotic to an equilibrium.

Our conclusion stresses again the question of whether the population genetic process of mutation — selection is by itself sufficient to account for macroevolution. As before, this seems not so surprising, as even speciation, the divergence of a single species into different species, seems to require external environmental factors. Again, quoting Smith [12, p.275], "It is widely agreed that the differences between species usually originate during geographical isolation." The isolation can be physical geographic isolation or any factor that inhibits reproduction between two groups of organisms. In terms of evolutionary genetics dynamics models, creating of a new species (let alone new anatomy) seems to require an external dialing of the fitness parameters by a changing external environment. We discuss additional conclusions in Section 4.

2. Evolutionary Models and Dynamical Systems

The primary laws governing the interactions between genotype, phenotype, fitness, and the resulting variation over time can be described by mathematical dynamical systems [1]. A mathematical dynamical system is any system that changes over time with governing rules for change that depend on previous states of the system, possibly including external factors that may be deterministic or stochastic.

The two primary classes of mathematical models for evolutionary dynamics are discrete systems (iterated maps) and continuous systems (differential equations). In either case we have a state space, X, which is the space of all possible states of the system. In evolutionary dynamics, the state space usually incorporates the number of organisms of each genotype. That is, if we are considering a system with n different possible genotypes then X is n-dimensional Euclidean space, points (or states) $\vec{x} = (x_0, x_1, \ldots, x_n)$ are vectors of length n with x_i being the number (or proportion) of organisms with genotype i and n being the number of genotypes being considered. We use $\vec{x}$ when we want to emphasize the vector nature of this variable or just x otherwise.

For discrete systems we have a function (or map) f such that if x is the state of the system at a given time then $f(x)$ is the state one unit of time later. Thus, in discrete systems time passes in discrete steps — that is in jumps. If our units of time are say years, then state of the system two years later will be $f(f(x)) = f^2(x)$, and n years later it will be $f(f(\ldots f(x) \ldots)) = f^n(x)$.

Continuous dynamical systems typically arise as solutions to differential equations. The state space X still constitutes the space of all possible states. For a state x, the state that will occur t time units later will be written as either $\varphi(t, x)$ or $x(t)$. If the system is governed by a differential equation, we begin an equation $x' = f(x)$ and then $\varphi(t, x)$ is the solution with initial condition x (that is, $\frac{d\varphi}{dt}(t,x) = f(x)$ and $\varphi(0, x) = x$.)

There is an efficient mathematical framework for simultaneously treating continuous and discrete dynamical systems. A *mathematical dynamical system* is a state space X together with a time space T (T is either the real numbers or integers) and a continuous group action (or semi-group action) φ: $T \times X \rightarrow X$. For a differential equation, φ (t, x) is the solution with initial condition x. For a discrete dynamical systems defined by iteration of a map f: $X \rightarrow X$ the group action is φ $(n, x) = f^n$ (x). In either case, the system inputs a state (given by x) and a time (given by either t or n in T) and outputs that state after the allotted time has passed. Treating dynamical systems in such general terms enables us to focus on the topological and geometric phenomena that are true in general instead of what is only true for a given formulation.

The class of dynamical systems described above includes all deterministic dynamical systems (ie differential equations and iterated maps), those systems where the future is determined by the current state and time. Non-deterministic systems will be treated separately, although these often 'average out' to deterministic ones when many organisms are involved as with the quasi-species equation. (See Basener [7] for a treatment of topology in general; see Strogatz [5], Devaney [13] and Robinson [14] for dynamical systems; and see Novak [1] for dynamical systems as models of evolution). Like all mathematical models, the system can be simple or complex, depending on the number of simplifying assumptions.

2.1. *Simple Population Models*

Some simple models incorporate only the competition between populations, and thus focus on the competition-selection portion of evolution. Such models include the Malthusian logistic single species $y' = ay(1 - y)$ and competing species model $x' = r_1x(1 - bc - cy)$, $y' = r_2y(1 - fx - gy)$ which can lead to survival of one or both species. Nonlinear systems model more complex interactions, and can result in finite time extinction of one or more of the species.

The theory of mathematical dynamical systems can be applied to general formulations of these types of equations. In 1936 A. N. Kolmogorov gave conditions under which equations of the form

$$x' = xF(x, y)$$
$$y' = yG(x, y)$$

has either a stable limit cycle or equilibrium. This has broad implications for biological systems — see May [15]. (A *limit cycle* is either a periodic orbit or a sequence of equilibria, $p_1, p_2, \ldots, p_n$ with heteroclinic trajectories connecting p_i to $(p_{i+1 \ mod \ n})$.) More generally, the Poincare-Bendixson Theorem says that any bounded solution to a 2-dimensional system of differential equations is asymptotic to either an equilibrium or a limit cycle [16]. These examples illustrate the power of the dynamical systems approach; geometric or topological theorems restrict the potential behavior of a system even if the governing laws/equations are only partially known.

Discrete systems in any dimension and continuous systems in more than 2-dimensions can exhibit more complex behavior. For example, an orbit in the discrete 2-dimensional system for a simple ecosystem with two organisms

$$P_{n+1} = P_n + aP_n(1 - P_n/R_n)$$
$$R_{n+1} = R_n + cR_n(1 - R_n/M) - hP_n$$

is shown in Figure 1 for three sets of parameters. This system was used in Basener *et al.* [17] to model the rise and fall of the civilization on Rapa Nui (Easter Island). The mathematics of chaotic and recurrent behavior is discussed in Section 3.

2.2. *Simple Mutation-Selection Models*

Simple models may also focus solely on the genetic aspect of evolution. The METHINKSITISAWEASEL system, created by Dawkins in 1989, is commonly used to illustrate evolution by mutation and natural selection as in Smith [12]. The state space X is the space of all strings of 19 letters. Topologically, X is a discrete space with $26^{19} \approx 7.66 \times 10^{26}$ points. Iteration of the system involves making ten copies of a parent state x in which each letter of the copy has a 0.99 probability of being the same as the corresponding letter in the parent. The fitness of a state is equal to the hamming distance from the sequence METHINKSITISAWEASEL;

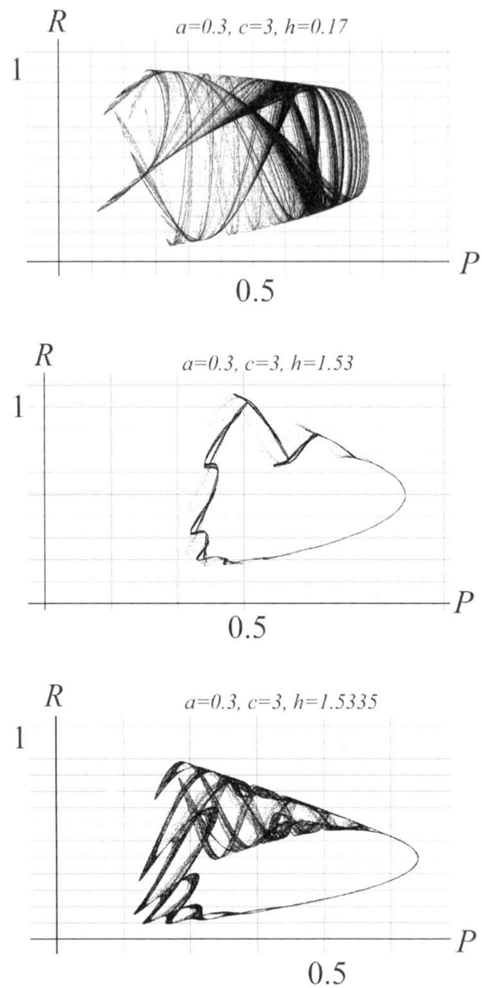

Fig. 1. Three chaotic attractors for discrete dynamical system modeling a simple two species ecosystem.

that is, the number of letters in the correct location with this goal sequence. The child with the highest fitness level is chosen as the new parent in the next generation. Although the system is not deterministic, for any initial condition the probability that the resulting sequence of iterations will reach the goal sequence is equal to 1.

This system is not an accurate model of evolution (see [12]), but it is useful to illustrate the simple description of evolution with mutation and natural selection. It also has typical evolutionary behavior; the 'genotype' undergoes modification over generations and then stabilizes at an equilibrium. This is the same behavior as with the RNA molecules in the Qβ replicase. It is worth considering more general systems of this variety. In Theorem 1 we show that if the state space is finite with a simple model of evolution, then evolution will run its process and then cease. Once maximal fitness is achieved, all subsequent mutations are irrelevant for fitness.

To describe a general situation of an evolutionary process, we will use X to denote a state space which could in general be any topological space. We assume that there are some laws governing the process of evolution on X, although they may in general be continuous differential equations, a discrete map, stochastic, or nondeterministic. If the system is discrete, the process of evolution will result in a sequence of points

$$x_0 = x,\ x_1 = f(x),\ x_2 = f(f(x)) = f^2(x),\ \ldots,\ x_n = f^n(x)$$

according to the governing laws. If it is continuous, the process of evolution will result in a path $x(t)$ in X. In the first case, we call $x_0, x_1, \ldots$ a (discrete) evolutionary trajectory and in the second we call $x(t)$ a (continuous) evolutionary trajectory. By a fitness function on X we mean a continuous function F: $X \to R$ (the domain is X and the range is R, the real numbers) that is nondecreasing on evolutionary trajectories. (Either $F(x_i) \leq F(x_j)$ for $i < j$ in the first case, or $F(x(s)) \leq F(x(t))$ for $s < t$ in the second.). Our first theorem, Theorem 1, shows that the behavior of the METHINKSITISAWEASEL system is the only possible behavior for a system with only finitely many states.

THEOREM 1. *Let X be any finite state space with a fitness function* F: $X \to R$. *Suppose* $x_0, x_1, \ldots$ *is a discrete evolutionary trajectory. Then there exists an N such that* $F(x_n) = F(x_N)$ *for all* $n > N$.

The proof is very simple; the set $F(\{x_0, x_1, \ldots\})$ is finite, being a subset of the finite space X, and therefore attains a maximum at some x_N. Since this is the maximum on the sequence $F(x_n) \leq F(x_N)$ for all (x_n) and F is nondecreasing $F(x_N) \leq F(x_n)$ for $(N < n)$, we have $F(x_n) = F(x_N)$ for all $n > N$.

It is clear from Theorem 1 that this type of a system — either a deterministic or nondeterministic progression of increasing fitness of a genotype in a sequence space — by itself does not result in an ongoing increase in fitness of organisms.

Related models can be constructed incorporating multiple organisms as well as spatial distributions. As long as the state space is compact (such as any a closed and bounded subset of Euclidean space, as is the case for any system with a finite area in which the organisms live), a similar theorem holds for systems with a fitness function that does not decrease over time. To work with continuous and discrete dynamical systems, with stochastic and deterministic ones, and with cases where the system is chaotic, we make some general terminology. If X is a state space, F is a fitness function on X and $x_0, x_1, \ldots$ is a sequence of points in X resulting from a model of evolution on X for which F is nondecreasing, we will refer to $x_0, x_1, \ldots$ as a *discrete evolutionary trajectory* in X. Similarly, a path $x(t)$ in X on which F is nondecreasing will be called an *continuous evolutionary trajectory* in X.

THEOREM 2. *Let X be any compact state space with a fitness function* $\boldsymbol{F}: \boldsymbol{X} \to \boldsymbol{R}$. *If* $\boldsymbol{x}_0, \boldsymbol{x}_1, \ldots$ *is a discrete evolutionary trajectory, then there exists an* $\boldsymbol{F}_*$ *such that* $\boldsymbol{F}(\boldsymbol{x}_n) \to \boldsymbol{F}_*$ *as* $n \to \infty$. *If* $x(t)$ *is a continuous evolutionary trajectory, then there exists an* F^* *such that* $\boldsymbol{F}(\boldsymbol{x}(\boldsymbol{t})) \to \boldsymbol{F}_*$ *as* $\boldsymbol{t} \to \infty$.

Proof. Since X is compact, F is bounded on X. In the first case, $\boldsymbol{F}(\{\boldsymbol{x}_0, \boldsymbol{x}_1, \ldots\})$ is a bounded subset of R, and thus has a supremum F^*. Since $\boldsymbol{F}(\boldsymbol{x}_n)$ is nondecreasing, it goes monotonically to F^*. In the second case, $\boldsymbol{F}(\{\boldsymbol{x}(\boldsymbol{t}) \mid \boldsymbol{t} \varepsilon \boldsymbol{R}\})$ is a bounded subset of R, and thus has a supremum F^*. As before, $\boldsymbol{F}(\boldsymbol{x}(\boldsymbol{t}))$ is nondecreasing, and so it goes monotonically to F^*.

The sequence $x_0, x_1, \ldots$ in Theorem 2 can be the solution to either a stochastic or deterministic discrete system on X, and the path $x(t)$ can be the solution to either a stochastic or deterministic continuous system on X. Observe that this theorem states that evolution will run its course until some point after which increase in fitness is inconsequential. (Specifically, for any small positive number ε there is a time after which the increase in fitness is less than ε.)

It may seem counterintuitive that Theorem 2 would apply to systems with chaos; for chaos has often been suggested as a mechanism for producing very complex structures. We address chaotic dynamics in Section 3, where it is proven that fitness never increases on chaotic sets.

2.3. *Population Models with Mutation-Selection*

To construct a more accurate model of evolution, we need to consider more aspects of genetics, mutations, populations and ecology. In this section we consider quasispecies, which is are ensemble of similar genomic sequences generated by a mutation-selection process, a notion developed by Manfred Eigen and Paul Schuster [18].

As before, we take our genotype information in a sequence space, say $X = \{A,T,C,G\}^N$ which is the set of all sequences in the letters A, T, C and G of length N. There are 4^N different organisms that can have their genotype in this space. Imagine a large population of such organisms. We denote the fraction of the total population consisting of genotype i by x_i, for $i = 1, \ldots, N$. So each x_i is in [0, 1] and $\Sigma_i x_i = 1$. Our state space X is the set of all $\vec{x} = (x_0, x_1, \ldots, x_N)$ satisfying $\Sigma_i x_i = 1$, which is the unit simplex in R^{N+1}. Observe that the state space X is compact.

Let $f_i > 0$ be the fitness of species i. For now, assume that the fitness corresponds to the growth rate. (It is common practice to equate fitness with growth rate. This seems sufficient in the short-term. However, organisms with a high fitness, resulting in a high reproduction rate, can overpopulate their ecosystem, destroying their food source and subsequently themselves as a population. This behavior is the main topic in the study of the collapse of ancient human civilizations in Basener and Ross [19] and Basener *et al.* [20].) The state space X together with the fitness function $\vec{f} = (f_0, f_1, \ldots, f_N)$ is called a *fitness landscape*.

Let Q be the matrix such that q_{ij} is the probability of mutation from genotype i to genotype j. (The rows of the square matrix sum to 1.) The *quasispecies equation* is then the differential equation

$$\vec{x}' = \sum_{j=0}^{N} x_j f_j q_{ij} - \phi x_i$$

where $\phi = \Sigma_i f_i x_i$ is the average fitness. The first term provides for reproduction and mutation, while the second term maintains $\Sigma_i x_i = 1$. If we let $\vec{W} = \vec{f}Q$, then the equation becomes

$$\vec{x} = W\vec{x} - \phi\vec{x}$$

which has a (generically stable) equilibrium at the solution to the eigenvector equation $W\vec{x} = \phi\vec{x}$.

For quasispecies, the fitness function determines the fitness of each genomic sequence, not the fitness of the quasispecies. Because individuals with more fit genomic sequences continually produce mutations with lower fitness, the quasispecies equation does not maximize an overall fitness. For modest mutation rates, quasispecies will appear as a peak centered on the genomic sequence with the greatest fitness. For this reason, we cannot apply Theorem 2 directly using the given fitness function. However, generically the conclusion still holds — evolution runs its course to the equilibrium.

Stochastic systems — systems in which mutations occur from each genotype to other genotypes at prescribed mutations rates and with some approximately deterministic rules governing population change over time for various genotypes — can all be modeled as a dynamical system on the same state space as the quasispecies equation. The following theorem says that even though fitness is not strictly increasing in these systems, regardless of the rules governing the population change the net effect of evolution over time is not much different than in previous theorems. Instead of going to an equilibrium, the system has a point which it will get close to, then may move away and will come back to again even closer, and then repeat the process of closer and closer approaches, infinitely often. To an observer, the system will continue to repeat (or return to) some state infinitely often.

THEOREM 3. *Let X be any compact state space. Then for any trajectory of a (discrete or continuous) dynamical system on X, there is a state x_* such that the orbit comes repeatedly close to x_* as time goes to ∞, as follows. If $x_0, x_1, \ldots$ is a discrete evolutionary trajectory, then there exists a state $x^* \varepsilon X$ and a subsequence $x_{i(1)}, x_{i(2)}, \ldots$ such that $x_{i(k)} \rightarrow x_*$ as $k \rightarrow \infty$. If $\varphi(t, x)$ is a continuous evolutionary trajectory, then there exists an x_* and a sequence of times $t_1, t_2, \ldots$ such that $\varphi(t_k, x) \rightarrow x_*$ as $k \rightarrow \infty$.*

Proof. If $x_0, x_1, \ldots$ is a sequence of points, since X is compact the collection of sets $\overline{\{x_i\}_{i=1}^{\infty}} \supseteq \overline{\{x_i\}_{i=2}^{\infty}} \supseteq \overline{\{x_i\}_{i=3}^{\infty}} \ldots$ is a nested sequence of compact sets. Thus the intersection $\bigcap_{n=1}^{\infty}\overline{\{x_i\}_{i=n}^{\infty}}$ is nonempty. Then let x_* be any point in $\bigcap_{n=1}^{\infty}\overline{\{x_i\}_{i=n}^{\infty}}$ and x_* is the desired point.

If $\varphi(t, x)$ is a path in X, since X is compact the collection of sets $\overline{\{\varphi(t,x)\}_{t>\alpha}}$ is a nested sequence of compact sets. Thus the intersection $\bigcap_{\alpha \in R}\overline{\{\varphi(t,x)\}_{t>\alpha}}$ is nonempty. Then let x_* be any point in $\bigcap_{\alpha \in R}\overline{\{\varphi(t,x)\}_{t>\alpha}}$ and x_* is the desired point.

3. Chaos and Recurrent Behavior

A dynamical system $\varphi: T \times X \rightarrow X$ is said to be *chaotic* on an infinite subset $A \subseteq X$ if

(i) Periodic orbits are dense in A.
(ii) There exists one orbit in A which is dense.
(iii) If X is a metric space then the system has *sensitive dependence of initial conditions*: There exists an $\varepsilon > 0$ such that for any point $x \varepsilon A$ and any neighborhood N of x, there exists a $y \varepsilon N$ *and a* $t > 0$ such that $d(\varphi(t, x), \varphi(t, x)) > \varepsilon$.

(See Basener [7] for mathematical terms, Robinson [14] for details on the dynamical systems in this section and Strogatz [5] for applications). Note that chaotic subsets are necessarily compact and in variant. It has been shown that the first two conditions are sufficient to imply the third (See Banks et al. [21] and Basener [7]), although in practice sensitive dependence is often taken alone as a definition of chaos because it is easy to compute experimentally.

For real-life systems, the periodic orbits in chaotic sets are less observable that the tendency to wander around the set; the small perturbations that occur in any real system will prevent it from actually being periodic. However, the behavior both in theory and practice has periodic-like aspects. Specifically, for any point $x \varepsilon A$, there is a sequence of times $t_0, t_1, \ldots$ with $t_i \to \infty$ such that $\varphi(t_i, x) \to x$ as $i \to \infty$. To an observer, the system appears to repeatedly return to its initial state forever. Hence, whether there is a fitness function that is nondecreasing on trajectories as with mutation-selection models, or if there is a fitness function that is defined on species but is not optimized in general as with the quasispecies model, on a chaotic set the system will continue to repeat a given state, and thus a given level of fitness, repeatedly.

More can be said if we assume that the fitness is nondecreasing with time; in this case, the fitness level is constant on a chaotic set.

THEOREM 4. *Let $\varphi: T \times X \to X$ be any dynamical system with a fitness function $F: X \to R$ such that $F(\varphi(s, x)) < F(\varphi(t, x))$ for any $s < t$. If A is a subset of X upon which φ is chaotic then F is constant on A. That is, there is no increase in fitness for orbits in A.*

Proof. Since φ is chaotic on A, there is a sequence of times t_0, $t_1, \ldots$ with $t_i \to \infty$ such that $\varphi(t_i, x) \to x$. Then, since F is continuous, $F(\varphi(t_i, x)) \to F(x)$. Since F is nondecreasing on orbits, F is constant.

4. Conclusions

Our first conclusion is that chaos and nonlinear dynamical systems contribute nothing to the ongoing increase in complexity or evolutionary fitness of biological systems. Statements such as that quoted earlier from Novak [1, p.9], suggesting that complexity of life results from nonlinear chaotic systems, are contrary to mathematics.

Second, the evolutionary process driven by mutation-selection, in both mathematical models and directly observed behavior, is that of a system going to an equilibrium and staying there. It seems the discussion of evolution in biology is that of an ongoing process but the study of mathematical models of evolution is that of equilibrium dynamics. There is nothing inherent in the fitness-driven mathematical system that leads to ongoing progress; to the contrary, mathematical systems, both those which are specific models such as the quasispecies equation and very general classes of models, have limits on the amount of increase in fitness that occurs. This is really well-known, as speciation is believed to occur only when driven by geographical isolation [12, p.275].

We have determined certain means of evolutionary progress to be impossible, and some of these means, for example the idea that chaos can lead to extreme evolutionary progress, have in the past been used as hypothetical possibilities for evolutionary dynamics. This leads us to ask what is left?

The space of all possible genotypes, while a compact space (assuming we disallow genotypes of unbounded length), is still enormous. The potential fitness, while bounded, is still extremely high. We can imagine this space as an enormous dimensional space, and imagine every viable species as a point in this space. We can image a line segment connecting every pair of viable genotypes if there is a reasonable probability that mutation from one to the other, as suggested in Figure 2. The result is an enormous network amenable to analysis by mathematical

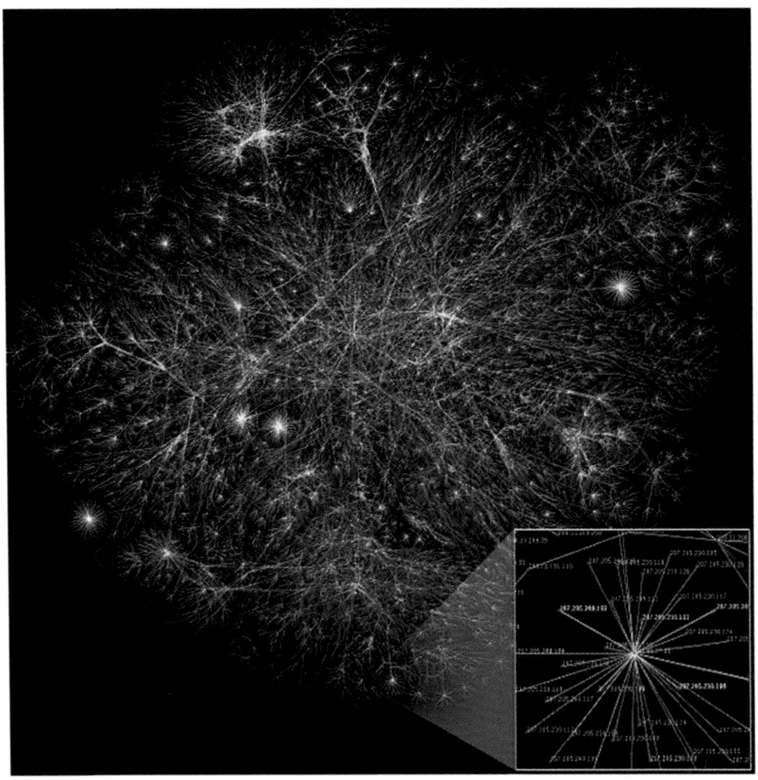

Fig. 2. A large network with sparsely connected groups. The question we pose is whether the genotype network is connected like this, or if there are many disconnected islands. This image shows a partial map of the internet based on the January 15, 2005 data found on opte.org. Each line is drawn between two nodes, representing two IP addresses. The length of each line are indicates the delay between its endpoint nodes. See [22].

network theory. The quasispecies equation provides the local equilibrium dynamics in this space, and there is no mathematical reason to expect generally other than the equilibrium state naturally from the system; stability is what we observe experimentally and from well-supported equations.

In the genotype network described above, each quasispecies lives within a group of highly interconnected points, called a community or clique in social network theory. If environmental conditions change, the quasispecies shifts within this group. In most cases, if the environment shifts to far (or at least too quickly) then the quasispecies is pushed to the edge of its local group, to points with low fitness, and then goes extinct. This decrease in fitness near the boundary of a local group can be observed in selective breeding; if too many desired properties in an animal or vegetation are attempted to be optimized through selective breeding, the simultaneous optimization becomes difficult and the species becomes less fit as a whole.

A question for evolution is to determine the structure of this genotype network. Are there bridges between groups of interconnected genotypes? How can we tell? What is the density of the network? How populated must a group be in order to support a quasispecies? Can the dimension of a local group be inferred, for example as the number of properties of a species that can be simultaneously optimized through selective breeding?

References

1. Novak MA (2006) Evolutionary Dynamics, Exploring the Equations of Life. Harvard University Press, Cambridge, MA.
2. James IM (ed)(2006) A History of Topology. Elsevier, Amsterdam.
3. McGehee R (1999) Celestial Encounters. Princeton University Press, Princeton, NJ.
4. Li TY, Yorke JA (1975) Period Three Implies Chaos. American Mathematical Monthly 82: 985–92.
5. Strogatz S (2001) Nonlinear Dynamics and Chaos: With Applications to Physics, Biology, Chemistry, and Engineering. Westview Press, Cambridge, MA.
6. Osborne M (2003) An introduction to game theory. Oxford University, Oxford.
7. Basener W (2007) Topology and Its Applications. Wiley, Hoboken, NJ.
8. Mermin ND (1979) The Topological Theory of Defects in Ordered Media. Rev Mod Physics 51(3): 591–648.
9. Kleman M, Lavrentovich OD (2003) Soft Matter Physics. Springer, New York.
10. Sethna J (2006) Entropy, Order Parameters, and Complexity. Oxford University Press.
11. Weeks JR (2001) The Shape of Space. Marcel Dekker, New York.
12. Smith JM (1998) Evolutionary Genetics. Oxford University Press, Oxford.
13. Devaney R, Hirsch M (2003) Differential Equations, Dynamical Systems, and an Introduction to Chaos. Academic Press.

14. Robinson C (1999) Dynamical Systems, Symbolic Dynamics and Chaos. CRC Press.
15. May R (2001) Stability and Complexity of Model Ecosystems. Princeton University Press, Princeton, NJ.
16. Perko L (2006) Differential Equations and Dynamical Systems. Princeton University Press, Springer.
17. Basener WF, Brooks BP, Radin M, Wiandt T, (2008) Dynamics of a Discrete Population Model for Extinction and Sustainability in Ancient Civilizations, Nonlinear Dynamics. Psychology and Life Sciences 12(1): 29–53.
18. Eigen M, Schuster P (1979) The Hypercycle: A Principle of Natural Self-Organization. Springer, Berlin.
19. Basener W, Ross D (2004) Booming and crashing populations and Easter Island. SIAM J. Appl. Math. 65(2): 684–701.
20. Basener WF, Brooks B, Radin M, Wiandt T (2008) Rat Instigated Human Population Collapse on Easter Island, Nonlinear Dynamics. Psychology and Life Sciences 12 (3): 227–240.
21. Banks J, Brooks J, Cairns G, Davis G, Stacey P (1992) On Devaneys definition of chaos. Am. Math. Month 99(4):33234.
22. The OPTE Project, www.opte.org/maps/.

Tierra: The Character of Adaptation

Winston Ewert[1]*, William A. Dembski[2] and Robert J. Marks II[1]

[1]*Electrical & Computer Engineering, One Bear Place #97356, Baylor University, Waco, TX 76798-7356.* [2]*Discovery Institute, 208 Columbia Street, Seattle, WA 98104. (*Corresponding author: evoinfo@winstonewert.com)*

Abstract

Tierra is a digital simulation of evolution for which the stated goal was the development of open-ended complexity and a digital "Cambrian Explosion." However, Tierra failed to produce such a result. A closer inspection of Tierran evolution's adaptations show very few instances of adaptation through the production of new information. Instead, most changes result from removing or rearranging the existing pieces within a Tierra program. The open-ended development of complexity depends on the ability to generate new information, but this is precisely what Tierra struggles to do. The character of Tierran adaptation does not allow for open-ended complexity but is similar to the character of adaptations found in the biological world.

Key words: Adaptive loss, artificial life, complexity, novelty, open-ended evolution, simulation, Tierra

1. Introduction

Tierra, a digital evolution simulation, was originally developed by Thomas Ray in 1989 [1]. Some such simulations attempt to accomplish a specific task or to solve a particular problem. Examples include finding a phrase [2], logic function synthesis [3], and designing an antenna [4,5]. While such simulations take inspiration from the concepts of natural selection and random mutation, they differ from Darwinian processes in a significant way. Such examples of evolutionary computation have a predetermined goal, while biological evolution, as commonly understood, does not. Tierra does not define such a predetermined goal; instead, the intent is simply to observe the outcome of the evolutionary process. As Ray states: "The creatures invent their own fitness functions" [6].

This is not to say that research using Tierra has no goal. In fact, Tierra's goal is much more ambitious. Ray's intent with Tierra was nothing less than to simulate the genesis of complexity and open-ended evolution, analogous to the Cambrian explosion:

> While the origin of life is generally recognized as an event of the first order, there is another event in the history of life that is less well known but of comparable significance: the origin of biological diversity and macroscopic multicellular life during the Cambrian explosion 600 million years ago [6].

The Cambrian explosion is an event recorded in the fossil record during which there was a relatively sudden shift in the evolution of life on earth. Prior to this point, biological life was almost entirely composed of single-celled organisms. However, in a brief period of geological time, there was an "explosion" of biological forms in which most of the phyla now in existence appeared suddenly in the fossil record. The causes behind this geological event are debated within biological circles [7].

Why is the goal to produce a Cambrian explosion in artificial life? The underlying intent is to produce countless forms through an evolutionary process similar to what is found in biology. The potential of this process in biology appears to have been unleashed during the Cambrian explosion. If artificial evolution could be unleashed in the same way, we might also be able to produce a plethora of fascinating forms analogous to those found in biology. Essentially, once evolution (whether biological or artificial) has produced a Cambrian explosion, the rest of evolution should proceed easily.

Ray's view was that the complexity needed to reach a critical mass. Once past this point, evolution's creativity would be unleashed. In the case of biological life, this happened during the Cambrian explosion. Tierra was Ray's attempt to give evolution the critical mass it needed. In fact, there were three different versions of Tierra each starting with more complexity in an attempt to kick start the evolutionary process.

Tierra produced a variety of interesting phenomena, including parasites, hyperparasites, social behavior, cheating, and loop unrolling. However, twenty years after the introduction of Tierra, the conclusion is that Tierra did not produce a Cambrian explosion or open-ended evolution. Though Ray described Tierran evolution as generating "rapidly diversifying communities of self-replicating organisms exhibiting open-ended evolution by natural selection" [6], others disagree:

> Artificial life systems such as Tierra and Avida produced a rich diversity of organisms initially, yet ultimately peter out. By contrast, the Earth's biosphere appears to have continuously generated new and varied forms throughout the 4×10^9 years of the history of life [8].
>
> These strong increasing trends imply a directionality in biological evolution that is missing in the artificial evolutionary systems [9].

Ray has recently stated that he regards Tierra as having failed to reach its goal. He describes the evolution seen within Tierra as transitory. He no longer considers himself part of the artificial life community, and is now studying biological questions rather than those of artificial evolution [10].

The absence of a Cambrian explosion in artificial life demands an explanation. If biological evolution produced a Cambrian explosion, why does artificial evolution not do the same? Our inability to mimic evolution in this regard suggests a deficiency in our understanding of it. In the words of Feynman: "What I cannot create, I do not understand" [11].

Tierran evolution can be characterized as an initial period of high activity producing a number of novel adaptations followed by barren stasis. It would appear that Tierra easily produced the novel information required for a variety of adaptations. Why did it cease? If Tierra could produce novel information, it should continue to do so as long as it was run. However, if Tierra was incapable of producing such information, it should not have been able to produce the adaptations that it did.

A closer look at Tierran evolution reveals an important characteristic of the adaptations. Tierra started with a designed ancestor to seed the population. In other words, it presupposed something like an origin of life and was concerned with the development of complexity after that point. The ancestor provides initial information to Tierra. Adaptations primarily consist of rearranging or removing that information. Open-ended evolution requires adaptations which increase information. However, such adaptations are rare in Tierra. Tierra's informational trajectory is reversed from what evolution requires. It is dominated by loss and rearrangement with only minimal new information instead of being dominated by the production of new information with minimal cases of removal or rearrangement of information. Long term evolutionary progress is dependent on the generation of new information.

If Tierra is capable of generating new information even in small amounts, does this not provide evidence that Darwinism can account for new information? Many small gains will eventually accumulate into a large amount of information. However, if this were true, we would see evidence of it within Tierra. There ought to be a steady stream of information gaining adaptations, rather than stasis actually observed.

The purpose of this paper is to review the published results of Tierran evolution. By investigating these results, we elucidate the characteristics of adaptations found within this system. In particular, we demonstrate that Tierran programs adapt primarily through loss and rearrangement. Tierra initially appeared to hold great promise as a model of biological evolution displaying open-ended evolution. However, we see that the character of Tierran developments was never that which could produce open-ended evolution.

2. Description of Tierra

2.1 *Programs*

Tierra seeks to create artificial life within a computer. In some cases similar evolutionary simulations are meant to model biology [12,13]. As a result, the rules of the system are derived from a simplification of biological reality. Other cases seek to use the evolutionary process to solve a particular problem [3–5]. The rules of the system are derived from the nature of the problem being solved. In contrast, Tierra seeks to use the underlying rules of computer systems, trusting the evolutionary process to make use of whatever medium it finds itself in.

However, in developing Tierra, Ray did not maintain perfect fidelity to the design of computer hardware. Instead, the design of Tierra was also influenced by the design of biological systems. He was concerned, based partially on the results of previous similar experiments, that computer code would be too "brittle," prompting him make design decisions to make code more evolvable [10]. He realized that random modifications to the computer code would too easily break existing functionality and make it difficult to evolve new behaviors.

Tierran programs can be considered similar to proteins. A Tierra program is a sequence of instructions in much the same way that a protein is a sequence of amino acids. Both of these can be compared to English sentences. The function of a sentence, Tierran program, or protein is determined in some way by the sequence which makes it up. The meaning of a sentence is determined by the letters which make up the sentence. If different letters are substituted into the sentence or the letters are rearranged, a different sentence with a completely different meaning will likely result. In a similar way, the structure and function of a protein is determined by the sequence of amino acids that make up the protein. The behavior of a Tierra program is also determined by the sequence of instructions that make up the program.

Programs need to refer to locations inside themselves. This is especially true for Tierra as the program must copy itself. In actual computer systems, this is typically done through the use of numerical offsets, e.g. a reference to position 5 in the program. The problem with such a technique is that adding or removing instructions will tend to change all of the position numbers in the program. This will leave all the position numbers incorrect, thereby breaking the program. This is a primary cause of the brittleness that Ray was trying to avoid.

When biological proteins need to interact with other biological entities, they make use of binding sites. A binding site is a particular region on a protein to which other molecules bind. Which molecules will bind depends on the exact binding site properties. As a result, changing the binding site will change how the protein interacts with other molecules and thus possibly its function.

Tierra borrows this idea by having some of the instructions function as labels. A label consists of a sequence of nop0 and nop1 instructions, which are considered complementary to one another. Each label "binds" to another label with the complementary instructions. That is, a label nop1, nop1, nop0 will bind to the label nop0, nop0, nop1. Figure 1 shows the use of labels within the ancestor program. This solves the problem of referencing different parts of the program with specific position numbers, because the program can refer to the label itself, a referencing technique that will still work if the label is relocated.

English sentences do not have a precise analog to biological binding sites. The sites can, however, be considered roughly similar to punctuation. A binding site or label is useless by itself, as it has no actual function except to bind other things together. As such, binding sites modify the rest of system in useful ways, while lacking intrinsic functionality. Punctuation acts much the same way in English sentences. Consider the difference between, "No price too high," and "No, price too high." None of the words in the phrase have been modified; nevertheless, the meaning has been changed significantly.

Tierra programs contain instructions. The exact sequence of instructions specifies the operation of the program. Some of the instructions form labels which are like binding sites. Binding sites perform no tasks in isolation, but manipulate the functions of other instructions in the program.

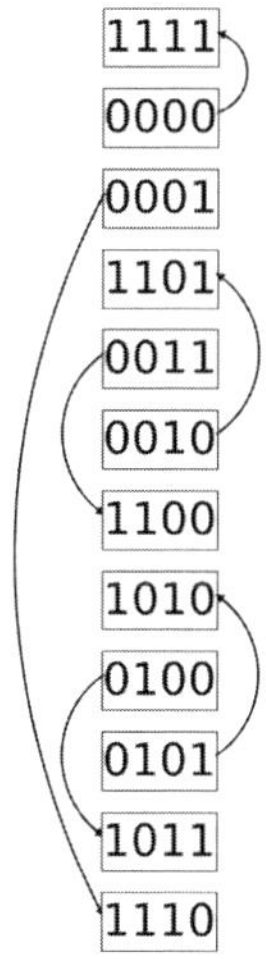

Fig. 1. A depiction of the use of labels in the Tierran ancestor.

2.2 *Ancestor*

Tierra runs by simulating many different programs running inside a computer. As time goes on, older programs are killed off. As the programs run, they make copies of themselves to produce new programs. Some of these programs have mutations and are thus slightly different from their predecessors. These mutations randomly replace, insert, or remove instructions like similar mutations in a DNA sequence. There is a selective force present, as those programs which are able to replicate more times before they die will leave more offspring and thus dominate the population through a process like natural selection.

This is similar to the idea of a soup of self-replicating proteins. In terms of sentences, it is as if the computer simulating Tierra is reading sentences and following their instructions. In this case, the sentence reads something like, "make a copy of this sentence." Thus as long as the simulation is kept running, more and more copies are made. If some sentences provide better instructions for making copies, they will tend to dominate the population.

In all of these situations, an ancestor is needed, i.e., the initial self-replicating program, protein, or sentence. Tierra starts with a program that is capable of replicating itself. This is equivalent to a self-replicating protein or the sentence, "copy this sentence."

A depiction of the structure of the original program can be found in Figure 1. The ancestor is important in the case of Tierra because the adaptations mostly derive from rearranging the information contained in that ancestor.

2.3 *Parallel Tierra*

Further development on the Tierra program produced a version which made use of parallelism [14–16]. Modern computers have the ability to run different code at the same time, that is, in parallel. By taking a large task and dividing it into smaller tasks which can be run at the same time, it is possible to perform the whole task more quickly. An analogy is drawn between these parallel "threads" of execution and cells in a biological organism [15]. The developers were able to produce "significant increases in parallelism" [15] in this version of Tierra.

2.4 *Network Tierra*

A later version of Tierra was developed known as Network Tierra [17,18]. Results using this version have been published, but much of the data produced remains unanalyzed. The papers published about the result of Network Tierra did suggest interesting results [19]. A particular portion of the code in the Tierra program was duplicated and, "While the two copies are initially identical, gradually, the two copies diverge in their structure and function" [19]. However, no actual code was presented and details as to what exactly is meant by divergence of structure and function were lacking. The lack of presented code prevents an analysis, and thus further discussion about Network Tierra will not take place here.

3. Looking for complexity

Tierra produced a number of adaptations. However, in order to produce a Cambrian explosion, adaptation alone is insufficient. It is necessary that new information is produced. Adaptations can lose or rearrange existing information and thus provide benefit without new information.

There is a parallel to this idea in biology. Fish found in dark environments can lack functioning eyes. Since the eyes do not work in the dark, they are useless if not deleterious in that environment. As a result, the process of natural selection works to eliminate the eyes. Thus we have a clear example of a biological adaptation being brought about through changes in the environment. However, this change has been produced by removing something rather than adding it, and therefore constitutes an example of reductive evolution. Could humans have evolved from a bacteria-like organism by successively disabling features? Clearly not.

Biological experiments have been performed in which insects have undergone changes due to mutations that produce extra sets of wings or eyes [20]. This does not appear to have been a beneficial change for the insect; however, it does show the ability to produce novel features due to relatively minor mutations. In this case, we are only observing the repeated expression of what the insect was already capable of producing. Clearly, the insect already contained instructions (genes) needed to construct the eyes and the wings. Mutations have simply caused those instructions to be repeated. Such duplications, modified expressions, or rearrangements of the genetic information can produce significant results. But many repetitions of this will not explain the origin of eyes or wings in the first place.

A similar idea can be seen in English sentences. Consider the sentence, "the quick brown fox jumps over the lazy dog." We can easily obtain a new valid sentence by omitting the word "quick" and obtaining "the brown fox jumps over the lazy dog." In this case, we have eliminated something. On the other hand, suppose that we add the word "blind," and obtain "the quick brown fox jumps over the lazy blind dog." There is a completely new word in place. It is much easier to remove a word than it is to add a new word. The letters in the new word must be selected at random, which is a relatively difficult task. While removing words is easier, it is clearly a very limited approach, as there are only so many words that can be removed.

For a biologist to determine if new information is produced in an adaptation can be difficult. Because we have a limited understanding of biological systems, the nature of a biological adaptation can be difficult to determine. In an artificial system such as Tierra this is not the case. We have a complete understanding of Tierra and thus can determine how any adaptation functions.

Tierra produced a number of adaptations. But did Tierra produce new information? What would new information look like inside of Tierra? It would be in the form of new functional code within Tierra programs. Of course, it is easy to produce new code by inserting extra instruction into a Tierra program. However, it is difficult to produce functional code. In order to be considered information, the code must be beneficial — not neutral or detrimental.

In some cases parts of Tierra programs are duplicated or moved. It does not make sense to count these as new information because the evolutionary process did not produce the code in question. The code was already given in the ancestral program; it has merely been relocated. However, by duplicating and moving individual instructions it is possible to construct any program. It only makes sense to appeal to a duplication or movement event when explaining a sequence of instructions. In terms of the English sentences, it only makes sense to consider words being moved and duplicated, not individual letters. As such, a word formed by rearranging the letters of another word is a completely new word not a rearrangement of the old one.

Tierra contains labels that are analogous to binding sites. These control the "expression" of the program. They changed within the time frame of Tierran evolution, and these changes caused many of the adaptations observed. However, since the labels are inert in and of themselves, they are not solely responsible for the behaviors they produce. Rather like the extra wings or eyes on an insect, they are manipulating the expression of other information. Clearly, change that can be produced by manipulating expression is limited. As such, we should not consider such changes as new information.

In some cases, a mutation will be neutral. The program with the mutation performs exactly the same as a program without the mutation. This is not new information because it has no adaptive benefit. In other cases, a given instruction may perform no useful task. It can be replaced by almost another instruction and the program will execute in the same way. Due to the lack of specificity such instructions do not carry informational content.

The importance of new information is due to its being both necessary and difficult. Without new information, evolution is restricted to rearrangements of existing information. But there is only a limited number of ways to rearrange existing information. In order to avoid stasis, evolution must produce new information. Obtaining new information is difficult because it depends on improbable random events. In the case of Tierra, the improbability derives from having to select particular sequences of instructions with functionality. However, this difficulty depends on the length of the sequence. It should be expected that short sequences of new instructions can arise. The difficulty of selecting the correct instructions grows exponentially as the number of instructions is increased.

What we find in Tierra is that most of the changes do not produce new information. In various ways, they rearrange the code already present in the ancestor. There are cases were new information, that is functional code, is produced. Such cases consist of only small pieces of code. That is, we see a few scattered instructions not blocks of new code.

But if these small changes can be combined, is it not possible to gain a large amount of information? Darwinism depends on precisely this point to explain all information found within biological life. Nevertheless, Tierra does not support the Darwinist contention. Despite the substantial amount of time spent running Tierra simulations, this predicted repeated information gain did not occur. It never gained more than a small amount of information. On the other hand, we do observe significant adaptations making use of deletion or rearrangement. Tierra does show new information; however, it fails to vindicate Darwinian theory's expectations of that information.

Ray sought to produce a digital Cambrian explosion. It initially seemed to work but ultimately stalled. A closer inspection shows that even during that initial period, the process could not be characterized by an increase in information. The trajectory of Tierra was never correct for open-ended evolution or unbounded complexity.

4. Examples

This section will look at the individual programs produced by Tierra to show what kinds of changes were necessary to bring them into existence. Most of the actual code is taken from the Tierra distribution available from the Tierra website and discussed in the Tierra manual [21]. In some cases, code that is considered is taken from other papers published about Tierra. This section deals with a high-level overview of the adaptations observed in these programs. A look at the precise code involved can be found in Appendix 6.

4.1 *Parasite*

Tierra's first interesting adaptation was parasitism. These programs were called parasites because they were unable to make copies of themselves on their own. However, they could replicate inside the Tierra simulation because they made use of the code in nearby ancestors. The parasite was shorter than the ancestor because it did not contain all of the code necessary to self-replicate. This allowed the parasite to replicate more quickly and more often, giving it a competitive advantage against the ancestors. Such parasites came to dominate Tierra; however, they required the presence of an ancestor in order to replicate, and thus never completely replaced the ancestors.

The ability to make use of the code belonging to another program would, at first glance, appear to be a fairly complex task. However, this was not the case within Tierra. As Figure 3 shows, the original ancestor was written divided into two parts. The first was the main loop and controlled the operation of the program. The second was a copy loop procedure; it was responsible for actually copying one block of memory to another. It was used to copy the parent's code in memory to the location of the child. This is analogous to the procedure used for DNA replication in biology. The sole difference between the parasite and the ancestor was that the parasite did not contain a copy procedure. However, because the copy procedure is located using the label addressing technique, Tierra looked for the copy procedure in nearby code. Typically, it found one in a nearby ancestor and thus executed that code, thereby allow the parasite to self-replicate even without a copy loop procedure.

Figure 2 shows the label references as they differ between the parasite and the ancestor. The parasite is simply a truncated version of the original ancestor. The jump into the copying code is still present, but does not point anywhere within the program. Instead it points into a nearby program which it will use to make copies.

A complete comparison of the code in the ancestor and the parasite can be found in Section 6.1. The only changes found are the removed block of code and a change to a label, which was the original cause behind the removal of that block of code. Neither of these changes qualifies as new information.

4.2 *Immunity*

Some Tierra research indicates that the ancestors develop immunity to parasites [16]. Neither the papers nor the official Tierra distribution appear to provide the actual code of a program which exhibits such immunity. Nevertheless, the method of immunity is described as follows: "Immune hosts cause their parasites to loose[sic] their sense of self by failing to retain the information on size and location" [16]. Such

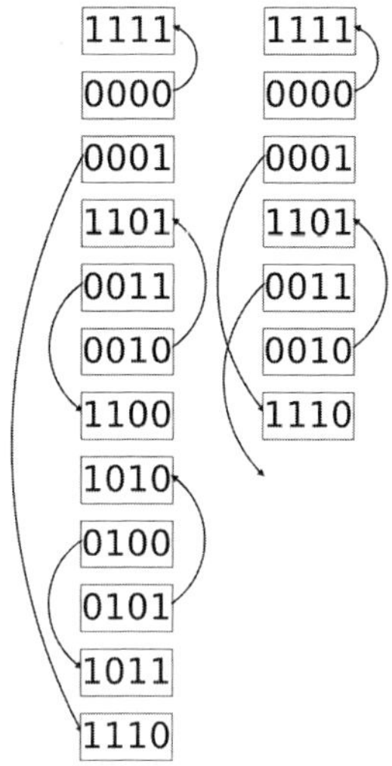

Fig. 2. Labels compared between the ancestor and the parasite.

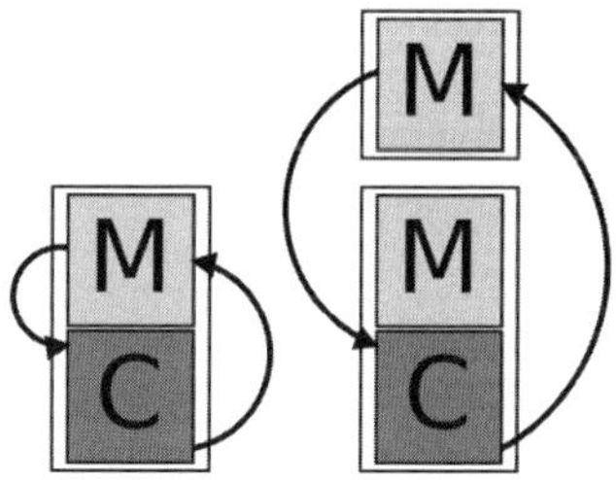

Fig. 3. The structure of the original Tierra ancestor compared with that of the parasite. The image on the left is a regular ancestor. On the right a parasite is depicted using the copy loop of a nearby ancestor.

behavior can be caused by having a subset of the adaptations of the hyper-parasite. See Section 6.2 for further discussion. See Section 4.3 for details on the changes producing the hyper-parasite.

4.3 *Hyper-parasites*

The evolutionary response to the parasites was hyper-parasites. They were termed hyper-parasites because they acted as a parasite on a parasite. While the original parasites used the code of other programs to replicate, the hyper-parasites tricked parasites into copying the code of the hyper-parasite. This technique worked because the parasite was executing code inside the hyper-parasite allowing the hyper-parasite to take control of it.

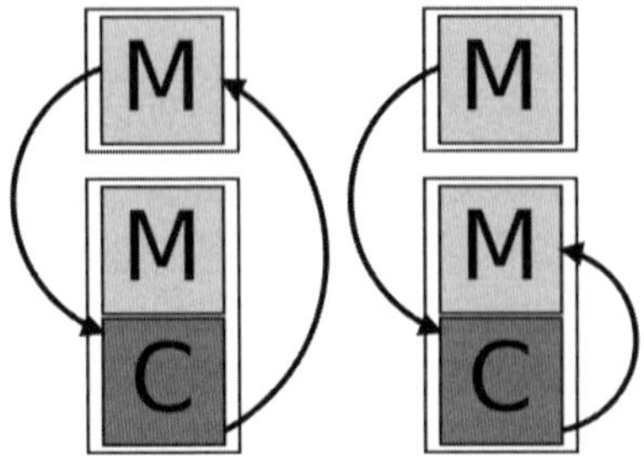

Fig. 4. The operation of a parasite and a hyper-parasite. The left side shows the typical parasitical interaction, but the right side shows how the hyper-parasite traps the parasite's CPU.

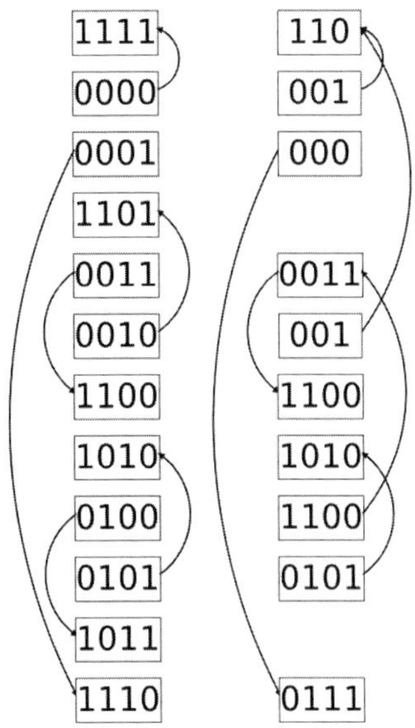

Fig. 5. Labels compared between the ancestor and the hyper-parasite.

As Figure 4 shows, the original ancestor returns back to the calling code once the copying is done. This behavior is used by the parasite in order to use another program's copying code. However, the hyper-parasite has mutated so that it no longer gives control back to the calling code, instead maintaining control itself. This alone was not actually enough; that change alone would still have continued to produce parasites because the internal state of the program would still be that which was configured by the parasite. The hyper-parasite managed to avoid this by always resetting the state of the program after a copy has been made.

Figure 5 compares the use of labels between the ancestor and the hyper-parasite. Some of the actual labels have changed, but those changes are not important.

For the most part, the same activity can be seen in the ancestor and the parasite. There are two significant changes, shown by arrows now pointing to different locations. The changed arrow in the lower half of the figure shows the change necessary to keep control of the CPU instead of returning it to the parasite. The other changed arrow corresponds to the change necessary to reset thc state of the process so that it copies the hyper-parasite instead of the parasite.

See Section 6.3 for details on the exact code changes involved. By the time hyper-parasites arise in the simulation, there have been a large number of changes to Tierran genomes. However, most of these have no actual effect and none of them consist of new functional code.

4.4 *Social behavior*

The Tierran programs eventually developed social behavior. A program was deemed to be social if it cannot replicate without being surrounded by similar creatures. Once a program has finished replicating it must return to the beginning of the program in order to make a second replication. In the case of social programs, the program jumped into the end of a previous program and then fell off into the start of the current program. This is depicted in Figure 6. Figure 7 shows the underlying labels being used here. The only significant change is that the jump that had previously gone to the first part of the program now jumps into memory behind it.

Social behavior was an interesting development but with one major caveat. The program exhibiting the social behavior does not appear to gain any benefit for doing so. A program is deemed social by the fact that it cannot reproduce except

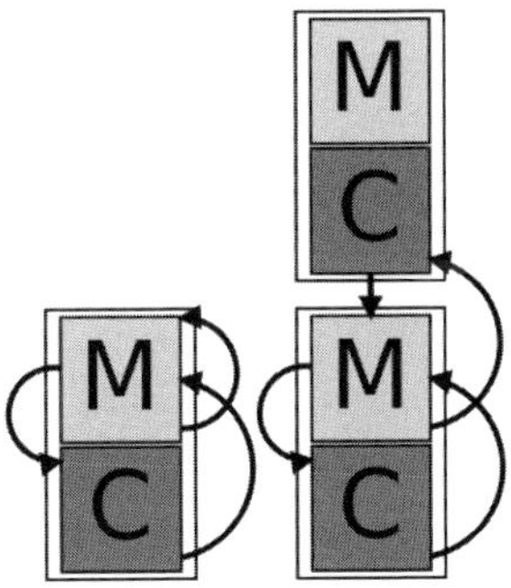

Fig. 6. Comparison of the mechanics of the ancestor and a social creature. On the left we see a typical ancestor which jumps back to the beginning of its main loop when a copy is finished. On the right a social creature jumps into the end of the creature before it and trails into the copy loop.

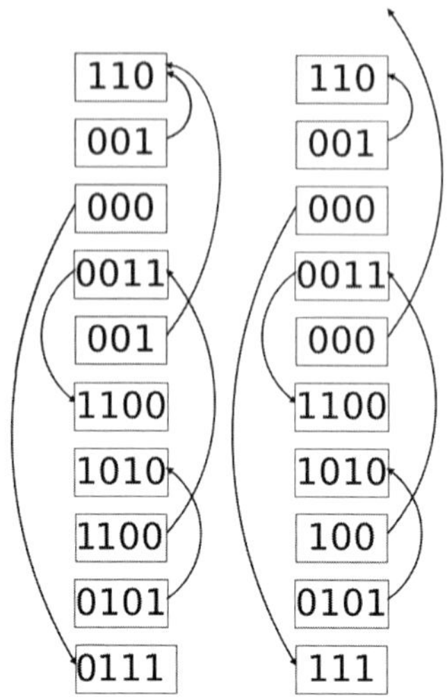

Fig. 7. Labels compared between the hyper-parasite and the social program.

in aggregate groups. It has lost the ability to replicate alone. In fact replication will be slightly slower because it must execute some code belonging to a neighbor before actually reaching its own code.

Ray gives his reasoning for the evolution of sociality:

> It appears that the selection pressure for the evolution of sociality is that it facilitates size reduction. The social species are 24% smaller than the ancestor. They have achieved this size reduction in part by shrinking their templates from four instructions to three instructions. This means that there are only eight templates available to them, and catching each others[sic] jumps allows them to deal with some of the consequences of this limitation as well as to make dual use of some templates [6].

It is true that the social species were considerably smaller than the ancestor. However, they were not considerably, or at all, smaller than similar creatures which did not exhibit "social" behavior. The social programs did not have a size advantage over the non-social creatures that dominated at the time of their arrival. Ray's explanation of selection pressure for sociality does not work

We propose another explanation. These social programs were produced by nearly neutral deleterious mutations which became fixed in the population. Once Tierra's population filled the available space, Tierra programs very rarely produced more than one child. It took a long time to make a copy of a program in memory. A program would typically die while in the process of making its second child. The result of this is that there was very little selective pressure on the code responsible for performing the transition for a second replication. Social behavior was a degradation of performance in this area, but it was not large enough to be selccted against.

Section 6.4 demonstrates the code differences between the hyper-parasite and the social program. The interesting changes are to the labels; everything else involves removal of code or changes with no effect on behavior.

4.5 *Cheater*

Eventually a cheater arose which took advantage of the programs exhibiting the social behavior. As Figure 8 shows, a truncated program was created which sits between two social programs. When the social program attempted to jump into its predecessor's end, it ends up running into the cheater's code instead of its own. The cheater then uses the captured CPU to make additional copies of itself.

As with the parasite this ability derives from having deleted a large portion of the genome. Figure 9 depicts the resulting program structure. See Section 6.5 for an actual look at the code. The only change which is not a deletion is neutral.

4.6 *Shorter program*

The shortest self-replicating program reported to evolve was 22 instructions in length. Interestingly, this was shorter than either of the parasitic designs. It was a

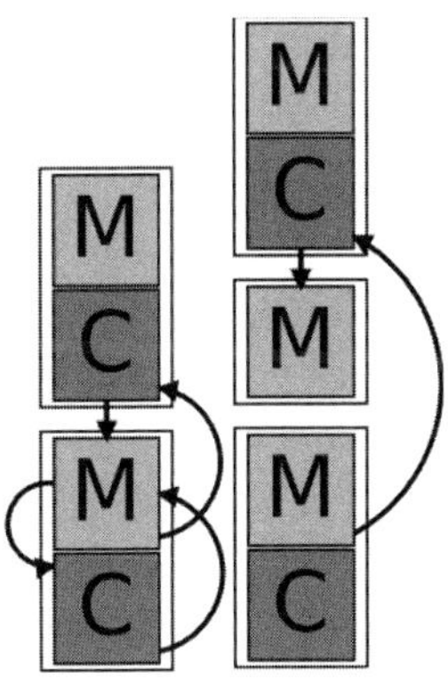

Fig. 8. Comparison of the mechanics of the social program and the cheater. The left hand side shows the typical behavior of a social creature, whereas the right shows a cheater taking advantage of this.

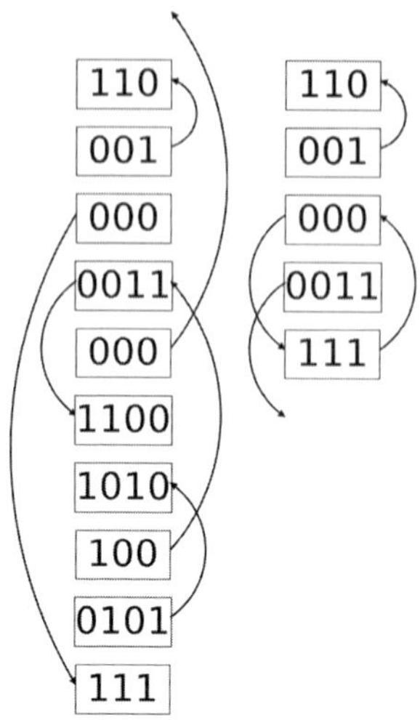

Fig. 9. Labels compared between the social program and the cheater.

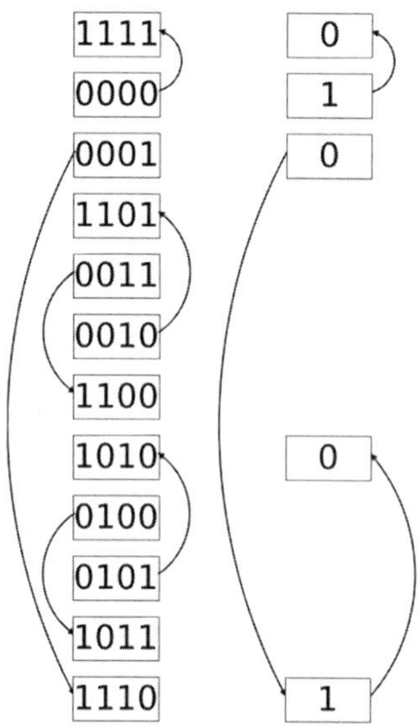

Fig. 10. Labels compared between the ancestor and the short program.

very substantial reduction from the 80 instructions of the original ancestor. However, as Figure 10 shows, the structure was a subset of the original. As one might guess, the construction of this short program was largely done through the removal of instructions. However, as discussed in Section 6.6, there was an exception. The short program was generated mostly by code elimination, but two of the instructions of new code were inserted which helped replace longer code. i.e. The new instructions perform the same task as the original but with less instructions required.

4.7 *Loop unrolling*

An optimization known as loop unrolling also evolved in Tierra. This arose in a version of Tierra operating on slightly different rules. In this case, longer programs were rewarded for their length in order to discourage the development of shorter and shorter programs. Normally a shorter program had an advantage in terms of the time it takes to make a copy, simply due to being shorter. Rewarding longer programs removed that advantage. As a result, this scheme is known as size neutrality. Under these rules, Tierra removes the incentive to shrink genomes and instead promotes the development of techniques to copy existing instructions faster. The evolutionary process managed to implement an optimization known as unrolling a loop.

Ray presented this an example of an intricate adaptation:

> The astonishing improbability of these complex orderings of instructions is testimony to the ability of evolution through natural selection to build complexity [22].

However, Ray's perspective does not hold up to scrutiny. In fact this adaption results from a duplication of the code inside the program. Loop unrolling is an optimization which works through duplicating code in a loop. To repeat an action, such as copying an instruction, a program must jump backwards in the code so as to re-execute the instructions. This jump takes time and thus constitutes overhead cost. By repeating the contents of the loop, it is possible jump half as often thereby reducing this extra cost, leading to more efficient replication.

Ray stated that "unrolling did not occur through an actual replication of the complete sequence." This claim was derived from the idea that the copies of the loop in the unrolled version differ in instruction order. However, as Section 6.7 discusses, most of the instructions were in a consistent order. In fact, they remained in the same order as in the original loop. Since the instructions can be reordered in several ways without affecting the operation of the program, this consistency strongly implies that the new loop was generated through a duplication event.

New functional code did show up; however, it was not directly related to the unrolled loop. Instead, the program "lied" about its length, causing it to receive a larger bonus. Ordinarily, this bonus would have been counteracted by the need to execute a longer program. However, this program neither executed nor copied the instructions in the second half. This means that it managed to gain the benefits of doubling the program length without any of the drawbacks. Doing so required introducing four new instructions.

Contrary to the claims of Ray, this is not an example of an astonishingly improbable sequence of instructions. The program results mostly from duplication

5. Summary

The author of Tierra sought to create a digital Cambrian explosion whereby the power of the evolutionary process was unleashed. It is agreed that Tierra did not succeed in accomplishing this feat. Rather, the evolutionary activity within Tierra dies after only a transitory period. No Cambrian explosion occurs.

Furthermore, the evolutionary activity that occurred was not of the sort that can be used as the basis for the ongoing evolution of novel information. Most change in Tierra was created by rearranging the existing code in the system, not by producing new code. Some cases did produce new code; however, the amount of change produced in this fashion is very small compared to change produced in other ways. What information gain existed only manages to tweak the existing system. The trajectory of Tierra was wrong, it is dominated by the wrong category of adaptation.

The observation that evolution consists largely of adaptations that remove or manipulate existing information, rather than adaptations producing new information, is not restricted to Tierra. Many observed adaptations in biology are in fact derived from changes which break existing systems [23]. Studies of biological adaptations have shown that they proceed via the elimination of unnecessary and costly functions [24,25]. A survey of lab experiments showed that the adaptations found in such scenarios fit the same picture [26]. Further discussion of adaptation by loss in biological scenarios can be found within these proceedings [27].

Unlike many artificial life simulations, Tierra followed Darwinism by not imposing an external artificial fitness. Tierran programs were not rewarded for performing calculations or solving problems. Rather in Tierra there was only survival and replication. As a result Tierra paralleled biology more closely on this point. As discussed, the pattern of observed adaptation is similar between Tierra and biology. Rather than being a system which fails to imitate biology closely enough to produce a Cambrian explosion, Tierra is a system which manages to imitate the character of directly observed biological adaptations.

Some other evolutionary systems do show an increase in complexity and the production of new functional code. Avida is one such example, in which a sequence of instructions is generated which computes the bitwise EQU (XNOR) operation [28]. However, Avida's ability to generate such sequences of instructions is derived from its use of stair step active information [29]. Avida rewards the development of partial implementations of its target, thereby helping the programs to evolve [3]. Essentially, action was taken in Avida to make it easier for evolution to find new valid code sequences, enabling it to succeed. Whereas Tierra's primary source of information is the ancestral program, Avida's primary source of information is in the design of the "environment" in which Avida programs are run.

Tierra also derives some information from the environment in which it runs. Ray was concerned about the brittleness of machine code [10], and accordingly made specific design decisions. Additionally, the original instruction set was created by choosing exactly the instructions which were used in the ancestor [14]. This results in the Tierra instruction set being specifically tuned to the problem it faces. This work has not attempted to investigate the implications of these decisions, but is our opinion that the Tierran evolution is substantially assisted through them.

Almost any design-based view of biological origins allows the existence of some variation occurring by Darwinian mechanisms while remaining skeptical that such mechanisms can explain all of biology. Defenders of Darwinism claim that the distinction is artificial and that minor variation will necessarily eventually add up to large scale variation. Tierra provides evidence for the design position. Tierra demonstrates adaptation, but also demonstrates that the adaptation fails to add up to open-ended complexity. It shows that minor variation does imply major variation.

Tierra did not succeed in producing open-ended evolution and a Cambrian-like explosion as was hoped. Changes were dominated by loss or rearrangement rather than the production of new functional code. The character of Tierran evolution never held promise for long term evolutionary growth. However, it did manage to replicate something of the character of actual biological change. Biological adaptations also often make use of loss or rearrangement of existing information. As such, the models of evolution like Tierra may well provide insights into biological change. However, it fails to demonstrate evolution of the sort that could explain the innovations of the Cambrian explosion or the development of the biological world.

Acknowledgments

The authors thank Dr. Paul Brown whose thoughts on the character of adaptation led to our interpretation of Tierra. They also are grateful for the comments of anonymous reviewer especially for reminding the authors that not everyone is as familiar with the technical concepts involved as the authors themselves.

6. Appendix: Tierra program comparisons

Prior to this point, we have attempted to explain the content in a generally accessible manner. This appendix seeks to provide detailed backup for the claims made in the rest of the paper. It is necessarily technical. The reader is assumed to have good grasp on the mechanics of computer machine code. As such, technical computer terminology will be used without explanation in this appendix.

6.1 *Ancestor and parasite*

Table 3 shows the difference in code between the ancestor and the parasite. The most significant change is that a substantial portion of the code has been removed. The only other change is on instruction 43, where a label is changed. That change actually causes the loss of the code because it makes that part of the program look

TABLE 3: Comparison of the code of the ancestor and the parasite. (Bold indicates changes in the program code)

	Ancestor	Parasite
1	nop1	nop1
2	nop1	nop1
3	nop1	nop1
4	nop1	nop1
5	zero	zero
6	not0	not0
7	shl	shl
8	shl	shl
9	movDC	movDC
10	adrb	adrb
11	nop0	nop0
12	nop0	nop0
13	nop0	nop0
14	nop0	nop0
15	subAAC	subAAC
16	movBA	movBA
17	adrf	adrf
18	nop0	nop0
19	nop0	nop0
20	nop0	nop0
21	nop1	nop1
22	incA	incA
23	subCAB	subCAB
24	nop1	nop1
25	nop1	nop1
26	nop0	nop0
27	Nop1	nop1
28	Mal	mal
29	call	call
30	nop0	nop0
31	nop0	nop0
32	nop1	nop1
33	nop1	nop1
34	divide	divide
35	jmpo	jmpo
36	nop0	nop0
37	nop0	nop0
38	nop1	nop1
39	nop0	nop0
40	ifz	ifz
41	nop1	nop1
42	nop1	nop1
43	**nop0**	**nop1**
44	nop0	nop0
45	pushA	pushA
46	**pushB**	
47	**pushC**	
48	**nop1**	
49	**nop0**	
50	**nop1**	
51	**nop0**	
52	**movii**	
53	**decC**	
54	**ifz**	
55	**jmpo**	
56	**nop0**	
57	**nop1**	
58	**nop0**	
59	**nop0**	
60	**incA**	
61	**incB**	
62	**jmpo**	
63	**nop0**	
64	**nop1**	
65	**nop0**	
66	**nop1**	
67	**ifz**	
68	**nop1**	
69	**nop0**	
70	**nop1**	
71	**nop1**	
72	**popC**	
73	**popB**	
74	**popA**	
75	**ret**	
76	**nop1**	
77	**nop1**	
78	**nop1**	
79	**nop0**	
80	**ifz**	

the same as the end. This confuses the copying code, resulting in a partial copy, thus the truncated code.

6.2 *Immunity*

The functionality of at least one form of immunity to parasites is described as "failing to retain the information on size and location." The original ancestor stores its size in the CX register and its location in the BX register. When running in the copy loop, these registers are used for other purposes. The original values are saved by pushing them onto the stack before running the copying code and popping them back off the stack afterwards. The only reason that the program needs to maintain those values is in order to make additional copies of the program. However, by jumping to the beginning of the program rather than its originally specified location, the main program can recalculate the values each time. At this point it can remove or break the pushing and popping code without ill effects. However, the parasite assumes that the pushing and popping code is still active and thus becomes confused.

The hyper-parasite does this same thing with an additional twist. The hyper-parasite jumps back into its main loop rather than returning back into the parasite. This means that the hyper-parasite maintains control of the parasite's CPU and thus uses it to make new hyper-parasites.

6.3 *Ancestor and hyper-parasite*

Table 4 shows the differences between a hyper-parasite and the ancestor. A substantial number of changes are made. As discussed, changes to labels and the removal of code do not constitute new code. The following discusses each case that might otherwise be considered new code:

- 21 This jump instruction does nothing as there is no label after it.
- 22 This sets the CX register to 0, but the CX register is reset by the next instruction, leaving it with no effect.
- 35 The two jump instructions, jmpo and jmpb, will both have the same effect here.
- 39–40 These two instructions will never be executed because the jump instruction at position 35 will have already taken effect.
- 64 The two jump instructions, jmpo and jmpb, will both have the same effect here.
- 69–77 This code is dead and is no longer being executed.

TABLE 4: Comparison of the code of the ancestor and a hyper-parasite

	Ancestor	Hyper-parasite
1	nop1	nop1
2	nop1	nop1
3	**nop1**	**nop0**
4	**nop1**	
5	**zero**	
6	**not0**	
7	**shl**	
8	**shl**	
9	**movDC**	
10	adrb	adrb
11	nop0	nop0
12	nop0	nop0
13	**nop0**	**nop1**
14	**nop0**	
15	subAAC	subAAC
16	movBA	movBA
17	adrf	adrf
18	nop0	nop0
19	nop0	nop0
20	nop0	nop0
21	**nop1**	**jmpb**
22	**incA**	**zero**
23	subCAB	subCAB
24	**nop1**	
25	**nop1**	
26	**nop0**	
27	**nop1**	
28	mal	mal
29	call	call
30	nop0	nop0
31	nop0	nop0
32	nop1	nop1
33	nop1	nop1
34	divide	divide
35	jmpo	jmpb
36	nop0	nop0
37	nop0	nop0
38	nop1	nop1
39		**jmpo**
40		**nop1**
41	nop0	nop0
42	ifz	ifz
43	nop1	nop1
44	nop1	nop1
45	nop0	nop0
46	nop0	nop0
47	pushA	pushA
48	pushB	pushB
49	pushC	pushC
50	nop1	nop1
51	nop0	nop0
52	nop1	nop1
53	nop0	nop0
54	movii	movii
55	decC	decC
56	ifz	ifz
57	jmpo	jmpo
58	**nop0**	**nop1**
59	nop1	nop1
60	nop0	nop0
61	nop0	nop0
62	incA	incA
63	incB	incB
64	**jmpo**	**jmpb**
65	nop0	nop0
66	nop1	nop1
67	nop0	nop0
68	nop1	nop1
69	**ifz**	**jmpb**
70	nop1	nop1
71	nop0	nop0
72	**nop1**	**popB**
73	nop1	nop1
74	popC	popC
75	popB	popB
76	**popA**	**popB**
77	ret	ret
78		**nop0**
79	nop1	nop1
80	nop1	nop1
81	nop1	nop1
82	**nop0**	
83	**ifz**	

Thus all new instructions introduced have no actual affect on the execution of the program. The only interesting changes are to the labels which produced the hyperparasitism effect.

6.4 *Hyper-parasite and social program*

Table 5 presents the difference between a hyper-parasite and a social program. The actual change making the program social is to instruction 27. None of the other changes produce interesting effects. Most of those changes relate to code which no longer serves a purpose.

- 14 Sets the CX register to zero, which is repeated by the next instruction.
- 30 This code is never executed.
- 37 The values pushed on the stack are no longer being used, so this has no effect on the program.
- 58 Code is not executed due to the hyper-parasite change.

TABLE 5: Comparison of the code of a hyper-parasite and a social program.

	Hyper-parasite	Social
1	nop1	nop1
2	nop1	nop1
3	nop0	nop0
4	adrb	adrb
5	nop0	nop0
6	nop0	nop0
7	nop1	nop1
8	subAAC	subAAC
9	movBA	movBA
10	adrf	adrf
11	nop0	nop0
12	nop0	nop0
13	nop0	nop0
14	**jmpb**	**zero**
15	zero	zero
16	subCAB	subCAB
17	mal	mal
18	call	call
19	nop0	nop0
20	nop0	nop0
21	nop1	nop1
22	nop1	nop1
23	divide	divide
24	jmpb	jmpb
25	nop0	nop0
26	nop0	nop0
27	**nop1**	**nop0**
28	jmpo	jmpo
29	nop1	nop1
30	**nop0**	**subCAB**
31	ifz	ifz
32	nop1	nop1
33	nop1	nop1
34	nop0	nop0
35	nop0	nop0
36	pushA	pushA
37	pushB	pushC
38	pushC	pushC
39	nop1	nop1
40	nop0	nop0
41	nop1	nop1
42	nop0	nop0
43	movii	movii
44	decC	decC
45	ifz	ifz
46	jmpo	jmpb
47	**nop1**	
48	nop1	nop1
49	nop0	nop0
50	nop0	nop0
51	incA	incA
52	incB	incB
53	jmpb	jmpb
54	nop0	nop0
55	nop1	nop1
56	nop0	nop0
57	nop1	nop1
58	**jmpb**	**ifz**
59	**nop1**	
60	**nop0**	
61	popB	popB
62	**nop1**	
63	**popC**	
64	**popB**	
65	**popB**	
66	**ret**	
67	**nop0**	
68	nop1	nop1
69	nop1	nop1
70	nop1	nop1

6.5 *Social program and cheater*

Table 6 compares the code of the first hyper-parasite with that of the cheater. As can be seen, a large section of code has been removed. The only other change is at position 15 which repeats the action of the previous instruction and as a result makes no lasting change on the state of the program.

6.6 *Ancestor and short code*

Table 7 shows the changes between the ancestor and a short self-replicator.

- 12 The divide instruction is used, a new functional instruction
- 55 The ret instruction is used, a new functional instruction

TABLE 6: Comparison of the code of a social program and a cheater.

	Social	Cheater
1	nop1	nop1
2	nop1	nop1
3	nop0	nop0
4	adrb	adrb
5	nop0	nop0
6	nop0	nop0
7	nop1	nop1
8	subAAC	subAAC
9	movBA	movBA
10	adrf	adrf
11	nop0	nop0
12	nop0	nop0
13	nop0	nop0
14	zero	zero
15	**zero**	**subCAB**
16	subCAB	subCAB
17	mal	mal
18	call	call
19	nop0	nop0
20	nop0	nop0
21	nop1	nop1
22	nop1	nop1
23	divide	divide
24	**jmpb**	
25	**nop0**	
26	**nop0**	
27	**nop0**	
28	jmpo	jmpo
29	**nop1**	
30	**subCAB**	
31	**ifz**	
32	**nop1**	
33	**nop1**	
34	**nop0**	
35	**nop0**	
36	**pushA**	
37	**pushC**	
38	**pushC**	
39	**nop1**	
40	**nop0**	
41	**nop1**	
42	**nop0**	
43	**movii**	
44	**decC**	
45	**ifz**	
46	**jmpb**	
47	**nop1**	
48	**nop0**	
49	**nop0**	
50	**incA**	
51	**incB**	
52	**jmpb**	
53	**nop0**	
54	**nop1**	
55	**nop0**	
56	**nop1**	
57	**ifz**	
58	**popB**	
59	nop1	nop1
60	nop1	nop1
61	nop1	nop1

- 62 There is no difference between jmpb and jmpo in this case.
- 65 This code is never executed.

We see that the divide and ret instructions are new. As such, both of these instructions do indicate a degree of novelty in the system. Both instructions were in the original ancestor, and one might be inclined to argue that they are not new code as they have merely been moved. However, since both are only single instructions rather than sequences appealing to a code movement event is not justified.

TABLE 7: Comparision of the code of the ancestor and a short self-replicator.

	Ancestor	Short
1	nop1	nop0
2	**nop1**	
3	**nop1**	
4	**nop1**	
5	**Zero**	
6	**not0**	
7	**shl**	
8	**shl**	
9	**movDC**	
10	adrb	adrb
11	nop0	nop1
12	nop0	divide
13	**nop0**	
14	**nop0**	
15	subAAC	subAAC
16	movBA	movBA
17	adrf	adrf
18	nop0	nop0
19	**nop0**	
20	**nop0**	
21	**nop1**	
22	incA	incA
23	subCAB	subCAB
24	**nop1**	
25	**nop1**	
26	**nop0**	
27	**nop1**	
28	mal	mal
29	**call**	
30	**nop0**	
31	**nop0**	
32	**nop1**	
33	**nop1**	
34	**divide**	
35	**jmpo**	
36	**nop0**	
37	**nop0**	
38	**nop1**	
39	**nop0**	
40	**ifz**	
41	**nop1**	
42	**nop1**	
43	**nop0**	
44	**nop0**	
45	**pushA**	
46	pushB	pushB
47	**pushC**	
48	**nop1**	
49	**nop0**	
50	**nop1**	
51	nop0	nop0
52	movii	movii
53	decC	decC
54	ifz	ifz
55	**jmpo**	**ret**
56	**nop0**	
57	**nop1**	
58	**nop0**	
59	**nop0**	
60	incA	incA
61	incB	incB
62	jmpo	jmpb
63	**nop0**	
64	nop1	nop1
65	nop0	movii
66	**nop1**	
67	**ifz**	
68	**nop1**	
69	**nop0**	
70	**nop1**	
71	**nop1**	
72	**popC**	
73	**popB**	
74	**popA**	
75	**ret**	
76	**nop1**	
77	**nop1**	
78	**nop1**	
79	**nop0**	
80	**ifz**	

6.7 *Loop unrolling*

Table 8 shows the changes between the ancestor and the optimized program.

- 12 This divide instruction is new.
- 23 There is no label; this piece of code has no affect.

TABLE 8: Comparison of the code of the ancestor and a unrolled loop.

	Ancestor	Unrolled
1	nop1	nop1
2	**nop1**	
3	**nop1**	
4	**nop1**	
5	**zero**	
6	**not0**	
7	**shl**	
8	**shl**	
9	**movDC**	
10	adrb	adrb
11	nop0	nop0
12	nop0	divide
13	**nop0**	
14	**nop0**	
15	subAAC	subAAC
16	movBA	movBA
17	adrf	adrf
18	nop0	nop0
19	**nop0**	
20	**nop0**	
21	**nop1**	
22	incA	incA
23		**call**
24	subCAB	subCAB
25	**nop1**	**pushB**
26	**nop1**	**shl**
27	**nop0**	
28	**nop1**	
29	mal	mal
30	**call**	
31	**nop0**	
32	**nop0**	
33	**nop1**	
34	**nop1**	
35	**divide**	
36	**jmpo**	
37	**nop0**	
38	**nop0**	
39	**nop1**	
40	**nop0**	
41	**ifz**	
42	**nop1**	
43	**nop1**	
44	**nop0**	
45	**nop0**	
46	**pushA**	
47	**pushB**	
48	**pushC**	
49	**nop1**	
50	**nop0**	
51	**nop1**	
52	nop0	nop0
53	movii	movii
54	decC	decC
55	ifz	decC
56	jmpo	jmpb
57	**nop0**	**decC**
58	**nop1**	
59	**nop0**	
60	**nop0**	
61	incA	incA
62	incB	incB
63	**jmpo**	**movii**
64	**nop0**	**decC**
65	**nop1**	**incA**
66	**nop0**	**incB**
67		**movii**
68		**decC**
69		**not0**
70		**ifz**
71		**ret**
72		**incA**
73		**incB**
74		**jmpb**
75	nop1	nop1
76	ifz	ifz
77	**nop1**	
78	**nop0**	
79	**nop1**	
80	**nop1**	
81	**popC**	
82	**popB**	
83	**popA**	
84	**ret**	
85	**nop1**	
86	**nop1**	
87	**nop1**	
88	**nop0**	
89	**ifz**	

- 25, 29 These two instructions were both in the ancestor. The order has been switched, but this has no effect on the program.
- 26 This shl instruction is new.
- 49–75 This section is result of a loop unrolling.

Instructions 49–75 derive from a three-fold duplication of the original version of that code. Table 9 compares a repeated version of the ancestor loop with the optimized version.

- 7 A decrement CX instruction.
- 8 No label; has no function.
- 9 A decrement CX instruction.
- 28 Has the effect of decrementing CX.
- 30 New instruction.
- 37 Neutral change.

It is obvious that the code was produced by a straightforward duplication of the original loop. There are three features which have been added.

1. The same changes to ret/divide from the very short program.
2. The program requests twice as much space as it needs, and counts down twice as fast to make up for it.
3. The loop has been unrolled.

TABLE 9: Comparison of a repeated ancestor copy loop and the unrolled loop.

	Ancestor	Unrolled
1	**nop1**	
2	**nop0**	
3	**nop1**	
4	nop0	nop0
5	movii	movii
6	decC	decC
7	**ifz**	**decC**
8	**jmpo**	**jmpb**
9	**nop0**	**decC**
10	**nop1**	
11	**nop0**	
12	**nop0**	
13	incA	incA
14	incB	incB
15	**jmpo**	
16	movii	movii
17	decC	decC
18	**ifz**	
19	**jmpo**	
20	**nop0**	
21	**nop1**	
22	**nop0**	
23	**nop0**	
24	incA	incA
25	incB	incB
26	movii	movii
27	decC	decC
28		**not0**
29	ifz	ifz
30	**jmpo**	**ret**
31	**nop0**	
32	**nop1**	
33	**nop0**	
34	**nop0**	
35	incA	incA
36	incB	incB
37	jmpo	jmpb
38	**nop0**	
39	nop1	nop1
40	**nop0**	
41	**nop1**	

This change was discussed in Section 6.6. The second change introduces the shl instruction as well as the inserted instructions in the copy loop aside from the ret. This change consisted of four instructions. Three of the four instructions do the same thing, i.e., decrement CX.

6.8 *Parallel*

Table 10 shows the differences between the threaded ancestor and the optimized version.

- 19–20 Since CX is never zero here, these instructions have no effect.
- 40 This instruction lacks a label and has no effect.
- 50–74 These instructions are part of a duplicated section.
- 90–91 These instructions have been moved from earlier in the program.

TABLE 10: Comparison of a parallel ancestor with the increased parallelism program.

	Ancestor	Developed
1	**nop1**	**nop0**
2	**nop1**	
3	**nop1**	
4	**nop1**	
5	Adrb	adrb
6	**nop0**	**nop1**
7	**nop0**	
8	**nop0**	
9	**nop0**	
10	subAAC	subAAC
11	MovBA	movBA
12	Adrf	adrf
13	nop0	nop0
14	nop0	nop0
15	**nop0**	
16	nop1	
17	incA	
18	subCAB	subCAB
19	**nop1**	ifz
20	nop1	ifz
21	nop0	
22	nop1	
23	mal	mal
24	**zeroD**	
25	zeroD	
26	split	split
27	**call**	
28	nop0	
29	nop0	
30	nop1	
31	nop1	
32	join	
33	divide	
34	jmpo	
35	nop0	
36	nop0	
37	nop1	
38	nop0	
39	ifz	ifz
40	**nop1**	**adrb**
41	**nop1**	
42	**nop0**	
43	**nop0**	
44	**pushA**	
45	**pushB**	
46	pushC	
47	shr	shr
48	offAACD	offAACD
49	offBBCD	offBBCD
50	**nop1**	**zeroD**
51	**nop0**	**ifz**
52		adro
53		**ifz**

54		**split**
55		**split**
56		**shr**
57		**offAACD**
58		**pushB**
59		**offBBCD**
60		**zeroD**
61		**split**
62		**movii**
63		**shr**
64		**offAACD**
65		**offBBCD**
66		**incC**
67		**zeroD**
68		**split**
69		**not0**
70		**ifz**
71		**ifz**
72		**shr**
73		**offAACD**
74		**offBBCD**
75	nop1	nop1
76	nop0	nop0
77	movii	movii
78	decC	decC
79	ifz	ifz
80	jmpo	jmpo
81	nop0	nop0
82	**nop1**	
83	**nop0**	
84	**nop0**	
85	incA	incA
86	incB	incB
87	jmpb	jmpb
88	nop0	nop0
89	nop1	nop1
90	**nop0**	**join**
91	nop1	divide
92	**ifz**	
93	**nop1**	
94	**nop0**	
95	**nop1**	
96	**nop1**	
97	**popC**	
98	**popB**	
99	**popA**	
100	ret	ret
101	nop1	nop1
102	nop1	nop1
103	**nop1**	
104	**nop0**	
105	**ifz**	

We can gain a better idea of the changes in the duplicate section by comparing a duplicated version of the ancestor's code as in Table 11. The original section that was duplicated includes a large section which was removed either before or after the duplication. This section has not been included in the comparison.

- 4–5 adrb has no label, so these instructions have no effect.
- 11 adro has no label, it has no effect.
- 12–13 CX is not zero, so these instructions have no effect.
- 17 This instruction was preserved since the ancestor. It is the lone surviving instruction from the section removed from the comparison.
- 22 This copies an instruction which is simply recopied later; it is thus useless.
- 26 This instruction is actually novel and useful
- 29 manipulates CX, but effect is lost by rounding
- 30–31 Since CX is not zero these instructions have no function.

TABLE 11: Comparison of a duplicated threaded ancestor with the increased parallelism program.

	Ancestor	Developed
1	**zeroD**	
2	**zeroD**	
3	split	split
4		**ifz**
5		**adrb**
6	shr	shr
7	offAACD	offAACD
8	offBBCD	offBBCD
9	zeroD	zeroD
10	**zeroD**	**ifz**
11		**adro**
12		**ifz**
13		**split**
14	split	split
15	shr	shr
16	offAACD	offAACD
17		**pushB**
18	offBBCD	offBBCD
19	zeroD	zeroD
20	**zeroD**	
21	split	split
22		**movii**
23	shr	shr
24	offAACD	offAACD
25	offBBCD	offBBCD
26	**zeroD**	**incC**
27	zeroD	zeroD
28	split	split
29		**not0**
30		**ifz**
31		**ifz**
32	shr	shr
33	offAACD	offAACD
34	offBBCD	offBBCD

References

1. Ray T (2001) Overview of tierra at atr. In: Technical Information, No. 15, Technologies for Software Evolutionary Systems.
2. Ewert W, Montañez G, Dembski W, Marks R (2010) Efficient per query information extraction from a hamming oracle. In: System Theory (SSST), 2010 42nd Southeastern Symposium on, pp. 290 –297. DOI 10.1109/SSST.2010.5442816
3. Ewert W, Dembski WA, Marks II RJ (2009) Evolutionary Synthesis of Nand Logic: Dissecting a Digital Organism. In: Proceedings of the 2009 IEEE International Conference on Systems, Man, and Cybernetics. San Antonio, TX, USA - October 2009, pp. 3047–3053.
4. Lohn JD, Linden DS, Hornby GS, Rodriguez-Arroyo A, Seufert S, Blevins B, Greenling T (2005) Evolutionary design of a single-wire circularly-polarized X-band antenna for NASAs Space Technology 5 mission. In: Antennas and Propagation Society International Symposium, 2005 IEEE, vol. 2.
5. Lohn JD, Linden DS, Hornby GS, Kraus WF, Rodriguez-Arroyo A (2003) Evolutionary design of an X-band antenna for NASA's space technology 5 mission. In: EH '03: Proceedings of the 2003 NASA/DoD Conference on Evolvable Hardware, p. 155. IEEE Computer Society, Washington, DC, USA.
6. Ray TS (1992) An approach to the synthesis of life. In: Langton CG, Taylor C, Farmer JD, Rasmussen S (eds.) Artificial Life II, pp. 371–408. Addison Wesley Publishing Company.

7. Simon CW (2000) The Cambrian "explosion": slow-fuse or megatonnage. Proc Natl Acad Sci USA 97(9): 4426–4429.
8. Standish RK (2003) Open-ended artificial evolution. International Journal of Computational Intelligence and Applications 3(2): 167–175.
9. Bedau MA, Snyder E, Brown CT, Packard NH (1997) A comparison of evolutionary activity in artificial evolving systems and in the biosphere. In: Proceedings Of The Fourth European Conference On Artificial Life, pp. 125–134. MIT Press, Cambridge.
10. Ray T, Barbalet T (2009) Biota live #56, Tom Ray on twenty years of tierra. podcast http://www.biota.org/podcast/live.html\#56.
11. Hawking S (2001) The universe in a nutshell, Bantam
12. Jefferson D, Collins R, Cooper C, Dyer M, Flowers M, Korf R, Taylor C, Wanq A (1991) Evolution as a theme in artificial life: The genesys/tracker system.
13. Collins RJ, Jefferson DR (1991) Antfarm: Towards simulated evolution. In: Artificial Life II, pp. 579–601. Addison-Wesley
14. Thearling K, Ray T (1994) Evolving multi-cellular artificial life. In: Artificial life IV: Proceedings of the Fourth International Workshop on the Synthesis and Simulation of Living Systems, p. 283. The MIT Press, Cambridge.
15. Thearling K, Ray T (1997) Evolving parallel computation. Working Papers 97–05–040, Santa Fe Institute. URL http://ideas.repec.org/p/wop/safiwp/97–05–040.html
16. Ray T (1996) Evolution of parallel processes in organic and digital media. In: Natural & artificial parallel computation: Proceedings of the Fifth NEC Research Symposium, p. 69. Soc for Industrial & Applied Math.
17. Ray T (1994) A Proposal to Create a Network-Wide Biodiversity Reserve for Digital Organisms. unpublished. See also: Science 264: 1085.
18. Charrel, A (1995) Tierra network version.
19. Ray TS (1998) Selecting naturally for differentiation: preliminary evolutionary results. Complex. 3: 25–33. DOI 10.1002/(SICI)1099–0526(199805/06)3:5⟨25::AID-CPLX5⟩3.3.CO;2-O. URL http://portal.acm.org/citation.cfm?id=295404.295413.
20. Wells J. (2002) Icons of Evolution: Science or Myth? Regnery Publishing, Washington, DC.
21. Ray T, Xu C, Charrel A, Kimezawa T, Yoshikawa T, Chaland M, Uffner T (2000) Tierra Documentation.
22. Ray TS (1994) Evolution, complexity, entropy and artificial reality. Physica D: Nonlinear Phenomena 75(1–3): 239–263.
23. Behe M (2008) The Edge of Evolution. Free Press, New York.
24. Cooper VS, Lenski RE (2000) The population genetics of ecological specialization in evolving Escherichia coli populations. Nature 407(6805): 736–739.
25. Cooper VS, Schneider D, Blot M, Lenski RE (2001) Mechanisms causing rapid and parallel losses of ribose catabolism in evolving populations of Escherichia coli B. J Bacteriol 183(9): 2834–41.

26. Behe M (2010) Experimental evolution, loss-of-function mutations, and the first rule of adaptive evolution. The Quarterly Review of Biology 85(4): 419–445.
27. Behe M (2013) Getting there first: an evolutionary rate advantage for adaptive loss-of-function mutations. In: Marks II RJ, Behe MJ, Dembski WA, Gordon B, Sanford JC (eds) Biological Information – New Perspectives. World Scientific, Singapore, pp. 450–473.
28. Lenski RE, Ofria C, Pennock RT, Adami C (2003) The evolutionary origin of complex features. Nature 423(6936): 139–144.
29. Dembski WA, Marks RJ (2009) Conservation of information in search: measuring the cost of success. Trans Sys Man Cyber Part A 39(5): 1051–1061.

Multiple Overlapping Genetic Codes Profoundly Reduce the Probability of Beneficial Mutation

George Montañez[1], Robert J. Marks II[2], Jorge Fernandez[3] and John C. Sanford[4§]

[1]*Computer Science Department, One Bear Place #97356, Baylor University, Waco, TX 76798–7356.* [2]*Electrical & Computer Engineering, One Bear Place #97356, Baylor University, Waco, TX 76798–7356.* [3]*FMS Foundation, 7160 Stone Hill Rd. Livonia, NY 14487.* [4]*Department of Hort Sci, Hedrick Hall, NYSAES, Cornell University, Geneva, NY 14456, USA.* ([§]*Corresponding author: jcs21@cornell.edu*)

Abstract

There is growing evidence that much of the DNA in higher genomes is poly-functional, with the same nucleotide contributing to more than one type of code. Such poly-functional DNA should logically be multiply-constrained in terms of the probability of sequence improvement via random mutation. We describe a model of this relationship, which relates the degree of poly-functionality and the degree of constraint on mutational improvement. We show that: a) the probability of beneficial mutation is inversely related to the degree that a sequence is already optimized for a given code; b) the probability of beneficial mutation drastically diminishes as the number of overlapping codes increases. The growing evidence for a high degree of optimization in biological systems, and the growing evidence for multiple levels of poly-functionality within DNA, both suggest that mutations that are unambiguously beneficial must be especially rare. The theoretical scarcity of beneficial mutations is compounded by the fact that most of the beneficial mutations that do arise should confer extremely small increments of improvement in terms of total biological function. This makes such mutations invisible to natural selection. Beneficial mutations that are below a population's *selection threshold* are effectively neutral in terms of selection, and so should be entirely unproductive from an evolutionary perspective. We conclude that beneficial mutations that are unambiguous (not deleterious at any level), and useful (subject to natural selection), should be extremely rare.

Key words: beneficial mutation, probability, multiple codes, overlapping codes, ENCODE, poly-functional DNA, selection threshold

1. Introduction

It is almost universally acknowledged that beneficial mutations are rare compared to deleterious mutations [1–10]. However, there is controversy regarding just how rare beneficial mutations actually are. It appears that beneficial mutations may be too rare to actually allow the accurate measurement of how rare they are [11]. For decades it has been widely thought that beneficial mutations might be as rare as one in a million [12, 13]. However, more recently some have argued that beneficial mutations might be much more common [14, 15].

The actual rate of beneficial mutation is a crucial question, because if determines both the speed and the direction of genetic change. If beneficial mutations are extremely rare, this profoundly limits the *rate and range* of all forward genetic change. Furthermore, to the extent that beneficial mutations may be extremely rare, the question arises — "how can there be any net gain in total biological fitness?" This question arises because it is widely recognized that in large genomes most mutations should have very small effects, and so large numbers of low-impact deleterious mutations should not be subject to purifying selection [16-33]. This means that over time large numbers of such deleterious mutations should accumulate continuously, leading to ever-increasing genetic load [29-33]. In order to halt such genetic deterioration, one must invoke the continuous amplification of a large number of beneficial mutations to fully compensate for all the accumulating deleterious mutations [34–36].

Fisher addressed the problem of the rarity of beneficial mutations as long ago as 1930 [37]. He argued that beneficial mutations might be quite common. He used the illustration of focusing a microscope. A random change in focal length has a nearly equal chance of either improving or diminishing the focus, assuming three things: a) the microscope is significantly out of focus, b) the change in focus is very small, and c) focus is just a one dimensional trait (a single knob — turned either up or down). We now know that Fisher's three necessary conditions do not apply to the real biological world. Biological systems are highly optimized (the microscope is not substantially out of focus), a beneficial mutation must be subject to selection, so its biological effect must not be too small (so very tiny changes in focus are not feasible), and fitness is extremely multi-dimensional (there is much more to biological functionality than optimizing a single parameter such as focal length).

Fisher acknowledged that focusing a microscope just involves optimization in a single dimension, and conceded that to the extent that fitness is not a simple one-dimensional trait, his analogy would break down. He went on to show that as the number of "dimensions" of fitness increased, the probability of beneficial mutation should rapidly decrease. This insight was profound, yet in his day he could not have realized how extremely multi-dimensional biological fitness really is. Fisher lived before the revolution in biology — he knew nothing of cell biology, molecular biology, or molecular genetics. We now know that total biological fitness is multi-dimensional in the extreme. In a sense, every functional nucleotide within a genome adds another dimension to the fitness equation. So in a sense Fisher's allegorical "microscope" has millions of knobs that must be focused simultaneously and interactively.

In the last decade, we have discovered still another aspect of the multi-dimensional genome. We now know that DNA sequences are typically "poly-functional" [38]. Trifanov previously had described at least 12 genetic codes that any given nucleotide can contribute to [39,40], and showed that a given base-pair can contribute to multiple overlapping codes simultaneously. The first evidence of overlapping protein-coding sequences in viruses caused quite a stir, but since then it has become recognized as typical. According to Kapronov et al., "it is not unusual that a single base-pair can be part of an intricate network of multiple isoforms of overlapping sense and antisense transcripts, the majority of which are unannotated" [41]. The ENCODE project [42] has confirmed that this phenomenon is ubiquitous in higher genomes, wherein a given DNA sequence routinely encodes multiple overlapping messages, meaning that a single nucleotide can contribute to two or more genetic codes. Most recently, Itzkovitz et al. analyzed protein coding regions of 700 species, and showed that virtually all forms of life have extensive overlapping information in their genomes [43]. So not only are there many "knobs" in Fisher's microscope analogy, each one can affect multiple traits simultaneously and interactively.

In light of these new developments, it is timely to reexamine the question of the probability of beneficial mutation, the utility of Fisher's model, Fisher's Theorem, and Fisher's insight about multiple fitness dimensions. This paper examines the probability of a selectable beneficial mutation arising within a DNA sequence that is functional (hence must be significantly optimized), and contains multiple overlapping codes.

2. Method and Results

2.1 *The Model*

For illustration, in Figure 1 we show a hypothetical 100 base pair sequence, which participates in 12 partially overlapping codes.

Starting Assumptions:

1. We only consider here a "functional sequence". We assume this sequence is not primarily "junk DNA", but that for the most part it encodes information, yet we allow for rare nucleotide sites within the functional sequence that are perfectly neutral.
2. Each nucleotide within the functional genome is classified by level (L_1–L_{12}), depending on how many codes it contributes to. A nucleotide that does not contribute to a given code is considered neutral relative to that code. A nucleotide which does not contribute to any of the codes is considered perfectly neutral and will be designated L_0.

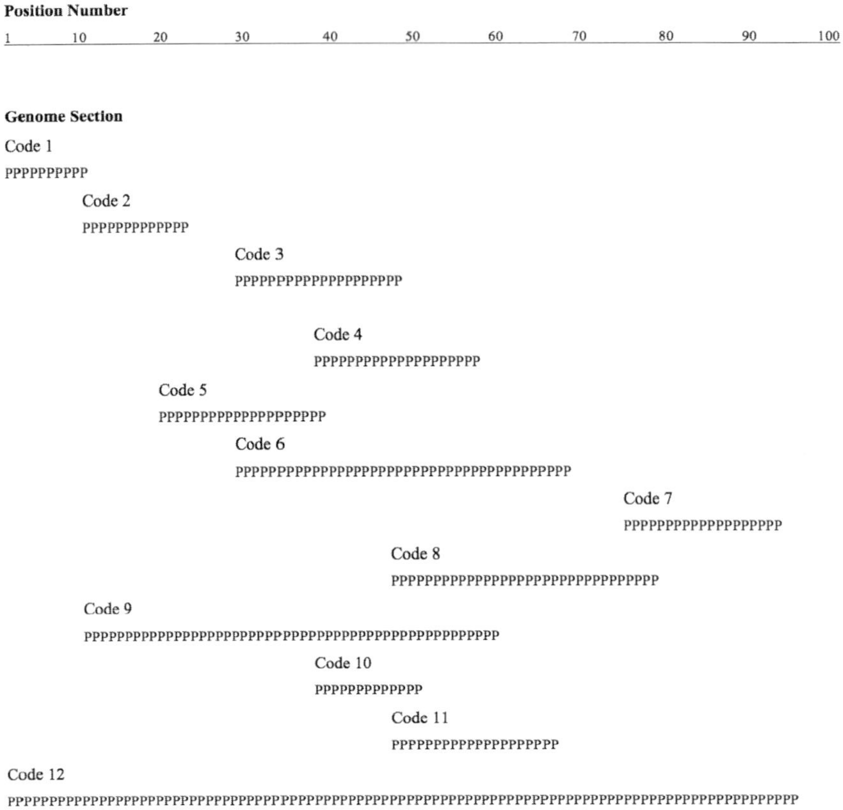

Fig. 1. A model nucleotide sequence of 100 bases that encodes 12 partially overlapping codes. Each sub-section represents the positions of the Genome Section that participate in that particular code. For example, only the first 10 positions of the Genome Section participate in Code 1 whereas all except the last 5 positions of the Genome Section participate in Code 12. Nucleotide positions of the Genome Section that do not fall into any code are considered entirely neutral with respect to those codes, since they play no part in what the function of those codes may be. In that regard, these neutral positions are not part of the functional genome (at least with respect to those specific codes).

3. Consistent with commonly used evolution models [41, 44–46], we assume the optimization of a composite organism is determined by a single fitness function. The contribution of each code to fitness is assumed to arise by aggregation of constraint commonly found in multi-objective optimization [47-50].
4. We assume a high degree of optimization within each code, although this assumption can be relaxed, and is a tunable parameter within the model. For the analysis and discussion we assume 99.9% of the nucleotide positions defining a code are already an optimal nucleotide.

5. For those nucleotides that are part of a given code but are not yet the optimal nucleotide for that code, we assume that only one of the three alternative nucleotides will be an improvement relative to the existing sub-optimal nucleotide. Mutations at such sites will therefore have one-third chance of being beneficial, but will still have a two-thirds probability of being deleterious. Another way to say this is that for a given site, relative to a given code, there is a hierarchy of most desirable nucleotide (ranked first) to least desirable nucleotide (ranked fourth), and as a rule the non-optimal nucleotide is ranked second, rather than third or fourth. This reflects the idea that even if non-optimal, the existing nucleotide is not truly random.
6. We assume independence in the designation of different optimal codes, such that the nucleotide deemed optimal for a given code (at a position) is chosen independently of the other nucleotides that are optimal for codes that may overlap at that position. In other words, for position 1, the first code may view G as the optimal nucleotide, whereas the second code may consider T the optimal nucleotide, or both may consider C optimal, etc. Although the nucleotide at a position may be shared by several codes (in the case of overlap), we assume that a nucleotide for an optimal code sequence is chosen only with respect to other nucleotides within that same code, and not with respect to other codes which may or may not overlap with it on the genome section currently or in the future. Modeling these optimal code sequence decisions as independent gives rise to the Bernoulli model presented here.
7. Lastly, we make the simplifying assumption that beneficial and deleterious mutations have "unit magnitude" effects, such that if one of each is present, their combined effects effectively cancel out (See Discussion).

2.2 *Analyses*

We analyzed how overlapping codes affect the probability of beneficial mutation in three ways. The first analysis involved a very simple calculation of how multiple overlapping codes affect the theoretical probability of an "unambiguously beneficial mutation". We define an unambiguously beneficial mutation as a mutation that causes a benefit in at least one code, without causing any deleterious effect in any other code. The second analysis is more involved, and examines the probability of a "net-effect beneficial mutation". A net-effect beneficial mutation is a mutation that improves more codes than it disrupts. The last analysis involves an empirical analysis of how overlapping English words (i.e. as in a crossword puzzle), affect the probability of creating a new valid word.

2.2.1 *First Level of Analysis:*

When we consider poly-functional nucleotide sites, it is relatively simple to calculate the probability of mutations which are "unambiguously beneficial" (i.e., beneficial in one code, and not deleterious in any other code). For example, let us assume all codes are 99.9% optimized, (such that 99.9% of all mutations will be deleterious for any given code). Even for that one-in-a-thousand site which is sub-optimal, on average only about half of the nucleotide substitutions at such a site will be an improvement (which in this simple analysis we can ignore). For L_1 nucleotides, the rate of unambiguously beneficial mutations will be at best one in 10^3, for L_2 nucleotides this rate will be at best one in 10^6, and for L_3 nucleotides this rate will be at best one in 10^9. Generalized, for a L_n nucleotide, the rate will be at best one in 10^{3n}. Overlapping codes, by their very nature, make unambiguous mutations vanishingly rare. This means that within all poly-functional nucleotide sites, essentially all "beneficial mutations" will at best be *ambiguously* beneficial, being beneficial at just one level, but simultaneously being deleterious at one or more levels. Therefore at any poly-functional nucleotide site, a "beneficial" mutation will almost always still consistently have deleterious effects, systematically eroding the total amount of information in the entire information system.

2.2.2 *Second Level of Analysis:*

We can calculate the probability of a net-effect beneficial mutation for each nucleotide level (L_1–L_{12}) as described below.

Within a given code, assume that sequences are highly optimized. We use p(optimal) = 99.9% = 0.999 of all nucleotides being optimal in our recurring example. In the case of optimal nucleotide bases, any change is deleterious, assuming no neutral changes. Therefore, only $r = 1 - p(\text{optimal}) = 0.1\% = 0.001$ are subject to beneficial mutation. There are no absolutely neutral positions in any given code, because by definition such a position is not part of that code. The conditions for net beneficial or net deleterious changes, therefore, are as follows:

<u>To be a net-beneficial mutation:</u>

- The current nucleotide in that position must be non-optimal AND
- The change must be to a beneficial nucleotide, which occurs with a 1/3 probability, denoted as

$$p(\text{beneficial} \mid \text{non-optimal})$$

<u>To be a net-deleterious mutation:</u>

- The current nucleotide in that position can be optimal OR
- The change must be to a deleterious nucleotide, which occurs with a 2/3 probability, denoted as

$$(1 - p(\text{beneficial} \mid \text{non-optimal}))$$

Given these assumptions, we calculate the probability that for a single code a mutation at a uniformly random chosen position is beneficial as follows, according to the law of total probability:

$$\begin{aligned} p(B) &= p(\text{non-optimal}) \times p(\text{benefical} \mid \text{non-optimal}) + p(\text{optimal}) \times p(\text{benefical} \mid \text{optimal}) \\ &= p(\text{non-optimal}) \times p(\text{benefical} \mid \text{non-optimal}) + 0 \\ &= (1 - p(\text{optimal})) \times p(\text{benefical} \mid \text{non-optimal}) \\ &= r \times p(\text{benefical} \mid \text{non-optimal}) \end{aligned}$$

Given the stated assumptions, for any single code, a mutation at a position chosen at random that mutates has a probability of being a *beneficial* (**B**) mutation equal to $p(B) = (1/3)r = 0.00033$. This, in turn, means that a random position that mutates has a probability of being a *deleterious* (**D**) mutation equal to $1 - p(B) = 0.99967$.

A mutation occurring to a single nucleotide may be beneficial or deleterious for any given code (as per previous discussion, neutral cases are excluded). Let's consider a few specific cases before generalizing:

(1) If the nucleotide is a L_1 nucleotide then there is only one possibility: a mutation will be either beneficial (**B**) or deleterious (**D**) with $p(B) = 0.00033$ and $1 - p(B) = 0.99967$.

(2) If the nucleotide is a L_2 nucleotide then there will be four possibilities: 1) a mutation may be beneficial for both codes (**B,B**); 2) a mutation may be beneficial to the first code and deleterious to the second code (**B,D**); 3) a mutation may be deleterious to the first code and beneficial to the second code (**D,B**) or, 4) a mutation may be deleterious to both codes (**D,D**). For such nucleotide positions, there is a value for each code, each of which is either beneficial or deleterious. We will make the simplifying assumption that where there is a beneficial effect in one code and a deleterious effect in another code, these effects will essentially cancel, leaving a neutral effect. Therefore (**B,B**) will be beneficial, (**D,D**) will be deleterious, while (**D,B**) and (**B,D**) will be neutral. In this case,

$p(\text{beneficial}) = p(\text{B,B}) = p(\text{B})^2 = 1.11 \times 10^{-7}$
$p(\text{neutral}) = p(\text{B,D}) + p(\text{D,B}) = 2 \times p(\text{B}) \times (1-p(\text{B})) = 6.66 \times 10^{-4}$
$p(\text{deleterious}) = p(\text{D,D}) = (1-p(\text{B}))^2 = 0.99933$

(3) In all other cases, where more than two codes are involved, there can be more than two factors to consider. For example, for L_3 positions, there are three levels of mutational effect.

(4). If the nucleotide is a L_N nucleotide, there will be 2^N possibilities. To generalize:

Let L_i be the level of a particular nucleotide. Combining all of the above, and formulating the binomial within our model parameters, if there are *N* codes and an *L*-level nucleotide, then the probability of a beneficial mutation for this *L*-level nucleotide, $p(B)_L$, is obtained with the binomial distribution [42]

$$p(B)_L = \sum_{k=\left\lceil \frac{L+1}{2} \right\rceil}^{L} \binom{L}{k} p(B)^k (1-p(B))^{L-k} \tag{1}$$

where *L* is the number of codes, $\left\lceil \frac{L+1}{2} \right\rceil$ is the minimum number of codes that constitute a majority (with the brackets denoting the ceiling function), and

$$p(B) = (1-p(\text{optimal})) \times p(\text{beneficial} \mid \text{non-optimal})$$

with $p(\text{optimal})$ denoting the probability that a nucleotide is already optimal.

In similar fashion, the probability of a deleterious mutation for this *L*-level nucleotide, $p(D)_L$, is obtained with:

$$p(D)_L = \sum_{k=\left\lceil \frac{L+1}{2} \right\rceil}^{L} \binom{L}{k} p(B)^{L-k} (1-p(B))^{k} \tag{2}$$

In general, the probability of a neutral mutation is

$$\begin{aligned} p(neutral)_L &= 1-(p(B)_L + p(D)_L) \\ &= \binom{L}{\frac{L}{2}} (p(B)(1-p(B))^{\frac{L}{2}} \times \delta(L \text{ is even}) \end{aligned} \tag{3}$$

where δ (*L* is even) is one when *L* is even and is zero otherwise. When *L* is even, $p(B)_L = 1-p(D)_L$. For $p(B) << 1$ (in other words, when $p(B)$ is near zero), this becomes approximately true for large odd *L*.

The value of $p(B)_L$ (the probability of a beneficial mutation) in equation (1) rapidly goes to zero for increasing *L* when $p(B) << 1$. Because differentiating between of probabilities like 10^{-11} and 10^{-22} is intuitively challenging, we propose use of the information measure [51, 52, 53]

$$I\,[L_+] = -\log_2(p(B)_L)$$

$I[L_+]$ measures the probability in terms of flips of a fair coin. If $I[L_+] = 3$ bits, for example, the corresponding probability is the same as forecasting the result of three flips of a fair coin, i.e. $p = (½)^3 = 0.125$. $I[L_+] = N$ bits corresponds to a probability of $p = (½)^N$. To place this measure in perspective, there are 10^{15} square millimeters in an area of 1000 square kilometers. The probability of two people choosing the same square millimeter is thus 10^{-15}. Since $-\log_2(10^{-15}) = 50$ bits, the success probability is the same as the probability of predicting 50 sequential outcomes of the flipping of a fair coin.

A plot of $I[L_+]$ is shown in Figure 2 as a function of L for various values of r, where $r = 1-p(\text{optimal})$. The plots rapidly approach improbable values. For $r = 0.001$, a value of $L = 12$, $p(B)_L = 4.15 \times 10^{-22}$ or $I[L_+] = 71$ bits. The chance of choosing the same millimeter twice in a distance of 100 light years (10^{-21}) is more probable.

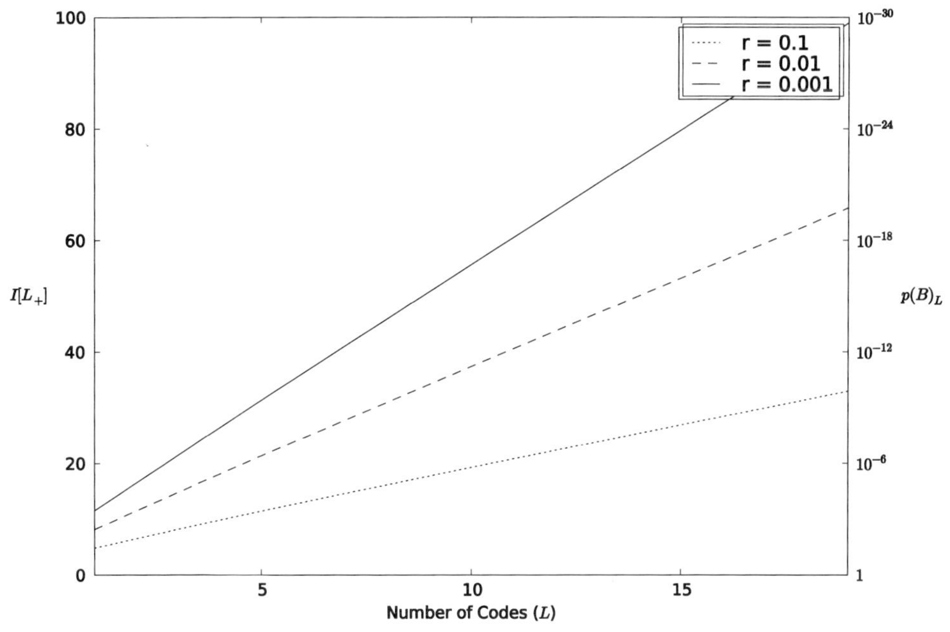

Fig. 2. Plot of $I[L_+]$ (information, in bits) versus L for various values of $r = 1$ — p(optimal). Even numbered codes are omitted for clarity. Since the probability of a beneficial mutation, $p(B)_L$, decreases exponentially with increasing L, the logarithmic information measure $I[L_+]$ increases linearly with increasing L. The right-hand scale indicates the probability of net beneficial mutation, using standard scientific notation. The three lines represent the cases where the overlapping codes are weakly optimized (10% of nucleotides are sub-optimal), moderately optimized (1% of nucleotides are sub-optimal), and highly optimized (0.1% of nucleotides are sub-optimal).

Our analysis suggests that increasing either the number of overlapping codes or the degree of optimization has negative effects on the probability of producing a beneficial mutation. A high degree of optimization makes beneficial mutations unlikely — even when considering just one code. As more codes are considered, the probability of beneficial mutation diminishes rapidly, as is shown in Figures 3, 4 and 5. The ratio of beneficial to deleterious mutations decreases so rapidly that for L_3 nucleotides in highly optimized sequences, the number of deleterious mutations expected before the first beneficial arose would be greater than the genome size of a typical bacterium. For L_5 nucleotides, the number of deleterious mutations expected before the first beneficial arose would be greater than the genome size of a typical mammal. While relaxing the optimization assumption reduces the severity of the problem (as can be seen in Figure 4), increasing the number of overlapping codes diminishes the likelihood of attaining a net beneficial mutation even for weakly optimized systems. If we allow, within a functional sequence, for overall optimization values as low as 50%, deleterious mutations remain roughly a thousand times more likely than beneficial mutations in the presence of twelve overlapping codes. As the organism becomes more optimized, the probability of receiving an overall beneficial mutation falls rapidly.

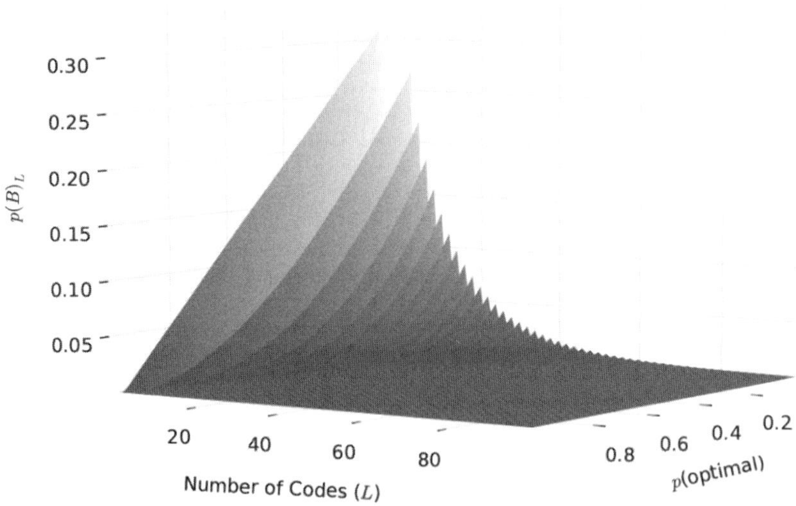

Fig. 3. Number of Codes (L) and p(optimal), plotted against $p(B)_L$, for one to one-hundred codes, showing the general behavior of the model as L increases. The probability of an overall beneficial mutation, $p(B)_L$, decreases exponentially with increasing L.

(Note: The spikes on the surface of the plot, visible near the rear plane of the figure, result from the difference between the majority of an even number of codes and the majority of an odd number of codes. For example, six is the majority for ten codes (60% of total); whereas six is also the majority for eleven codes (only 54% of total). The disparity declines with increasing L.)

We are forced to conclude that the poly-functionality of DNA profoundly affects the expected rate of beneficial mutations. The growing evidence for poly-functional DNA therefore suggests that unambiguously beneficial mutations should be vanishingly rare.

2.2.3 *Third Level of Analysis:*

To further test the effect of multiple constraints on the appearance of beneficial mutations, we constructed a simple poly-constrained artificial system based on English crossword puzzles. Crossword puzzles, for our purpose, are simply collections of words with overlapping, shared letters among some of the words. Figure 6 contains an illustration of such puzzles. We are most familiar with two-dimensional crossword puzzles, where up to two words may share a single letter, but crossword puzzles can be extended to many dimensions. An L-dimensional crossword puzzle is here defined as a collection of words, such that up to L words may share a single, overlapping letter, for one or more letters in the puzzle. Each

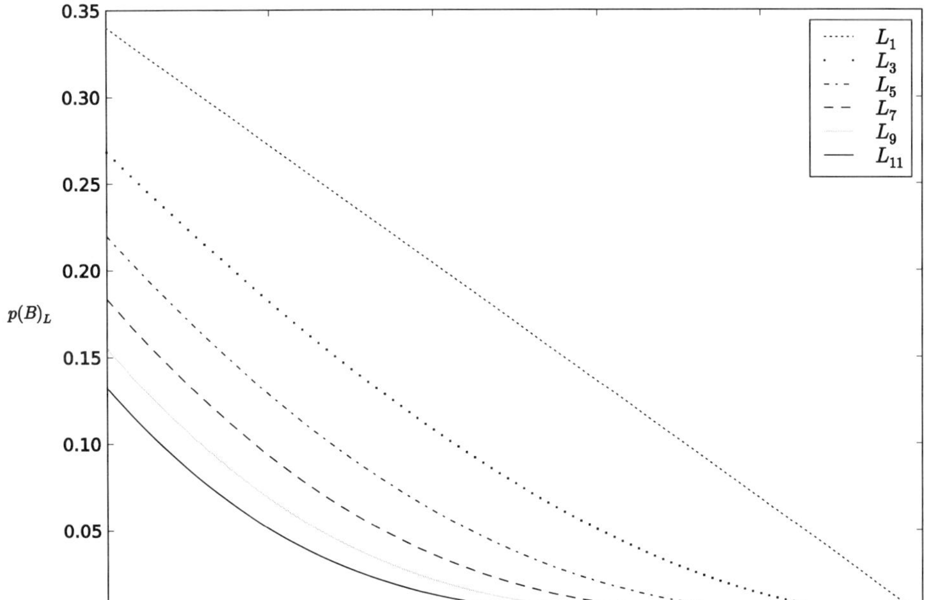

Fig. 4. $p(B)_L$ for different p(optimal) using a fixed p(beneficial | non-optimal) value of 0.34. Even numbered codes are omitted for clarity. If more than 80% of nucleotides are optimized, the probability of a beneficial mutation is near zero for $L \geq 5$.

randomly selected English word would result in another word was roughly 1.65% (using a *p*(optimal) value of 0.0). As we increase either the level of optimization or the number of overlapping words (*L*), this probability drops as expected. If more than five overlapping words are present, then the probability of making a change that is beneficial for the majority of words on the shared letter is empirically less than one in 10^7. For eleven overlapping words (similar to eleven overlapping codes in our biological model), we were unable to find a single example of an overall beneficial change during our tests. Therefore, we find the same dearth of unambiguously beneficial mutations in simple poly-constrained systems such as crossword puzzles, due to constraints imposed by the presence of interlocking, mutually dependent systems.

2.2.4 *Summary of Results:*

Having overlapping genetic codes profoundly reduces the probability of beneficial mutation. This is most dramatically seen when we consider unambiguous

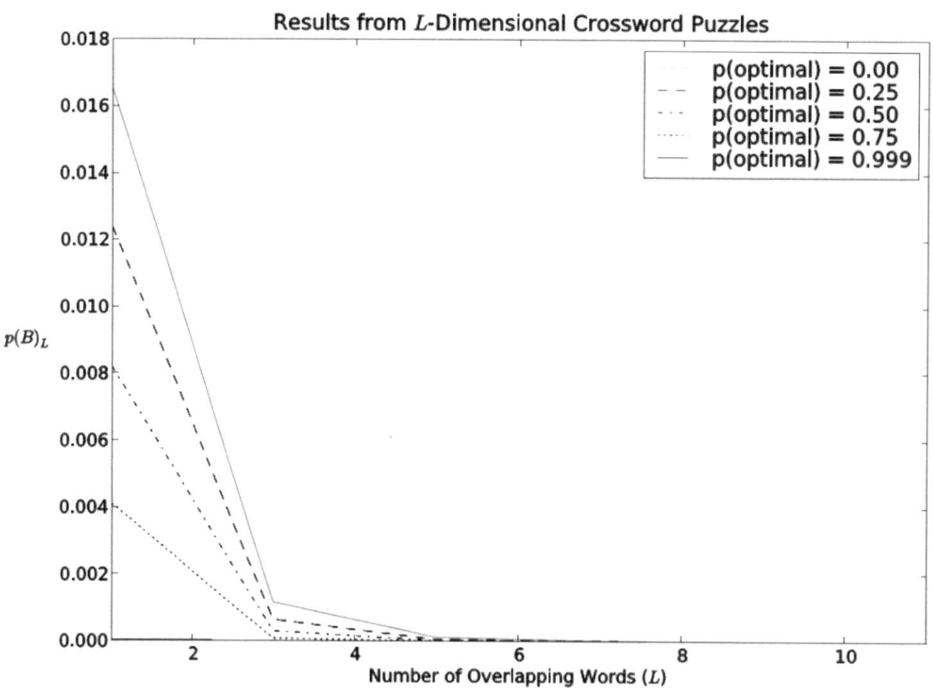

Fig. 7. Empirical results from ten-million trials, plotting the probability of achieving an overall beneficial mutation, $p(B)_L$, when mutating a shared letter among L words. Beneficial mutations were defined as changing a non-optimal word (with probability determined by *p*(optimal)) to another English word. Graph contains data points for odd numbered L only. The line for *p*(optimal) = 0.999 is indistinguishable from the horizontal axis.

beneficial mutations — which are not deleterious for any one of the overlapping codes. For example, for those nucleotides that contribute to just three different overlapping codes, assuming each code is 99.9% optimized, less than one in a billion mutations will be unambiguously beneficial. For net-effect beneficial mutations, having three overlapping codes still reduces the probability of beneficial mutation down to less than one per 10^6. When we experimentally test our basic model using a real information system (overlapping English words in the context of a crossword puzzle), we see empirical confirmation of our genetic analysis (even though our only requirement is that a letter substitution creates a new valid English word). Assuming no optimization (namely p(optimal) = 0.0), the probability of having a productive letter substitution within a single word is 1.65%, but when a letter substitution occurs where just three words overlap and p(optimal) $\geq$ 0.75, the probability drops to 7.64×10^{-5}. For nine overlapping words and p(optimal) $\geq$ 0.75 it drops to less than 10^{-7}. Our results clearly show that overlapping codes reduce the potential for beneficial mutation in a most profound way, even for moderately optimized systems.

3. Discussion

Beneficial mutations in nature appear to be so rare that after decades of research we still cannot empirically determine just how rare they are [11]. This suggests they are very rare indeed. There are many reasons to believe that beneficial mutations must be very rare. A mutation is a component of an organism's genetic specifications. Specifications are, by definition, specific. For life to be life requires an exquisite degree of specification — optimization that is hard for us to understand, involving global integration of thousands of systems which have hundreds of thousands of interactions [54]. What is being specified are all the instructions for the establishment, maintenance, and operation of a network of countless biological functions. These functions are integrated into a single elaborate system that is more complex than anything man has ever designed. Each biological specification is encoded by strings of characters (nucleotides or amino acids) that are very specific (and hence very unlikely), with each character having meaning only in the *context* of many other characters — like letters in a book or like the binary bits comprising a computer code. Any random change in such a set of specifications causes some loss of useful information — with a very high degree of probability. The more that each character is contextually interactive with other characters, the less feasible it becomes to improve a set of specifications via random character changes, because each character is multiply *constrained* by its many contextual relationships.

It has often been argued that life's specifications must be very *unconstrained*, citing "junk DNA", synonymous sites in protein coding regions, and the general concept of "bad design". However since the ENCODE project the term "junk DNA" has been largely abandoned [42,55]. "Synonymous mutations" have been shown to be biologically very important [56]. Arguments of bad design have assumed we understand every possible design constraint for a given biological component — which seems unreasonable in light of evidence for poly-functionality of most biological components.

It is now clear that biological systems are very robust and can tolerate much genetic damage. While many in the past have argued that this is due to a general lack of specificity (many sequences will do), this no longer seems reasonable. It now seems more likely that biological systems are robust because of many levels of auto-regulation, self-correction, and countless back-up systems. The new field of systems biology informs us of near-optimality in biological systems, and this appears to be ubiquitous. Such ubiquitous optimality is only conceivable given extremely specific (hence extremely constrained) genetic specifications. Such nearly-optimal genetic specifications should inherently be very difficult to improve, especially when limited to changes which only arise as rare, random, and isolated events.

The discovery of ubiquitous poly-functional DNA is profound, and forces us to reassess our understanding of the degree of genetic specificity and the probability of beneficial mutation. Trifanov pioneered the concept that genomes have a multiplicity of codes and such codes can overlap [40,41]. He showed that a given nucleotide site can participate in multiple genetic codes (with the standard protein code being just being one such code). This is the basic meaning of "poly-functional DNA" [38]. Regrettably, Trifanov's profound discovery generated limited interest. However the ENCODE project has validated the importance of his ideas, and has shown that poly-functional DNA appears to be ubiquitous in higher genomes.

To illustrate how a single nucleotide pair can participate in many different codes, let us consider some of the multiple functions a given nucleotide can participate in (each of these modes of functionality has its own code). A given nucleotide could be: 1) part of an isochore structure; 2) part of a nucleosome binding site; 3) part of a cohesion binding site; 4) part of a transcriptional promoter or enhancer; 5) part of numerous forward-strand RNA transcripts, each with its own transcriptional start and stop points; 6) part of numerous reverse-strand RNA transcripts, each within its own transcriptional start and stop points; 7) part of an mRNA splice site; 8) part of an antisense RNA; 9) part of a nucleo-protein complex; 10) part of several alternately-spliced proteins within the source genic region; 11) part of several alternately-spliced proteins between different genic regions; 12) part of the genome which regulates alternative splicing of proteins; 13) part of the 3-dimensional organization of the chromosome; 14) part of the 3-dimensional organization of the entire genome; 15) part of the machinery which transports genic regions to active regions of transcription within the nucleus; 16) part of a site for attachment to the nuclear membrane; and 17) part of other undiscovered coding structures.

Given that a single nucleotide pair can potentially participate in so many different codes simultaneously, it should be obvious that this allows data amplification without increasing genome size, and so reflects a very sophisticated form of data compression. One interesting requirement of overlapping codes is that each code must be partially "degenerate" (imperfect) to create the "flexibility" required to allow other overlapping codes. Such degeneracy might appear to the casual observer as an example of bad design, but would actually reflect extreme optimization.

Poly-functional DNA has several implications. Firstly, it is difficult to understand how poly-functional DNA could arise through random isolated mutations. In illustration, when we write, it is difficult to compose a good paragraph (although with training our minds accomplish this with apparent ease). It involves a great deal of optimization because the letters interact, the words interact, the sentences interact, and the ideas interact. But imagine if it was required that such a paragraph had to also have several other messages, using different languages, embedded within it (i.e., using every-other-letter codes, or by reading parts of the message backwards). It would obviously be vastly more difficult to compose a coherent paragraph. The chance of random letter changes creating these types of overlapping messages (in multiple languages) seems incredible, and the chance that natural selection could sort out all the possible interactions also seems incredible.

Given an existing poly-functional DNA sequence, it would seem inordinately difficult to improve it via random mutation. This is at the heart of this paper's analysis. Poly-functional DNA by its very nature is ultra-specific, highly-optimized, and hence highly-constrained. This paper shows that when a nucleotide participates in more than one code, a mutation at that site is going to almost certainly be deleterious relative to the first code, and even when a mutation is beneficial in the first code, it will still almost certainly be deleterious in one or more of the other codes. Hence a mutation at a poly-functional site will at best be only "ambiguously beneficial" — still being deleterious at one or more other levels. The exact degree to which nucleotides participate in two or more codes is still unknown, but if it is at all common, it should profoundly reduce the probability of mutations which are unambiguously beneficial.

Mutations that affect more than one code are pleiotropic, in that they have multiple biological effects. This is consistent with what geneticists have known for many decades — most known mutations are pleiotropic at some level — affecting more than one biological trait. In the case of most human genetic pathologies, the multiple effects of a mutation are usually all negative. In the rare case of an ambiguously beneficial mutation, a certain beneficial effect will be combined with one or more deleterious effects (for example, carriers of the mutation for sickle cell anemia are more resistant to malaria — but suffer from impaired hemoglobin function and reduced red blood cell counts).

In our analysis we have for simplicity assumed that if a mutation has a single beneficial effect and a single deleterious effect, it is counted as neutral. However this is not realistic because we can logically expect most such ambiguous mutations to have a net deleterious effect. This is because, not only is it more likely for a random change to damage an optimized system than improve it, the nature of that damage will tend to be more pronounced than any potential improvement. Within a highly optimized genetic system, mutational damage can range from very slight to lethal — but improvements will consistently be only very slight. For example, certain spelling errors in a plane's assembly manual could cause the plane to fly twice as slow, but no spelling error can be expected to cause the plane to fly twice as fast. Therefore selection for the ambiguous beneficial mutation is especially problematic — the positive and negative effects will tend to cancel out, but the deleterious effect will tend to overshadow the beneficial effect.

The analysis in this paper provides strong evidence that the discovery of multiple overlapping codes requires us to re-adjust downward our estimates of the rate of beneficial mutation. At the same time, the newly emerging field of systems biology strongly points to a very high degree of optimization in all biological systems, and this also requires us to adjust downward our estimate of the rate of beneficial mutation. Lastly, there is clearly a selection threshold [57], wherein below a certain limit, all low impact beneficial mutations must become invisible to natural selection. Using realistic biological conditions, it appears that in a large genome, at least 99% of all beneficial mutations should be so subtle as to be un-selectable [57]. So the rate of *useful* beneficial mutations should be at least two orders of magnitude less that the rate of actual beneficial mutations. Taking this into consideration, this suggests we should reduce the probabilities reported in this analysis by another two orders of magnitude. Although we do not quantitatively analyze the problem of drift in this paper, it is important to note that the vast majority of beneficial mutations that do arise, and are above the selection threshold of the population, are still lost due to genetic drift.

Logic and mathematical analysis persuade us that unambiguous beneficial mutations should be extremely rare. This is consistent with the apparent absence of documented mutations that are unambiguously beneficial (i.e., beneficial at one or more levels, while not deleterious on any level). To our knowledge there is no case of a mutation which is unambiguously beneficial and which has been shown to distinctly improve the inner workings or an organism. Certainly there are numerous documented cases of simple adaptations to an external environment factor, but these special cases have little bearing on how most of the information within a genome arose — because most of a genome's information specifies life's internal workings.

The long-term *E. coli* experiments of Lenski *et al.* [58] have been widely acclaimed as "proof of evolution before our very eyes". Such evolution would suggest that numerous beneficial mutations were arising. It is useful to examine these claims more carefully. The *E. coli* in these long-term experiments (which involved vast numbers of cells over vast numbers of generations), did not appear to evolve any new functions. The only changes that were observed involved adaptations to the specific artificial growth medium. This type of adaptive change to an external factor is only a superficial improvement — it does not explain how the *E. coli* genome arose, nor how the information specifying the bacteria's internal workings arose. Moreover, those studies failed to show any specific mutation which was unambiguously beneficial. In fact, it is clear that most of the adaptive mutations involved loss of function mutations — including deletions of genetic material [59]. It should be obvious that genetic material not essential for a given environment, if inactivated or deleted, can decrease metabolic load, and so can allow more total growth in that given medium. But all such broken genes and deletions clearly involve a net loss of information, and there is no question that the resulting bacteria became less "fit" in the broader and truer sense. Such strains of bacteria would immediately go extinct in virtually any natural environment.

In that enormous evolutionary experiment, the closest instance to an unambiguously beneficial mutation was a mutation that allowed the bacteria to utilize citrate from the artificial medium [60]. However, this did not actually involve evolution of a new function — the *E. coli* already had all the machinery needed for metabolizing citrate, but the citrate could not normally pass through the bacteria's external membrane. In light of the work of Behe [61], in such a case the most likely explanation for this mutant strain would be a loss-of-function mutation that would result in a leaky membrane. Certainly no exhaustive research was done to prove that the mutation in question had zero deleterious effect.

3.1 *Possible Objections*

Contrary to the thesis of this paper, some scientists have argued that beneficial mutations might be extremely common — even approaching 50% of all non-neutral mutations [14,15,37]. The concept that beneficial mutations might be extremely common traces back to some simple mental constructs suggested by

Fisher [37]. Fisher's most famous illustration was the example of focusing a microscope. If the microscope is significantly out of focus and one makes a small random adjustment, there is roughly a 50% chance of improving the focus (this would only be true for extremely small adjustments). Fisher argued that in the same way, a very low impact mutation might have roughly a 50% chance of improving fitness (in his day the near-neutral mutation problem had not yet been identified, and he apparently did not consider that such a low-impact mutation might be inherently un-selectable). When Fisher developed this illustration, DNA had not yet been discovered, genes seemed to be very simple (beads on a string), and the nature of mutation was unknown. With the advent of molecular genetics it is now evident why this analogy simply is not applicable.

Fisher knew mutations happened, but he did not know what they really were. We now know mutations are essentially spelling errors in the assembly manual of the cell. There are some small isolated parts of the genome (such as gene promoters), which can act like an electric rheostat or like a microscope's focusing knob. Mutations within these special regions can raise of lower a gene's expression level — and in this special case mutations that can increase expression can conceivably be almost as common as those that decrease expression. For example, mutations in the promoter region of the growth hormone gene might cause either giants or dwarfs. These special variable switches within DNA appear to function for the purpose of fine-tuning a trait such as height. But these special cases do not reflect the true nature of total fitness (total biological functionality), and do not reflect the way most of the genome functions. A change in height can only result in two possibilities — taller or shorter. But overall biological fitness is inherently multi-dimensional, it involves a multitude of separate traits and is contingent upon millions of nucleotides, and requires very precise genetic specifications. When a single trait is defined by just 100 functional nucleotides, that trait's genetic optimum is an extremely specific set of 100 base pairs (one specific set of 4^{100} sets, or one in 10^{60}). If that trait is anywhere near its optimum, then there are a multitude of mutations which can make the trait worse, but there are very few opportunities to make the trait better. This is analogous to a random letter change in a text that results in a superior text. As a message becomes more and more complex and refined, a text change must be more and more specific in order to enhance that message, and hence the greater the constraint for achieving improvement via any random change. As this paper shows, the recent discovery of poly-functional DNA vastly compounds this problem. To his credit, Fisher acknowledged that the chance of improvement via a random change must approach zero — either when the focus is already nearly optimized, as the size of the change in focus grows larger, or as the number of dimensions defining the trait (i.e., overall fitness) becomes larger [37].

There is another aspect of Fisher's theoretical work, which arose because he did not understand that genes specify information and that mutations are just errors in genetic specifications. Fisher imagined that all biological variation arose symmetrically. In the case of the focusing knob on a microscope, the knob turns equally well both ways, and Fisher imagined this would be equally true for mutations affecting any biological trait — such as height or vigor. There would be just as many mutations that increased performance as diminished it. This is the error underlying Fisher's famous "Fundamental Theorem of Natural Selection" [37]. Given a population with performance levels following a bell-shaped curve, he reasoned that any level of selection will always remove at least some of the underperformers and will favor at least some of the higher performing individuals. This would consistently yield higher mean performance in the next generation. He then assumed new mutations would arise creating new variation *symmetrically* around the new mean. This is what led Fisher to believe he had a mathematical proof that continuous evolutionary improvement was unavoidable. But we now know that mutations are essentially word-processing errors in the DNA, so new variation will be *extremely asymmetrical* and will be almost exclusively deleterious. So, for example, apart from a small set of mutations within its promoter region, mutations deleterious for a gene's function will be much more common than mutations for enhanced function — invalidating Fisher's Theorem, and negating his simple microscope analogy.

When we consider the organism as an integrated whole, we conclude beneficial mutations should generally be very rare for the reasons discussed above. We can only rationalize that beneficials might be common when considering one tiny component of fitness at a time, such as height. When we do this we artificially make fitness seem one-dimensional — analogous to Fisher's example of focusing his microscope. Within this very limited context, most of the constraints on what constitute a "beneficial" mutation disappear. For example, in terms of malaria resistance, a deleterious mutation in the hemoglobin gene can be defined as "beneficial", even though it is actually a semi-lethal mutation. Under this type of very limited one-dimensional analysis, the rate of beneficial mutation can appear much higher than it really is. This is especially true in the case of those rare mutations that strongly interact with major environmental factors that are external to the organism (i.e., antibiotic resistance). Relative to just that single component of the entire biological system, one can expect a reasonable probability of beneficial mutation. This is because any genetic change that interacts with that specific external factor has a nearly equal probability of making that factor's impact either better or worse. This allows biological fine-tuning for a single isolated trait, relative to a single external factor. In these special cases Fisher's microscope analogy has some validity, so that relative to that single trait (or within a single code), random mutations can have a reasonable probability of being beneficial. This may explain why most examples of beneficial mutation involve a form of adaptation to a local condition. However, most genomic information does not involve adaptation to specific high-impact external factors, but rather specifies a labyrinth of complex, integrated, and optimized biological functions internal to a living system. The important distinction between adaptation to some local external condition versus maintenance of total genomic integrity is illustrated by a recent study. That study showed that specific adaptive mutations within a mutagenized population, when tested in a particular environment, obscured, but did not halt genetic degeneration [62].

A few recent studies have inferred extremely high beneficial mutations rates, based on data from mutation accumulation (MA) experiments [14,15]. These MA experiments have significant problems. No actual mutations were actually seen, the beneficial and deleterious mutation rates were only inferred based upon the differential growth rates of a limited number of isolated strains. These experiments were not capable of identifying the vast majority of subtle mutations that arose in the populations. They could only detect those few mutations that had large effects and affected a single trait (growth rate on a given medium) making inferences about total mutation rates entirely unwarranted. The observed effects in these two studies could be attributed to a specific one-dimensional adaptation, which could arise due to a specific mutational hotspot, or could even be due to an epigenetic effect. Lastly, unintentional selection could not be rigorously precluded.

Given the one-dimensional nature of these MA experiments, a relatively high rate of beneficial mutation is not unexpected because only one trait was measured, making fitness appear one-dimensional (like Fisher's microscope), or like a simple one-dimensional trait such as height. In both of these studies, fitness was measured only in a very narrow sense and in a very specific and unnatural environment. Instead of total fitness, what was being measured was the degree of biological fine-tuning to a very specific and very artificial circumstance. In one case [14], the researchers tested the ability of yeast strains that were initially grown under minimal selection conditions (to allow mutations to accumulate), to then grow slightly faster than the source strain in the same artificial medium where the mutations had been accumulating. In that study 5.75% of the derived lines grew faster than the parental strain, under those specific conditions. In a very similar yeast experiment [15], the researchers again minimized selection to allow mutation accumulation, and then tested derived strains for ability to compete with the parental genotype in artificial medium. In the second study 25% of the derived lines out-grew the parental strain. In both cases the researchers used extremely narrow and unnatural criteria for measuring "fitness", and the singular traits they focused on might easily have been affected (for better or worse) by very simple genetic or epigenetic variations. However, natural selection, as it occurs in the natural world, must act on "fitness" in a much fuller sense — it must involve all heritable traits, all functional nucleotides, all codes, all relevant environments, and all phases of the life cycle. The authors of one of these two studies freely acknowledge these types of limitations on the interpretation of their study (including the possibility of unintentional selection) and state: "the large proportion of beneficial mutations observed in our experiment may in part reflect a combination of factors: the ancestor's distance from the fitness optimum, yeast's recent genome duplication, our examination of only a single environment and life-history stage, and the recessive nature of deleterious mutations" [14].

The two isolated reports mentioned above, which claim very high rates of beneficial mutation, are inconsistent with a much broader range of observations. For example, the *net effect* of currently observed human mutation is universally recognized as being distinctly deleterious, and hence clearly represents a serious problem in terms of public health. This is made obvious by the fact that there are thousands of Mendelian pathologies documented in man, in spite of the tendency for natural selection to eliminate such mutations from the population. Conversely, there are only a handful of putative beneficial mutations commonly cited for man, despite the tendency for natural selection to amplify such mutations. Moreover, the "benefit" of most such mutations is typically equivocal, usually being defined as beneficial in only a very narrow sense (as in the case of sickle cell anemia).

Another possible argument against the thesis of this paper might be that it is contradicted by a substantial volume of scientific literature that uses DNA sequence comparisons to infer historical positive selection events for great numbers of putative beneficial mutations. It is important to realize that the vast majority of the putative beneficial mutations claimed in these papers are just observed alternative nucleotides — with no known biological function (the presumed benefits being inferred, not being in any way understood or observed). We naturally acknowledge the operation of selection for beneficials in the past, but argue that such selection is severely constrained by the reality of very low rates of beneficial mutations, as this study and common sense both demand. It is noteworthy that a significant part of this body of literature that claims proof of so much positive selection in the past (based upon observed sequence variability in the present), may suffer from systematic error and is now being challenged [43,54,55]. Inferences of specific positive selection events in the past, based solely upon sequence data and allele frequencies, are mere historical inferences. The observed sequence variations might be explained using alternative mechanisms such as differential mutation rates or ordinary statistical fluctuations.

A final possible argument against the thesis of this paper might be that our analysis involved point mutations, but did not consider duplications. Some might

6. Bataillon T (2000) Estimation of spontaneous genome-wide mutation rate parameters: whither beneficial mutations? Heredity 84:497–501.
7. Imhof M, Schlotterer C (2001) Fitness effects of advantageous mutations in evolving *Escherichia coli* populations. Proc Natl Acad Sci USA 98:1113–1117.
8. Orr H (2003) The distribution of fitness effects among beneficial mutations. Genetics 163: 1519–1526.
9. Keightley P, Lynch M (2003) Toward a realistic model of mutations affecting fitness. Evolution 57:683–685.
10. Barrett R, et al (2006) The distribution of beneficial mutation effects under strong selection. Genetics 174:2071–2079.
11. Bataillon T (2000) Estimation of spontaneous genome-wide mutation rate parameters: whither beneficial mutations? Heredity 84:497–501.
12. Elena S, et al (1998) Distribution of fitness effects caused by random insertion mutations in *Escherichia coli*. Genetica 102/103: 349–358.
13. Gerrish P, Lenski R N (1998) The fate of competing beneficial mutations in an asexual population. Genetica 102/103:127–144.
14. Joseph S, Hall D (2004) Spontaneous mutations in diploid *Saccharomyces cerevisiae*: More beneficial than expected. Genetics 168:1817–1825.
15. Dickinson W (2008) Synergistic fitness interactions and a high frequency of beneficial changes among mutations accumulated under relaxed selection in *Saccharomyces cerevisiae*. Genetics 178:1572–1578.
16. Muller HJ (1964) The relation of recombination to mutational advance. Mutat Res 1:2–9.
17. Kimura M (1968) Evolutionary rate at the molecular level. Nature 217:624-626.
18. Ohta T (1973) Slightly deleterious mutant substitutions in evolution. Nature 246: 96–98.
19. Ohta T (1974) Mutational pressure as the main cause of molecular evolution and polymorphism. Nature 252:351–354.
20. Ohta T (1992) The nearly neutral theory of molecular evolution. Ann Rev Ecol Syst 23:263–286.
21. Ohta T (2002) Near-neutrality in evolution of genes and gene regulation. Proc Natl Acad Sci USA 99:16134–16137.
22. Kimura M (1979) Model of effectively neutral mutations in which selective constraint is incorporated. Proc Natl Acad Sci U S A 76:3440–3444.
23. Kimura M (1983) Neutral Theory of Molecular Evolution. Cambridge Univ. Press, New York.
24. Kondrashov AS (1995) Contamination of the genome by very slightly deleterious mutations: why have we not died 100 times over? J Theor Biol 175:583–594.
25. Lynch M, Conery J, Burger R (1995) Mutation accumulation and the extinction of small populations. Am Nat 146:489-518.

26. Lynch M, Conery J, Burger R (1995) Mutational meltdown in sexual populations. Evolution 49(6):1067–1080.
27. Higgins K, Lynch M (2001) Metapopulation extinction caused by mutation accumulation. Proc Natl Acad Sci USA 98:2928–2933.
28. Loewe L (2006) Quantifying the genomic decay paradox due to Muller's ratchet in human mitochondrial DNA. Genet Res 87:133–159.
29. Sanford J, Baumgardner J, Gibson P, Brewer W, ReMine W (2007) Using computer simulation to understand mutation accumulation dynamics and genetic load. In 7th International Conference on Computational Science, Beijing, China, May 27–30, 2007 Proceedings, Part II, LNCS 4488. Edited by Shi Y, van Albada GD, Dongarra J, Sloot PMA, Berlin/Heidelberg: Springer-Verlag: 386–392.
30. Gibson P, Baumgardner JR, Brewer WH, Sanford JC (2013) Can purifying natural selection preserve biological information? In: Marks II RJ, Behe MJ, Dembski WA, Gordon BL, Sanford JC (eds) Biological Information — New Perspectives. World Scientific, Singapore, pp. 232–263.
31. Brewer WH, et al (2013) Using numerical simulation to test the "mutation-count" hypothesis. In: Marks II RJ, Behe MJ, Dembski WA, Gordon BL, Sanford JC (eds) Biological Information — New Perspectives. World Scientific, Singapore, pp. 298–311.
32. Baumgardner JR, et al (2013) Can synergistic epistasis halt mutation accumulation? Results from numerical simulation. In: Marks II RJ, Behe MJ, Dembski WA, Gordon BL, Sanford JC (eds) Biological Information — New Perspectives. World Scientific, Singapore, pp. 312–337.
33. Nelson CW, et al (2013) Computational evolution experiments reveal a net loss of genetic information despite selection. In: Marks II RJ, Behe MJ, Dembski WA, Gordon BL, Sanford JC (eds) Biological Information — New Perspectives. World Scientific, Singapore, pp. 338–368.
34. Ohta T (1973) Slightly deleterious mutant substitutions in evolution. Nature 246: 96–98.
35. Hirsch A, Fraser H (2001) Protein dispensability and rate of evolution. Nature 411:1046–1049.
36. Poon A, Otto S (2000) Compensating for our load of mutations: freezing the meltdown of small populations. Evolution 54: 1467–1479.
37. Fisher R (1930) The genetical theory of natural selection. Clarendon Press, Oxford. Pages 38–41.
38. Sanford J (2008) Genetic Entropy and the Mystery of the Genome. FMS Publications, NY. Pages 131–142.
39. Trifonov EN (1989) Multiple codes of nucleotide sequences. Bull of Mathematical Biology 51:417–432.
40. Trifanov EN (1997) Genetic sequences as products of compression by inclusive superposition of many codes. Mol Biol 31:647–654.

41. Kapranov P, et al (2005) Examples of complex architecture of the human transcriptome revealed by RACE and high density tiling arrays. Genome Res 15:987–997.
42. Birney E, et al (2007) Encode Project Consortium: Identification and analysis of functional elements in 1% of the human genome by the ENCODE pilot project. Nature 447:799–816.
43. Itzkovitz S, Hodis E, Sega E (2010) Overlapping codes within protein-coding sequences. Genome Res. 20:1582–1589.
44. Schneider, TD (2000) Evolution of biological information. Nucleic Acids Res 28 (14): 2794–2799.
45. Montañez G, Ewert W, Dembski WA, Marks II RJ (2010). Vivisection of the ev Computer Organism: Identifying Sources of Active Information. BIO-Complexity, 2010 (3):1–6
46. Lenski RE, Ofria C, Pennock RT, Adami C (2003) The evolutionary origin of complex features. Nature 423:139–144
47. Ewert W, Dembski WA, Marks II RJ (2009) Evolutionary Synthesis of Nand Logic: Dissecting a Digital Organism,. Proceedings of the 2009 IEEE International Conference on Systems, Man, and Cybernetics. San Antonio, TX, USA: 3047–3053.
48. Goh CK, Tan KC (2009) A Competitive-Cooperative Coevolutionary Paradigm for Dynamic Multiobjective Optimization. IEEE Transactions on Evolutionary Computation, 13 (1): 103–127
49. Rachmawati L, Srinivasan, D (2009) Multiobjective Evolutionary Algorithm With Controllable Focus on the Knees of the Pareto Front. IEEE Transactions on Evolutionary Volume: 13 , Issue 4: 810–824.
50. Farina, M., Deb, K., Amato, P. (2004) Dynamic multiobjective optimization problems: test cases, approximations, and applications. IEEE Transactions on Evolutionary Computation, Vol 8 , Issue 5: 425–442
51. Marks II RJ (2009) Handbook of Fourier Analysis and Its Applications. Oxford University Press.
52. Cover TM. Thomas JA (2006) Elements of Information Theory, 2nd edn. Wiley-Interscience, New York.
53. Dembski WA, Marks II RJ (2009) Conservation of Information in Search: Measuring the Cost of Success. IEEE Transactions on Systems, Man and Cybernetics A, Systems and Humans: 39(5): 1051–1061.
54. Stumpf M, et al (2008) Estimating the size of the human interactome. Proc Natl Acad Sci USA 105:6959–6964.
55. Wells J (2013) Not junk after all: non-protein-coding DNA carries extensive biological information. In: Marks II RJ, Behe MJ, Dembski WA, Gordon BL, Sanford JC (eds) Biological Information — New Perspectives. World Scientific, Singapore, pp. 210–231.

56. Kudla G., Murray AW, Tollervey D, Plotkin JB (2009) Coding-sequence determinants of gene expression in *Escherichia coli*. Science 324:5924.
57. Sanford JC, Baumgardner JR, Brewer WH (2013) Selection threshold severely constrains capture of beneficial mutations. In: Marks II RJ, Behe MJ, Dembski WA, Gordon BL, Sanford JC (eds) Biological Information — New Perspectives. World Scientific, Singapore, pp. 264–297.
58. Lenski RE, Travisano M (1994) Dynamics of adaptation and diversification: A 10,000-generation experiment with bacterial populations. Proc Natl Acad Sci USA 91:6808-6814.
59. Cooper VS, Lenski RE (2000) The population genetics of ecological specialization in evolving *Escherichia coli* populations. Nature 407:736–739.
60. Blount ZD, Borland CZ, Lenski RE (2008) Historical contingency and the evolution of a key innovation in an experimental population of *Escherichia coli*. Proc Natl Acad Sci USA 105:7899–7906.
61. Behe MJ (2013) Getting there first: an evolutionary rate advantage for adaptive loss-of-function mutations. In: Marks II RJ, Behe MJ, Dembski WA, Gordon BL, Sanford JC (eds) Biological Information — New Perspectives. World Scientific, Singapore, pp. 450–473.
62. Springman R, Keller T, Molineux I, Bull J (2009) Evolution at a high imposed mutation rate: adaptation obscures the load in phage T7. Genetics (26 October 2009), genetics.109.108803.
63. Hughes A (2007) Looking for Darwin in all the wrong places: the misguided quest for positive selection at the nucleotide level. Heredity 99:364–373.
64. Hughes A (2008) The origin of adaptive phenotypes. Proc Natl Acad Sci USA 105: 13193–13194.
65. Kryukov G, Schmidt S, Sunyaev S (2005) Small fitness effect of mutations in highly conserved non-coding regions. Human Molecular Genetics 14: 2221–2229.

Entropy, Evolution and Open Systems

Granville Sewell

*Mathematics Department, University of Texas at El Paso,
El Paso, TX 79968, USA. sewell@utep.edu*

Abstract

It is commonly argued that the spectacular increase in order which has occurred on Earth is consistent with the second law of thermodynamics because the Earth is not an isolated system, and anything can happen in a non-isolated system as long as the entropy increases outside the system compensate the entropy decreases inside the system. However, if we define "X-entropy" to be the entropy associated with any diffusing component X (for example, X might be heat), and, since entropy measures disorder, "X-order" to be the negative of X-entropy, a closer look at the equations for entropy change shows that they not only say that the X-order cannot increase in an isolated system, but that they also say that in a non-isolated system the X-order cannot increase faster than it is imported through the boundary. Thus the equations for entropy change do not support the illogical "compensation" idea; instead, they illustrate the tautology that "if an increase in order is extremely improbable when a system is isolated, it is still extremely improbable when the system is open, unless something is entering (or leaving) which makes it not extremely improbable." Thus unless we are willing to argue that the influx of solar energy into the Earth makes the appearance of spaceships, computers and the Internet *not* extremely improbable, we have to conclude that at least the basic principle behind the second law has in fact been violated here.

Key words: Entropy, Second Law of Thermodynamics

1. Compensation

It is probably fair to say that the majority view of science today holds that physics explains all of chemistry, chemistry explains all of biology, and biology completely explains the human mind; thus, physics alone explains the human mind, and all it does.

In fact, since there are only four known forces of physics (the gravitational, electromagnetic and strong and weak nuclear forces), this means that these four forces must explain everything that has happened on Earth, according to this majority view. For example, Peter Urone, in *College Physics* [1], writes "One of the most remarkable simplifications in physics is that only four distinct forces account for all known phenomena."

In my 2000 *Mathematical Intelligencer* article, "A Mathematician's View of Evolution" [2], I argued against this view, asserting that the increase in order which has occurred on Earth seems to violate the underlying principle behind the second law of thermodynamics, in a spectacular way. I wrote:

> I imagine visiting the Earth when it was young and returning now to find highways with automobiles on them, airports with jet airplanes, and tall buildings full of complicated equipment, such as televisions, telephones and computers. Then I imagine the construction of a gigantic computer model which starts with the initial conditions on Earth 4 billion years ago and tries to simulate the effects that the four known forces of physics would have on every atom and every subatomic particle on our planet. If we ran such a simulation out to the present day, would it predict that the basic forces of Nature would reorganize the basic particles of Nature into libraries full of encyclopedias, science texts and novels, nuclear power plants, aircraft carriers with supersonic jets parked on deck, and computers connected to laser printers, CRTs and keyboards? If we graphically displayed the positions of the atoms at the end of the simulation, would we find that cars and trucks had formed, or that supercomputers had arisen? Certainly we would not, and I do not believe that adding sunlight to the model would help much.

Anyone who has made such an argument is familiar with the standard reply: the Earth is not an isolated system, it receives energy from the sun, and entropy can decrease in a non-isolated system, as long as it is "compensated" somehow by a comparable or greater increase outside the system. For example, Isaac Asimov, in the *Smithsonian* journal [3] recognizes the apparent problem:

> You can argue, of course, that the phenomenon of life may be an exception [to the second law]. Life on earth has steadily grown more complex, more versatile, more elaborate, more orderly, over the billions of years of the planet's existence. From no life at all, living molecules were developed, then living cells, then living conglomerates of cells, worms, vertebrates, mammals, finally Man. And in Man is a three-pound brain which, as far as we know, is the most complex and orderly arrangement of matter in the universe. How could the human brain develop out of the primeval slime? How could that vast increase in order (and therefore that vast decrease in entropy) have taken place?

But Asimov concludes that there is no conflict with the second law here, because

> Remove the sun, and the human brain would not have developed. ... And in the billions of years that it took for the human brain to develop, the increase in

> entropy that took place in the sun was far greater; far, far greater than the decrease that is represented by the evolution required to develop the human brain.

Similarly, Peter Urone, in *College Physics* [1], writes:

> Some people misuse the second law of thermodynamics, stated in terms of entropy, to say that the existence and evolution of life violate the law and thus require divine intervention. ... It is true that the evolution of life from inert matter to its present forms represents a large decrease in entropy for living systems. But it is always possible for the entropy of one part of the universe to decrease, provided the total change in entropy of the universe increases.

Some other authors appear to feel a little silly suggesting that increases in entropy anywhere in the universe could compensate for decreases on Earth, so they are careful to explain that this "compensation" only works locally; for example in *Order and Chaos* [4], the authors write:

> In a certain sense the development of civilization may appear contradictory to the second law. ... Even though society can effect local reductions in entropy, the general and universal trend of entropy increase easily swamps the anomalous but important efforts of civilized man. Each localized, man-made or machine-made entropy decrease is accompanied by a greater increase in entropy of the surroundings, thereby maintaining the required increase in total entropy.

2. The Equations of Entropy Change

Of course the whole idea of compensation, whether by distant or nearby events, makes no sense logically: an extremely improbable event is not rendered less improbable simply by the occurrence of "compensating" events elsewhere. According to this reasoning, the second law does not prevent scrap metal from reorganizing itself into a computer in one room, as long as two computers in the next room are rusting into scrap metal — and the door is open. (Or the thermal entropy in the next room is increasing, though I am not sure how fast it has to increase to compensate computer construction![1])

[1]Daniel Styer, however, in an *American Journal of Physics* article [5], apparently *has* figured out how fast thermal entropy needs to increase to compensate evolution. Assuming that "each individual organism is 1000 times more improbable than the corresponding individual was 100 years ago" (a "very generous" assumption) and using a generous estimate for the number of organisms on Earth, he calculates that the rate of decrease of entropy due to evolution is very small, only about 302 Joules

To understand where this argument comes from, we need to look at the equations for entropy change, as given in Appendix D of my 2005 John Wiley book [6], and previously in my 2001 *Mathematical Intelligencer* article [7], "Can ANYTHING Happen in an Open System?"

Consider the diffusion (conduction) of heat in a solid, R, with absolute temperature distribution U(x, y, z, t). The first law of thermodynamics (conservation of energy) requires that

$$Q_t = -\nabla \bullet \mathbf{J} \tag{1}$$

where Q is the heat energy density ($Q_t = c\rho U_t$) and J is the heat flux vector. The second law requires that the flux be in a direction in which the temperature is decreasing, i.e.

$$\mathbf{J} \bullet \nabla\mathbf{U} \leq 0 \tag{2}$$

Equation 2 simply says that heat flows from hot to cold regions — because the laws of probability favor a more uniform distribution of heat energy.

"Thermal entropy" is a quantity that is used to measure randomness in the distribution of heat. The rate of change of thermal entropy, S, is given by the usual definition as

$$S_t = \iiint_R Q_t / U \, dV \tag{3}$$

Using (3) and the first law (1), after doing a (multidimensional) integration by parts, we get

$$S_t = \iiiint_R -(\mathbf{J} \bullet \nabla U)/U^2 \, dV - \iint_{\partial R} (\mathbf{J} \bullet \mathbf{n})/U \, dA \tag{4}$$

per degree Kelvin per second! He concludes, "Presumably the entropy of the Earth's biosphere is indeed decreasing by a tiny amount due to evolution and the entropy of the cosmic microwave background is increasing by an even greater amount to compensate for that decrease." It should be noted that if one is dealt a given poker hand, then replaces some cards, according to Styer we can compute the resulting entropy decrease in the universe, in units of Joules per degree Kelvin (!), as $k_B \log(N)$, where k_B is the Boltzmann constant, if the new hand is N times more improbable than the first. It should also be noted that if organisms become 1000 times more improbable every century, that would imply that organisms today are, on the average, about $10^{30000000}$ times "more improbable" than organisms a billion years ago, but, according to Styer, there is no conflict with the second law as long as something (anything, apparently!) is happening outside the Earth which, if reversed, would be even more improbable.

where **n** is the outward unit normal on the boundary ∂R. From the second law (2), we see that the volume integral is nonnegative, and so

$$S_t \geq - \iint_{\partial R} (J \bullet n)/U \, dA \tag{5}$$

From (5) it follows that $S_t \geq 0$ in an isolated system, where there is no heat flux through the boundary ($\mathbf{J} \bullet \mathbf{n} = 0$). Hence, in an isolated system, the entropy can never decrease. Since thermal entropy measures randomness (disorder) in the distribution of heat, its opposite (negative) can be referred to as "thermal order," and we can say that the thermal order can never increase in an isolated system.

Since thermal entropy is quantifiable, the application of the second law to thermal entropy is commonly used as the model problem on which our thinking about the other, less quantifiable, applications is based. The fact that thermal entropy cannot decrease in an isolated system, but can decrease in a non-isolated system, was used to conclude that, in other applications, any entropy decrease in a non-isolated system is possible as long as it is compensated somehow by entropy increases outside this system, so that the total "entropy" (as though there were only one type) in the universe, or any other isolated system containing this system, still increases.

However, there is really nothing special about "thermal" entropy. Heat conduction is just diffusion of heat, and we can define an "X-entropy" (and an X-order = -X-entropy), to measure the randomness in the distribution of any other substance X that diffuses; for example, we can let U(x, y, z, t) represent the concentration of carbon diffusing in a solid, and use equation (3) again to define this entropy ($c\rho = 1$ now, so $Q_t = U_t$), and repeat the analysis leading to equation (5), which now says that the "carbon order" cannot increase in an isolated system.[2]

Furthermore, equation (5) does not simply say that the X-entropy cannot decrease in an isolated system; it also says that in a non-isolated system, the X-entropy cannot decrease faster than it is exported through the boundary, because the boundary integral there represents the rate at which X-entropy is exported across the boundary. To see this, notice that without the denominator U, the integral in (3) represents the rate of change of total X (energy, if X=heat) in the system; with the denominator it represents the rate of change of X-entropy. Without the denominator U, the boundary integral in (5) represents the rate at

[2] "Entropy" sounds much more scientific than "order," but note that in this paper, "order" is simply defined as the opposite of "entropy." Where entropy is quantifiable, such as here, order is equally quantifiable. Physics textbooks also often use the term "entropy" in a less precise sense, to describe the increase in disorder associated with, for example, a plate breaking or a bomb exploding (e.g., [8], p 651). In such applications, "order" is equally difficult to quantify!

which X (energy, if X=heat) is exported through the boundary; with the denominator therefore it must represent the rate at which X-entropy is exported. Although I am certainly not the first to recognize that the boundary integral has this interpretation (see [9], p. 202)[3], this has been noticed by relatively few people, no doubt because usually the special case of isotropic heat conduction or diffusion is assumed, in which case $J = -K\nabla U$, and then the numerator in the boundary integral is written as $-K\partial U/\partial n$, and in this form it is not obvious that anything is being imported or exported, only that in an isolated system, the boundary integral is zero. Furthermore, entropy as defined by (3) seems to be a rather abstract quantity, and it is hard to visualize what it means to import or export entropy.

Stated in terms of order, equation (5) says that the X-order in a non-isolated system cannot increase faster than it is imported through the boundary. According to (4), the X-order in a system can decrease in two different ways: it can be converted to disorder (first integral term) or it can be exported through the boundary (boundary integral term). It can increase in only one way: by importation through the boundary.

3. A Tautology

The second law of thermodynamics is all about probability; it uses probability at the microscopic level to predict macroscopic change.[4] Carbon distributes itself more and more uniformly in an isolated solid because that is what the laws of probability predict when diffusion alone is operative. Thus the second law predicts that natural (unintelligent) causes will not do macroscopically describable things which are extremely improbable from the microscopic point of view. The reason natural forces can turn a computer or a spaceship into rubble and not vice versa is probability: of all the possible arrangements atoms could take, only a very small

---

[3]Dixon has a section "The Entropy Inequality for Open Systems," which contains the inequality, written out in words: "rate of change of entropy inside > rate of entropy flow in — rate of entropy flow out." In any case, even if one refuses to recognize that the boundary integral in (5) represents the (net) rate that entropy is exported, the tautology given in the next section is still illustrated by this application, because this boundary integral still represents the "something" that is crossing the boundary that makes the decrease in entropy not extremely improbable.

[4]In *Classical and Modern Physics*, Kenneth Ford [8] writes "There are a variety of ways in which the second law of thermodynamics can be stated, and we have encountered two of them so far: (1) For an isolated system, the direction of spontaneous change is from an arrangement of lesser probability to an arrangement of greater probability. (2) For an isolated system, the direction of spontaneous change is from order to disorder."

percentage could add, subtract, multiply and divide real numbers, or fly astronauts to the moon and back safely.

Of course, we must be careful to define "extremely improbable" events to be events of probability less than some very small threshold: if we define events of probability less than 1% to be extremely improbable, then obviously natural causes *can* do extremely improbable things.[5] But after we define a sufficiently low threshold, everyone seems to agree that "natural forces will rearrange atoms into digital computers" is a macroscopically describable event that is still extremely improbable from the microscopic point of view, and thus forbidden by the second law — at least if this happens in an isolated system. But it is not true that the laws of probability only apply to isolated systems: if a system is not isolated, you just have to take into account what is crossing the boundary when deciding what is extremely improbable and what is not. What happens in an isolated system depends on the initial conditions; what happens in a non-isolated system depends on the boundary conditions as well.

The "compensation" counter-argument was produced by people who generalized the model equation for isolated systems, but forgot to generalize the equation for non-isolated systems. Both equations are only valid for our simple models, where it is assumed that only heat conduction or diffusion is going on; naturally in more complex situations, the laws of probability do not make such simple predictions. Nevertheless, in "Can ANYTHING Happen in an Open System?" [7], I generalized the equations for non-isolated systems to the following tautology, which is valid in all situations:

> If an increase in order is extremely improbable when a system is closed, it is still extremely improbable when the system is open, unless something is entering which makes it not extremely improbable.

---

[5] If we repeat an experiment 2^k times, and define an event to be "simply describable" (macroscopically describable) if it can be described in m or fewer bits (so that there are 2^m or fewer such events), and "extremely improbable" when it has probability $1/2^n$ or less, then the probability that *any* extremely improbable, simply describable event will *ever* occur is less than $2^{k+m}/2^n$. Thus we just have to make sure to choose n to be much larger than $k + m$. If we flip a billion fair coins, any outcome we get can be said to be extremely improbable, but we only have cause for astonishment if something extremely improbable and simply describable happens, such as "all heads," or "every third coin is tails," or "only every third coin is tails." Since there are 10^{23} molecules in a mole of anything, for practical purposes anything that can be described without resorting to an atom-by-atom accounting (or coin-by-coin accounting, if there are enough coins) can be considered "macroscopically" describable.

The fact that order is disappearing in the next room does not make it any easier for computers to appear in our room — unless this order is disappearing *into* our room, and then only if it is a type of order that makes the appearance of computers not extremely improbable, for example, computers. Importing thermal order into a system may make the temperature distribution less random, and importing carbon order may make the carbon distribution less random, but neither makes the formation of computers more probable.

My conclusion, from "Can ANYTHING Happen in an Open System?" [7] is the following:

> Order can increase in an open system, not because the laws of probability are suspended when the door is open, but simply because order may walk in through the door.... If we found evidence that DNA, auto parts, computer chips, and books entered through the Earth's atmosphere at some time in the past, then perhaps the appearance of humans, cars, computers, and encyclopedias on a previously barren planet could be explained without postulating a violation of the second law here.... But if all we see entering is radiation and meteorite fragments, it seems clear that what is entering through the boundary cannot explain the increase in order observed here.

4. The Common Sense Law of Physics

I was discussing the second law argument with a friend recently, and mentioned that the second law has been called the "common sense law of physics." The next morning he wrote:

> Yesterday I spoke with my wife about these questions. She immediately grasped that chaos results in the long term if she would stop caring for her home.

I replied:

> Tell your wife she has made a perfectly valid application of the second law of thermodynamics.[6] In fact, let's take her application a bit further. Suppose you and

[6] Isaac Asimov [3] writes, "We have to work hard to straighten a room, but left to itself, it becomes a mess again very quickly and very easily.... How difficult to maintain houses, and machinery, and our own bodies in perfect working order; how easy to let them deteriorate. In fact, all we have to do is nothing, and everything deteriorates, collapses, breaks down, wears out — all by itself — and that is what the second law is all about."

> your wife go for a vacation, leaving a dog, a cat and a parakeet loose in the house (I put the animals there to cause the entropy to increase more rapidly, otherwise you might have to take a much longer vacation to see the same effect). When you come back, you will not be surprised to see chaos in the house. But tell her some scientists say, "but if you leave the door open while on vacation, your house becomes an open system, and the second law does not apply to open systems... you may find everything in better condition than when you left." I'll bet she will say, "If a maid enters through the door and cleans the house, maybe, but if all that enters is sunlight, wind and other animals, probably not."

Imagine trying to tell my friend's wife that, provided her house is an open system, the fact that chaos is increasing in the rest of the universe — or on the sun, provided sunlight enters through the door — means that chaos could decrease in her house while she is gone. Even if the door is left open, it is still extremely improbable that order in the house will improve, unless something enters that makes this not extremely improbable — for example, new furniture or an intelligent human.

Suppose we take a video of a tornado sweeping through a town, and run the video backward. Would we argue that although a tornado turning rubble into houses and cars represents a decrease in entropy, tornados derive their energy from the sun, and the increase in entropy outside the Earth more than compensates the decrease seen in the video, so there is no conflict with the second law? Or would we argue that what we were seeing was too difficult to quantify, so we can't be sure there is a problem? Some things are obvious even if they are difficult to quantify.

In *Signature in the Cell* [10], Stephen Meyer appeals to common sense in applying the second law to information:

> [M]ost of us know from our ordinary experience that information typically degrades over time unless intelligent agents generate (or regenerate) it. The sands of time have erased some inscriptions on Egyptian monuments. The leak in the attic roof smudged the ink in the stack of old newspapers, making some illegible.... Common experience confirms this general trend — and so do prebiotic simulation experiments and origin-of-life research.

A recent article by Andy McIntosh [11] in *International Journal of Design & Nature and Ecodynamics* takes a detailed and technical look at the relationship between entropy and biological information, but also includes appeals to common sense such as:

> Both Styer [5] and Bunn [12] calculate by slightly different routes a statistical upper bound on the total entropy reduction necessary to 'achieve' life on earth…

these authors are making the same assumption — viz. that all one needs is sufficient energy flow into a [non-isolated] system and this will be the means of increasing the probability of life developing in complexity and new machinery evolving. But as stated earlier this begs the question of *how* a local system can possibly reduce the entropy without existing machinery to do this… machines need to be pre-existing to enable an increase in order and complexity to take place.

5. Conclusions

Of course, one can still argue that the spectacular increase in order seen on Earth is consistent with the underlying principle behind the second law because what has happened here is not really extremely improbable. One can still argue that once upon a time, on a special planet called Earth, a collection of atoms formed by pure chance that was able to duplicate itself, and these complex collections of atoms were able to pass their complex structures on to their descendants generation after generation, even correcting errors. One can still argue that, after a long time, the accumulation of genetic accidents resulted in greater and greater information content in the DNA of these more and more complex collections of atoms, and eventually something called "intelligence" allowed some of these collections of atoms to design cars and trucks and spaceships and nuclear power plants. One can still argue that it only seems extremely improbable, but really isn't, that under the right conditions, the influx of stellar energy into a planet could cause atoms to rearrange themselves into computers and laser printers and the Internet.[7]

But one would think that at least this would be considered an open question, and those who argue that it really *is* extremely improbable, and thus contrary to the basic principle underlying the second law of thermodynamics, would be given a measure of respect, and taken seriously by their colleagues, but we aren't.

6. References

1. Urone PP (2001) College Physics. Brooks/Cole, Pacific Grove, CA.
2. Sewell G (2000) A Mathematician's View of Evolution. The Mathematical Intelligencer 22 (4): 5–7.

[7] For further development of the ideas in this paper, see Chapter 5 of my book *In the Beginning and Other Essays on Intelligent Design*, Discovery Institute Press, 2010. (www.discoveryinstitutepress.com/sewell).

3. Asimov I (1970) In the Game of Energy and Thermodynamics, You Can't Even Break Even. Smithsonian **1**, August 1970, p4.
4. Angrist S, Hepler L (1967) Order and Chaos. Basic Books, New York.
5. Styer D (2008) Entropy and Evolution. Am J of Physics **76**(11): 1031–1033.
6. Sewell G (2005) The Numerical Solution of Ordinary and Partial Differential Equations, second edition, John Wiley & Sons, Hoboken, NJ.
7. Sewell G (2001) Can ANYTHING Happen in an Open System? The Mathematical Intelligencer 23(4): 8–10.
8. Ford K (1973) Classical and Modern Physics. Xerox College Publishing, Lexington, MA.
9. Dixon J (1975) Thermodynamics I: An Introduction to Energy. Prentice-Hall, Englewood Cliffs, NJ.
10. Meyer S (2009) Signature in the Cell. Harper One, New York, NY.
11. McIntosh A (2009) Information and Entropy — Top-down or Bottom-up Development in Living Systems? International Journal of Design & Nature and Ecodynamics 4(4): 351–385.
12. Bunn E (2009) Evolution and the Second Law of Thermodynamics. *Am J of Physics* **77**(10): 922–925.

Information and Thermodynamics in Living Systems

Andy C. McIntosh

School of Process, Environmental and Materials Engineering, University of Leeds, LEEDS, U.K. LS2 9JT, a.c.mcintosh@leeds.ac.uk

Abstract

Are there laws of information exchange? And how do the principles of thermodynamics connect with the communication of information?

We consider first the concept of information and examine the various alternatives for its definition. The reductionist approach has been to regard information as arising out of matter and energy. In such an approach, coded information systems such as DNA are regarded as accidental in terms of the origin of life, and it is argued that these then led to the evolution of all life forms as a process of increasing complexity by natural selection operating on mutations on these first forms of life. However scientists in the discipline of thermodynamics have long been aware that organisational systems are inherently systems with low local entropy, and have argued that the only way to have consistency with an evolutionary model of the universe and common descent of all life forms is to posit a flow of low entropy into the earth's environment and in this second approach they suggest that islands of low entropy form organisational structures found in living systems.

A third alternative proposes that information is in fact non-material and that the coded information systems (such as, but not restricted to the coding of DNA in all living systems) is not defined at all by the biochemistry or physics of the molecules used to store the data. Rather than matter and energy defining the information sitting on the polymers of life, this approach posits that the *reverse* is in fact the case. Information has its definition outside the matter and energy on which it sits, and furthermore *constrains* it to operate in a highly non-equilibrium thermodynamic environment. This proposal resolves the thermodynamic issues and invokes the correct paradigm for understanding the vital area of thermodynamic/organisational interactions, which despite the efforts from alternative paradigms has not given a satisfactory explanation of the way information in systems operates.

Starting from the paradigm of information being defined by non-material *arrangement* and *coding*, one can then postulate the idea of laws of information exchange which have some parallels with the laws of thermodynamics which undergird such an approach. These issues are explored tentatively in this paper, and lay the groundwork for further investigative study.

Keyword: Information, thermodynamics, free energy, organisation, entropy, open systems, machines, biopolymers

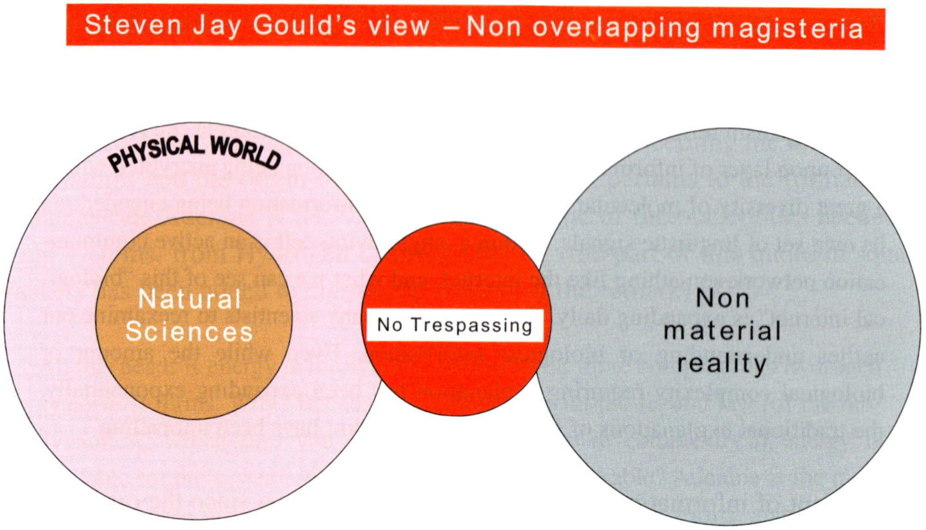

Fig. 2. Stephen Jay Gould's non-overlapping magisteria (NOMA) view of reality.

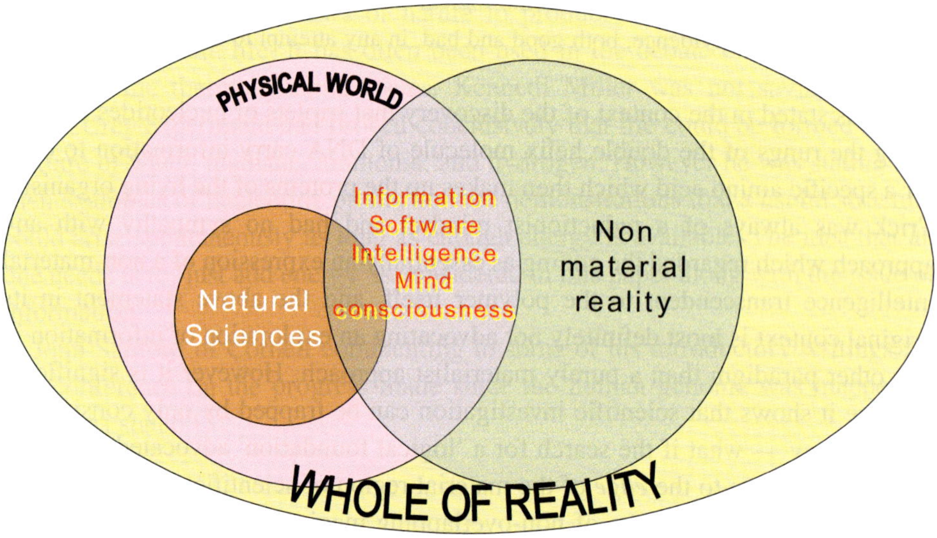

Fig. 3. The view of reality advocated in this paper which defines information as constraining the matter and energy it sits on, but not defined by it.

in recent years. The antithesis to Gould's approach is illustrated in Figure 3. It is argued that there is a legitimate realm where information, mind and consciousness lie — this area is undoubtedly interacting with the physical realm but is not entirely controlled by it. Though this clearly can have metaphysical implications, we are here not talking about religious matters, but simply the area where thoughts, logic and mind exists, and where the importance of *arrangement* rather than *matter* itself is dominant — as for instance in the sequencing of the nucleotides in DNA.

The paradigm adopted here is the assumption that information is essentially defined as non-material but profoundly influences the material in which it is found, in a similar way that software is essentially coded non-material instructions but nevertheless controls the hardware of a computer. It should be emphasised that this is not a license for any lazy thinking, whereby anything which cannot be understood is put metaphorically into a box labelled 'non-material and not to be further investigated'. This is no 'god of the gaps' thesis. Indeed, once adopted, this approach opens out a whole raft of new research routes which properly explain the control of living systems. A far more profound methodology is in view. What is being advocated here is an entirely different paradigm whereby the non-material message is accepted as being of an origin outside the area of physical investigation, but that its effect can readily be seen in the organisation of the molecular machinery in living organisms. Rather than the material and energy forming the information system as advocated by evolutionary philosophy, the non-material informational message expressed in the coded ordering of nucleotides is actually the mechanism of constraining the material itself. In this paradigm, it is the information which *organises* and *constrains* the biopolymers. It is a known feature of living systems that they are information-rich and it is this that is more and more being recognised as the cause of their great efficiency [6]. Rather than the intricate machinery for such systems evolving from simpler systems, it is the thesis of this paper that the message/information itself is sitting on biochemical molecular bonds which are in a significantly raised free energy state. Understanding the thermodynamics of this machinery shows that it is thermodynamically impossible both to form such machinery (abiogenesis) without intelligence, and that the laws of thermodynamics prohibit any formation of new machinery which is not there already or latently coded for in the DNA (evolutionary development). A hierarchical system of information is involved in living systems (see Figure 4).

Furthermore recent research has confirmed that non-coding parts of DNA previously thought to be 'junk DNA' are in fact not to be regarded as such [7]. More research is now coming to light [8] that the very folding of proteins carries with it a separate form of information transfer. This intertwining of information and matter lies at the heart of what is life itself, and fundamentally changes our view of how to understand living systems.

(Adenine (A), Thymine (T), Cytosine (C) and Guanine (G)) connecting the two deoxyribose sugar-phosphate strands (see Figure 5a). This is summarised in Figure 5b where the excision and repair of a damaged nucleotide base is shown. First the enzyme Glycolsylase finds the damaged site and the nucleotide, and then endonuclease removes neighbouring bases. Then the protein DNA polymerase manufactures the appropriate nucleotide and the enzyme ligase encircles the damaged DNA, and the replacement nucleotide base is put in place. This means that mutations are generally corrected (Jackson [14] and de Laat *et al.* [15]), so that even if speciation does occur due to slight modifications and adaptations of the phylogeny, any serious departures in the genetic information would be acted against by the DNA's own repair factory. Mutations generally do not increase information content — rather the reverse is true.

The second obstacle is a more fundamental issue. At the molecular level, the principles of thermodynamics do not permit the formation of new machinery from that which is already set up or coded for in a particular function performed by the cells of living organisms. There is in fact an 'uphill' gradient in the formation of any of the molecular bonds in the nucleotides and most of the proteins, since they want to pull apart. Consequently there is no natural chemical pathway to form these, rather there is a move away from their formation to equilibrium. In the following sections we examine the thermodynamic principles governing living systems.

2.1 *Thermodynamics and isolated systems*

One form of the statement of the second law of thermodynamics is "The amount of energy available for useful work in a given isolated system is decreasing. The entropy (dissipated energy per degree Kelvin which can no longer be used to do work) is *always* increasing."

Thus according to the second law, heat always flows from hot to cold. In the process it can be made to do work but always some energy will be lost to the environment, and that energy cannot be retrieved. Water flows downhill and loses potential energy which is changed into kinetic energy. This can again be made to do work (as in a hydroelectric power plant). However some energy will be lost such that if one was to use all the energy generated to pump the same water back up to its source, it would not reach the same level. The difference of original potential energy to that corresponding to the new level, divided by the temperature (which in that case is virtually constant) is the entropy of the system. Such a measure will always give an entropy *gain*.

There is no known system where this law does not apply. The fact that the entropy of a given isolated system increases, effectively brings with it an

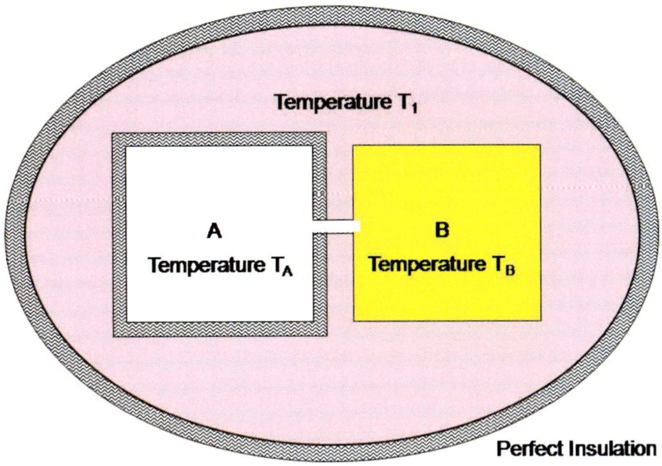

Fig. 6. Non-isolated system A and B

inevitable eventual decline in usefulness of all sub-systems within such an isolated system.

2.2 *Non isolated systems*

In that the second law of inevitable entropy increase applies to an isolated system, some have maintained that with a closed (boundaries open to heat transfer) or open system (boundaries open to heat and mass transfer) one could have entropy decreasing in one area while the overall entropy of the two systems together is increasing. An illustration would be of two ice boxes A and B (Figure 6) where there is an allowance for some small contact between them but with (perfect) insulation round the rest of the cube A and poor insulation round cube B. Systems A and B are both then non-isolated systems (technically closed as heat can pass the boundaries but not mass), as is the system A and B together (referred to as A+B), but system A and B with the surrounding region 1, (that is the complete system) is isolated. The entropy of the overall complete system then must increase with time. That is, there will eventually be equilibrium throughout every region.

2.2.1 *Entropy deficiency*

Suppose we start with Temperature T_1 appreciably hotter than T_A and T_B. Thus for instance we could have $T_1 = 100°C$ and T_A and T_B both at -10°C. Initially as time progresses the original equal temperatures T_A and T_B become different. T_A will stay close to the original −10°C, but T_B will begin to move to a higher value (say +5°C) due to there being good conduction of heat into ice box B (as against the insulated ice box A). Now consider system A and B together (A+B). One now has an open system with decreasing entropy called an *entropy deficiency*, in that useable energy transfer between the two ice boxes is possible, and work can be achieved where before in that system, treated in isolation, none was possible. However one notes two things. First that this is possible only for a finite time — eventually the temperature difference will reach a maximum (when T_B gets close to T_1) and at this point system A+B will have a minimum entropy condition. After this system A+B will then experience a rising entropy condition. Secondly one must insert some device between A and B before use can be made of this energy flow. This demonstrates the reality of how the underlying principles of energy flow and its use to do useful work, still apply in open systems. Extra energy is of no use *unless there is a mechanism to use it.*

2.2.2 *Open systems and machinery*

In the debate concerning origins where thermodynamic considerations are in view, much is made of the fact that the earth is an open system receiving energy and some mass transfer from extra-terrestrial sources. The main source of energy of course is the Sun. When one considers non-isolated systems where heat transfer can take place at the boundary, some have argued that by adding energy in to the original system then one should be able to reverse the overall trend of entropy increase. But this is not the case [10]. Adding energy without an existing mechanism which can make use of that additional energy, generally leads to simply the heating up of the surroundings faster than would otherwise have been the case. There can be cases where differential heating can occur (in the atmosphere or in the earth where rock and soil have thermal conductivity differences) following the same principle as outlined in Figure 6. Locally the entropy ($\Delta Q/T$ where ΔQ is the heat gained by the system being considered and T is temperature) can increase at different rates and give rise to a deficiency in entropy in one location compared to another. This can then potentially give rise to free energy which can do work. Thus for instance Freske [16] considers the entropy deficiency that sometimes can occur in a pile of bricks or rubble receiving energy from the sun, and that a device could make use of that energy supply

> .... under the given conditions, an entropy deficiency is in fact generated in the pile. After several hours of exposure to the sun, the temperature will be higher at the top than at the bottom. If we were to measure the temperatures throughout the

> pile, it would be a fairly simple matter to calculate the entropy deficiency. Useful energy could actually be extracted from the pile by means of a thermocouple, for example.

The last sentence concerning energy extraction actually demonstrates that the point at issue is not so much whether deficiency in entropy can take place and thus useful energy can be made to do work, so much as *the capacity to use the energy available*. Whether it is capturing directly the energy input from the sun, or harvesting the differential energy flow due to entropy deficiency, *a mechanism for making use of that energy flow is essential*. Without the thermocouple in Freske's illustration, very little will happen without directed purpose behind it.

In Section 3.2 below, we define a *machine* as a functional device that can do work [10], and it then follows that only by having in existence such a mechanism for capturing the incoming energy, can further useful work be achieved.

2.3 *Can negative entropy be harvested from somewhere else?*

Prigogine [17] and others (see for instance Styer [18]) have proposed that there is information in the non-material arrangement and organisation of systems and refer to an organisational entropy or 'logical' entropy. They propose the addition of other entropies which could then feed negative entropy into a given (non-isolated) system. Consequently the total entropy is considered to be

$$ds = ds_T + ds_{\text{logical}} \tag{1}$$

where ds_{logical} represents the ordering principle or 'logical' negative entropy which gradually seeps in to the system. Thus even though ds overall is increasing with the thermal entropy ds_T positive, the presence of ds_{logical} coming in at the boundary ensures locally the low entropy needed to spark evolutionary development. Styer [18] speaks of a net entropy flux at the earth which would then be the source of evolution of early prokaryotes (cells reckoned to be primitive with no nuclei) to eukaryotic (cells with nuclei) individuals.

Thus complexity and the ordering principle is predicated on the notion that information can gradually increase from a random state. Again this is flawed for two reasons:

(i) Firstly as stated above in Section 2.2.2, no flux of energy from outside the system can be made to do work within the system unless there is the necessary machinery to capture this energy [10].
(ii) Secondly the information itself (that is the message and meaning behind the communication) is not defined in purely thermodynamic terms or even in any ordered code such as in DNA when considering biological systems. Gitt [19] has shown that information is hierarchical in at least five levels. Two important levels are code (or language) and message which uses the coded communication system. Neither of these can actually be thought of as arising simply from a flux of entropy locally. Rather the reverse of this is the reality, viz. that non-material information (that is arrangement and coded instructions) sits on a material substrate in living systems and *the non-material information arrangement/coding causes a thermodynamic effect.*

3. Free energy and Machines

In order to propose an alternative understanding of the information in living systems, one of the key parts of the argument concerns the availability of energy to do work, coupled with the mechanism for harnessing of that energy.

3.1 *Free energy*

The Gibbs free energy g is defined as the net energy available to do work. It effectively takes away the unusable lost energy (associated with entropy) from the enthalpy h (which can be regarded as the total thermodynamic energy available). Thus

$$g = h - Ts, \text{ and } \Delta g = \Delta h - T\Delta s \tag{2a,b}$$

3.2 *Machines and raised free energies*

As a consequence of the principles of thermodynamics applied to non-isolated systems [9,10] one can state that the following applies concerning the spontaneity of chemical reactions:

$$\begin{array}{l} \Delta g < 0 \text{ Favoured reaction – Spontaneous} \\ \Delta g = 0 \text{ Reversible reaction – Equilibrium} \\ \Delta g > 0 \text{ Disfavoured reaction – Non-spontaneous} \end{array} \tag{3}$$

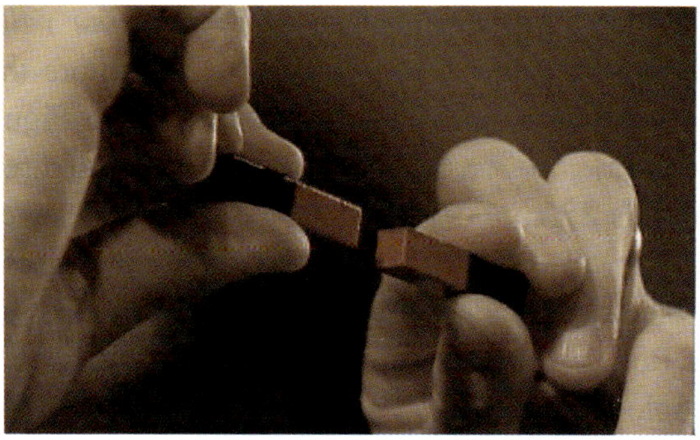

Fig. 7. All natural molecule formations are like magnets with the same pole facing each other such that if one lets the system 'go' they would pull apart: $\Delta g < 0$ (due to $g \equiv h - Ts > 0$). To set this system up — that is to keep the opposing magnets together work needs to be put in — the free energy change to bring them together is positive. In a similar way to bring the molecules together which form living polymers requires an *initial input* of ordered energy by another machine.

Consequently a positive free energy device cannot arise spontaneously. It always requires another operational machine to enable the free energy to be loaded/'primed', ready to do work. This can be illustrated in the example above (Figure 7) of two magnets with the same pole facing each other. Work needs to be put into the system to hold the opposing magnets together — the free energy change is positive — it is non-spontaneous, and the magnets want to pull apart. In a similar way to bring the molecules together which form living polymers requires an *initial input* of ordered energy to cause them to stay together. Δg is positive.

This leads to a definition:

We define a *machine* as a device which can locally raise the free energy to do useful work.

Even if material exchange was involved (and one had a completely open system), no amount of matter or energy exchange without information exchange would alter the fundamental finding (eqn (3)) concerning the spontaneity of chemical reactions.

Thus the free energy argument applies both to isolated systems with no contact with the surroundings and non-isolated systems. The latter include open systems where heat and mass can cross the boundary, as well as closed systems where just heat is allowed to cross.

One can now consider what happens if energy is added to a non-isolated system (as in Section 2.2) and it is evident that without a machine, the free energy to do

phosphate bonds, the three other nucleotides, and the weak Hydrogen bonds which couple the paired nucleotides together (e.g. Thymine to Adenine) — all requiring positive free energy steps [21] (that is they will break apart if left to themselves). On top of this one has the homochirality issue. When Harold Urey and Stanley Miller [1] in 1953 managed to produce amino acids by passing an electric charge through a reducing atmosphere of methane, ammonia, water and hydrogen, they did produce amino acids with some (but by no means all) of the standard 20 amino acids which need to be coded for, in order to make the proteins for life. But the most important difficulty was that they produced both left handed and right handed chirality molecules in the expected 50:50 ratio. However, living systems have entirely left handed versions of these molecules (i.e. homochirality) which otherwise are identical in chemical behaviour. Living systems are not just to do with chemicals, but with the *shape* and *positioning* of the chemicals. Stanley Miller acknowledged that the difficulties were indeed formidable when in 1991 he stated to Scientific American [22] that 'the problem of the origin of life has turned out to be much more difficult than I, and most other people, imagined ...'.

Furthermore the latest work in DNA studies [23] has produced some astounding discoveries that Hoogsteen base pairing (where a different part of the nucleotide bases is temporarily used to connect the coding and complimentary parts of the DNA) constitute a second transient layer of the genetic code. They state

> ...the DNA double helix intrinsically codes for excited state Hoogsteen base pairs as a means of expanding its structural complexity beyond that which can be achieved based on Watson-Crick base-pairing.

That is, there is already evidence that there is a further layer of information transfer in evidence — this again requires control of a suite of thermodynamic raised free energies by a different information system!

We stated in Section 3.2 that biopolymers could not change radically such that a low entropy and sustained sequence of free energies would emerge. To alter the DNA constituents from one stable state say to another representative state with a distinct improvement, cannot be done by natural means alone without additional *information*. The laws of thermodynamics are against such a procedure.

Put another way the carrier molecules of the information in living systems are actually kept in a non-equilibrium state by *the very presence of the coded information*. They would fall apart to a disordered equilibrium state were it not for the information in the system making them act in this way.

4. A different paradigm: Thermodynamics constrained by functional information

We thus propose a different treatment which quantifies the effect of functional information in a system. This approach recognizes Gitt's important deductions concerning real information systems being impossible to define in terms of matter and energy alone [19]. However one can recognise the effect of machines/ information systems (that is teleonomy) being present in exactly the same way as a digitally controlled machine (i.e. a computer) is operated by software. The high level program controls a set of electronic switches on a micro chip which are set in a certain predefined pattern (see right hand part of diagram in Figure 4). Thus the logical negative entropy (the switching of the micro chip in the analogy) rather than being the *source* of the information should be thought of as the *effect* of information carrying systems.

Only with the presence of a free energy device (a machine already existing) will an energy flux outside the system do useful work and enable a local lowering of the entropy of the system. This is illustrated for photosynthesis in Figure 9 whereby it is evident that the machinery of the production of chlorophyll in the leaf acts as an important system for taking in Carbon Dioxide and forming sugars

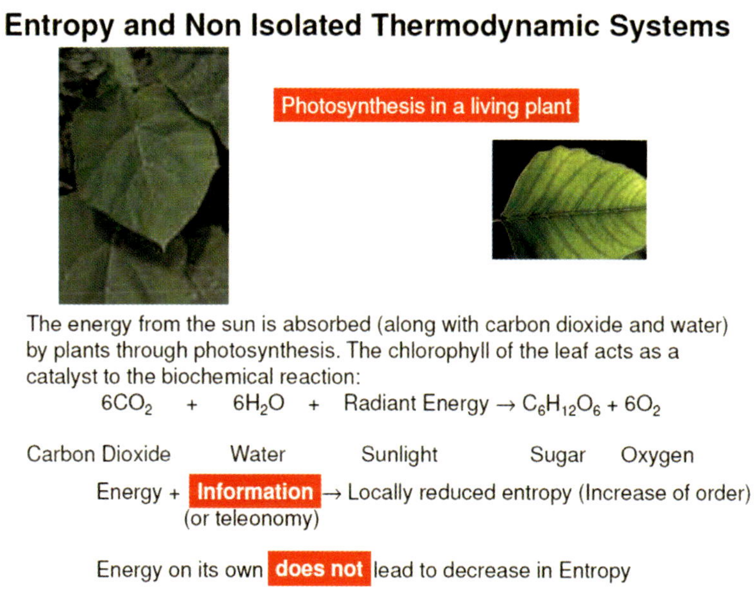

Fig. 9. Photosynthesis in a living plant requires energy input, but the energy flux on its own would do nothing unless there was a machine already present (a free energy device) to enable the system to do work using the sunlight.

and Oxygen. The energy flux on its own would do nothing unless there was a machine already present (a free energy device) to enable the system to do work using the sunlight.

In this approach it is expected that there will be levels of information, and in particular language (code) and semantics (meaning). This is a very different paradigm to that which is currently adopted, and leads to the proposition that machinery and information are closely intertwined in living systems [10,24,25], in a very similar way that software in a digital computer is intertwined with the electronic hardware and the very small but precise energy levels used in the computer registry, memory and storage facilities.

For a pure materialist there may be a natural reticence to adopting such an approach because of their presuppositions, but the evidence of the thermodynamics of living systems supports the view that it is *information* in living systems that controls the thermodynamics, and not the other way round.

4.1 *A different paradigm: Information definitions*

In order to construct a new approach to information exchange in living systems it is becoming evident that a new set of definitions is needed to set up this very valuable line of research. The following are suggested, and have come from valuable discussions with John Sanford [13] of Cornell University.

Information: That which is communicated through symbolic language.

Intelligent agent: An entity with the ability to create information and communicate it (i.e. — a human being).

Agency of intelligence: A secondary entity which is capable of being used to communicate information deriving from a higher source (i.e. — a computer).

Language: The symbolic medium through which information is communicated (i.e. Spanish).

Communication: The transmission of meaningful information through symbolic language.

4.2 *A different paradigm: principles of information and thermodynamics*

The following are suggested principles to understand the nature of how non-material information is transferred and communicated. In both the realm of thermodynamics and non-material information there are principles of conservation and degeneration. The following principles of information exchange are similar to the first two laws of thermodynamics and the thermodynamic law of non-isolated systems referred to in Section 3.3.

4.2.1 *Principles of information exchange*

We postulate the following principles of information exchange:

The First Principle concerning information, language, and communication
Apart from creative intelligence, information cannot be derived from nothing. There has to be a precursor bank of such information.

This is a parallel principle to the principle of conservation of mass and energy (first law of thermodynamics).

The Second Principle concerning information, language, and communication
Apart from a sustaining intelligence, all information degenerates in terms of its functional utility. Information will corrupt unless it is sustained by an intelligent agent.

This principle is a parallel to the second law of thermodynamics which effectively states that in a given isolated system, the energy available for doing useful work is diminishing — there is a principle of decay.

The principle of Information gain
The information content in a system is never greater than the total of that which was there already and that coming in through the boundary of the system.

This principle mirrors the thermodynamic law of non-isolated systems (Section 3.3).

4.2.2 *Principles of information interaction with energy and matter in biological systems*

We now summarize two further important principles which have been the main subject of this paper concerning the interaction of information with energy and matter in biological systems:

The First Principle of information interaction with matter in biological systems
Information in biological systems is expressed as coded instructions and is not defined by the energy levels or by the matter it resides in. It is not defined by the properties of that matter and is transcendent to it.

Comment: This principle is best exemplified by the fact that software in a computer is not defined by the hardware.

The Second Principle of information interaction with matter in biological systems
Information in biological systems constrains the matter and energy to be in a non-equilibrium state.

Corollary to second principle of information interaction
In biological systems all information sits on a substrate where a series of free energies are kept in disequilibrium. Thus information in biological systems relies on machines — that is on devices which raise the free energy.

Comment: This principle can be referred to as the 'top-down' principle — that is the information organises the thermodynamics of the system. The information does not arise out of the matter and energy.

Third principle of degeneration in living systems
Consequently the second principle of information interaction combined with the principle of thermodynamics decay, implies that degeneration, and in particular information corruption (mutations), will inevitably take place.

5. Conclusions

Three views of informational reality (ontology) are considered in this paper. The first is that matter and energy is all there is. This is the materialistic view of information (Dawkins (Oxford), Jones (University College, London), Atkins (Oxford) and others). Such authors argue that functional non-material information and design are an illusion. In their view matter and energy is all that there is in the Universe. Patterns only have meaning in a reductionist sense and do not carry any non-material 'value'. The second scenario is a variation of the bottom up approach. In this view information is regarded as non-material but has arisen out of matter and energy. This is the view of Prigogine [17], Yockey [26], Wicken [27] and Kenneth Miller [2,28,29] and many other authors.

Both these approaches are flawed on two counts. Firstly they ignore the fact that real information systems are *not* defined by the codes and languages they use and that the arrangement of the physical objects used in the system (for DNA, this would be the nucleotide triplets) has to be in a specified order. So even non-materialists such as Prigogine, Yockey, Wicken or Kenneth Miller have insuperable hurdles with such a system. By proposing an evolutionary model of the bottom up approach, they do not have the means to derive the specificity [30] in the ordering arrangement of the nucleotides in DNA. These issues are discussed in the work of Abel and Trevors [31, 32]. Secondly a more subtle point, but a very important one, is that there is an impossible thermodynamic barrier to such an approach. The information in living systems is mounted on molecules with a raised free energy such that the carriers of information would fall apart into equilibrium chemistry were it not for the information present. It is this barrier which shows that a top down approach is the only way to understand information in living systems.

The third view then that we have proposed in this paper is the top down approach. In this paradigm, the information is non-material and *constrains* the local thermodynamics to be in a non-equilibrium state of raised free energy. It is the information which is the active ingredient, and the matter and energy are passive to the laws of thermodynamics within the system.

As a consequence of this approach, we have developed in this paper some suggested principles of information exchange which have some parallels with the laws of thermodynamics which undergird this approach. They also have some profound implications concerning the inevitable decay in genetic material and the uniqueness of information in the beginning.

Acknowledgement

The author is indebted to many useful and profitable exchanges with Professor John Sanford of Cornell University. He has in particular suggested ideas for the principles on information exchange outlined in Section 4.

References

1. Miller SL (1953)Production of amino acids under possible primitive earth conditions. Science 117(3046):528–529.
2. Miller K.R., NCSE, The Miller/Morris Debate (1981), http://ncse.com/creationism/general/miller-morris-debate-1981 accessed January 2011.
3. Sanford J (2013) Biological Information and genetic theory — introductory comments. In: Marks II RJ, Behe MJ, Dembski WA, Gordon B, Sanford JC (eds) Biological Information — New Perspectives. World Scientific, Singapore, pp. 203–209.
4. Crick F "On the Genetic Code", Nobel Lecture, December 11, 1962 and also recorded in, Science 139(3554), 8 February 1963.
5. Gould SJ (1999) Non over-lapping magisteria. Skeptical Inquirer, July August, p55–61.

6. Vincent JFV, Bogatyreva OA, Bogatyrev NR, Bowyer A, Pahl AK (2006) Biomimetics: Its practice and theory. J Royal Society. Interface 3:471–482. In particular see discussion on p. 477 of nature's vast resources of information and small energy useage by comparison with man-made technology.
7. Birney E, *et al* (2007) Identification and analysis of functional elements in 1% of the human genome by the ENCODE pilot project. Nature 447:799–816.
8. Tramontano A (2005) Problem 4 Protein Structure Prediction. In: The ten most wanted solutions in protein bioinformatics. Chapman and Hall/CRC, Boca Raton, FL.
9. McIntosh AC (2006) Functional Information and Entropy in living systems. In: Brebbia CA, Suchrov LJ, Pascolo P (eds) Design and Nature III: Comparing Design in Nature with Science and Engineering. WIT Press, UK.
10. McIntosh AC (2009) Information and Entropy — Top-down or bottom-up development in living systems? Int. J. of Design & Nature and Ecodynamics 4(4):351–385.
11. Dawkins R (2003) The Information Challenge. Chapter 2.3 in: A Devil's Chaplain; Selected Essays by Richard Dawkins, Ed. Latha Menon, Phoenix.
12. Shannon CE (1948)The mathematical theory of communication. The Bell System Technical Journal 27: 379–423, 623–656.
13 Sanford JC (2005) Are random mutations good? In: Genetic Entropy and the Mystery of the Genome. Ivan Press, Lima NY, pp. 15–32.
14. Jackson SP (2002) Sensing and repairing DNA double strand breaks. Carcinogenesis 23(5):687–696, OUP.
15. de Laat WL, Jaspers NGJ, Hoeijmakers JHJ (1999) Molecular mechanism of nucleotide excision repair. Genes and Devel 13(7):768–785.
16. Freske S (1981) Creationist Misunderstanding, Misrepresentation, and Misuse of the Second Law of Thermodynamics. Creation Evol J 2(2): 8–16.
17. Nicolis G and Prigogine I (1989) Exploring Complexity: An Introduction. W.H. Freeman, New York.
18. Styer DF (2008) Entropy and Evolution. Am J Phys 76(11):1031–1033.
19. Gitt W (1989) Information: the third fundamental quantity. Siemens Review 56 (6):36–41.
20. Sewell, G (2000) A mathematician's view of evolution. Mathematical Intelligencer 22(4):5–7.
21. Thaxton CB, Bradley WL & Olsen RL (1992) Thermodynamics and the origin of life. In: The Mystery of Life's Origin. Lewis and Stanley, Dallas.
22. Horgan J (1991) In the beginning. Scientific American 264:116–125. Section concerning Stanley Miller, p. 117.
23. Nikolova EN, Kim E, Wise AA, O'Brien PJ, Andricioaei I, Al-Hashimi HM (2011) Transient Hoogsteen base pairs in canonical duplex DNA, Nature 470(7335): 498–502.

24. Trevors JT, Abel DL (2004) Chance and necessity do not explain the origin of life. Cell Biology International 28:729–739.
25. Abel DL (2009) The Capabilities of Chaos and Complexity. Int J Mol Sci 10: 247–291.
26. Yockey H (1992) Information Theory and Molecular Biology. Cambridge University Press, Cambridge.
27. Wicken JS (1987) Evolution, Thermodynamics and Information: Extending the Darwinian Program. Oxford University Press, Oxford.
28. Miller KR (2000) Finding Darwin's God: A Scientist's Search for Common Ground between God and Evolution ISBN 0–06–093049–7
29. Miller KR (2008) Only a Theory: Evolution and the Battle for America's Soul ISBN 978–0670018833
30. Meyer SC (2004) The origin of biological information and the higher taxonomic categories. Proc Biol Soc of Washington 117(2):213–239.
31. Abel DL & Trevors JT (2005) Three subsets of sequence complexity and their relevance to biolpolymeric information. Theor Biol & Med Mod 2:29.
32. Abel DL, (2009) The capabilities of chaos and complexity. Int J Mol Sci 10:247–291.

Section Two — Biological Information and Genetic Theory: Introductory Comments

John C. Sanford — Section Chairman

In the 21st century, *biological information* has become the over-arching theme which unifies the life sciences. In the 19th century, Charles Darwin and his colleagues did not yet have the notion of biological information. Indeed Darwin completely misunderstood the nature of inheritance, which he pictured to be Lamarckian in nature. One of Darwin's contemporaries, Gregor Mendel, discovered that the determinants of certain biological traits are transmitted from generation to generation in discrete packages (this work was ignored for a generation). Mendel probably had some vague notion that these genetic packages somehow might contain a very simple type of "biological information". But he could never have guessed that these genetic units which he observed were actually precisely-specified instructions, encoded by language, with each gene being comparable in complexity to a book. When the early population geneticists developed their models, they employed over-simplified mathematical models to try to describe their understanding of genetic change, but at that time genes were considered to be merely "beads on a string."

When DNA was discovered, it finally became clear that genetic information is very much like human written information — an extensive array of language-encoded strings of text. Where did all these text strings come from? For most biologists the already-ruling Darwinian paradigm seemed to be sufficient — they assumed that all biological information must arise merely by random letter changes in the text, combined with some reproductive filtering. In the last 60 years, many thousands of scientists have made a truly monumental effort to try to explain the entire biosphere, just in terms of random mutations which are filtered by natural selection. Has this effort been successful? It has certainly been successful in a sociological sense — this view is now faithfully upheld by the large majority in the academic community. The neo-Darwinian paradigm literally saturates the content of most biological journals. In fact any deviation from this view is generally regarded as academic treason — often being characterized as a threat to science itself. Yet in this section of our proceedings (Biological Information and Genetic Theory), we will show that there are huge genetic problems which bring this reigning paradigm into serious question.

As the figure below graphically illustrates, a paradigm shift appears to be imminent. This is because the amount of biological information which demands explanation is exploding, even while the explanatory power of Darwin's mechanism of

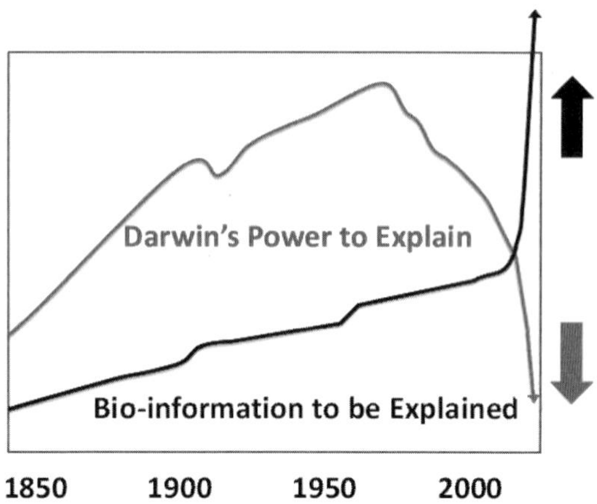

natural selection is virtually collapsing. This section of our symposium focuses on these two things — the explosion and the collapse.

The first problem is the explosion in the amount of biological information which requires explanation. We now realize that the last century's simplistic concept of biological information ("DNA makes RNA makes protein makes life") was incredibly naïve. We are just beginning to understand that biological information is profoundly multidimensional and moves in all directions through elaborate communication networks. The many layers of biological information are not only dynamic, they are globally integrated — overwhelming the previous generation's understanding of information (a gene encodes a protein). This will be clearly demonstrated by **Wells** in the first paper in this section, and is further developed by **Seaman** and **Johnson** in the last two papers of this section. Seaman and Johnson both correctly characterize the cell as being more like a network of computers than a set of books. These papers by Wells, Seaman, and Johnson act as the 'bookends' for this collection of research papers.

We need to better grasp the full scope of what "biological information" really is. It is a serious error to think of biological information as simply the genome. As discussed by Seaman, we can best understand the genome as the hard drive of the cell — it largely reflects *stored static information*. In that light, we should see that the RAM or active memory of the cell is that galaxy of RNAs and proteins which comprises the active communication network within the cell. These RNAs/proteins are actually the *active information* which makes life alive. As discussed by Johnson, RNA and proteins can be viewed as actively operating algorithms, specifying their own folding, their own transport, their own operation, and their various communication links with other molecules. Countless messages are continually being transmitted in both directions between the hard drive (the genome), and the

RAM (RNA and proteins). There is also continuous information being exchanged between different parts of the genome, and between RNAs and proteins, so there is a continuous interchange of information between all components. All this information which is continuously being exchanged within a single cell has been termed the "interactome", and it is vastly more complex than the genome itself. Such interactions within a living cell are beyond counting — and might best be compared to an internet system. The entire cell can be considered to be an extensive communication network. Above and beyond the individual cell, there is still more biological information being regularly communicated between cells, between tissues, and between individuals. Lastly, there is the biological information network that constitutes the brain/mind — which dwarfs everything else we have spoken of. With all this in mind, in this section we will primarily focus our attention on just the simplest level of biological information — the genome.

For decades it was believed that there is just one genetic code, and that only the protein-coding sequences within the genome were functional (less than 2% of the human genome). Essentially all other sequences were designated "junk DNA". This concept has been dramatically reversed in the last ten years, as revealed by Wells in this section's first paper. It is now clear that most of the non-protein-coding genome is functional. This means two things — firstly it means there is a lot more information in the genome that needs to be explained, and secondly it means there are many codes other than the amino acid code.

The implications of having many languages (genetic codes) in the same genome are staggering, and the fact that these codes overlap extensively is breath-taking (see Montanez *et al.*, in the previous section of these proceedings — Biological Information and Information Theory). In addition to the basic protein code, other codes associated with the conventional gene concept include the 12 codes of Trifanov, the transcription codes, the alternative splicing codes, and the RNA folding/processing codes. On an entirely different level, there are genome-wide codes that transcend the gene concept. These include the isochore codes, the nucleosome-positioning codes, the topological 3-D codes, and the epigenetic codes. Even the tiny but super-abundant Alu elements in the human genome, the most famous class of "junk DNA", are now known to contain multiple codes. These include transcription-regulating code, protein-binding code, and also a special 'pyknon' (small RNA) code. Some, but not all, of these codes are described in more detail by Wells. It should be obvious that more codes are waiting to be discovered. In the second to last paper in this section, Seaman, discloses very exciting new evidence for repeat-based codes in the genome, which have an uncanny resemblance to the repeat structures characteristic of executable computer code.

How many genes are in the human genome? The textbooks still suggest there are just over 20,000 human genes — because they have not yet acknowledged the paradigm shift ushered in by the ENCODE project. We now know that what we used to call a gene was a gross over-simplification. What we used to call a gene, we now know is actually a complex of many functional elements, encoding multiple proteins and many RNAs. If we define each of these functional elements as a gene, there must be hundreds of thousands of genes. Since there is now strong evidence that SINES and LINES are themselves functional elements, we should also recognize these as genes — so depending on how we define a gene, there may be over a million genes in the genome. Our awareness of biological information, just within the genome, is truly exploding. In the following section of this symposium (Biological Information and Molecular Biology), Dent and Wells each present papers proposing additional new types of biological information which entirely transcend the genome. If validated, each of these will clearly require its own language or code. I am convinced that none of us has yet fully absorbed the significance of what is emerging, in terms of the richness and depth of biological information. There has simply never been a more exciting time to be a biologist.

The second problem is the collapse of the Darwinian mechanism, in terms of its power to explain how all this biological information arose and is sustained. This will be clearly demonstrated by the papers of Gibson *et al.*, Sanford *et al.*, Nelson *et al.*, Brewer *et al.*, and Baumgardner *et al.* Natural selection obviously works, the problem is it does not appear to be capable of performing as advertised. These papers show that, most fundamentally, the Darwinian mechanism cannot consistently create a net gain of information. This is because even as rare beneficial mutations arise (only some of which can be selectively amplified), many more deleterious mutations must be accumulating continuously. Certainly this should result in "genetic change over time" — but the change should primarily be downward. If mutation/selection causes genomes to primarily go down, not up, then the Darwinian mechanism cannot explain the origin of genomes, or even their maintenance. Consequently, the explanatory power of the Darwinian mechanism appear to be limited to the trivial and the mundane (i.e., minor superficial adaptations in response to environmental change — mere fine-tuning). This is clearly documented in the following papers.

Gibson *et al.* summarize their extensive numerical simulation research addressing the problem of deleterious mutation accumulation — as affected by the *selection threshold* phenomenon. They have developed what is clearly the most advanced numerical simulation for modeling mutation accumulation within populations ("Mendel's Accountant"). They use numerical simulation to demonstrate that given biologically realistic conditions, natural selection fails to selectively remove the large majority of deleterious mutations. They show that there are various reasons why this happens, but the most important reason in that each population has a certain characteristic selection threshold, and mutations which have very small fitness effects fall below this threshold, and hence will become essentially invisible to natural selection. Gibson *et al.* show that when biologically realistic conditions are modeled for a higher organism, the selection threshold is especially high, such that the vast majority of deleterious mutations are not selectable, and hence accumulate continuously. If the mutation/selection process is really all that is happening, then this means that all higher organisms should be continuously accumulating deleterious mutations at a high rate, even when there is strong natural selection pressure — which would logically lead to eventual extinction.

Sanford *et al.* have also studied the selection threshold problem, but they examine how it affects the accumulation of beneficial mutations. They use numerical simulation (again, Mendel's Accountant) to demonstrate that there is a very clear selection threshold for beneficial mutations, and that only a very tiny fraction of all beneficial mutations have a large enough effect to be able to respond to selection. They show that the selection threshold problem is even more severe for beneficial mutations than it is for deleterious mutations. Because it is clear that beneficial mutations are very rare anyway, the fact that only a very tiny fraction of them are selectable means that selectable beneficial mutations should be vanishingly rare. When rare beneficial mutations do occur which are above the selection threshold, they respond to selection beautifully and can be rapidly amplified. This reflects the type of response we see when there is strong selection for something like a bacterial mutation for antibiotic resistance. But these types of rare and isolated events can only explain what is known as microevolution (mere adaptation). Clearly, this type of fine-tuning to some specific environmental factor has no bearing on how genomes might be created or sustained. Sanford *et al.* raise the important question — "What mechanism could have established the hundreds of millions of very low-impact nucleotide sites within any higher genome?"

Nelson *et al.* use the well-known Avida simulation program to show that when Avida is run using biologically realistic parameters, the results are remarkably similar to when similar parameters are used in Mendel's Accountant. For example, when a realistic distribution of mutation effects is employed (the Mendel default setting), both programs show no forward evolution at all, but rather a rapid loss of whatever genetic information was initially present. Conversely, when all mutations have very large fitness effects (the Avida default setting), both simulation programs demonstrate explosive forward evolution. Avida, like Mendel's Accountant, when run with biologically reasonable parameters, shows reverse evolution. Nelson *et al.* go on to use Avida to illustrate something that Mendel's Accountant fails to demonstrate — that there is a clearly defined threshold for establishing irreducible complexity via the selective process, given reasonable probabilistic resources.

The profound difficulties with the classic Darwinian mechanism, as described in the preceding papers, have been known within the population genetics community for decades. The standard response to these problems has been either to ignore them, or to invoke ad hoc models to explain away the problems. These ad hoc models have never been critically examined or properly tested. There are two primary models used to explain why genetic change over time might primarily be upward, rather than downward. The first is what can be called the *mutation count mechanism* and the second is the *synergistic epistasis mechanism.*

Brewer *et al.* use numerical simulation to test the mutation count mechanism model. This model suggests that if selection is strongly directed specifically against those individuals with higher mutation counts, deleterious mutation accumulation can be halted. The numerical simulations of Brewer *et al.* show that this mechanism actually can work, but only when mutation effects are relatively uniform, when there is truncation selection, and where there is sexual recombination. However, numerical simulations clearly show the mutation count mechanism becomes ineffective when *any* of the following are true: 1) there is a distribution of mutation effects which is realistically broad; 2) probability selection is operating; 3) a species reproduces clonally. Few if any situations occur in nature where *none* of three conditions are present, hence the mutation count mechanism cannot be operational in any general sense. Therefore, Brewer *et al.* have effectively falsified the mutation count hypothesis.

Baumgardner *et al.* use numerical simulation to test the synergistic epistasis hypothesis. This hypothesis proposes that as mutations accumulate continuously — they will amplify each other's deleterious effect, so that genetic damage does not increase linearly but rather increases exponentially. It is thought that at some critical point, just one or a few additional mutations will create a profoundly deleterious effect ("the straw that broke the camel's back"). In this way selection might be focused more strongly against those individuals who have a higher mutation count (just as with the mutation count mechanism). This hypothesis is highly problematic, is entirely ad hoc, and it is entirely incompatible with all the normal population genetics assumptions. None the less, this hypothetical mechanism is rigorously tested using numerical simulation by Baumgardner *et al.* It is shown that the synergistic epistasis mechanism fails to halt deleterious mutation accumulation, and consistently accelerates mutational degeneration, just as common sense would dictate.

Given all the theoretical evidence that the mutation/selection should yield to a net loss in functional information, it's very reasonable to ask if there are living systems that actually show this might be happening. This is generally difficult to demonstrate experimentally, because most biological systems change very slowly, especially on the level of the whole genome. **Brewer, Smith, and Sanford** have chosen to study RNA viruses, which have short replication cycles and extremely high mutation rates, and so they can change rapidly in short intervals of time. They examine such viruses to better understand loss of information in real biological systems. They examine pandemic histories which suggest that some human pandemics involving RNA virus may come to an end because of mutation accumulation leading to *natural genetic attenuation* of the virus. They then do a series of numerical simulations that confirm that based upon known RNA viral mutations rates and based upon the biology of pandemics, a significant fitness decline in the virus should be expected during the course of a typical pandemic. These authors then go on to use numerical simulations to examine what factors might accelerate such natural genetic attenuation. They show that use of pharmaceuticals that are known to enhance the viral mutation rate should be highly effective in reducing both the extent and the duration of pandemics. Other practices which should accelerate genetic attenuation would include reducing inoculum levels during disease transmission (stronger bottlenecking), and reducing titer levels in the infected host (lower selection efficiency).

These papers, along with many other lines of evidence (i.e., see Behe's paper in the following section "Biological Information and Molecular Biology"), clearly show that the explanatory power of the classic Darwinian mechanism is suddenly collapsing. This is happening at exactly the same time that we are being overwhelmed with evidence that the actual amount of biological information that requires explanation is vastly deeper and richer than we could have imagined. Surely this is an exciting time to be a biologist!

Not Junk After All: Non-Protein-Coding DNA Carries Extensive Biological Information

Jonathan Wells

Discovery Institute, 208 Columbia Street, Seattle, WA 98104, USA. jonwells2001@comcast.net

Abstract

In the 1950s Francis Crick formulated the Central Dogma of molecular biology, which states (in effect) that DNA makes RNA makes protein makes us. By 1970, however, biologists knew that the vast majority of our genome does not encode proteins, and the non-protein-coding fraction became known as "junk DNA." Yet data from recent genome projects show that most nuclear DNA is transcribed into RNAs, many of which perform important functions in cells and tissues. Like protein-coding DNA, non-protein-coding regions carry multiple overlapping codes that profoundly affect gene expression and other cellular processes. Although there are still many gaps in our understanding, new functions of non-protein-coding DNA are being reported every month. Clearly, the notion of "junk DNA" is obsolete, and the amount of biological information in the genome far exceeds the information in protein-coding regions.

Key words: Central Dogma, Sequence Hypothesis, junk DNA, non-protein-coding DNA, non-protein-coding RNA, chromatin, centromere, inverted nuclei

1. Introduction

James Watson and Francis Crick's 1953 discovery that DNA consists of two complementary strands suggested a possible copying mechanism for Mendel's genes [1,2]. In 1958, Crick argued that "the main function of the genetic material" is to control the synthesis of proteins. According to the "Sequence Hypothesis," Crick wrote that the specificity of a segment of DNA "is expressed solely by the sequence of bases," and "this sequence is a (simple) code for the amino acid sequence of a particular protein." Crick further proposed that DNA controls protein synthesis through the intermediary of RNA, arguing that "the transfer of information from nucleic acid to nucleic acid, or from nucleic acid to protein may be possible, but transfer from protein to protein, or from protein to nucleic acid, is impossible." Under some circumstances RNA might transfer sequence information to DNA, but the order of causation is normally "DNA makes RNA makes protein." Crick called this the "Central Dogma" of molecular biology [3], and it is sometimes stated more generally as "DNA makes RNA makes protein makes us."

The Sequence Hypothesis and the Central Dogma imply that only protein-coding DNA matters to the organism. Yet by 1970 biologists already knew that much of our DNA does not code for proteins. In fact, less than 2% of human DNA is protein-coding. Although some people suggested that non-protein-coding DNA might help to regulate gene expression, the dominant view was that non-protein-coding regions had no function. In 1972, biologist Susumu Ohno published an article wondering why there is "so much 'junk' DNA in our genome" [4].

In 1976, Oxford biologist Richard Dawkins wrote: "The amount of DNA in organisms is more than is strictly necessary for building them: A large fraction of the DNA is never translated into protein. From the point of view of the individual organism this seems paradoxical. If the 'purpose' of DNA is to supervise the building of bodies, it is surprising to find a large quantity of DNA which does no such thing. Biologists are racking their brains trying to think what useful task this apparently surplus DNA is doing. But from the point of view of the selfish genes themselves, there is no paradox. The true 'purpose' of DNA is to survive, no more and no less. The simplest way to explain the surplus DNA is to suppose that it is a parasite, or at best a harmless but useless passenger, hitching a ride in the survival machines created by the other DNA" [5].

If one assumes that only protein-coding regions of DNA matter to the organism, and non-protein-coding DNA is just parasitic junk, it makes sense also to assume that only protein-coding regions would be transcribed into RNA. Why would an organism engaged in a struggle for survival waste precious internal resources on transcribing junk? Yet it turns out that organisms *do* transcribe most of their DNA into RNA — and there is growing evidence that much (perhaps even most) of this RNA performs essential functions in cells and tissues.

2. Widespread Transcription Into RNAs That Are Probably Functional

Even before the Human Genome Project was completed in 2003 [6] there had been reports of the widespread transcription of non-protein-coding DNA. In 2002, the Japanese FANTOM Consortium (for **F**unctional **AN**no**T**ation **O**f the **M**ammalian Genome) identified 11,665 non-protein-coding RNAs in mice and concluded that "non-coding RNA is a major component of the transcriptome" [7]. Other scientists reported that transcription of two human chromosomes resulted in ten times more RNA than could be attributed to protein-coding exons [8].

In 2003, the ENCODE project (for **ENC**yclopedia **O**f **D**NA **E**lements) set out to identify all the functional elements in the human genome. It soon became obvious that most of the mammalian genome is transcribed into RNA [9,10]. Preliminary data provided "convincing evidence that the genome is pervasively transcribed, such that the majority of its bases can be found in primary transcripts, including non-protein-coding transcripts" [11].

The ENCODE Project and FANTOM Consortium showed that RNAs are transcribed from *both* strands of DNA, and that antisense RNA is a major component of the mammalian transcriptome [12-15]. Not only is some RNA transcribed from the antisense strand, but RNAs can also be transcribed from several different start sites within an open reading frame. So a single open reading frame can carry multiple overlapping codes that specify both protein-coding RNAs and non-protein-coding RNAs [16-20].

Widespread transcription suggests probable function; so does sequence conservation. In 2004 and 2005, several groups of scientists identified non-coding regions of DNA that were *completely identical* in humans and mice. They called these "ultra-conserved regions (UCRs)" and noted that they were clustered around genes involved in early development. The researchers concluded that the long non-coding UCRs act as regulators of developmentally important genes [21-24].

In 2006, a team studying endothelial cells (which line the inside of human blood vessels) reported that "conserved non-coding sequences" — some within introns — were enriched in sequences that "may play a key role in the regulation of endothelial gene expression" [25]. Oxford geneticists comparing large non-protein-coding RNAs in humans, rats and mice reported conserved sequences that "possess the imprint of purifying selection, thereby indicating their functionality" [26]. And in 2009, a team of American scientists found "over a thousand highly conserved large non-coding RNAs in mammals" that are "implicated in diverse biological processes" [27].

3. Direct Evidence for Some Specific Functions of Non-Protein-Coding RNAs

There is also direct evidence for specific functions of non-protein-coding RNAs. Paraspeckles are domains inside the nuclei of mammalian cells that play a role in gene expression by retaining certain RNAs within the nucleus [28]. Several non-protein-coding RNAs are known to be essential constituents of them [29,30], binding to specific proteins to form ribonucleoproteins that stabilize the paraspeckles and enable them to persist through cell divisions even though they are not bounded by membranes [31,32].

Non-protein-coding RNAs are also involved in alternative splicing. When a eukaryotic gene is transcribed into RNA, its non-protein-coding introns are removed and the protein-coding exons are then spliced together before being translated into protein. In the great majority of cases (80-95%), the exons can be "alternatively spliced," which means that the resulting transcripts can lack some exons or contain duplicates of others [33,34]. Alternative splicing plays an essential role in the differentiation of cells and tissues at the proper times during embryo development, and many alternatively spliced RNAs occur in a developmental-stage- and tissue-specific manner [35-37].

Although introns do not code for proteins, the RNAs transcribed from them contain specific codes that regulate alternative splicing [38-40]. The mammalian thyroid hormone receptor gene produces two variant proteins with opposite effects, and the alternative splicing of those variants is regulated by an intron [41]. An intronic element plays a critical role in the alternative splicing of tissue-specific RNAs in mice [42], and regulatory elements in introns control the alternative splicing of growth factor receptors in mammalian cells [43].

In 2007, Italian biologists reported that intronic sequences regulate the alternative splicing of a gene involved in human blood clotting [44]. In 2010, a team of Canadian and British scientists studying splicing codes in mouse embryonic and adult tissues — including the central nervous system, muscles, and the digestive system — found that introns are rich in splicing-factor recognition sites. It had previously been assumed that most such sites are close to the affected exons — leaving long stretches of DNA not involved in the process of alternative splicing — but the team concluded that their results suggested "regulatory elements that are deeper into introns than previously appreciated" [45].

Introns encode other functional RNAs, as well. Short non-protein-coding RNAs are known to regulate gene expression [46], and in 2004 British scientists identified such RNAs within the introns of 90 protein-coding genes [47]. In 2007, Korean biologists reported that in humans a "majority" of short non-protein-coding RNAs originate "within intronic regions" [48]. One of these, according to American medical researchers, is involved in regulating cholesterol levels in humans [49]. Introns also encode many of the small RNAs essential for the processing of ribosomal RNAs, as well as the regulatory elements associated with such RNA-coding sequences [50,51].

Chromatin organization profoundly affects gene expression. Non-protein-coding RNAs are essential for chromatin organization [52,53], and non-protein-coding RNAs have been shown to affect gene expression by modifying chromatin structure [54,55]. A recent study of chromatin-associated RNAs in some human cells revealed that almost two-thirds of them are derived from introns [56].

concluded: "Given the scale dependence of nucleotide function, large amounts of 'junk DNA', contrary to common belief, must be assumed to contribute to the complexity of gene interaction systems and of organisms" [82]. In 2007, Zuckerkandl (with Giacomo Cavalli) wrote that "SINEs and LINEs, which have been considered 'junk DNA,' are among the repeat sequences that would appear liable to have teleregulatory effects on the function of a nearby promoter, through changes in their numbers and distribution" [83].

Since enhancers can be tens of thousands of nucleotides away from the genes they regulate, bringing together enhancers and promoters that are on the same chromosome requires chromosome "looping" [84,85]. The size of a chromosome loop depends on the length of the DNA. For physical reasons, a loop consisting only of DNA must be at least 500 nucleotides long, while a loop consisting of chromatin (because of its greater stiffness) must be at least 10,000 nucleotides long [86]. In such cases it may be the sheer length of the DNA that matters, not whether it encodes RNAs.

4.2 *Chromatin Organization*

Because DNA is packaged into chromatin, and because RNA polymerase must have access to the DNA to transcribe it, the structure of chromatin is all-important in gene regulation. In many cases, various proteins and RNAs mediate the attachment of RNA polymerase to the DNA by interacting with specific sequences of nucleotides, but in some cases a mere change in the three-dimensional conformation of chromatin can activate transcription by exposing the DNA to RNA polymerase [87].

In 2007, scientists in Massachusetts produced a genome-scale, high-resolution three-dimensional map of DNA and found similar conformations that were independent of the underlying nucleotide sequences. They concluded that "considerably different DNA sequences can share a common structure" due to their similar chromatin conformation, and some transcription factors may be "conformation-specific … rather than DNA sequence-specific" [88].

Two years later, scientists reported that functional non-protein-coding regions of the human genome are correlated with chromatin-related "local DNA topography" that can be independent of the underlying sequence. "Although similar sequences often adopt similar structures," they wrote, "divergent nucleotide sequences can have similar local structures," suggesting that "they may perform similar biological functions." The authors of the report concluded that "some of the functional information in the non-coding portion of the genome is conferred by DNA structure as well as by the nucleotide sequence" [89].

The clearest example of a chromatin-level function that can be independent of the exact DNA sequence is the "centromere," a special region on a eukaryotic chromosome that serves as the chromosome's point of attachment to other structures in the cell. For example, before a eukaryotic cell divides it makes a duplicate of each chromosome, and the duplicate copies of each chromosome are joined together at their centromeres until they separate and move to daughter cells.

Centromeres can form only on a foundation provided by the chromosome. Yet centromeres are built upon long stretches of repetitive DNA that some biologists have regarded as junk [90]. Although much of the DNA that underlies centromeres is now known to be transcribed into RNAs that perform a variety of functions [91-96], it turns out that centromere formation is to a great extent independent of the exact nucleotide sequence.

The DNA sequences of centromere regions vary significantly from species to species, though all centromeres function similarly [97]. If the chromosome region containing a centromere is artificially deleted and replaced by synthetic repetitive DNA, a functional centromere can form again at the same site [98]. Extra centromeres (called "neo-centromeres") can also form abnormally elsewhere on a chromosome that already has one, or on a chromosome fragment that has separated from the part bearing a centromere [99,100]. It seems that centromeres can form at many different places on a chromosome, regardless of the underlying DNA sequence.

Nevertheless, the underlying chromatin must have certain characteristics that make centromere formation possible. For example, there is evidence that some aspects of the DNA sequence are conserved [101,102]. In humans and other primates, centromere activity is normally associated with repeated blocks of 171- nucleotide subunits termed alpha-satellite DNA. (Researchers in the 1960s discovered that a fraction of DNA consisting of millions of short, repeated nucleotide sequences produced "satellite" bands when DNA was centrifuged to separate it into fractions with different densities.) Every normal human centromere is located on alpha-satellite DNA [103–105].

Human neo-centromeres form on parts of a chromosome that do *not* consist of alpha-satellite DNA, though the neo-centromere DNA still has special characteristics — most notably, an unusually high proportion of LINEs [106]. These non-protein-coding segments apparently play a role in localizing proteins that are required for the formation of the centromere and kinetochore [107,108].

In the 1980s, biologists identified several proteins associated with centromeres and called them CENPs (for **CEN**tromere **P**roteins) [109]. Subsequent research revealed that one of these, CENP-A, takes the place of some of the histones in chromatin [110]. The incorporation of CENP-A makes chromatin stiffer and provides a foundation for assembling the other components of centromeres

[111,112]. In fact, centromeres in all organisms are associated with CENP-A, which must be present for a centromere to form, though CENP-A by itself is not sufficient [113,114].

The modification of chromatin by CENP-A and other centromere-specific proteins can be passed down from generation to generation. Indeed, the location of a centromere on a particular chromosome can persist for thousands of generations. From the perspective of the Central Dogma and Sequence Hypothesis (i.e., the view that DNA sequences determine the essential features of organisms by encoding proteins), centromeres are an enigma because they show that a cell can impose an essential and heritable structure on its DNA that is independent of the precise nucleotide sequence.

4.3 *Chromosome Arrangement in the Nucleus*

Between cell divisions, chromosomes are not randomly distributed in the nucleus; instead, they occupy distinct domains [115]. Chromosome domains affect gene regulation, in part, by bringing together specific regions of chromosomes and facilitating interactions among them [116,117]. Different cell and tissue types in the same animal can have different three-dimensional patterns of chromosomes in their nuclei, which account for at least some differences in gene expression [118,119].

One notable feature of nuclear domains is their radial arrangement [120]. In 1998, biologists in New York reported that chromatin localized to the periphery of the nucleus in yeast cells tends to be "transcriptionally silent" [121]. In 2001, British biologists wrote that "most gene-rich chromosomes concentrate at the centre of the nucleus, whereas the more gene-poor chromosomes are located towards the nuclear periphery" [122]. In 2008, Dutch biologists reported that human chromosome domains associated with the periphery of the nucleus "represent a repressive chromatin environment" [123]. The same year, several teams of researchers reported independently that they could suppress the expression of specific genes by relocating them to the nuclear periphery [124–126].

These data are consistent with the observation that in most nuclei the gene-rich euchromatin is concentrated near the center while the gene-poor heterochromatin is situated more peripherally. An important exception to this radial arrangement, however, occurs in the retinas of nocturnal mammals (Fig. 1).

The retina of a vertebrate eye contains several different kinds of light-sensing cells. Cone cells detect colors and function best in bright light; rod cells are more numerous and more sensitive to low light. Nocturnal animals such as mice need to see under conditions of almost no light, so they need exceptionally sensitive rod

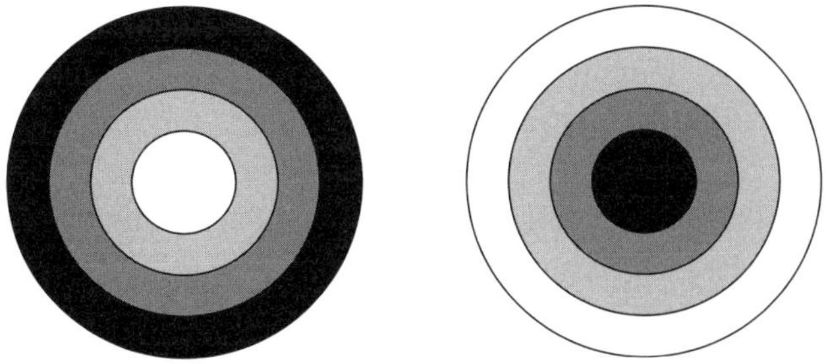

Fig. 1 Left: A simplified view of the internal arrangement of chromatin in most eukaryotic nuclei. Gene-poor heterochromatin (black) is on the periphery, and the gene content of the chromatin increases toward the center, which consists of gene-rich euchromatin (white). Right: A simplified view of the inverted chromatin arrangement found in rod cells in the retinas of nocturnal mammals. Gene-rich euchromatin is on the periphery, while gene-poor heterochromatin is in the center. The centrally located heterochromatin acts as a liquid-crystal lens that focuses the few photons available at night onto the light-sensitive outer segments of the rod cells.

cells. In 1979, medical researchers examined mouse retinas with an electron microscope and found that the heterochromatin in cone cells was located near the periphery of the nucleus, as in most other eukaryotic cells, but the heterochromatin in rod cells was concentrated in "one large, central clump" [127].

Another team of medical researchers used mice to study the genetic mutation responsible for an inherited human disease that causes nerve degeneration [128]. The team found that the mutation causes blindness in mice by altering the arrangement of the chromatin in rod cells. Instead of containing "a single, large clump of heterochromatin surrounded by a spare rim of euchromatin," the rod cells in mutant mice "showed a dramatic chromatin decondensation" and "resembled cone nuclei" [129].

Clearly, the unique localization of heterochromatin in the center of rod cells in mouse retinas is essential for normal vision in these animals. In 2009, European scientists called the unusual pattern of centrally located heterochromatin "inverted," and they reported finding an inverted pattern in the rod cell nuclei of various other mammals that are primarily nocturnal (including cats, rats, foxes, opossums, rabbits and several species of bats) but not of mammals that are primarily active in daylight (such as cows, pigs, donkeys, horses, squirrels, and chipmunks). These scientists observed that the centrally located heterochromatin had a high refractive index — a characteristic of optical lenses — and by using a two-dimensional computer simulation they showed that a main consequence of the inverted pattern was to focus light on the light-sensitive regions of rod cells [130].

In 2010, molecular biologists in France reported that the organization of the central heterochromatin in the rod nuclei of nocturnal mammals is consistent with a "liquid crystal model" [131], and British biophysicists improved upon the 2009 study by using a new computer simulation to show that "the focusing of light by inverted nuclei" in three dimensions is "at least three times as strong" as it is in two dimensions [132].

So evidence for the functionality of non-protein-coding DNA comes from several sources: pervasive transcription of the genome, including transcription from antisense DNA and from multiple start sites within open reading frames; conservation of a substantial fraction of non-protein-coding sequences; particular sequence-dependent functions of RNAs transcribed from introns, pseudogenes, repetitive DNA (much of which is *not* conserved, but species-specific); and functions that are to a large extent independent of the exact nucleotide sequence, such as the influence of intron length on transcription timing, the role of chromatin topology in gene expression and centromere placement, and the light-focusing property of heterochromatin in inverted nuclei. Clearly, it is no longer reasonable to maintain that the vast majority of our DNA is "junk."

5. Conclusion: Multiple Levels of Biological Information

The concept of information as applied to a linear sequence — such as letters in an English sentence or nucleotides in a DNA molecule — has been extensively analyzed [133-143]. Although protein-coding DNA constitutes less than 2% of the human genome, the amount of such information in such DNA is enormous. Recent discoveries of multiple overlapping functions in non-protein-coding DNA show that the biological information in the genome far exceeds that in the protein-coding regions alone.

Yet biological information is not limited to the genome. Even at the level of gene expression — transcription and translation — the cell must access information that is not encoded in DNA. Many different RNAs can be generated from a single piece of DNA by alternative splicing, and although some splicing codes occur in intronic DNA there is no empirical justification for assuming that *all* of the information for tissue- and developmental-stage-specific alternative splicing resides in DNA. Furthermore, even after RNA has specified the amino acid sequence of a protein, additional information is needed: Protein function depends on three-dimensional shape, and the same sequence of amino acids can be folded differently to produce proteins with different three-dimensional shapes [144–147]. Conversely, proteins with different amino acid sequences can be folded to produce similar shapes and functions [148,149].

Many scientists have pointed out that the relationship between the genome and the organism — the genotype-phenotype mapping — cannot be reduced to a genetic program encoded in DNA sequences. Atlan and Koppel wrote in 1990 that advances in artificial intelligence showed that cellular operations are not controlled by a linear sequence of instructions in DNA but by a "distributed multilayer network" [150]. According to Denton and his co-workers, protein folding appears to involve formal causes that transcend material mechanisms [151], and according to Sternberg this is even morc evident at higher levels of the genotype-phenotype mapping [152].

So non-protein-coding regions of DNA that some previously regarded as "junk" turn out to encode biological information that greatly increases the known information-carrying capacity of DNA. At the same time, DNA as a whole turns out to encode only part of the biological information needed for life.

Addendum

Due to a delay in the publication of these proceedings, the material in this chapter is now (2013) over two years old. Yet it is still accurate. Indeed, the fact that most non-protein-coding DNA serves biological functions was dramatically confirmed in September 2012 by 37 papers published by the ENCODE Project in Nature, Genome Research, Genome Biology, The Journal of Biological Chemistry, *and* Sciencc *[153-189]. The Project concluded that 80% of the genome is linked to biological functions, but Project Coordinator Ewan Birney pointed out that this conclusion was based on analyses of only 147 cell types, and "the human body has a few thousand." As more cell types are studied, Birney said, "It's likely that 80 percent will go to 100 percent." [190] A commentary accompanying the papers in* Nature *described the ENCODE results as "dispatching the widely held view that the human genome is mostly 'junk DNA'." [191] A commentary published at the same time in* Science *announced "ENCODE Project writes eulogy for junk DNA." [192]*

Acknowledgments

The author gratefully acknowledges the assistance of Richard v. Sternberg, helpful comments from reviewers, and the financial support of the Discovery Institute.

References

1. Watson JD, Crick FHC (1953) Molecular structure of nucleic acids: a structure for deoxyribose nucleic acid. Nature 171:737–738

2. Watson JD, Crick FHC (1953) Genetical implications of the structure of deoxyribonucleic acid. Nature 171:964–967
3. Crick FHC (1958) On protein synthesis. Symp Soc Exp Biol 12:138–163
4. Ohno S (1972) So much 'junk' DNA in our genome. Brookhaven Symp Biol 23: 366–70
5. Dawkins R (1976) The selfish gene. Oxford University Press, New York
6. National Human Genome Research Institute (2003) International consortium completes human genome project. http://www.genome.gov/11006929 Accessed 2013 Mar 15
7. Okazaki Y, et al (2002) Analysis of the mouse transcriptome based on functional annotation of 60,770 full-length cDNAs. Nature 420:563–573
8. Kapranov P, et al (2002) Large-scale transcriptional activity in chromosomes 21 and 22. Science 296:916–919
9. Carninci P, et al (2005) The transcriptional landscape of the mammalian genome. Science 309:1559–1563
10. Pheasant M, Mattick JS (2007) Raising the estimate of functional human sequences. Genome Res 17:1245–1253
11. Birney E, et al (2007) Identification and analysis of functional elements in 1% of the human genome by the ENCODE pilot project. Nature 447:799–816
12. Katayama S, et al (2005) Antisense transcription in the mammalian transcriptome. Science 309:1564–1566
13. Engström PG, et al (2006) Complex loci in human and mouse genomes. PLoS Genet 2:e47
14. He Y, et al (2008) The antisense transcriptomes of human cells. Science 322:1855–1857
15. Morris KV, et al (2008) Bidirectional transcription directs both transcriptional gene activation and suppression in human cells. PLoS Genet 4:e1000258
16. Gustincich S, et al (2006) The complexity of the mammalian transcriptome. J Physiol 575:321–332
17. Kapranov P, Willingham AT, Gingeras TR (2007) Genome-wide transcription and the implications for genomic organization. Nat Rev Genet 8:413–423
18. Carninci P (2007) Constructing the landscape of the mammalian transcriptome. J Exp Biol 210:1497–1506
19. Wu JQ, et al (2008) Systematic analysis of transcribed loci in ENCODE regions using RACE sequencing reveals extensive transcription in the human genome. Genome Biol 9:R3
20. Itzkovitz S, Hodis E, Segal E (2010) Overlapping codes within protein-coding sequences. Genome Res 20:1582–1589
21. Bejerano G, et al (2004) Ultraconserved elements in the human genome. Science 304:1321–1325

22. Sandelin A, et al (2004) Arrays of ultraconserved non-coding regions span the loci of key developmental genes in vertebrate genomes. BMC Genomics 5:99
23. Woolfe A, et al (2005) Highly conserved non-coding sequences are associated with vertebrate development. PLoS Biol 3:e7
24. Siepel A, et al (2005) Evolutionarily conserved elements in vertebrate, insect, worm, and yeast genomes. Genome Res 15:1034–1050
25. Bernat JA, et al (2006) Distant conserved sequences flanking endothelial-specific promoters contain tissue-specific DNase-hypersensitive sites and over-represented motifs. Hum Mol Genet 15:2098–2105
26. Ponjavic J, Ponting CP, Lunter G (2007) Functionality or transcriptional noise? Evidence for selection within long noncoding RNAs. Genome Res 17:556–565
27. Guttman M, et al (2009) Chromatin signature reveals over a thousand highly conserved large non-coding RNAs in mammals. Nature 458:223–227
28. Bond CS, Fox AH (2009) Paraspeckles: nuclear bodies built on long noncoding RNA. J Cell Biol 186:637–644
29. Clemson CM, et al (2009) An architectural role for a nuclear noncoding RNA: NEAT1 RNA is essential for the structure of paraspeckles. Mol Cell 33:717–726
30. Sasaki YTF, et al (2009) MENepsilon/beta noncoding RNAs are essential for structural integrity of nuclear paraspeckles. Proc Natl Acad Sci USA 106:2525–2530
31. Sasaki YTF, Hirose T (2009) How to build a paraspeckle. Genome Biol 10:227
32. Souquere S, et al (2010) Highly-ordered spatial organization of the structural long noncoding NEAT1 RNAs within paraspeckle nuclear bodies. Mol Biol Cell 21:4020–4027
33. Wang ET, et al (2008) Alternative isoform regulation in human tissue transcriptomes. Nature 456:470–476
34. Sultan M, et al (2008) A global view of gene activity and alternative splicing by deep sequencing of the human transcriptome. Science 321:956–960
35. Blencowe BJ (2006) Alternative splicing: new insights from global analyses. Cell 126:37–47
36. Prokunina-Olsson L, et al (2009) Tissue-specific alternative splicing of *TCF7L2*. Hum Mol Genet 18:3795–3804
37. Mercer TR, et al (2010) Regulated post-transcriptional RNA cleavage diversifies the eukaryotic transcriptome. Genome Res 20:1639–1650
38. Ladd AN, Cooper TA (2002) Finding signals that regulate alternative splicing in the post-genomic era. Genome Biol 3:reviews0008
39. Hui J, et al (2005) Intronic CA-repeat and CA-rich elements: a new class of regulators of mammalian alternative splicing. EMBO J 24:1988–1998
40. Nakaya HI (2007) Genome mapping and expression analyses of human intronic noncoding RNAs reveal tissue-specific patterns and enrichment in genes related to regulation of transcription. Genome Biol 8:R43

41. Hastings ML, Wilson CM, Munroe SH (2001) A purine-rich intronic element enhances alternative splicing of thyroid hormone receptor mRNA. RNA 7: 859–874
42. Nakahata S, Kawamoto S (2005) Tissue-dependent isoforms of mammalian Fox-1 homologs are associated with tissue-specific splicing activities. Nucleic Acids Res 33:2078–2089
43. Wagner EJ, et al (2005) Characterization of the intronic splicing silencers flanking FGFR2 exon IIIb. J Biol Chem 280:14017–14027
44. Marcucci R, Baralle FE, Romano M (2007) Complex splicing control of the human thrombopoietin gene by intronic G runs. Nucleic Acids Res 35:132–142
45. Barash Y, et al (2010) Deciphering the splicing code. Nature 465:53–59
46. Mendes Soares LM & Valcárcel J (2006) The expanding transcriptome: the genome as the 'Book of Sand', EMBO J 25:923–931
47. Rodriguez A, et al (2004) Identification of mammalian microRNA host genes and transcription units. Genome Res 14:1902–1910
48. Kim Y-K, Kim VN (2007) Processing of intronic microRNAs. EMBO J 26: 775–783.
49. Najafi-Shoushtari SH, et al (2010) MicroRNA-33 and the SREBP host genes cooperate to control cholesterol homeostasis. Science 328:1566–1569
50. Hoeppner MP, et al (2009) Evolutionarily stable association of intronic snoRNAs and microRNAs with their host genes. Genome Biol Evol 2009:420–428
51. Monteys AM, et al (2010) Structure and activity of putative intronic miRNA promoters. RNA 16:495–505
52. Lavrov SA, Kibanov MV (2007) Noncoding RNAs and chromatin structure. Biochemistry (Mosc) 72:1422–1438
53. Rodríguez-Campos A, Azorín F (2007) RNA is an integral component of chromatin that contributes to its structural organization. PLoS One 2:e1182
54. Malecová B, Morris KV (2010) Transcriptional gene silencing through epigenetic changes mediated by non-coding RNAs. Curr Opin Mol Ther 12:214–222
55. Caley DP, et al (2010) Long noncoding RNAs, chromatin, and development. ScientificWorldJournal 10:90–102
56. Mondal T, et al (2010) Characterization of the RNA content of chromatin. Genome Res 20:899–907
57. Meister G, Tuschl T (2004) Mechanisms of gene silencing by double-stranded RNA. Nature 431:343–349
58. Watanabe T, et al (2008) Endogenous siRNAs from naturally formed dsRNAs regulate transcripts in mouse oocytes. Nature 453:539–543
59. Tam OT, et al (2008) Pseudogene-derived small interfering RNAs regulate gene expression in mouse oocytes. Nature 453:534–538
60. Franco-Zorrilla JM, et al (2007) Target mimicry provides a new mechanism for regulation of microRNA activity. Nat Genet 39:1033–1037

61. Piehler AP, et al (2008) The human ABC transporter pseudogene family: Evidence for transcription and gene-pseudogene interference. BMC Genomics 9:165
62. Poliseno L, et al (2010) A coding-independent function of gene and pseudogene mRNAs regulates tumour biology. Nature 465:1033–1038
63. Garcia-Perez JL, et al (2007) Distinct mechanisms for trans-mediated mobilization of cellular RNAs by the LINE-1 reverse transcriptase. Genome Res 17:602–611
64. Shepard EA, et al (2007) Alternative promoters and repetitive DNA elements define the species-dependent tissue-specific expression of the FMO1 genes of human and mouse. Biochem J 406:491–499
65. Spadafora C (2008) A reverse transcriptase-dependent mechanism plays central roles in fundamental biological processes. Syst Biol Reprod Med 54:11–21
66. Tang YA, et al (2010) Efficiency of Xist-mediated silencing on autosomes is linked to chromosomal domain organization. Epigenetics Chromatin 3:10
67. Chow JC, et al (2010) LINE-1 activity in facultative heterochromatin formation during X chromosome inactivation. Cell 141:956–969
68. Vansant G, Reynolds WF (1995) The consensus sequence of a major *Alu* subfamily contains a functional retinoic acid response element. Proc Natl Acad Sci USA 92:8229–8233
69. Mariner PD, et al (2008) Human *Alu* RNA is a modular transacting repressor of mRNA transcription during heat shock. Mol Cell 29:499–509
70. Häsler J, Strub K (2006) *Alu* RNP and *Alu* RNA regulate translation initiation in vitro. Nucleic Acids Res 34:2374–2385
71. Häsler H, Strub K (2006) *Alu* elements as regulators of gene expression. Nucleic Acids Res 34:5491–5497
72. Häsler J, Samuelsson T, Strub K (2007) Useful 'junk': Alu RNAs in the human transcriptome. Cell Mol Life Sci 64:1793–1800
73. Gu TJ, et al (2009) Alu-directed transcriptional regulation of some novel miRNAs. BMC Genomics 10:563
74. Barak M, et al (2009) Evidence for large diversity in the human transcriptome created by *Alu* RNA editing. Nucleic Acids Res 37:6905–6915
75. Walters RD, Kugel JF, Goodrich JA (2009) InvAluable junk: the cellular impact and function of *Alu* and B2 RNAs. IUBMB Life 61:831–837
76. van Driel R, Fransz PF, Verschure PJ (2003) The eukaryotic genome: a system regulated at different hierarchical levels. J Cell Sci 116:4067–4075
77. Misteli T (2007) Beyond the sequence: cellular organization of genome function. Cell 128:787–800
78. Gubb D (1986) Intron-delay and the precision of expression of homeotic gene products in *Drosophila*. Dev Genet 7:119–131
79. Thummel CS (1992) Mechanisms of Transcriptional Timing in *Drosophila*. Science 255:39–40

80. Swinburne IA, Silver PA (2008) Intron delays and transcriptional timing during development. Dev Cell 14:324–330
81. Zuckerkandl E (1997) Junk DNA and sectorial gene repression. Gene 205:323–343
82. Zuckerkandl E (2002) Why so many noncoding nucleotides? The eukaryote genome as an epigenetic machine. Genetica 115:105–129
83. Zuckerkandl E, Cavalli G (2007) Combinatorial epigenetics, 'junk DNA', and the evolution of complex organisms. Gene 390:232–242
84. Celniker SE, Drewell RA (2007) Chromatin looping mediates boundary element promoter interactions. BioEssays 29:7–10
85. Palstra RJ (2009) Close encounters of the 3C kind: long-range chromatin interactions and transcriptional regulation. Brief Funct Genomic Proteomic 8:297–309
86. Göndör A, Ohlsson R (2009) Chromosome crosstalk in three dimensions. Nature 461:212–217
87. Angermayr M, et al (2002) Transcription initiation in vivo without classical transactivators: DNA kinks flanking the core promoter of the housekeeping yeast adenylate kinase gene, AKY2, position nucleosomes and constitutively activate transcription. Nucleic Acids Res 30:4199–4207
88. Greenbaum JA, Pang B, Tullius TD (2007) Construction of a genome-scale structural map at single-nucleotide resolution. Genome Res 17:947–953
89. Parker SCJ, et al (2009) Local DNA topography correlates with functional noncoding regions of the human genome. Science 324:389–392
90. Gardiner K (1997) Clonability and gene distribution on human chromosome 21: reflections of junk DNA content? Gene 205:39–46
91. Reinhart BJ, Bartel DP (2002) Small RNAs correspond to centromere heterochromatic repeats. Science 297:1831
92. Lu J, Gilbert DM (2007) Proliferation-dependent and cell cycle regulated transcription of mouse pericentric heterochromatin. J Cell Biol 179:411–412
93. Wong LH, et al (2007) Centromere RNA is a key component for the assembly of nucleoproteins at the nucleolus and centromere. Genome Res 17:1146–1160
94. O'Neill RJ, Carone DM (2009) The role of ncRNA in centromeres: a lesson from marsupials. Progr Mol Subcell Biol 48:77–101
95. Ferri F, et al (2009) Non-coding murine centromeric transcripts associate with and potentiate Aurora B kinase. Nucleic Acids Res 37:5071–5080
96. Eymery A, Callanan M, Vourc'h C (2009) The secret message of heterochromatin: new insights into the mechanisms and function of centromeric and pericentric repeat sequence transcription. Int J Dev Biol 53:259–268
97. Sullivan BA, Blower MD, Karpen GH (2001) Determining centromere identity: cyclical stories and forking paths. Nat Rev Genet 2:584–596
98. Harrington JJ, et al (1997) Formation of de novo centromeres and construction of first-generation human artificial microchromosomes. Nat Genet 15:345–355

99. du Sart D, et al (1997) A functional neo-centromere formed through activation of a latent human centromere and consisting of non-alpha-satellite DNA. Nat Genet 16:144–153
100. Warburton PE (2004) Chromosomal dynamics of human neocentromere formation. Chromosome Res 12:617–626
101. Grady DL, et al (1992) Highly conserved repetitive DNA sequences are present at human centromeres. Proc Natl Acad Sci USA 89:1695–1699
102. Jiang J, et al (1996) A conserved repetitive DNA element located in the centromeres of cereal chromosomes. Proc Natl Acad Sci USA 93:14210–14213
103. Waye JS, Willard HF (1985) Chromosome-specific alpha satellite DNA: nucleotide sequence analysis of the 2.0 kilobasepair repeat from the human X chromosome. Nucleic Acids Res 13:2731–2743
104. Heller R, et al (1996) Mini-chromosomes derived from the human Y chromosome by telomere directed chromosome breakage. Proc Natl Acad Sci USA 93:7125–7130
105. Murphy TD, Karpen GH (1998) Centromeres take flight: Alpha satellite and the quest for the human centromere. Cell 93:317–320
106. Lo AW, et al (2001) A 330 kb CENP-A binding domain and altered replication timing at a human neocentromere, EMBO J 20:2087–2096
107. Chueh AC, et al (2005) Variable and hierarchical size distribution of L1-retroelement-enriched CENP-A clusters within a functional human neocentromere. Hum Mol Genet 14:85–93
108. Chueh AC, et al (2009) LINE retrotransposon RNA is an essential structural and functional epigenetic component of a core neocentromeric chromatin. PLoS Genet 5:e1000354
109. Palmer DK, et al (1987) A 17-kD centromere protein (CENP-A) copurifies with nucleosome core particles and with histones. J Cell Biol 104:805–815
110. Yoda K, et al (2000) Human centromere protein A (CENP-A) can replace histone H3 in nucleosome reconstitution in vitro. Proc Natl Acad Sci USA 97:7266–7271
111. Black BE, et al (2007) An epigenetic mark generated by the incorporation of CENP-A into centromeric nucleosomes. Proc Natl Acad Sci USA 104:5008–5013
112. Torras-Llort M, Moreno-Moreno O, Azorín F (2009) Focus on the centre: the role of chromatin on the regulation of centromere identity and function. EMBO J 28:2337–2348
113. Van Hooser AA, et al (2001) Specification of kinetochore-forming chromatin by the histone H3 variant CENP-A. J Cell Sci 114:3529–3542
114. Vos LJ, Famulski JK, Chan GKT (2006) How to build a centromere: from centromeric and pericentromeric chromatin to kinetochore assembly. Biochem Cell Biol 84:619–639
115. Cremer T, Cremer M (2010) Chromosome territories. Cold Spring Harb Perspect Biol 2:a003889

116. Lanctôt C, et al (2007) Dynamic genome architecture in the nuclear space: regulation of gene expression in three dimensions. Nat Rev Genet 8:104–115
117. Joffe B, Leonhardt H, Solovei I (2010) Differentiation and large scale spatial organization of the genome. Curr Opin Genet Dev 20:562–569
118. Parada LA, McQueen PG, Misteli T (2004) Tissue-specific spatial organization of genomes. Genome Biol 5:R44
119. Sexton T, et al (2007) Gene regulation through nuclear organization. Nat Struct Mol Biol 14:1049–1055
120. Takizawa T, Meaburn KJ & Misteli T (2008) The meaning of gene positioning, Cell 135:9–13
121. Andrulis ED, et al (1998) Perinuclear localization of chromatin facilitates transcriptional silencing. Nature 394:592–595
122. Boyle S, et al (2001) The spatial organization of human chromosomes within the nuclei of normal and emerin-mutant cells. Hum Mol Genet 10:211–219
123. Guelen L, et al (2008) Domain organization of human chromosomes revealed by mapping of nuclear lamina interactions. Nature 453:948–951
124. Finlan LE, et al (2008) Recruitment to the nuclear periphery can alter expression of genes in human cells. PLoS Genet 4:e1000039
125. Reddy KL, et al (2008) Transcriptional repression mediated by repositioning of genes to the nuclear lamina. Nature 452:243–247
126. Ruault M, Dubarry M, Taddei A (2008) Re-positioning genes to the nuclear envelope in mammalian cells: impact on transcription. Trends Genet 24:574–581
127. Carter-Dawson LD, LaVail MM (1979) Rods and cones in the mouse retina. I. Structural analysis using light and electron microscopy. J Comp Neurol 188:245–262
128. Blackshaw S, et al., (2001) Comprehensive analysis of photoreceptor gene expression and the identification of candidate retinal disease genes. Cell 107:579-589
129. Helmlinger D, et al (2006) Glutamine-expanded ataxin-7 alters TFTC/STAGA recruitment and chromatin structure leading to photoreceptor dysfunction. PLoS Biol 4:e67
130. Solovei I, et al (2009) Nuclear architecture of rod photoreceptor cells adapts to vision in mammalian evolution. Cell 137:356–368
131. Kizilyaprak C, et al (2010) In vivo chromatin organization of mouse rod photoreceptors correlates with histone modifications. PLoS One 5:e11039
132. Kreysing M, et al (2010) Physical insight into light scattering by photoreceptor cell nuclei. Opt Lett 35:2639–2641
133. Shannon CE (1948) A mathematical theory of communication. Bell System Technical J 27:379–423,623–656
134. Brillouin L (1956) Science and information Theory, 2nd Edition. Academic Press, New York
135. Yockey HP (1992) Information theory and molecular biology. Cambridge University Press, Cambridge

136. Gitt W (1997) In the beginning was information. Christliche Literatur-Verbreitung; Bielefeld, Germany
137. Dembski WA (1998) The design inference. Cambridge University Press, Cambridge
138. Dembski WA (2002) No free lunch: Why specified complexity cannot be purchased without intelligence. Rowman & Littlefield; Lanham, MD
139. Meyer SC (2004) The origin of biological information and the higher taxonomic categories. Proc Biol Soc Wash 117:213–239
140. Sanford JC (2005) Genetic entropy & the mystery of the genome. Elim Publishing; Lima, NY
141. Dembski WA & Marks RJ II (2009) Conservation of information in search: Measuring the cost of success. IEEE Trans Syst Man Cybern A 5(5):1051–1061
142. Meyer SC (2009) Signature in the cell: DNA and the evidence for intelligent design. HarperCollins, New York
143. Ewert W, et al (2010) Efficient per query information extraction from a Hamming Oracle. Proceedings of the 42nd Meeting of the Southeastern Symposium on System Theory, IEEE, University of Texas at Tyler, March 7-9, 2010, pp. 290–297
144. Kimchi-Sarfaty C, et al (2007) A 'silent' polymorphism in the MDR1 gene changes subsrate specificity. Science 315:525–528
145. Newman SA, Bhat R (2007) Genes and proteins: dogmas in decline. J Biosci 32:1041–1043
146. Pezza JA, Serio TR (2007) Prion propagation: the role of protein dynamics. Prion 1:36–43.
147. Bryan PN, Orban J (2010) Proteins that switch folds. Curr Opin Struct Biol 20: 482–488
148. Perry AJ, et al (2006) Convergent evolution of receptors for protein import into mitochondria. Curr Biol 16:221–229
149. Nickson AA, Clarke J (2010) What lessons can be learned from studying the folding of homologous proteins? Methods 52:38-50
150. Atlan H, Koppel M (1990) The cellular computer DNA: program or data. Bull Math Biol 52:335-348
151. Denton MJ, Marshall CJ, Legge M (2002) The protein folds as Platonic forms: new support for the pre-Darwinian conception of evolution by natural law. J Theor Biol 219:325–342
152. Sternberg RV (2008) DNA codes and information: formal structures and relational causes. Acta Biotheor 56:205–232
153. Arvey A, et al (2012) Sequence and chromatin determinants of cell-type-specific transcription factor binding. Genome Res 22:1723–1734
154. Bánfai B, et al (2012) Long noncoding RNAs are rarely translated in two human cell lines. Genome Res 22:1646–1657

155. Boyle AP, et al (2012) Annotation of functional variation in personal genomes using RegulomeDB. Genome Res 22:1790–1797
156. Charos AE, et al (2012) A highly integrated and complex PPARGC1A transcription factor binding network in HepG2 cells. Genome Res 22:1668–1679
157. Cheng C, et al (2012) Understanding transcriptional regulation by integrative analysis of transcription factor binding data. Genome Res 22:1658–1667
158. Derrien T, et al (2012) The GENCODE v7 catalog of human long noncoding RNAs: analysis of their gene structure, evolution, and expression. Genome Res 22: 1775–1789
159. Djebali S, et al (2012) Landscape of transcription in human cells. Nature 489: 101–108
160. Dong X, et al (2012) Modeling gene expression using chromatin features in various cellular contexts. Genome Biol 13:R53
161. Dunham I, et al (2012) An integrated encyclopedia of DNA elements in the human genome. Nature 489:57–74
162. Euskirchen G, Auerbach RK, Snyder S (2012) SWI/SNF chromatin-remodeling factors: multiscale analyses and diverse functions. J Biol Chem 287:30897–30905
163. Frietze S, et al (2012) Cell type-specific binding patterns reveal that TCF7L2 can be tethered to the genome by association with GATA3. Genome Biol 13:R52
164. Gerstein MB, et al (2012) Architecture of the human regulatory network derived from ENCODE data. Nature 489:91–100
165. Hardison RC (2012) Genome-wide epigenetic data facilitate understanding of disease susceptibility association studies. J Biol Chem 287:30932–30940
166. Harrow J, et al. (2012) GENCODE: the reference human genome annotation for the ENCODE project. Genome Res 22:1760–1774
167. Howald C, et al (2012) Combining RT-PCR-seq and RNA-seq to catalog all genic elements encoded in the human genome. Genome Res 22:1698–1710
168. Kundaje A, et al (2012) Ubiquitous heterogeneity and asymmetry of the chromatin environment at regulatory elements. Genome Res 22:1735–1747
169. Ladewig E, et al (2012) Discovery of hundreds of mirtrons in mouse and human small RNA data. Genome Res 22:1634–1645
170. Lan X, Farnham PJ, Jin VX (2012) Uncovering transcription factor modules using one- and three-dimensional analyses. J Biol Chem 287:30914–30921
171. Landt SG, et al (2012) ChIP-seq guidelines and practices of the ENCODE and modENCODE consortia. Genome Res 22:1813–1831
172. Lee B-K, Iyer VR (2012) Genome-wide studies of CCCTC-binding factor (CTCF) and cohesin provide insight into chromatin structure and regulation. J Biol Chem 287:30906–30913
173. Maurano MT, et al (2012) Systematic localization of common disease-associated variation in regulatory DNA. Science 337:1190–1195

174. Natarajan A, et al (2012) Predicting cell-type-specific gene expression from regions of open chromatin. Genome Res 22:1711–1722
175. Neph S, et al (2012) An expansive human regulatory lexicon encoded in transcription factor footprints. Nature 489:83–90
176. Park E, et al (2012) RNA editing in the human ENCODE RNA-seq data. Genome Res 22:1626–1633
177. Pei B, et al (2012) The GENCODE pseudogene resource. Genome Biol 13:R51
178. Sanyal A, et al (2012) The long-range interaction landscape of gene promoters. Nature 489:109–113
179. Schaub MA, et al (2012) Linking disease associations with regulatory information in the human genome. Genome Res 22:1748–1759
180. Sindhu C, Samavarchi-Tehrani P, Meissne A (2012) Transcription factor-mediated epigenetic reprogramming. J Biol Chem 287:30922–30931
181. Spivakov M, et al (2012) Analysis of variation at transcription factor binding sites in Drosophila and humans. Genome Biol 13:R49
182. Thurman RE, et al (2012) The accessible chromatin landscape of the human genome. Nature 489:75–82
183. Tilgner H, et al (2012) Deep sequencing of subcellular RNA fractions shows splicing to be predominantly co-transcriptional in the human genome but inefficient for lncRNAs. Genome Res 22:1616–1625
184. Vernot B, et al (2012) Personal and population genomics of human regulatory variation. Genome Res 22:1689–1697
185. Wang H, et al (2012) Widespread plasticity in CTCF occupancy linked to DNA methylation. Genome Res 22:1680–1688
186. Wang J, et al (2012) Sequence features and chromatin structure around the genomic regions bound by 119 human transcription factors. Genome Res 22:1798–1812
187. Whiteld TW, et al (2012) Functional analysis of transcription factor binding sites in human promoters. Genome Biol 13:R50
188. Yip KY, et al (2012) Classification of human genomic regions based on experimentally determined binding sites of more than 100 transcription-related factors. Genome Biol 13:R48
189. Zentner GE, Scacheri PC (2012) The chromatin fingerprint of gene enhancer elements. J Biol Chem 287:30888–30896
190. Quoted in Yong E (2012) ENCODE: the rough guide to the human genome. Discover Magazine Blog. Available at http://blogs.discovermagazine.com/notrocket-science/2012/09/05/encode-the-rough-guide-to-the-human-genome/. Accessed 2013 Jan 14
191. Ecker JR (2012) Serving up a genome feast. Nature 489:52–53.
192. Pennisi E (2012) ENCODE project writes eulogy for junk DNA. Science 337: 1159–1161

Can Purifying Natural Selection Preserve Biological Information?

Paul Gibson[1]*, John R. Baumgardner[2], Wesley H. Brewer[3] and John C. Sanford[4]

[1]*Dept. Plant, Soil, and Agric. Syst., Southern Illinois University, Carbondale, IL 62901, USA.* [2]*Department of Earth and Environmental Sciences, Ludwig Maxmilians University, Theresienstrasse 41, 80333 Munich, Germany.* [3]*Fluid Physics International, Starkville, MS 39759, USA.* [4]*Dept. Hort. Sci., NYSAES, Cornell University, Geneva, NY 14456, USA. (*Corresponding author: pgibson@siu.edu)*

Abstract

Most deleterious mutations have very slight effects on total fitness, and it has become clear that below a certain fitness effect threshold, such low-impact mutations fail to respond to natural selection. The existence of such a selection threshold suggests that many low-impact deleterious mutations should accumulate continuously, resulting in relentless erosion of genetic information. In this paper, we use numerical simulation to examine this problem of selection threshold.

The objective of this research was to investigate the effect of various biological factors individually and jointly on mutation accumulation in a model human population. For this purpose, we used a recently-developed, biologically-realistic numerical simulation program, Mendel's Accountant. This program introduces new mutations into the population every generation and tracks each mutation through the processes of recombination, gamete formation, mating, and transmission to the new offspring. This method tracks which individuals survive to reproduce after selection, and records the transmission of each surviving mutation every generation. This allows a detailed mechanistic accounting of each mutation that enters and leaves the population over the course of many generations. We term this type of analysis *genetic accounting.*

Across all reasonable parameters settings, we observed that high impact mutations were selected away with very high efficiency, while very low impact mutations accumulated just as if there was no selection operating. There was always a large transitional zone, wherein mutations with intermediate fitness effects accumulated continuously, but at a lower rate than would occur in the absence of selection. To characterize the accumulation of mutations of different fitness effect we developed a new statistic, selection threshold (ST_d), which is an empirically determined value for a given population. A population's selection threshold is defined as that fitness effect wherein deleterious mutations are accumulating at exactly half the rate expected in the absence of selection. This threshold is mid-way between entirely selectable, and entirely unselectable, mutation effects.

Our investigations reveal that under a very wide range of parameter values, selection thresholds for deleterious mutations are surprisingly high. Our analyses of the selection threshold problem indicate that given even modest levels of noise affecting either the genotype-phenotype relationship or the genotypic fitness-survival-reproduction relationship, accumulation of low-impact mutations continually degrades fitness, and this degradation is far more serious than has been previously acknowledged. Simulations based on recently published values for mutation rate and effect-distribution in humans show a steady decline in fitness that is not even halted by extremely intense selection pressure (12 offspring per female, 10 selectively removed). Indeed, we find that under most realistic circumstances, the large majority of harmful mutations are essentially unaffected by natural selection and continue to accumulate unhindered. This finding has major theoretical implications and raises the question, "What mechanism can preserve the many low-impact nucleotide positions that constitute most of the information within a genome?"

Key words: deleterious mutation, genetic deterioration, mutation accumulation, near-neutral, population genetics, selection threshold, simulation

Introduction

More than forty years ago, Muller [1] concluded that there exists a class of low-impact mutations that are beyond the reach of natural selection. Kimura greatly expanded upon this theme, using mathematical modeling to study the problem [2]. Although Kimura initially described such mutations as 'neutral', Ohta [3–6] argued that such mutations should more accurately be termed 'nearly neutral', and Kimura later agreed [7, 8]. Kondrashov realized that very low impact mutations are not only inherently unselectable, but they also create a profound evolutionary paradox [9]. Later, Lynch *et al.* [10, 11] and Higgins and Lynch [12] provided evidence that accumulation of low-impact mutations plays an important role in the extinction process. Recently, Loewe [13] showed that accumulation of nearly neutral mutations is a theoretical problem even for haploid genomes as small as that of human mitochondria. His analysis suggests that accumulation of nearly-neutral mutations within the mitochondria alone could potentially lead to human extinction. Given that nearly-neutral mutations have such profound biological implications, it would seem important to understand better the primary factors that control the accumulation of low-impact deleterious mutations.

A useful way to conceptualize selection's ability to influence the accumulation of low-impact mutations is in terms of signal versus noise. 'Signal' corresponds to the level of influence a mutation has on its own transmission. 'Noise', by contrast, corresponds to various types of interference that reduce the correlation between a mutation's effect on functional fitness and its probability of transmission. When the signal is weak and the noise is sufficiently strong, the signal is obscured and selection breaks down. At that point the correlation between the mutation's effect on functional fitness and the likelihood of that mutation's transmission becomes too small for selection to affect the frequency of that mutation in the population in any significant way.

Kimura [7] was the first to attempt to quantify the threshold for selection breakdown. His calculations focused only on the influence of one source of 'noise' on the rate of mutation fixation, i.e., that of gametic sampling. Kimura found that the strength of this confounding effect on selection varies inversely with the effective population size, N_e. In small populations, a relatively small number of gametes are extracted to produce the next generation. This restricted gametic sampling results in sampling error that leads to random fluctuations in each allele's frequency within the population. These random fluctuations represent a type of noise that interferes with selection. It is well known that this type of genetic drift is strong in small populations and can override all but the strongest selection pressures. However, in larger populations the gametic sampling error is smaller, and thus the resulting random fluctuations in allele frequency are smaller. Therefore, selection for low-impact mutations can be more effective in larger populations. Restricting his analysis to this single source of noise, Kimura developed his now well-known approximation of the magnitude of the selection coefficient needed to overcome drift, expressed as $s = 1/(2N_e)$. This expression implies a direct relationship between the selection threshold and the effective population size N_e [7]. Most subsequent studies of nearly-neutral mutations and their accumulation have utilized this estimate for the point at which selection breaks down and genetic drift becomes predominant [9–13].

It is obvious, however, that there are other sources of biological noise besides gametic sampling. All of these other sources of noise should reduce the correlation between the magnitude of the effect (d_i) of a specific mutation on the functional fitness of an individual and the influence of that mutation on the individual's reproductive success. Lynch [14], for example, notes that small population size, large nucleotide distances between crossovers, and high mutation rates all synergistically reduce the efficiency of natural selection. To study some of these biological factors and to quantify how they affect the selection threshold beyond their predicted direct effect on the selection coefficient, *s*, we adopt a numerical simulation strategy using the program Mendel's Accountant (Mendel) [15, 16, http://www.MendelsAccountant.info]. This numerical approach affords us much flexibility to explore the biological complexity of the mutation-selection process, as it actually occurs in nature. Numerous other studies have explored mutation accumulation via simulation [17–19], including the consequences of a non-uniform distribution of mutational effects. We extend those explorations by including environmental variance, a range of different mutation rates, and various forms of selection (truncation, partial truncation, and standard probability selection).

The earliest reference to the idea of a selection threshold seems to be from Muller [1]. He stated, "*There comes a level of advantage, however, that is too small to be effectively seized upon by selection, its voice being lost in the noise, so to speak. This level would necessarily differ greatly under different circumstances (genetic, ecological, etc.), but this is a subject that has as yet been subject to little analysis… although deserving of it.*" Muller's recognition that there are deleterious mutations that are practically invisible to the selection process contributed to his overall concern about genetic deterioration. It also contributed to his concern about the problem of linkage-mediated deterioration in fitness ("Muller's ratchet"). The goal of this paper is to explore the biological circumstances (to which Muller alluded) that can make a large fraction of deleterious mutations immune to selection. Our results reveal that even modest degrees of either environmental variance or randomness in the selection process (probability selection) cause selection breakdown for most deleterious mutations, and this problem is compounded by high mutation rates.

Results

Conditions allowing perfect purifying selection

Several experiments were first conducted to discover the region of parameter space in which there is zero near-neutral mutation accumulation. We found that complete elimination of near-neutrals requires that all sources of noise be reduced to either extremely low levels or zero. As a general rule, this requires zero environmental variation (heritability = 1), perfect truncation selection, sufficiently high selection intensity, and sufficiently low mutation rates to maintain near-zero genetic variance. Only when these conditions were satisfied was selection effective enough to preclude accumulation of nearly neutral mutations. Under these special circumstances, low-impact mutations were eliminated just as if they were fully lethal. This was because under these conditions, selection becomes a matter of simply choosing between mutant versus non-mutant individuals. We obtained this result, for example, for the case of zero environmental variance, perfect truncation selection, a mutation rate of one mutation per individual per generation, and the default reproduction rate of six offspring per female (allowing for selection to eliminate 2/3 of all offspring, maintaining a constant population size). In this case, the Poisson distribution defining the number of new mutations assigned to each offspring yielded enough individuals with no mutations (37% on average) so that truncation selection against all mutations still allowed maintenance of the designated population size. This guaranteed elimination of all individuals with even a single mutation, regardless of how small the mutation's effect. As in all other experiments reported here, replicate experiments with different random number seeds produced no meaningful differences in outcome. Therefore for this and all following analyses, we will only report results from single representative runs.

Effects of high mutation rate and mutation-mutation interference

We next conducted a series of similar experiments, but with mutation rates of 5, 10, 20, and 40 per diploid genome per generation. For mutation rates greater than one new mutation per individual, a type of biological noise arises associated with selection interference among mutations. Results are summarized in Figure 1, which plots the mutation fitness effect versus mutation accumulation relative to the neutral expectation. While high-impact mutations had zero accumulation, extremely low-impact mutations displayed accumulation fractions approaching 1.0. The transition zone between these two extremes is characterized by an *S*-shaped curve. We define the selection threshold for deleterious mutations (ST_d) as the midpoint of this transition zone. More specifically, ST_d is the value of mutational fitness effect for which the accumulation fraction is 0.5, indicating that half as many mutations have accumulated as would be expected under complete neutrality (i.e., no selection). This can be visualized in Figures 1, 2, 3, and 4 as the intersection of the horizontal line corresponding to 0.5 on the y-axis and the curve that plots the fraction of mutational retention.

As shown in Figure 1, mutation rates greater than one per offspring resulted in accumulation of low-impact alleles. Increasing the mutation rate resulted in the accumulation of alleles with increasingly large fitness effects. In other words, higher mutation rates lead to progressively higher ST_d values. This means that increasing numbers of alleles that would otherwise have been selectable (to the left of the threshold) became unselectable (to the right of the threshold). With a mutation rate of 10, almost half of all deleterious mutations were retained, with a nearly constant accumulation rate of 4.5 mutations per individual per generation. The mean population fitness declined continuously, reflecting this accumulation of deleterious mutations, but the decline was very slow because the accumulating alleles had very small fitness effects. Figure 1 illustrates that an increased mutation rate, and consequent selection interference among alleles, led to ST_d values increasing from 6.8×10^{-9} for a mutation rate of 5; to 7.4×10^{-8} for a mutation rate of 10; to 5.2×10^{-7} for a mutation rate of 20; to 3.2×10^{-6} for a mutation rate of 40. At the highest mutation rate, 75% of the mutations were below the selection threshold, and hence were effectively unselectable.

Effects of environmental variance

We conducted a series of similar experiments, but instead of increasing mutation rate, we kept the rate at one per offspring and introduced environmental variance, quantified in terms of fitness heritability (i.e., genotypic variance/phenotypic

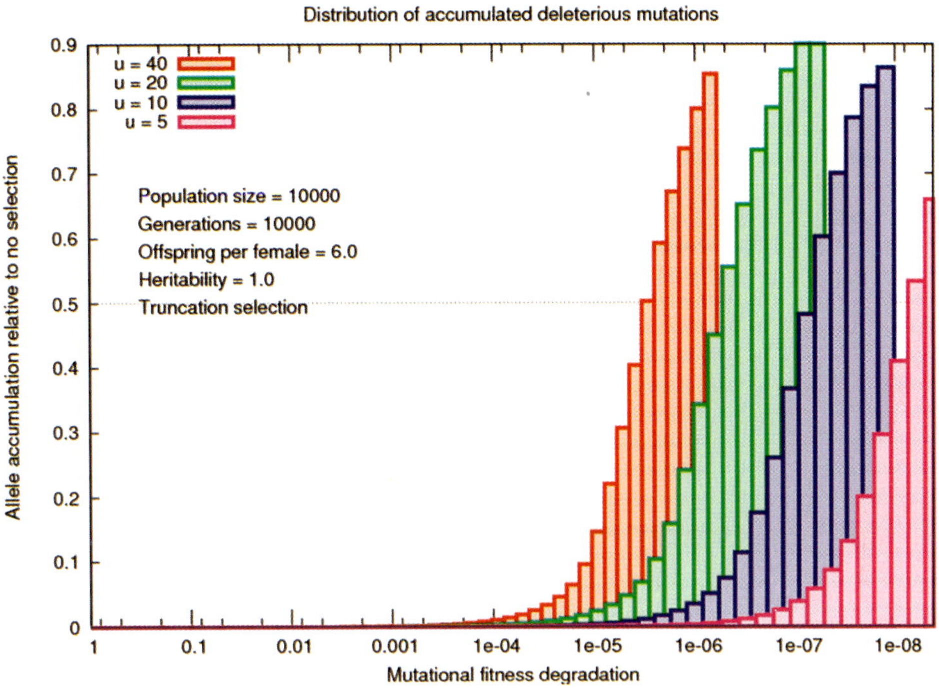

Fig. 1. Fractional retention of mutations as a function of fitness effect for various mutation rates. In these experiments, fitness heritability is 1.0 (i.e., there is no environmental noise), and truncation selection is chosen (i.e., there is no randomness in the selection process). Results for average mutation rates of 5, 10, 20, and 40 new mutations per offspring are displayed. Mutational fitness effect is shown using a log scale along the x-axis, with lethal mutations assigned the value of 1.0. Mutations of small effect are entirely unselectable, and have a fractional retention of 100% (y-axis value of 1.0), while mutations of large effect are eliminated entirely by selection and have a fractional retention of zero. The selection threshold (ST_d) is defined as that fitness effect class which has a fractional retention value of 0.5 (indicated by the dotted line). Note that selection breakdown becomes progressively worse as mutation rate increases. For a mutation rate of 1 per offspring on average, all mutations are selectively eliminated, so mutation accumulation is 0. With an average of 1 new mutation distributed in a Poisson manner and with four of every six offspring selectively eliminated, truncation selection is able to exclude every offspring that has one or more mutations. Because of the very large number of mutations accumulated in these experiments, given computer memory limitations, mutations with extremely small effects were not all tracked in detail, although their effects were fully accounted for. For this reason, the right edge of the distributions end at different fitness effect values.

variance ratio). To illustrate our findings we present three cases with fitness heritabilities of 0.4, 0.04, and 0.004 (Figure 2).

As can be observed in Figure 2, modest levels of environmental variance led to substantial ST_d levels. Heritability of fitness in nature has often been found to be very low, and such a fitness heritability value ($h^2 = 0.004$) yielded a high ST_d (2.6×10^{-5} after 10,000 generations). Given this level of environmental variance,

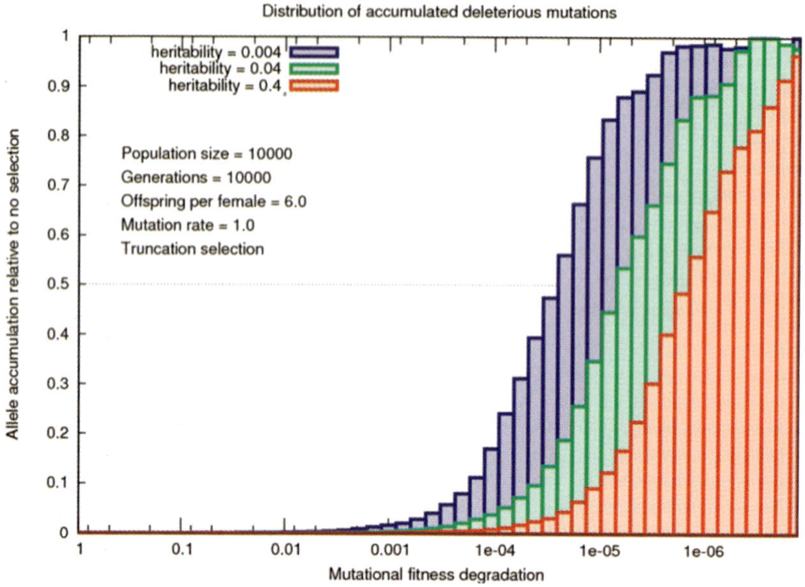

Fig. 2. Fractional retention of mutations as a function of fitness effect for various fitness heritabilities. In these experiments, the mutation rate is 1.0 per offspring on average, and truncation selection was applied (i.e., there was no randomness in the selection process). Results for fitness heritabilities of 0.4, 0.04, and 0.004 are displayed. Note that selection breakdown became progressively worse as heritability decreased (i.e., environmental variance increased). The selection threshold value for the lowest heritability value is 2.6×10^{-5}.

the average mutation count per individual increased at nearly a constant rate of 0.86 mutations per individual per generation. This means that 86% of all the newly arising mutations were below the selection threshold and were essentially unselectable, in spite of very intense selection pressure.

Effects of varying degrees of randomness within the selection process

In another series of experiments we examined the manner in which some randomness in the selection process itself (e.g., partial or complete probability selection) influences ST_d (Figure 3).

Figure 3 summarizes two experiments in which the only source of noise was a specified degree of randomness inherent to the selection process. These experiments were similar to the case that displayed zero mutation accumulation (that is, a mutation rate of one per offspring and zero environmental variance). However, instead of truncation selection, we applied two other forms of selection, i.e., probability selection and what we refer to as partial truncation (quasi-truncation)

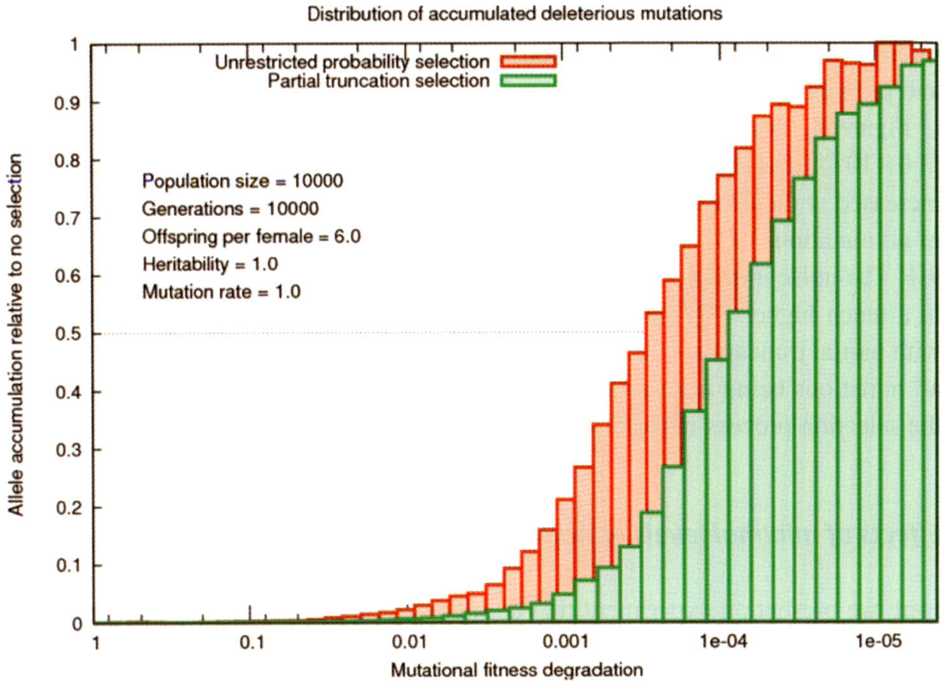

Fig. 3. Fractional retention of mutations as a function of fitness effect for various types of selection. In these experiments, the mutation rate was 1.0 per offspring and the fitness heritability was 1.0. Results are shown for three selection modes: truncation, partial truncation, and probability selection. Under truncation selection with this low mutation rate, all mutations are eliminated so that the fraction of mutations retained is zero for all fitness effect values (all bars in histogram have zero value). This occurs because, with new mutations distributed in a Poisson manner and with four of every six offspring selectively eliminated, truncation selection is able to exclude every offspring with one or more mutations. Note that selection breakdown becomes progressively worse as the level of randomness in the selection process increases. The transition from full truncation selection to partial truncation to probability selection results in increasing selection threshold (ST) values. The ST_d for probability selection is 3×10^{-4}.

selection. Under probability selection, the probability of an individual's reproduction is directly proportional to that individual's phenotypic fitness, such that even individuals with relatively low phenotypic fitness still have some likelihood of reproducing. It is generally understood that probability selection corresponds most closely to what occurs under natural circumstances. Probability selection contrasts strongly with truncation selection wherein there is no element of randomness. Under truncation selection, all individuals above a specific phenotypic value have a 100% probability of reproduction, while all individuals below that value have zero probability of reproduction. Such full truncation selection is almost never realized, even under the highly controlled conditions of artificial plant or animal

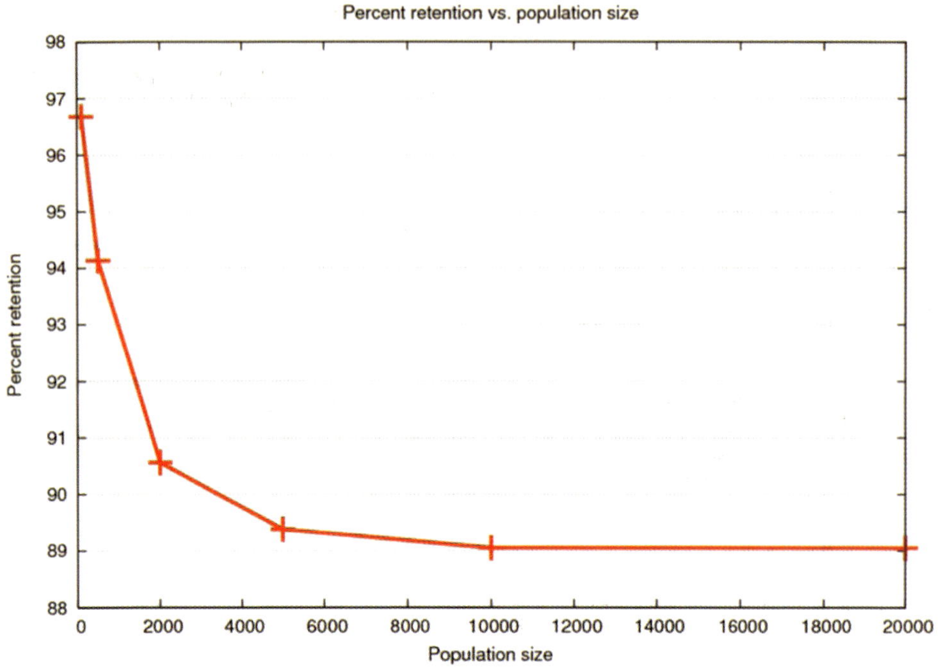

Fig. 9. Percent retention of deleterious mutations as a function of population size within 10,000 generations. The parameters for these experiments were the same as for figure 8, but with population sizes ranging from 100 to 20,000. Within the time frame of 10,000 generations, increasing population size beyond 5,000 resulted in no meaningful improvement in selection efficacy.

effects in Mendel's Accountant was adjusted to match Lynch's estimate of 27% of effects stronger than 10^{-2}, with the minimum fitness effect being adjusted upward to 10^{-6} by setting the genome size at 10^6, thus excluding from consideration the several other mutations per newborn, the effects of which might be less than 10^{-6} per mutation. The resulting distribution of fitness effects had a much higher mean fitness effect than the Mendel default distribution, and is a reasonable approximation of Lynch's distribution (ignoring all very low-impact mutations).

We ran two Mendel experiments using this new fitness effect distribution, employing a mutation rate of just 3 new deleterious mutations per newborn. The first experiment employed both partial truncation selection and a very high fitness heritability (0.4), as with the previous experiments. The second experiment used all the same parameters, except that it employed probability selection — which is much more realistic. Figure 10 shows the fitness history of these experiments. The result of using the Lynch fitness effect distribution was much faster degeneration than when using Mendel's default settings. The initial rate of fitness decline was approximately 5% per generation (data not shown), agreeing well with the fitness decline surmised by Lynch. However, over deeper time, as genetic variation for fitness built up, selection

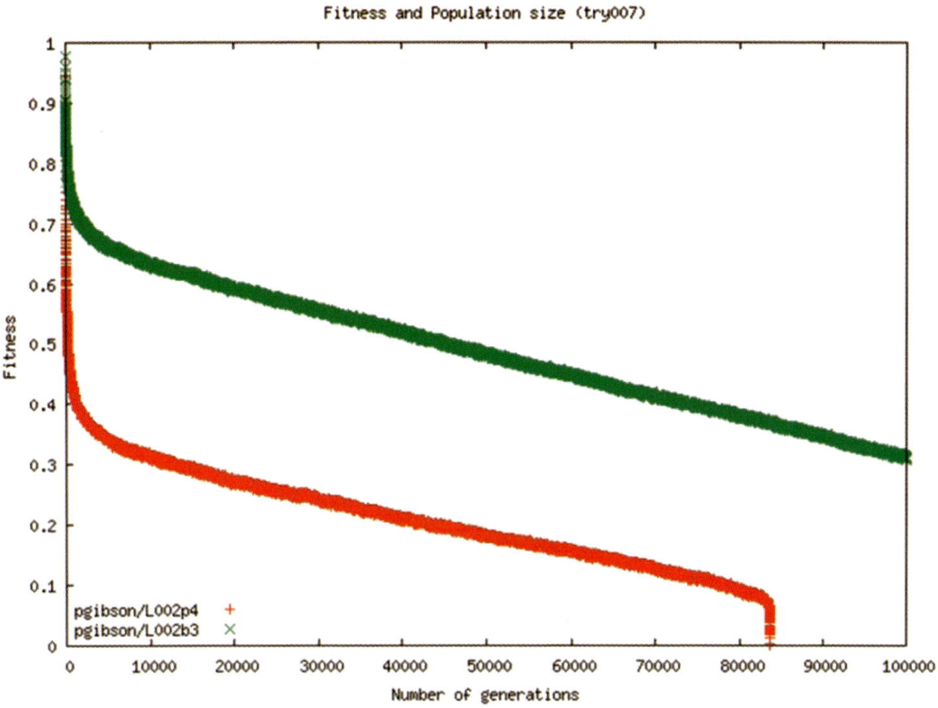

Fig. 10. Fitness history using the latest estimate of human mutation rate and fitness effect distribution, comparing partial truncation selection versus probability selection. The upper line (green) resulted from a run using partial truncation selection. The lower line (red) resulted from an identical run, but employing probability selection. In both cases, a fitness effect distribution was employed based upon Lynch [24], strongly skewed toward higher impact mutations. In both cases, the mutation rate was just 3, again in agreement with Lynch [24]. Population size was 10,000, fitness heritability was 0.4.

could act upon a wider range of variation and thus became more effective, slowing the decline, but not stopping it. The rate of fitness decline over time became extremely linear, with partial truncation selection resulting in a mean fitness of 0.31 after 100,000 generations. Using probability selection, the rate of fitness decline also became extremely linear over time, resulting in extinction at generation 83,647.

Discussion

General Implications

This study shows that, under conditions relevant to many mammalian populations, the large majority of deleterious mutations should escape purifying selection.

Given a specific population and specific circumstances, there must be a certain point where selection against low-impact mutations breaks down. Numerical simulation allows us to empirically determine this selection threshold, ST_d, for any particular set of conditions. We expand on previous work by showing that the value of ST_d is not a simple function of population size, but is affected by numerous variables. To our knowledge, the methodology used here (i.e., numerical simulation based on genetic accounting) provides the most biologically relevant treatment of the problem of germline mutation accumulation to date. The theoretical and practical implications of these results should be of wide interest.

For a typical mammalian model population (e.g. 10,000 individuals, genome size of 3 billion), our estimate for the lower limit of ST_d is in the range of 10^{-4} to 10^{-5}. Thus even with minimal levels of biological noise interfering with the phenotypic expression of the genotype, those deleterious mutations which reduce fitness by less than 10^{-4} to 10^{-5} will largely escape purifying selection and will accumulate linearly. We show that three important sources of noise which substantially increase the value of the selection threshold in large populations are: (1) selection interference between mutations; (2) environmental variance; and (3) any significant degree of probability selection (in contrast to truncation selection, which never occurs in nature). Our experiments show that depending on these variables, ST_d values for mammalian species may be as high as 10^{-3} or higher. Given Mendel's default fitness effect distribution, ST_d values in the range of 10^{-3} to 10^{-5} results in 82–97% of all deleterious mutations becoming effectively unselectable.

Our simulations indicate that the on-going accumulation of low-impact mutations results in continuous fitness loss. Consistent with the findings of others, our analyses reveal that the greatest contributor to this fitness loss is not the entirely unselectable mutations (having negligible fitness effects even in large numbers), but rather the accumulation of mutations with effects near the selection threshold. We observe that mutations in this zone accumulate more slowly than if there was no selection, yet still accumulate continuously and in large numbers. This transition zone between mutations that are entirely selectable and entirely un-selectable is often at least two orders of magnitude wide and typically encompasses fitness effects on the order of 0.001 to 0.00001. Accumulating alleles within this transition zone are primarily responsible for the reduction in fitness.

In view of the expected accumulation of low-impact mutations, it is important to estimate accurately the lower limit of effects that respond effectively to selection. Over the past several decades it has been tacitly assumed that population size is the primary determinant of this lower limit. This important assumption, explicit in Kimura's famous formula, $s = 1/(2Ne)$ [7], has been used by most investigators for defining the threshold for selection breakdown. However, our extensive investigations have indicated that mutation rate, environmental variance, selection mode, and time are all important variables that affect ST_d in addition to population size. In populations of 1000 or more, these other variables are often more important than population size. We consistently observed that, regardless of the mode of selection, increasing population size beyond 1,000 provided only modest gains in selection efficiency in the time frame of thousands of generations. The advantages of population sizes beyond 10,000 were only realized after tens of thousands of generations, and even that depended on the very questionable assumption that all selection coefficients could remain constant. It is clear that selection breakdown is not a simple function of population size. In other words, Kimura's famous formula represents an over-simplification of biological reality and the failure to consider other sources of noise can therefore lead to serious error and serious under-estimation of the selection threshold problem. This is especially true when mutation rates are above 1 per individual per generation (resulting in substantial selection interference between mutations), or when the effect of truncation or quasi-truncation selection is considered instead of simple probability selection. Although future studies should explore the behavior of larger populations in much deeper time (as greater computational power becomes available), the present results strongly suggest that population sizes larger than 10,000 will have a minimal effect on the effect on ST_d values.

The inability of natural selection to effectively remove large numbers of low-impact mutations has major implications regarding the long-term maintenance of the genetic integrity of populations. A substantial but unknown fraction of the many mutations in each eukaryotic individual must be deleterious. Yet this study indicates that most such deleterious mutations are too subtle to respond to natural selection. How can this be? Unless some entirely unknown mechanism is operating, it appears that net genetic deterioration is an inevitable aspect of the mutation/selection process, given known mutation rates and fitness effects. It is widely supposed that within any viable population, natural selection must be able to act effectively on deleterious mutations at millions of loci simultaneously, even though most such mutations have vanishingly small fitness effects and their selection is compromised by multiple levels of interfering biological noise. The results of the current study involving biologically realistic numerical simulation clearly show that selection simply cannot do this. This simple reality seems to be widely understood by leading population geneticists (e.g., see references [1–13]), yet it appears to be generally regarded as a matter of small significance judging by the lack of much serious investigation into factors influencing mutation accumulation. However, if natural selection cannot reasonably be expected to halt degeneration of genomic information, then there must be a profound problem with the present formulation of neo-Darwinian theory. We suggest this is a matter of great significance and should interest all serious scholars.

Robustness of Findings

The primary findings of this study are that the selection threshold problem is real and that it is more serious than generally recognized. These findings are very robust. Our basic conclusions do not depend on a narrow range of parameter settings; rather the same picture emerges under all reasonable biological settings, indicating that the basic phenomenon is fundamental. Our most realistic simulations (see Figures 7 and 10) still employed extremely conservative parameter settings, based upon the premise that most mutations are entirely neutral, the premise of partial truncation selection, and the premise of a very high fitness heritability. We do not believe any of these assumptions are reasonable--they were applied only to define the lower range of the deleterious selection threshold for a model human population. Simulations with what we consider to be more realistic parameter settings have indicated an even more serious erosion of genetic information than is presented here.

We suggest that, unlike many phenomena in the realm of physics, the biology of population dynamics is too complex to be reliably reduced to a small set of equations. The primary deficiency we observe in prior mutation accumulation studies is the extreme simplification that has been required both in mathematical formulations and in numerical simulations. Common simplifying restrictions include assuming that all mutation effects are equal and that environmental variance is zero; usually also assuming perfect probability selection or perfect truncation selection. These simplifications may be why previous analytical models have not fully illuminated the phenomenon of mutation accumulation. Such extreme simplification is no longer required. Today's rapidly expanding computational resources and much more sophisticated numerical simulations provide the capacity for comprehensive numerical simulations that can address population genetic systems in their entirety, simultaneously considering all the major variables that affect mutation accumulation.

Mendel's Accountant was programmed to be a comprehensive numerical simulation, reflecting biological reality as closely as possible for all the primary variables known to influence selection effectiveness [14, 15]. Mendel empirically and mechanistically tracks the basic biological processes of mutation, meiosis, crossover, gamete formation, mating, zygote formation, and selection. During the course of thousands of generations, millions of individuals are simulated, and hundreds of millions of mutations are tracked individually and continuously — an approach we call *genetic accounting*. This approach allows us to observe empirically how different biological factors interact as they influence selection efficiency, requiring far fewer prior assumptions and far less abstraction than the conventional algebraic analysis. We have repeatedly seen that, given parameter settings that correspond to the standard simplifying assumptions, Mendel supports the expectations of classic population genetic theory. However, in simulations that more realistically reflect the complexity of living populations (i.e., multiple sources of noise), Mendel's Accountant illuminates some fundamental problems in standard theory that were previously clouded by unrealistic simplifications. Mendel's Accountant thus marks a significant step forward in our ability to understand the problem of mutation accumulation, building upon the foundational work of Kimura and Ohta.

We have found these results to be highly reproducible. Replicated runs employing alternate random number seeds produce essentially identical results, creating only trivial variations. Other researchers can replicate the experiments reported here by downloading the Mendel's Accountant program along with its user manual at www.mendelsaccountant.info and by using the parameter settings listed in Appendix 1 for those parameters not presented in the specific experiments above.

Readers may ask whether we explored enough parameter space to enable us to reach the overall conclusions we claim. While the results of any given numerical experiment will, of course, depend on the specific parameter choices of the investigator, yet for each parameter, we included values that encompassed a range that was wider than seemed biologically reasonable, and explored an extensive number of combinations of the various parameters. These investigations revealed that a high selection threshold and continuous, nearly linear mutation accumulation are universal across all reasonable portions of parameter space. The results of these investigations cannot be summarized in any single paper, although our previous publications summarize many of our results [15, 16]. These extensive investigations have indicated that mutation rate, environmental variance, selection mode, and time are important variables that affect ST_d — in addition to population size. In populations of 1000 or more these factors are often more important than population size. For this reason we focused this paper on those specific variables, exploring the full range of their potential effects. In so doing we consistently find that the majority of deleterious mutations are not selectable, except within small and extremely unrealistic slivers of parameter space (e.g., the combination of less than 1 mutation per individual, no environmental variance, and full truncation selection). In this light, our conclusion that most deleterious mutations are beyond the reach of natural selection appears to be robust.

Potential Effects of Other Factors

Some will question the Mendel default settings for fitness effect distribution. We have tested other distributions and have not found them to produce fundamentally different results. In particular, in this paper we used Mendel to examine the latest incorporates all these factors, and suggests that the threshold for selection breakdown should be very substantial for most eukaryotic species. As stated by Keightley and Eyre-Walker "How humans and related species evade the effects of mutation load on an evolutionary time scale is also an open question" [32]. It is unclear what factors could realistically stop the decline of fitness due to mutation accumulation, although studies of the effects of bottlenecks, sub-populations, and other possible factors are underway using Mendel's Accountant. This issue deserves much more serious investigation, and Mendel's Accountant provides a biologically realistic simulation approach for such investigations.

Materials and Methods

We have applied Mendel's Accountant to simulate biological reality as closely as possible. Mendel introduces new mutations into the population every generation and tracks each mutation through the processes of recombination, gamete formation, mating, and transmission to the new offspring. This method tracks which individuals survive to reproduce after selection and records the transmission of each surviving mutation every generation. This allows a detailed mechanistic accounting of each mutation that enters and leaves the population over the course of many generations. We term this type of analysis *genetic accounting*, as reflected in the name of the program, Mendel's Accountant [15, 16]. Its inner workings are described in great detail elsewhere [15]. It meticulously records and tracks huge numbers of discrete genetic events over time. This discrete approach contrasts with the traditional approach that has been used by population geneticists for the past nine decades who have sought to represent the processes solely in terms of analytical equations and then to solve these equations. Like any accounting program, its primary limitations are the appropriateness of the input data, in this case a set of parameters that characterizes the particular biological circumstance the user wants to investigate, and the computer processing speed and memory.

Although Mendel is designed with the ability to model a broad spectrum of haploid and diploid organisms, for the sake of simplicity we have limited our consideration in this paper to sexual diploid organisms with large genomes. We use parameters appropriate for human populations because more is generally known about the relevant values in humans than in other complex eukaryotes. We start with a genetically-uniform population, approximating the relative genetic uniformity that follows a significant population bottleneck, and we initially assign each individual a fitness of 1. Across the experiments reported here, we keep all input parameters constant, except for the following: (1) mutation rate; (2) environmental variance; (3) selection mode; (4) population size; (5) number of linkage blocks; and (6) number of generations.

Mendel's calculations use a mutation's *effect on functional fitness (fitness effect)*, rather than its *selection coefficient,* in order to disentangle the genetic impact of a mutation on biological function from the selection process itself. In much of the population genetic literature, the selection coefficient and the influence of a given mutation on genetic fitness (fitness effect) have been equated by definition, which is true only when probability selection is combined with the multiplicative model of mutational effects and no other confounding factors occur. However, with other forms of selection and with the inclusion of other factors, a complex relationship emerges between a mutation's impact on functional fitness, its predicted selection coefficient, and its actual selectability [33, 34]. This actual selectability determines the change in allele frequencies, which by definition corresponds to the actual selection coefficient. *Functional fitness* is a concept integrating every element that influences survival and reproduction. We believe that the term functional fitness is both easily understood and conceptually useful. Our investigations show that numerous factors confound the correlation between a mutation's effect on functional fitness and its actual selectability.

Mendel outputs a new statistic we term *deleterious selection threshold* (ST_d), which marks the center of the transition zone in fitness effect between mostly selectable and mostly unselectable deleterious mutations. ST_d can be defined as the mutational fitness effect value at which the number of mutant alleles in the population is 50% of the number expected if there were no selection. The computed ST_d value lies at the mid-point of the transition zone separating large-effect, selectable mutations (that display nearly zero accumulation) and small-effect unselectable mutations (that display nearly 100% accumulation). This statistic provides, at any desired generation, a simple empirical basis for comparing selection effectiveness among cases involving different biological parameters. In this paper we restrict our discussion to only a few of the factors that influence this threshold, namely, mutation rate, environmental variation, selection scheme, population size, and degree of linkage.

The mutation rates we employ are based upon an estimate of approximately 100 new human mutations per person per generation [20, 21]. We adjust this estimate based on the fraction of the human genome assumed to be functional. We consider a minimal estimate of the functional genome to be 1% (yielding a functional mutation rate of 1), and a very conservative estimate to be 5% (yielding a functional mutation rate of 5). In light of increasing evidence of extensive genomic functionality [26, 27], we also examine functional mutation rates of 10, 20, or 40 new mutations per individual per generation, corresponding to a 10%, 20%, and 40% functional genome, respectively. By discounting the mutation rate based upon the circumstances, how large might such a selection threshold be? What are the biological implications of such a threshold?

Kimura [7] attempted to quantify the threshold for selection breakdown. His calculations focused on deleterious mutations and considered the influence of only one source of biological 'noise' on the rate of mutation fixation, that of gametic sampling. It is obvious, however, that there are other sources of biological noise besides gametic sampling. Except under strict probability selection (for which transmission of a gamete to the next generation is in strict proportion to the relative fitness of the parent), each of these other sources of noise should influence the selection threshold. Lynch [27], for example, notes that small population size, large nucleotide numbers between crossovers, and high mutation levels all synergistically reduce the efficiency of natural selection. To study some of these biological factors and to quantify how they affect the selection threshold, we have implemented a numerical simulation strategy using a program named Mendel's Accountant [28, 29]. Mendel's Accountant (Mendel) is freely available at http://www.MendelsAccountant.info. This numerical approach enables us to explore the biological complexity of the mutation-selection process as it actually occurs in nature in a way not before possible.

As early as 1964, Muller called for more research aimed at better understanding the selection threshold problem [1]. He stated, "*There comes a level of advantage, however, that is too small to be effectively seized upon by selection, its voice being lost in the noise, so to speak. This level would necessarily differ greatly under different circumstances (genetic, ecological, etc.), but this is a subject that has as yet been subject to little analysis...although deserving of it.*" The companion paper [14] does the very analysis which Muller felt was needed for deleterious mutations. The goal of this second paper is to describe the parallel analysis relative to the factors that affect the selectability of beneficial mutations.

Results

Conditions allowing optimal selection for beneficial mutations

To better understand the selection threshold phenomenon, we employed the same methodology described in our companion paper [14], conducting numerical simulation experiments using the genetic accounting program called "Mendel's Accountant". The details of how Mendel's Accountant works and how we conducted our experiments are given in the methods section at the end of this paper.

We first conducted experiments to see if there were any parameter settings that allowed selection to amplify beneficial allele frequencies across the full range of mutational fitness effects. We found that even under idealized selection conditions and zero biological noise, perfect selection for low-impact beneficial mutations never occurs. In this regard, beneficial mutations have a distinctly worse selection threshold problem than do deleterious mutations, because given the same biological parameters that allow all deleterious mutations to be selected away, a large fraction of beneficial mutations remain immune to selective amplification. Even with high selection intensity, minimal selection interference, zero environmental variation, and perfect truncation selection, we observe a significant ST_b, as seen in Figure 1.

Figure 1 displays the rate of accumulation of beneficial mutations as a function of mutational fitness effect, relative to the case of zero selection. Mutational fitness

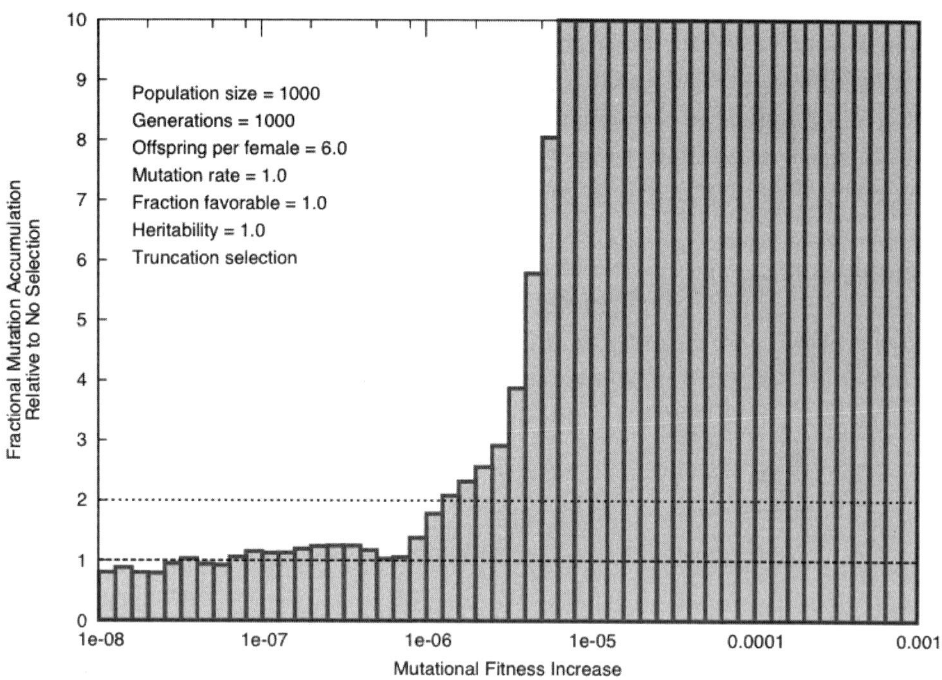

Fig. 1. Accumulation of beneficial mutations as affected by degree of benefit — optimal selection case. This experiment employed extremely unrealistic parameters chosen for maximal selection efficiency (low mutation rate, no deleterious mutations, 67% of all progeny were selected away every generation using truncation selection, with zero environmental variance). Beneficial mutation effects on fitness ranged from 3×10^{-8} to 1.0×10^{-3} (x axis). The height of the bins (y axis) reflects the relative rate of accumulation, compared to that expected when there is no selection. Bins at or near 1.0 are not responding to selection (see lower dotted line). Bins at or near 2.0 (see upper dotted line) are accumulating twice as fast as expected when there is no selection (we define this as the beneficial selection threshold — ST_b). Bins above 2.0 can be seen to be accumulating at increasingly rapid rates. Mutational effects falling in the first two orders of magnitude of mutational effect failed to respond to selection. Note that the vertical scale is clipped at a value of 10.

effect, shown on the x-axis using a logarithmic scale, ranges from a minimum non-neutral mutational value up to a maximal fitness effect. We define the minimal non-neutral mutation value as the reciprocal of the functional genome size (in this case we are considering a human population, and are assuming only 10% of the genome is functional). Each bin represents a fitness effect interval, and the height of the bin reflects the accumulation ratio of that class of mutations relative to the case of no selection. A height of 1.0, therefore, corresponds to the level of accumulation that occurs when selection is entirely ineffective (i.e., a mutation's frequency is influenced only by genetic drift). We define the beneficial selection threshold ST_b as the fitness effect value for which the distribution has the value 2.0. (i.e. the first fitness effect interval which displays twice the accumulation ratio expected in the absence of selection). This is in contrast to the deleterious selection threshold, ST_d, which is defined as the fitness effect where mutation accumulation is half of what is expected in the absence of selection. The beneficial selection threshold value can be seen visually in Figure 1 as the intersection point of the upper dotted line with the mutation distribution (at 1.34×10^{-6}). To the right of this selection threshold value, the heights of all bins increase rapidly because selection is highly effective in amplifying beneficial mutation frequency in this region.

Figure 1 reveals that, even under these idealized selection conditions, there is a fitness effect interval spanning more than two orders of magnitude, in which selection was exerting no meaningful influence on mutational frequency. This "zone of no selection" included all mutations from the smallest effect (3×10^{-8}), up to a value of just over 10^{-6} ($ST_b = 1.34\times10^{-6}$). This basic result was highly reproducible across multiple independent replicates that employed different random number seeds (data not shown). This method of representing the accumulating mutations is very useful, yet fails to convey the actual number of mutations in each bin, because the bin height represents merely a ratio of the actual mutation count versus the mutation count expected in the case of zero selection. It is important to realize that the mutation distribution is approximately exponential, so that the bins on the far left (i.e., low-impact mutations) contain the vast majority of beneficial mutations, while the bins on the right (i.e., high-impact mutations), even when filled, represent very few mutations. Even in this idealized selection experiment, given this mutation effect distribution, we actually observed that 92.7% of all beneficial mutations lay below the selection threshold. There will be occasional high-impact beneficial mutations that arise beyond the range of mutation effects of this experiment (above .001), but they will be so rare as to have very little effect on the fraction of mutations which are not selectable. As we will see, higher-impact beneficial mutations actually make the selection threshold problem worse, and need to be considered separately.

Effect of environmental variance

In the preceding experiment, parameters were chosen to maximize selection efficiency without any regard for biological realism. Two of the most unrealistic aspects of that experiment were the use of truncation selection and the assumption of zero environmental variation. To explore the influence of environmental variation, we conducted a series of experiments using identical parameters, except that we increased the level of environmental variance, quantified in terms of fitness heritability (the ratio of genotypic variance to total phenotypic variance). Figure 2 shows three cases, with fitness heritabilities (h^2) of 0.4, 0.04, and 0.004. Resulting ST_b values were 1.69×10^{-6}, 6.29×10^{-6}, and 1.4×10^{-5}, respectively. As can be observed, higher levels of environmental variance led to higher ST_b levels and a larger no-selection zone. The lowest fitness heritability value we used ($h^2 = 0.004$) is from Kimura [8], and is in keeping with the enormous impact environmental variance has on total phenotypic fitness under natural conditions. That particular heritability value yielded an ST_b approximately one order of magnitude higher

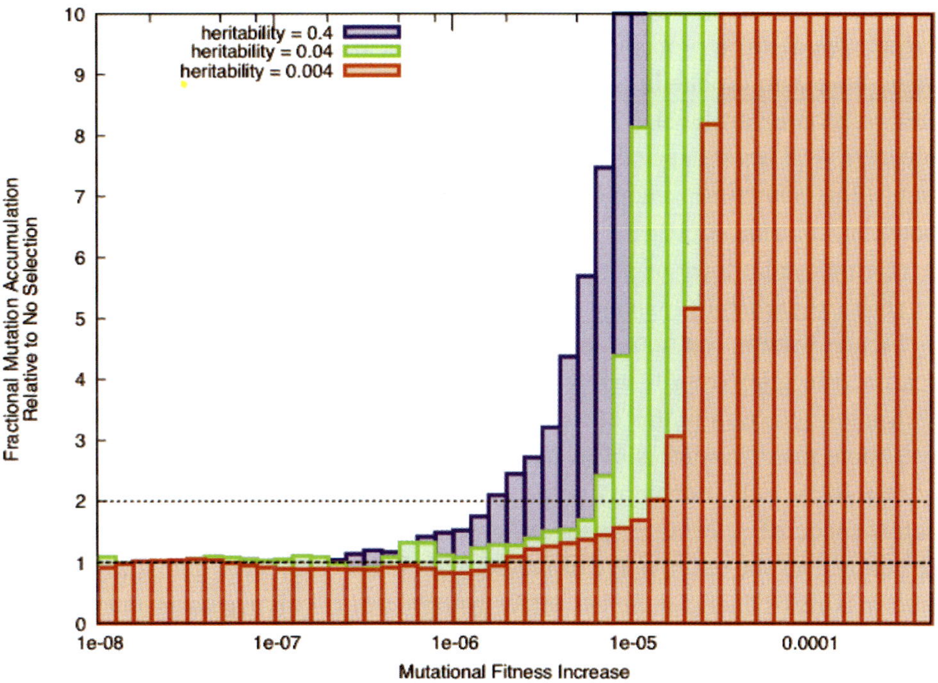

Fig. 2. Accumulation of beneficial mutations as affected by degree of benefit — introducing increasingly realistic levels of environmental variance. This figure combines the results of three experiments which employed the same unrealistic parameters as Figure 1, but simply introduced varying degrees of environmental variance (as reflected by heritability values less than 1.0). Heritability values (h^2) are shown in the figure. As can be seen, adding realistic levels of environmental variance increased the ST_b value by an order of magnitude.

than the zero environmental variance case (Figure 1), and we observed that in that instance 98.8% of the beneficial mutations fell below the selection threshold.

Introduction of probability into the selection process

In another series of experiments, we examined how more realistic modes of selection impact the beneficial selection threshold. Figure 3 contrasts our first experiment which employed truncation selection to more realistic cases employing partial truncation and probability selection. Figure 3 compares the results from the case shown in Figure 1 (red) with identical runs, but with partial truncation (green) or probability selection (blue).

It is well known that probability selection corresponds most closely to what occurs in nature. Under probability selection, the probability of an individual's reproduction is directly proportional to that individual's phenotypic fitness. Under this type of selection, even individuals with relatively low phenotypic fitness still have some likelihood of reproducing. Probability selection contrasts strongly with truncation selection, for which all individuals above a specific phenotypic fitness value have a 100% probability of reproduction, while all individuals below that value have zero probability of reproduction. Full truncation selection is an idealized version of artificial (conscious) selection, as employed by plant or animal breeders — it never happens in nature. The selection method we refer to as partial truncation (sometimes also referred to as "broken-line" selection) is intermediate between full truncation selection and probability selection. In this experiment we have employed a form of partial truncation representing an exact 50/50 blending of classical probability selection and full truncation selection.

Figure 3 shows that introducing even a modest degree of probability selection (partial truncation) results in markedly higher ST_b values. The ST_b value for partial truncation selection in this otherwise idealized selection experiment (2.54×10^{-4}) was more than two orders of magnitude larger than for pure truncation selection (1.68×10^{-6}). Full probability selection, which is commonly recognized as the actual mode of selection happening in nature, led to a complete breakdown of selection over the entire range of mutational effects considered in this experiment (the maximal beneficial fitness effect being 0.001). This indicates that the ST_b must have been greater than 0.001. We have consistently observed that the noise associated with the random aspects of probability selection leads to a greater increase in selection thresholds than any other source of noise we have examined. The only exception to this is in the case of extremely beneficial mutations, as will be described below. It is clear that even moderate levels of randomness in the selection process (i.e., a limited degree of "survival of the luckiest"), causes the vast majority of beneficial mutations to become un-selectable.

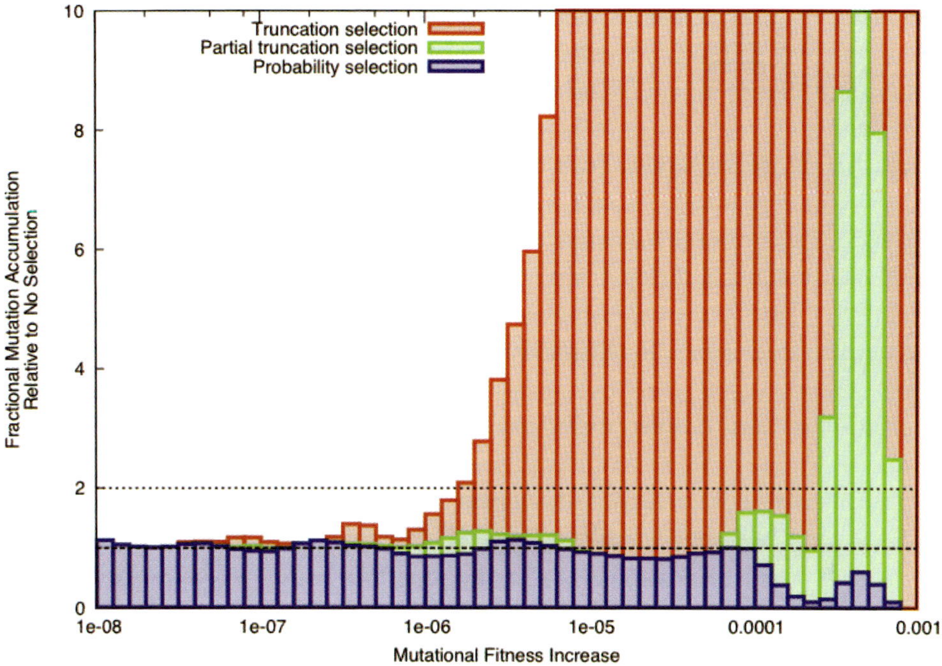

Fig. 3. Accumulation of beneficial mutations as affected by degree of benefit, employing three different modes of selection. Parameters are the same as in Figure 1, except that increasingly realistic forms of selection are introduced. Red: full truncation selection. Green: partial truncation selection (0.5). Blue: probability selection. As can be seen, introduction of probabilistic selection increased ST_b by roughly three orders of magnitude. The blue and green distributions become sparse on the right side of the figure because given an exponential distribution of mutational effects, alleles in this range were very rare apart from selective amplification.

Effect of high mutation rate and consequent selection interference among beneficial mutations

We next conducted a series of experiments, still using truncation selection and zero environmental variance, but with higher beneficial mutation rates, ranging from 5 to 40. As mutations accumulate, there arises a type of biological noise associated with *selection interference* among the mutations. Figure 4 summarized a series of experiments that reveal that increasing the rate of beneficial mutations lead to higher selection thresholds. This means that as mutation rate increases, more and more of the alleles that otherwise would be selectable escape selection. Increased mutation rate and the consequent selection interference among alleles resulted in ST_b values increasing from 1.68×10^{-6} for a mutation rate of 5; up to 5.84×10^{-6} for a mutation rate of 10; up to 1.00×10^{-5} for a mutation rate of 20; up to 1.46×10^{-5}

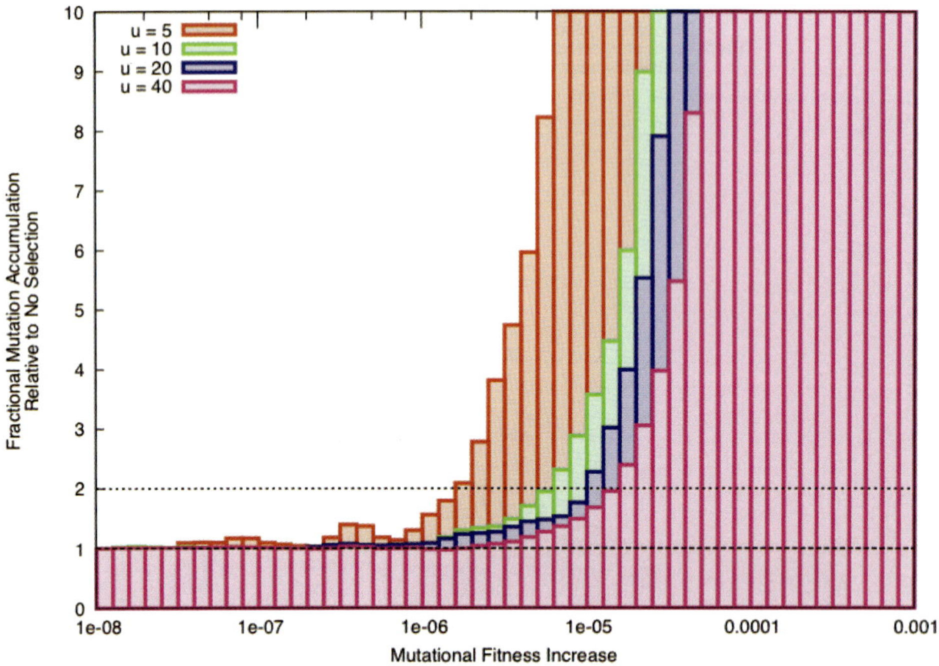

Fig. 4. Accumulation of beneficial mutations as affected by degree of benefit, employing four different mutation rates. Parameters are the same as in Figure 1, except that increasingly higher mutation rates (u) are introduced. As can be seen, higher mutation rates cause substantially higher ST_b values, due to selection interference.

for a mutation rate of 40. This last ST_b value for a mutation rate of 40 indicates that 98.8% of the beneficial mutations were below the selection threshold.

Effect of extremely beneficial mutations

Until this point, we have employed a ceiling value of 0.001 for beneficial mutational fitness effects. The rationale for this choice is given in the discussion section and was employed because very high-impact mutations need to be handled separately. We therefore conducted experiments with higher maximal fitness effect values, up to 1.0. When homozygous, a single beneficial mutation with a fitness effect of 1.0 will double the fitness of any individual, relative to the initial fitness value. We find that the inclusion of mutations with fitness values of 0.1 or greater have such a profound effect on the behavior of the whole population that we refer to them as "extremely beneficial" mutations. As can be seen in Figure 5, when we repeated the experiment illustrated in Figure 1, but merely extended the upper range of beneficial mutational effects up to 1.0, the result was a very dramatic

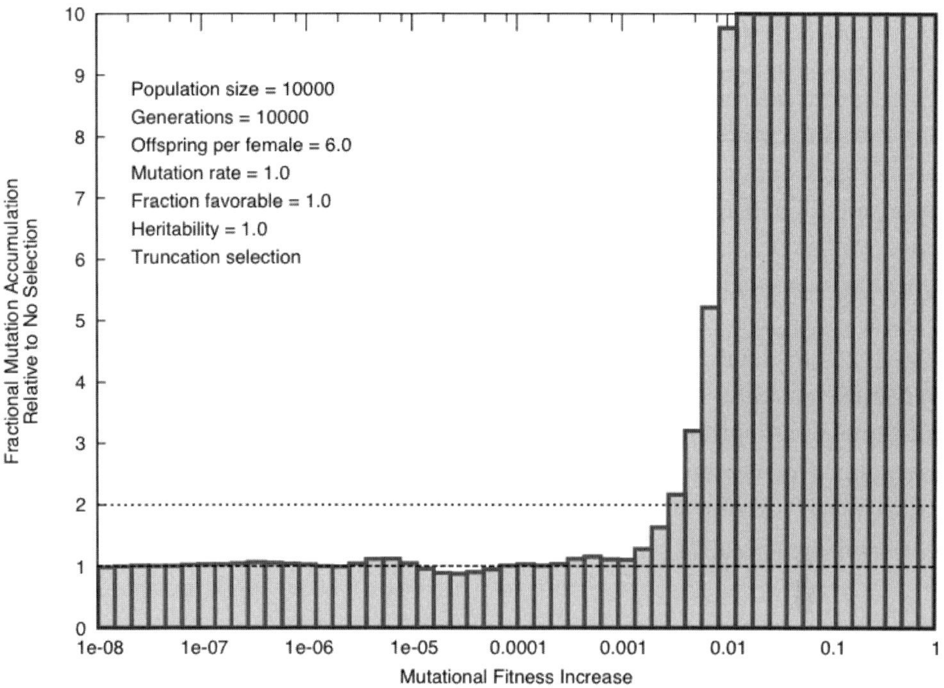

Fig. 5. Accumulation of beneficial mutations as affected by degree of benefit, when extremely beneficial mutations are allowed. Parameters are the same as in Figure 1, except that the maximal mutational fitness effect has been increased from .001 to 1.0. Allowing extremely beneficial mutations causes intense selection interference and raises the selection threshold more than 3 orders of magnitude.

increase in the ST_b value (2.96×10^{-3}). This was the single factor in our studies that by itself most dramatically increased the beneficial selection threshold.

The ST_b value seen in Figure 5 indicates that 98.0% of the beneficial mutations were below the selection threshold. This ST_b value is more than three orders of magnitude greater than what is seen in Figure 1 and is comparable to the increase we see when we switch from truncation selection to probability selection. Ironically, the effects of very high-impact beneficial mutations overshadow low-impact beneficial mutations so profoundly that it results in selection breakdown for all beneficial mutations with fitness effects less than approximately 0.001. This is true even when all other factors are chosen to minimize the selection threshold, including full truncation selection and zero environmental variance. These very high-impact beneficial mutations are in a sense "too selectable". The very rare alleles that are represented on the far right of Figure 5 dominate the selection process and exhaust almost all the selection potential available. This represents the most dramatic form of selection interference we have seen in over six years of experimentation with genetic accounting methodology.

Effect of adding deleterious mutations

The experiments described above show that increasing beneficial mutation rates leads to increased selection interference, and that introduction of extremely beneficial mutations leads to an especially profound type of selection interference. However, all experiments described thus far have involved only beneficial mutations. We know that, in reality, the majority of mutations are deleterious. To what extent do beneficial and deleterious mutations affect each other's relative selectability? To address this question we conducted an experiment similar to that of Figure 1 with truncation selection, zero environmental variance, and just one new beneficial mutation per offspring. In addition to the average of one beneficial mutation per offspring, we also added an average of one deleterious mutation per offspring. This experiment yielded a selection threshold for deleterious mutations of 2.30×10^{-6}, as shown in Figure 6. By contrast, the parallel case (as described in our companion paper [14]), with one new deleterious mutation per offspring but

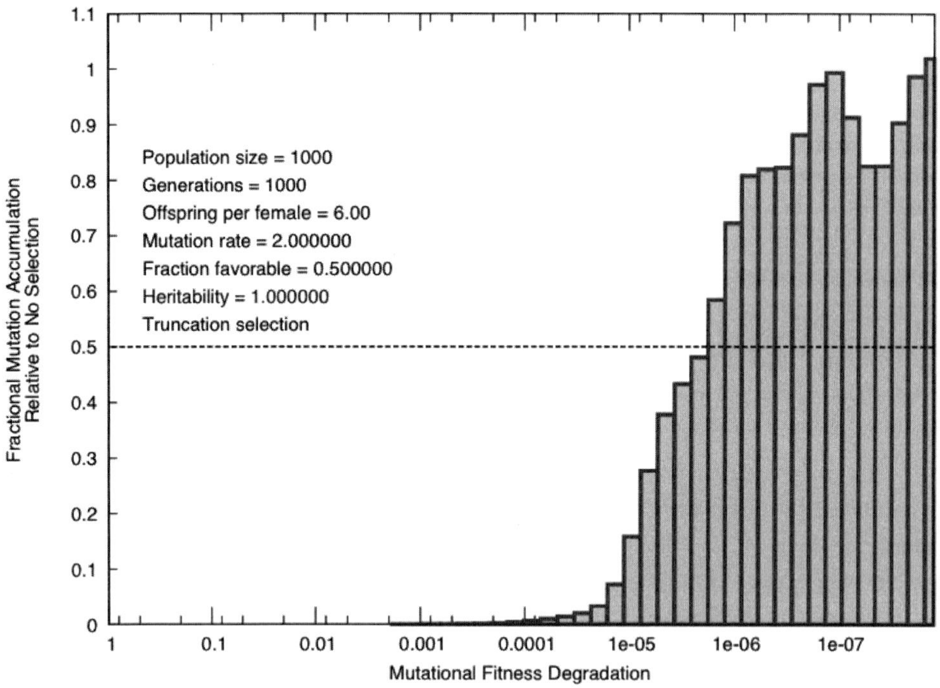

Fig. 6. Accumulation of deleterious mutations as affected by degree of harmfulness, given equal rates of deleterious and beneficial mutations. Parameters are the same as in Figure 1, except that an equal rate of deleterious mutation was added. Selection interference due to the accumulating beneficial mutations causes very significant accumulation of deleterious mutations under conditions where none would have accumulated otherwise (see companion paper [14]).

zero new beneficial mutations per offspring, gave the result of *zero* deleterious mutations accumulated.

The beneficial mutations clearly caused very serious selection interference in terms of the selectability of the deleterious mutations. However, the converse was not true. The accumulation of deleterious mutations only had a very modest effect on the accumulation of beneficial mutations. This can be seen by comparing Figure 7 ($ST_b = 2.00 \times 10^{-6}$) with Figure 1 ($ST_b = 1.34 \times 10^{-6}$). This asymmetrical aspect of selection interference between deleterious and beneficial mutations reflects a fundamental difference in dynamics between purifying selection versus positive selection. Purifying selection very effectively eliminates high-impact deleterious mutations, such that the remaining deleterious mutations are all low-impact, have a highly diffuse genetic effect, and constitute a minor source of noise relative to the selectability of the beneficial mutations. However, positive selection amplifies only the very high-impact beneficial mutations, which then very effectively "highjack" almost all the selection potential of the population, severely diminishing the effectiveness of purifying selection.

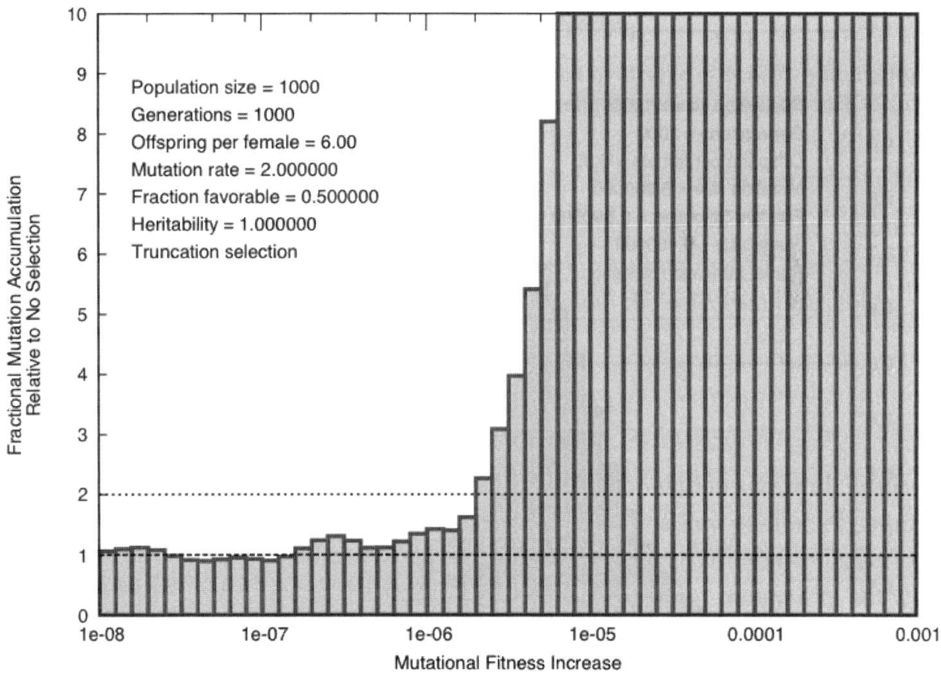

Fig. 7. Accumulation of beneficial mutations as affected by degree of benefit, given equal rates of deleterious and beneficial mutation. Parameters are the same as in Figure 1, except that an equal rate of deleterious mutation was included. When deleterious mutations are included, they have minimal effect on the selection threshold of beneficial mutations (contrast with Figure 1, where there were no deleterious mutations).

were chosen in order to approximate a more realistic lower limit for the beneficial selection threshold, as might be expected for a typical mammalian population. Results from this experiment are shown in Figures 10 and 11.

As seen in Figure 10, given a mixture of deleterious and beneficial mutations, combined with multiple sources of biological noise at modest levels, and with a larger population size, deleterious mutations again accumulated at very high rates, resulting in the highest ST_d value of this study, which was 4.96×10^{-3} (98.5% of all deleterious mutations were below the selection threshold).

Likewise, given a mixture of deleterious and beneficial mutations, combined with multiple sources of biological noise at modest levels, and with a larger population size, there was a failure to amplify the vast majority of beneficial mutations (Figure 11), resulting in the highest ST_b value of this study, which was 3.16×10^{-3} (99.6% of beneficial mutations were below the selection threshold).

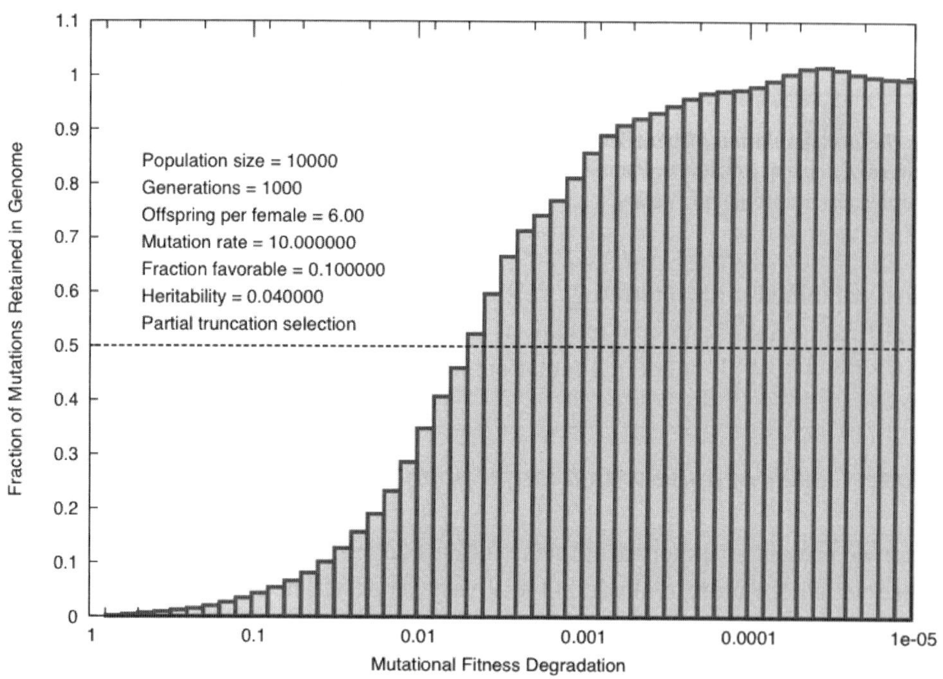

Fig. 10. Distribution of accumulating deleterious mutations, with multiple sources of noise at modest levels, larger population. Critical parameters: population size = 10000, generations = 1000, mutation rate = 10, fraction beneficial = 0.1, maximum beneficial effect = 0.1, fitness heritability = 0.04, partial truncation = 0.1. Multiple sources of noise, even at modest levels, and even with larger population size, cause very high ST_b values. Note: due to memory limits, in this experiment we used a tracking limit of 1.0×10^{-5}, and so could not plot the lowest three orders of low-impact mutations which would have been on far right.

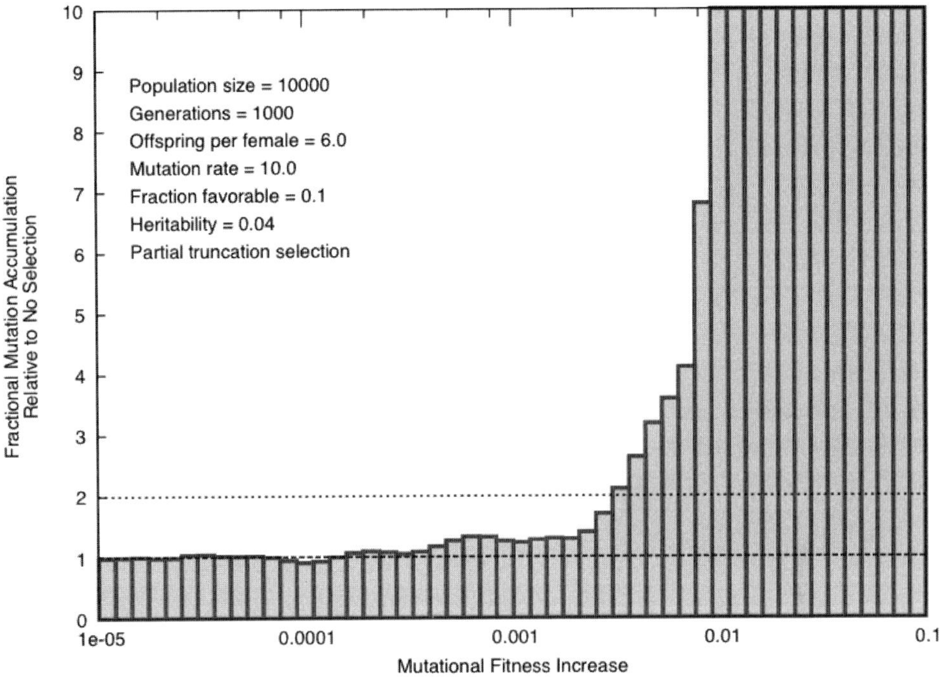

Fig. 11. Distribution of accumulating beneficial mutations, with multiple sources of noise at modest levels, larger population. Critical parameters: population size = 10000, generations = 1000, mutation rate = 10, fraction beneficial = 0.1, maximum beneficial effect = 0.1, fitness heritability = 0.04, partial truncation = 0.1. Multiple sources of noise, even at modest levels, and even with larger population size, cause very high ST_b values. Note: due to memory limits, in this experiment we used a tracking limit of 1.0×10^{-5}, and so could not plot the lowest three orders of low-impact mutations which would have been on far left.

The effect of time on ST_d and ST_b values

Here we present examples of how ST values can change over time. In all of our experiments where we begin with zero genetic variance, we see that ST values are initially exceptionally high, but rapidly decline as the population moves toward selection equilibrium, at which point ST values stabilize.

Figure 12 gives an example of this, where both beneficial and deleterious mutations are accumulating (population size = 1000, mutation rate = 5, fraction beneficial = 0.5, maximum benefit = 0.1, heritability = 0.4, partial truncation = 0.5). After 2000 generations, it can be seen that the beneficial mutations begin to approach selection equilibrium more rapidly than the deleterious mutations. After 5000 generations, ST values are very stable. In this experiment, ST_b stabilized at a slightly higher value than ST_d.

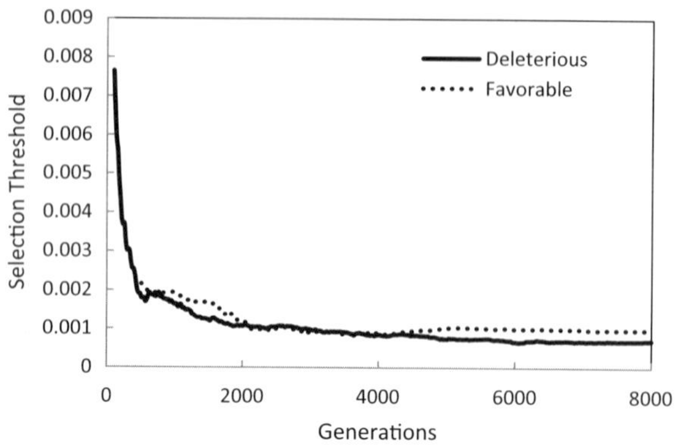

Fig. 12. Deleterious and beneficial selection thresholds plotted over time, with multiple sources of noise. It takes many generations to reach selection equilibrium. The beneficial and deleterious selection thresholds equilibrate at very nearly the same levels. Beneficial selection threshold cannot be plotted until about generation 500 (until then, there are too few beneficial mutations to produce meaningful data).

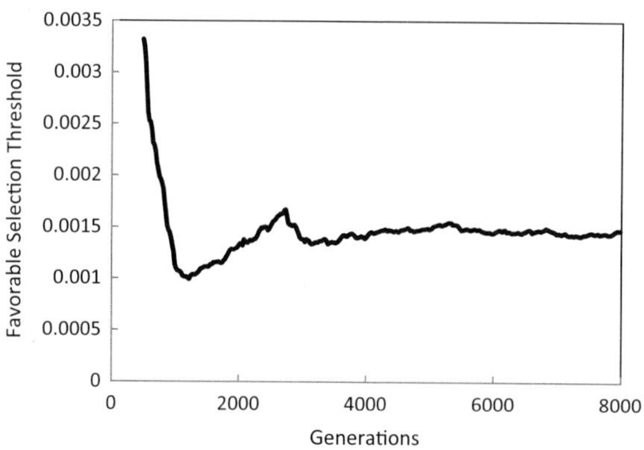

Fig. 13. Beneficial selection threshold plotted over time for a larger population, when extremely beneficial mutations are allowed. Critical parameters: population size = 10000, generations = 8000, mutation rate = 1, fraction beneficial = 1.0, maximum beneficial effect = 1.0, fitness heritability = 1.0, full truncation. Plotting started at generation 500.

Figure 13 shows how ST_b changes over time in the special case where there is a larger population (10,000), but extremely beneficial mutations are allowed (beneficial fitness effects up to 1.0), and all other parameters are optimized for selection efficiency (population size = 10,000; mutation rate = 1; fraction beneficial = 1,; heritability = 1; full truncation). Runs which include high-impact beneficial mutations tend to become limited by computer memory, because of the rapid amplification of those beneficial mutations. For that reason, longer-term experiments such as this require that all unnecessary tracking be suspended. Even with this accommodation, memory overflowed in this experiment after 8,000 generations. Adding extremely beneficial mutations, even under ideal conditions, greatly increases initial and ending ST_b values. ST_b values reach a minimum after roughly 1000 generations (about 2×10^{-3}) and then gradually increased due to growing selection interference as high-impact mutations increased.

Discussion

This analysis leaves no doubt that there must be a very significant selection threshold for beneficial mutations in higher organisms. This threshold is not a simple function of population size, but is affected by numerous factors. The reality of such a threshold has profound theoretical and practical implications. Our results show that the beneficial selection threshold for higher eukaryotes should be so large under realistic biological circumstances that nearly all beneficial mutations must be below that threshold. This constitutes a mystery. If the vast majority of beneficial mutations lie below the selection threshold and thus are not acted upon by selection, how can we explain the origin of low-impact functional nucleotides? Most functional nucleotides within a large genome must each make only an extremely small fractional contribution to total fitness, and therefore certainly must lie below the selection thresholds we are seeing. Simple logic therefore suggests that most functional nucleotides in large genomes could not have arisen via selection, at least not as natural selection is presently understood to operate.

There is substantial room for discussion regarding which parameter choices would be most appropriate for a given species and which choices might be most representative of a given natural circumstance. However, if we use extremely conservative estimates for all the relevant parameter choices that affect selection threshold, we should be able to estimate reasonably well the *lower limits* for mammalian ST_b values. The experiment summarized in Figures 10 and 11 does just this, yielding a ST_b value of approximately 3×10^{-3}. We have found that whenever we combine multiple sources of noise, even when using our most conservative parameter settings, we see ST_b values in this range. Therefore, we suggest that 10^{-3} is a reasonable approximation of the beneficial selection threshold for a typical mammalian population.

Even given extremely unrealistic selection parameters which confer the smallest possible selection threshold (Figure 1), we show that the large majority of beneficial mutations still lie below that threshold. When we introduce greater and greater levels of biological realism into our experiments, the selection threshold problem becomes progressively more severe (Figures 2–11). For example, our experiments show that when there are higher rates of mutations, or when there are just both classes of mutations (both beneficial and deleterious), this can cause strong selection interference, which further increases selection threshold values (Figures 4, 5, 6). This is seen when we increased beneficial mutations rates beyond one new mutation per offspring (Figure 4), or when we simultaneously allow both deleterious and beneficial mutations (Figure 6). We see this most dramatically (Figure 5), when we introduce very high-impact beneficial mutations, which strongly interfere with selection for all other mutations. The problem of selection interference has been casually recognized in several earlier papers [20, 26, 30], but no attempt has been made to quantify its effect under realistic circumstances, and the problem has largely been dismissed. Our studies suggest that selection interference is extremely important, and cannot be properly understood except by using biologically realistic genetic accounting programs such as Mendel's Accountant. This approach appears to bring the greatest clarity to the problem of selection interference and provides an excellent research tool for those who wish to study the problem further.

In a large genome (e.g., 10^8 functional nucleotides), non-neutral mutations must typically have very tiny fitness effects, with a lower limit of perhaps $\pm 10^{-8}$. Given that both deleterious and beneficial mutations have selection threshold magnitudes in the range of 10^{-3} or higher, it becomes clear that there exists a "zone of no-selection" which covers several orders of magnitude in fitness effect on either side of zero. We have previously shown that, when considering deleterious mutations by themselves, the large majority must fall within this "no selection zone" [14]. We here show that when high rates of beneficial mutations are included in the analysis, the selection breakdown for deleterious mutations becomes still worse (Figures 6, 8, 10). More importantly, we show that beneficial mutations themselves consistently have a very high selection threshold under all reasonable conditions (Figures 3, 5, 7, 9, 11). We show that given reasonable parameter settings, more than 99% all beneficial mutations are consistently un-selectable, leaving only a very small number of outlying high-impact beneficial mutations subject to selection. These findings raise a number of questions.

Can low-impact beneficial mutations contribute to genome building?

Building genomes without the use of low-impact nucleotides is very problematic. Since the time of Darwin it has been commonly thought that evolution must occur through an endless series of miniscule improvements (i.e. one nucleotide at a time). In light of our findings, this does not appear feasible. If beneficial mutations with fitness effects of less than 0.1% are not selectable, then evolution must only advance via larger and more discrete steps. For example, if fitness typically advances in increments of 1–10%, then only 10 to 100 mutational steps would be needed to double biological functionality. But the typical functional nucleotide in a large genome is generally assumed to carry a selection coefficient orders of magnitude smaller than this. How did such low-impact functional nucleotides arise? It is widely recognized that we each carry tens of thousands of deleterious mutations, yet we remain fairly robust, indicating that the damaged functional nucleotide sites in our genome must generally have each been conferring very tiny contributions to fitness. If selection cannot preserve such functional nucleotides, how could selection have put them in their place to begin with?

Can high-impact beneficial mutations explain the origin of the genome?

A very high-impact beneficial mutation (an extremely beneficial mutation), can obviously contribute to genome building, but only in a very limited sense. Indeed, we observe that, given high rates of high-impact beneficial mutations, net fitness can increase rapidly, even while a much larger number of deleterious mutations are continuously accumulating at a steady rate. Under these conditions we can see huge leaps in fitness scores, yet this improvement is entirely dependent upon only a handful of isolated, unlinked, non-complementary mutations. Under these conditions, selection can at best eliminate the worst deleterious mutations, while amplifying only the highest-impact beneficial mutations.

In terms of numerical scores within a simulation experiment, just a few extremely beneficial mutations can more than compensate for large numbers of low-impact deleterious mutations. But this leads to increasing "fitness" only in a narrow and artificial sense. In the broader sense, the whole genome is still degenerating, because, while a few nucleotide sites are being improved, large numbers are being degraded. This type of trade-off is not sustainable, as it results in a shrinking functional genome size. More and more nucleotide sites are losing their specificity, and hence their functionality. Taken to the extreme, this would eventually yield a biological absurdity — a functional genome consisting of a handful of high-impact nucleotide sites that somehow code for all of the organism's functionality.

Extremely beneficial mutations undoubtedly play an important role in adaptation to specific environmental circumstances, as in the case of microbial resistance to antibiotics, or in the case of human resistance to malaria. However, beyond this type of dramatic adaptation to some lethal external factor, extremely beneficial mutations seem to have very limited explanatory power in terms of genome building.

To the extent that extremely beneficial mutations are undergoing selection, our experiments show that they cause a sharp increase in both ST_d and ST_b values. This is a serious problem, because it means extremely beneficial mutations are hijacking most of the "selection power" inherent in the surplus population, and thus are contributing to selection breakdown for the vast majority of both deleterious and other beneficial mutations. Another way of expressing this is that the organism is being improved relative to only a few highly specific traits, but otherwise is "rusting out" in innumerable other ways. The actual fitness gain in such cases is generally no more than a transient response to a fluctuating environmental condition and so is fundamentally superficial, yet the cost is a continuously growing genetic load involving systematic, long-term, and irreversible decay of innumerable and essential internal functions.

Natural selection must explain more than just a few high-impact nucleotide sites. It needs to also explain all the low-impact nucleotide sites surrounding any given high-impact nucleotide site — because these create the proper context which gives the high-impact nucleotide its functionality. Because extremely beneficial mutations must be extraordinarily rare, there is a statistical necessity for extremely beneficial mutations to arise singly, unlinked, and with functional independence, and this profoundly limits their utility. They are self-limiting in that they can only accomplish the types of things that a single typographical error might achieve. Naturally, a single nucleotide change can readily destroy a function or interfere with some key interaction. But a single nucleotide change generally is not expected to create, *de novo*, any new complex functionality. If only high-impact nucleotide positions are selectable, where do the many low-impact nucleotides come from which create the context for the rare high-impact nucleotide?

Can equal-but-opposite compensating mutations stop degeneration?

One implication of high selection thresholds for beneficial mutations is that Ohta's hypothesis of compensating mutations [3,15,16] does not appear viable. A multitude of low-impact deleterious mutations cannot be systematically compensated by selection for equal-but-opposite beneficial mutations at other sites in the genome. Our analysis indicates that selection thresholds for beneficial mutations are comparable in amplitude to those for deleterious mutations, so equal-but-opposite beneficial mutations must be equally un-selectable, rendering such a stabilizing mechanism inoperative.

Can high-impact compensating beneficial mutations stop degeneration?

A single high-impact beneficial mutation can, in a limited sense, compensate for many low-impact deleterious mutations. If there were enough high-impact beneficial mutations, this might appear to solve the problem of genetic degeneration. We have conducted extensive analyses of this question using Mendel and find that stopping genetic degeneration is feasible only when the rate of high-impact beneficial mutations is sufficiently high.

A major unknown for any genetic simulation is the exact frequency of beneficial mutation. Beneficial mutations are generally considered much too rare to allow empirical determination of their exact rate. In this paper we used extremely high fractions of beneficial mutation (10–100%), not because we consider such high numbers to be realistic, but because it was necessary in order to obtain definitive estimates for ST_b. We needed to generate a relatively large number of beneficial mutations to define the selection thresholds in a reproducible manner. When we use rates of beneficial frequencies that are consistent with estimates of other investigators [19, 20] and that seem reasonable to us (e.g., less than one in 10,000), beneficial mutations have essentially no effect. In all our experiments where deleterious mutations outnumber beneficial mutations by 3–6 orders of magnitude, the beneficial mutations exert essentially no effect on fitness change over time (except rare and anomalous mutations which are extremely beneficial and create a short-term spike in fitness).

Might beneficial mutations be common?

Is it possible that the rate of beneficial mutations might actually be extremely high, such that random drift and just a little selection might fill the genome with functional nucleotides? This does not seem reasonable because it would imply that practically any sequence is equally functional and that functional sequence information requires little specificity. However, most biologists understand that functional information is very specific, and thus beneficial mutations must be very rare. Indeed, beneficial mutation rates have often been estimated to be in the range of only one in a million [19, 20]. A large majority of geneticists acknowledge the scarcity of beneficial mutations, and complain of the difficulty in studying them due to their scarcity [17–26]. However, a few scientists have argued that beneficial mutations might be extremely common, even approaching 50% of all non-neutral mutations [31, 32]. If applied to the written information within a given assembly manual, this concept would suggest that 50% of all typographical errors in a set of instructions will result in an improved product. This is obviously not reasonable, as it implies that almost any letter sequence will specify the same instruction. These issues are dealt with in more depth in another companion paper [33].

It is sometimes argued that genetic information must actually be quite non-specific, because many random changes have been thought to be perfectly neutral. This common misconception arose in part because of the casual use of the term "neutral mutation" to describe any low-impact mutation that escapes selection. However, on a functional level, the perfect neutrality of any mutation is neither testable nor logical. Every mutation should logically have some biological effect, no matter how small. Significantly, synonymous mutations, the long-standing paragon of neutral mutation, can no longer be assumed to be neutral. Synonymous mutations can be non-neutral because synonymous codon substitutions can profoundly affect RNA stability, protein translation rate, and even protein folding [34]. In a parallel development, the long-held paradigm of "junk DNA" is increasingly being challenged [35], undermining the other primary rationale for assuming that most mutations are perfectly neutral.

To address the issue of neutral mutation, Mendel's Accountant allows the mutation rate to be discounted by whatever fraction the user feels is a reasonable estimate of the rate of neutral mutation. For example, for the experiment summarized in Figures 10 and 11, we used a mutation rate of just 10, even though the actual human mutation rate is known to be in the range of 60–100. This reflects the premise that 90% of the genome is perfectly inert, and so 90% of all mutations are neutral, which we feel is extremely over-generous. We have earnestly sought to circumvent the confusion associated with the concept of neutral mutation by only considering mutations within the "functional genome" (as opposed to any junk DNA sequences). By focusing only on the functional genome, we feel we can focus just on those mutations within "functional sequences". To be functional, sequences must be specific, and so random changes within such sequences should very rarely increase their functionality.

For many reasons, unambiguously beneficial mutations must be very rare, and beneficial mutations above the selection threshold must be extraordinarily rare [33]. Invoking high rates of extremely beneficial mutations does not seem to offer a realistic solution to the selection threshold problem.

Possible criticisms

A possible criticism of this study might be that no one really knows the exact distribution of beneficial mutations. Therefore, some might claim that the Weibull distribution we used in these studies may be distorting our conclusions about selection threshold for beneficial mutations. However, our results do not depend on the precise shape of the distribution curve. As long as the distribution is approximately exponential, we get similar results and reach the same basic conclusions. There is essentially unanimous consent that the beneficial mutation distribution must be approximately exponential [17,23,24,26,36–43], with high-impact mutations being very rare and very low-impact mutations being the vast majority. Indeed various papers [38, 42, 44], contend that the Weibull distribution fits biological reality as well or better than the other variations on the basic exponential theme.

A second possible criticism of this study might be that our thesis is contradicted by a large volume of scientific literature that uses DNA sequence comparisons to infer historical positive selection events for great numbers of putative beneficial mutations. To the extent that theory and actual observations conflict, there arises a scientific paradox which demands a reexamination of either the standing theory, or the observed data, or both. We naturally acknowledge the operation of selection for beneficial mutations in the past, but argue that such selection is severely constrained by the reality of selection threshold, as this study and common sense both demand. Natural selection, as presently understood, simply cannot do what so many are attributing to it — at least relative to low-impact mutations. It is noteworthy that a significant part of this body of literature that claims proof of positive selection in the past (based upon observed sequence variability in the present), may suffer from systematic error and is now being challenged [45–47]. Authors arguing for ubiquitous positive selection in the past, based solely upon sequence data, need to explain why their observed sequence variations might not be explained just as readily using alternative mechanisms such as differential mutational rates or ordinary statistical fluctuations. At the same time, they rightfully should point to the findings of this study and include in their discussion the theoretical problems inherent in selecting simultaneously for a multitude of very low-impact mutations with both positive and negative effects.

A third possible criticism of this study might be that our results are unique to our program and that this program was specifically designed to give these results. Yet in truth we went to great lengths to design Mendel to best reflect biological reality, and it is in fact clear that Mendel's Accountant is the most biologically-realistic forward-time population genetics numerical simulation yet developed. Furthermore, apart from specific details, our observations are in good agreement with what sound population genetics and logic would predict, and our work reflects an expansion, not a reversal, of previous studies [1–29]. Moreover, in another paper in these proceedings [48], and also in a separate paper [49], it is shown that the digital genetics simulation program known as 'Avida' produces very similar results regarding selection threshold and selection breakdown as we report here — when Avida is run using realistic fitness effects.

consider (eight or more orders of magnitude). This function is similar to a gamma distribution but allows a wider range of fitness effect.

In regard to the parameters needed to characterize the Weibull distribution, for deleterious mutations we use $a = 3\times10^{-9}$ (reflecting the inverse of 3×10^{8} bp, a conservative estimate of the functional genome size in humans), which serves as the lower limit of the mutational effect for homozygous mutations in the model. Thus, the magnitude of homozygous deleterious mutational effects ranges from -1 (lethal) to -3×10^{-9}. With the Weibull-type distribution, mutations of small effect are much more frequent than those with large effect. To set the value of γ for the cases described in this study, we specify as high-impact mutations those with a homozygous deleterious fitness effect of at least 0.1 and fix their frequency at 0.001, reflecting an estimate that one in a thousand mutations in humans reduces fitness by ten percent. This parameterization generates almost no lethal mutations. Lethals have little effect on mutation accumulation, and thus are ignored in this analysis.

In this paper, when we specify the distribution of mutations, we must also include the beneficial mutations. Apart from their relative abundance, which is a user input, Mendel generates the distribution for deleterious and beneficial mutations in a very similar manner, such that they have the same basic shape to their distribution, except for their range. We take minimum magnitude for deleterious and beneficial mutations to be the same (one divided by the functional genome size). However, while the largest negative effect for deleterious mutations is always -1.0 (there can always be a few entirely lethal mutations), the maximum value Mendel allows for beneficial mutations is user-specified. While we believe a limiting value for beneficial effects in higher organisms should be on the order of a percent or less, we evaluate ST_b with values as large as $+1.0$. The distribution for beneficial mutation effects has the form $d = d_0 exp(ax^{\gamma})$, where d_0 is the limiting beneficial effect, a is the reciprocal of the product of the functional genome size and d_0, and γ is determined by the same parameters as deleterious mutations except that the cutoff value for "high-impact" mutations is scaled by the factor d_0.

Mendel outputs a statistic that we term *selection threshold* (ST), which marks the center of the transition zone in fitness effect between selectable and un-selectable mutations. For deleterious mutations, ST_d is defined as the mutational fitness effect value at which the number of mutant alleles in the population is exactly half of the number expected if there were no selection. The computed ST_d value lies at the mid-point of the transition zone separating large-effect, selectable mutations (that display essentially zero accumulation) and small-effect un-selectable mutations (that display essentially 100% accumulation). This statistic provides, at any desired generation, a simple empirical basis for comparing selection effectiveness among cases involving different biological parameters.

For beneficial mutations, a similar statistic, ST_b, can be defined as the mutational fitness effect value at which the number of mutant alleles in the population is exactly twice that of the number expected if there were no selection. This provides a very useful benchmark for tracking at what point selection for low-impact mutations breaks down, and has a basic symmetry with the deleterious selection threshold. The computed ST_b value lies at a critical point where beneficial mutation effects start to be strongly amplified. This marks the transition zone separating large-effect, extremely selectable mutations (which display greatly accelerated accumulation rates) and very small-effect un-selectable mutations that display accumulation rates consistent with random drift.

Our choice for mutation rate is informed by recent estimates that tend to fall in the range of 100 new human mutations per person per generation [52, 53]. We adjust this estimate based on the fraction of the human genome assumed to be functional. We consider a minimal estimate of the functional genome to be 1% (yielding a functional mutation rate of 1) and very conservative estimates to be 5% and 10% (yielding functional mutation rates of 5 and 10). In light of increasing evidence of extensive genomic functionality [35], we also examine functional mutation rates of 20 or 40 new mutations per individual per generation, corresponding to a 20% and 40% functional genome, respectively. By discounting the mutation rate based upon the size of the functional genome, we are postulating a very conservative mutation rate because we effectively remove from consideration all non-functional DNA. This also eliminates from consideration all mutations which are absolutely neutral.

In regard to environmental variance, we consider four cases: zero environmental variance (fitness heritability of 1.0), small variance (fitness heritability of 0.4), moderate variance (fitness heritability of 0.04), and large variance (fitness heritability of 0.004). While a heritability value of 0.04 would be very small for a simple phenotypic trait such as height, it is still about 10-fold higher than what is commonly estimated for total fitness heritability [8]. Indeed, heritability of overall fitness is often found to be too small to measure. Selection is always based on each individual's phenotypic fitness, which reflects the genotype fitness plus a random environmental effect. In Mendel, a given heritability is achieved by adding a random number to each individual's genotypic fitness to yield its phenotypic fitness value. These numbers are drawn from a zero-mean normal distribution of random numbers with just the right variance to produce the desired heritability.

We consider three relative frequencies of deleterious versus beneficial mutation: a) deleterious mutations are entirely absent; b) the deleterious mutation rate equals the beneficial mutation rate; and c) the deleterious mutations are 9-fold

19. Elena SF, Ekunwe L, Hajela N, Oden SA, Lenski RE (1998) Distribution of fitness effects caused by random insertion mutations in *Escherichia coli*. Genetica 102/103:349–358.
20. Gerrish P, Lenski R N (1998) The fate of competing beneficial mutations in an asexual population. Genetica 102/103:127–144.
21. Crow J (2000) The origins, patterns and implications of human spontaneous mutation. Nature Reviews 1:40–47.
22. Bataillon T (2000) Estimation of spontaneous genome-wide mutation rate parameters: whither beneficial mutations? Heredity 84:497–501.
23. Imhof M, Schlotterer C (2001) Fitness effects of advantageous mutations in evolving Escherichia coli populations. Proc Natl Acad Sci USA 98:1113–1117.
24. Orr H (2003) The distribution of fitness effects among beneficial mutations. Genetics 163: 1519–1526.
25. Keightley P, Lynch M (2003) Toward a realistic model of mutations affecting fitness. Evolution 57:683–685.
26. Barrett RDH, M'Gonigle LKM, Otto SP (2006) The distribution of beneficial mutation effects under strong selection. Genetics 174:2071–2079.
27. Lynch M (2006) The origins of eukaryotic gene structure. Mol Biol Evol 23(2):450–468.
28. Sanford J, Baumgardner J, Gibson P, Brewer W, ReMine W (2007) Mendel's Accountant: a biologically realistic forward-time population genetics program. Scalable Computing, Practice and Experience [http://www.scpe.org] 8(2):147–165.
29. Sanford J, Baumgardner J, Gibson P, Brewer W, ReMine W (2007) Using computer simulation to understand mutation accumulation dynamics and genetic load. In 7th International Conference on Computational Science, Beijing, China, May 27–30, 2007 Proceedings, Part II, LNCS 4488. Edited by Shi Y, van Albada GD, Dongarra J, Sloot PMA, Berlin/Heidelberg: Springer-Verlag:386–392.
30. Keightley P, Otto S (2006) Interference among deleterious mutations favors sex and recombination in finite populations. Nature 443:89–92.
31. Joseph S, Hall D (2004) Spontaneous mutations in diploid Saccharomyces cerevisiae: More beneficial than expected. Genetics 168:1817–1825.
32. Dickinson W (2008) Synergistic fitness interactions and a high frequency of beneficial changes among mutations accumulated under relaxed selection in Saccharomyces cerevisiae. Genetics 178:1572–1578.
33. Montañez G, Marks II RJ, Fernandez J, Sanford JC (2013) Multiple overlapping genetic codes profoundly reduce the probability of beneficial mutation. In: Marks II RJ, Behe MJ, Dembski WA, Gordon B, Sanford JC (eds) Biological Information — New Perspectives. World Scientific, Singapore, pp. 139–167.
34. Kudla G., Murray AW, Tollervey D, Plotkin JB (2009) Coding-Sequence Determinants of Gene Expression in Escherichia coli. *Science* 324:5924.

35. The Encode Project Consortium (2007): Identification and analysis of functional elements in 1% of the human genome by the ENCODE pilot project. Nature 447:799–816.
36. Piganeau G, Eyre-Walker A (2003) Estimating the distribution of fitness effects from DNA sequencing data: implications for the molecular clock. Proc Natl Acad Sci USA 100:10335–10340.
37. Estes S, Phillips PC, Denver DR, Thomas WK, Lynch M *et al.* (2004) Mutation accumulation in populations of varying size: The distribution of mutational effects for fitness correlates in *Caenorhabditis elegans*. Genetics 166:1269–1279.
38. Sanjuan R, Moya A, Elena S (2004) The distribution of fitness effects caused by single nucleotide substitutions in an RNA virus. Proc Natl Acad Sci 101:8396–8401.
39. Keightley P, Charlesworth B (2005) Genetic instability of C elegans comes naturally. Trends in Genetics 21:67–70.
40. Eyre-Walker A, Woolfit M, Phelps, T (2006) The distribution of deleterious amino acid mutations in humans. Genetics 173:891–900.
41. Kassen R, Bataillon T (2006) Distribution of fitness effects among beneficial mutations before selection in experimental populations of bacteria. Nature Genetics 38:484–488.
42. Martin G, Lenormand T (2008) The distribution of beneficial and fixed mutation effects close to an optimum. Genetics 179:907–916.
43. Lynch M, Sung W, Morris K, Coffey N, Landry CR, Dopman EB, Dickinson WJ, Okamoto K, Kulkarni S, Harti DL, Thomas WK (2008) A genome-wide view of the spectrum of spontaneous mutations in yeast. Proc Natl Acad Sci USA 105:9272–9277.
44. Peris JB, Davis P, Cuevas JM, Nebot MR, Sanjuán R (2010) Distribution of fitness effects caused by single-nucleotide substitutions in bacteriophage f1. Genetics 185:603–609.
45. Hughes A (2007) Looking for Darwin in all the wrong places: the misguided quest for positive selection at the nucleotide level. Heredity 99:364–373.
46. Hughes A (2008) The origin of adaptive phenotypes. Proc Natl Acad Sci USA 105:13193–13194.
47. Kryukov G, Schmidt S, Sunyaev S (2005) Small fitness effect of mutations in highly conserved non-coding regions. Human Molecular Genetics 14:2221–2229.
48. Nelson C, Sanford J (2013) Computational evolution experiments reveal a net loss of genetic information despite selection. In: Marks II RJ, Behe MJ, Dembski WA, Gordon B, Sanford JC (eds) Biological Information — New Perspectives. World Scientific, Singapore, pp. 338–368.
49. Nelson C, Sanford J (2011) The effect of low impact mutations in digital organisms. Theoretical Biology and Medical Modeling 8:9.
50. Milkman R (1977) Selection differentials and selection coefficients. Genetics 88:391–403.

51. Kimura M, Crow JF (1978) Effect of overall phenotypic selection on genetic change at individual loci. Proc Natl Acad Sci USA 75(12):6168–6171.
52. Weibull W (1951) A statistical distribution function of wide applicability. J Appl Mech-Trans ASME 18(3):293–297.
53. Kondrashov AS (2002) Direct Estimate of human per nucleotide mutation rates at 20 loci causing Mendelian diseases. Hum Mutat 21:12–27.
54. Nachman MW, Crowell SL (2000) Estimate of the mutation rate per nucleotide in humans. Genetics 156:297–304.
55. Lynch M (2010) Rate, spectrum, and consequences of human mutation. Proc Natl Acad Sci USA 107:961–968.

Appendix I: *Key parameter settings and their justification:*

Mutation rate = 1, 5, or 10 (unless otherwise specified). Although the human mutation rate is known to be roughly 100 new mutations per person per generation [53–55], we typically use the extremely conservative maximal value of 10. This presumes that at least 90% of the human genome is perfectly inert "junk", which is contrary to the mounting evidence indicating a substantial fraction of the human genome has function [35]. More realistic mutation rates only make the selection threshold problem worse.

Population size = 1,000 (unless otherwise specified). This default population size would be realistic for an isolated tribe or set of tribes. Population sizes larger than 1,000 do not significantly decrease ST values or change the percent of mutations which are un-selectable [14], but when we allow extremely beneficial mutations in larger populations, their rapid multiplication leads to overflow of memory.

Generations = 1000 (unless otherwise specified). We find that this is sufficient for ST_b to largely stabilize for the population sizes we have been studying.

Offspring per female = 6. In Mendel's default mode, all surplus progeny are selected away. Since two offspring per female are needed for population continuity, this setting causes two thirds of all progeny to be selected away and represents extremely intense selection.

Distribution of mutation effects = Weibull distribution, wherein 0.1% of all mutations reduce fitness by 10% or more. Altering the shape of the distribution to be either steeper or less steep, does not significantly affect the ST phenomenon.

Dominant versus recessive = co-dominance. Although Mendel allows some mutations to be partially or fully dominant, while others are partially or fully recessive, for simplicity we make all mutations in this paper co-dominant. We have observed that this parameter has only a minor impact on ST values.

Mutation effects combination method = additive. Mendel also allows use of the multiplicative model, but we feel the additive model is more realistic, and use of the multiplicative model does not significantly affect the ST phenomenon.

To reproduce these results: all other settings can be set to the normal Mendel default settings (Version 1.4.3).

Using Numerical Simulation to Test the "Mutation-Count" Hypothesis

Wesley H. Brewer[1], John R. Baumgardner[2] and John C. Sanford[3*]

[1]*Fluid Physics International, P.O. Box 4915, Mississippi State University, MS 39762, USA.* [2]*Department of Earth and Environmental Sciences, Ludwig Maximilians University, Theresienstrasse 41, 80333 Munich, Germany.* [3]*Dept. Hort. Sciences, NYSAES, Cornell University, Geneva, NY 14456, USA. (*corresponding author: jcs21@cornell.edu)*

Abstract

There is now abundant evidence that the continuous accumulation of deleterious mutations within natural populations poses a major problem for neo-Darwinian theory. It has been proposed that a viable evolutionary mechanism for halting the accumulation of deleterious mutations might arise if fitness depends primarily on an individual's "mutation-count". In this paper the hypothetical "mutation-count mechanism" (MCM) is tested using numerical simulation, to determine the viability of the hypothesis and to determine what biological factors affect the relative efficacy of this mechanism.

The MCM is shown to be very strong when given all the following un-natural conditions: all mutations have an equal effect, low environmental variance, and full truncation selection. Conversely, the MCM effect essentially disappears given any of the following natural conditions: asexual reproduction, or probability selection, or accumulating mutations having a natural distribution of fitness effects covering several orders of magnitude. Realistic levels of environmental variance can also abolish or greatly diminish the MCM effect.

Equal mutation effects when combined with partial truncation (quasi-truncation) can create a moderate MCM effect, but this disappears in the presence of less uniform mutation effects and reasonable levels of environmental variance.

MCM does not appear to occur under most biologically realistic conditions, and so is not a generally applicable evolutionary mechanism. MCM is not generally capable of stopping deleterious mutation accumulation in most natural populations.

Key words: mutation count mechanism, mutation accumulation, natural selection, neo-Darwinian theory, numerical simulation, Mendel's Accountant

Introduction

There is a significant body of literature, based upon both logic and mathematical modeling, which indicates that direct selection against deleterious mutations is insufficient to halt deleterious mutation accumulation [1–6]. Recent studies using numerical simulation have demonstrated this point [7–10]. A primary reason for this paradoxical mutation accumulation problem is that most deleterious mutations have extremely small biological effects, and thus are essentially invisible to selection [11–16].

It has been argued that this fundamental problem might be resolved by a form of selection not based directly upon the biological effect of each mutation, but instead upon an individual's "mutation count" [17–20]. We term this the "mutation-count mechanism" (MCM). In this paper we use numerical simulation to explore whether the MCM can realistically be expected to stop mutation accumulation. In a companion paper, numerical simulation is used to test a related concept, the synergistic epistasis hypothesis. That more elaborate hypothesis, also attempts to deal with the mutation accumulation problem by focusing selection specifically against high-mutation-count individuals [10].

The concept of selection based upon mutation count was first put forward by Muller [1], but has primarily been developed and expanded by Crow [17–20]. For decades, Crow, Muller, and others have acknowledged that deleterious mutations should logically accumulate continuously in populations, creating an evolutionary paradox. This is especially apparent when mutation rates are higher than one mutation per individual per generation [1]. Even when mutation rates are well below one per individual per generation, Ohta and others [11–16] have shown that most mutations have such small biological effects that they must be "nearly neutral" (effectively neutral), and must routinely escape the influence of selection, leading to continuous accumulation. The problem of continuous accumulation of deleterious mutations creates an evolutionary paradox, wherein populations should logically degenerate continuously, leading inevitably to extinction [1–6].

The idea of selection based upon an individual's mutation count was developed to address this theoretical problem of continuous genetic degeneration. The concept is that, when mutations accumulate to significant levels within a population, some individuals will have substantially more mutations than others due to random statistical fluctuations. If selection is strongly focused against those "high mutation count" individuals, elimination of single individuals might systematically eliminate proportionately more mutations. All this might be feasible if there were a strong correlation between mutation count and phenotypic fitness. Given a strong correlation, the MCM might progressively slow mutation accumulation and eventually even stop it. In such a case, the mean mutation count per individual would increase up to a maximum and then plateau, and mean fitness would cease its decline.

Numerical simulations using biologically reasonable parameters have consistently failed to show any evidence of the MCM, when using natural mutation distributions [8, 9]. This is most readily seen by plotting mean mutation count per individual over time. Using natural mutation distributions (wherein mutational effects vary over a wide range), the mutation count per individual consistently increases over time in a linear manner. This is seen even given intense selection, large populations, and many generations. In such experiments, no stabilization of mutation count is observed, and fitness declines continuously. This is because individuals are being selected based upon phenotypic fitness, as in nature, not based upon a contrived parameter such as an individual's "mutation count". Under realistic conditions, phenotypic fitness should have a weak correlation to mutation count within a natural population. Random sampling of gametes from within the same breeding population will have a strong statistical tendency toward producing similar mutation counts among all the progeny. Individual mutation counts will consistently track closely the population's mean mutation count. Not only will all individuals have approximately the same mutation count, the vast majority of the mutations within any individual will be nearly-neutral. Any meaningful genetic differences between individuals will be due to relatively few higher-impact mutations. These non-trivial mutations should strongly dominate the selection process, largely negating any correlation between an individual's mutation count and that individual's fitness. The correlation between an individual's mutation count and total fitness should logically be weak in most biological situations. This is exactly what is seen in careful numerical simulations; deleterious mutations invariably increase continuously at a constant rate.

Because the MCM hypothesis is a primary rationale for discounting pervasive genetic degeneration in nature, we desired to more carefully explore experimentally the potential for MCM using numerical simulation. For this purpose we employed the numerical forward-time population genetics program, *Mendel's Accountant* [7]. We modified this program so that selection could be based directly upon an individual's mutation count. This was achieved by specifying that all deleterious mutations have exactly the same fitness effect. The result is that an individual's reduction in genotypic fitness can correlate perfectly with its deleterious mutation count. This provided us with a research tool for evaluating the potential of MCM and allowed us to study various factors that affect the efficacy of this mechanism.

Methods

We apply the program *Mendel's Accountant* [7] (henceforth, 'MENDEL') to study the influence of MCM on mutation accumulation and genetic degeneration. This program was designed to study mutation accumulation [8–10], and we believe it is the first biologically-realistic population genetics program [7–10].

It is known that mutation accumulation is affected by many parameters. No set of equations solvable by hand can simultaneously account for all these interacting factors without introducing major simplifying assumptions. Of course, this limits both the scope and generality of such analyses. There is enormous biological complexity inherent in the mutation/selection process, especially when it is considered at the level of the whole genome and the whole population. Therefore it cannot be assumed that traditional analytical approaches are adequate for studying the consequences of hypotheses such as MCM. However, thanks to modern advances in scientific computing, complex systems of this type can now be analyzed reliably using numerical simulation. MENDEL tracks a complete biological system, starting with individual mutations, mutation-mutation interactions, linkage blocks, chromosomes, genotypes, phenotypes, mating/recombination events, sub-populations, and whole populations. Using MENDEL, all the primary known parameters that affect the selection/mutation process are accounted for, and can be specified by the program user, and so the computational processing can be faithful to our understanding of how genetic systems operate.

MENDEL can incorporate beneficial mutations, but for the sake of clarity in this paper we include only deleterious mutations. Except where indicated, we use MENDEL's human default parameters, as might reflect a small human population after a population bottleneck, with very intense selection (67% of progeny selected away every generation). Unless otherwise indicated, the most fundamental parameters were as follows: ploidy = diploid; reproduction = sexual; mating = random; linkage = dynamic recombination; new mutations per individual = 10; offspring per female = 6; mode of combining mutation effects = additive; population size = 1000; generations = 500; gene expression = co-dominance; fitness heritability = 1.0.

In these experiments we sometimes used "partial truncation", where selection was intermediate between full truncation selection and full probability selection. Mendel allows the user to specify the degree of partial truncation, with 0.1 specifying 10% truncation and 90% probability selection, while 0.5 specifies 50% truncation and 50% probability selection.

When either truncation or partial truncation selection are employed in our simulations, we have seen that it can result in un-naturally narrow genetic variance, and since we normally scale environmental variance to genetic variance (to specify a given heritability), this can result in a population that has an unreasonable narrow range of phenotypic variance. For this reason we established a non-scaling noise parameter where we can specify a minimal level of phenotypic variance, by adding some non-scaling environmental variance, to generate a reasonably heterogeneous phenotypic population even under truncation selection. In this study, whenever we select a heritability value less than one, we set the non-scaling noise at 0.05 (creating a minimum standard deviation of 0.05 for phenotypic fitness).

We begin by modeling the MCM using idealized conditions for optimal selection efficiency, and then investigate MCM in more depth by introducing more and more elements of realism.

Results

Our previous studies have clearly shown that given a natural distribution of mutational effects, mutations will accumulate continuously and at a constant rate [7–10]. Therefore, we already knew at the on-set of this research that one essential requirement for activation of MCM is some type of very narrow distribution of mutation fitness effects. For this reason all of the experiments done in this study employed either uniform mutations affects, or a relatively narrow range of fitness effects. This makes these experiments generally unrealistic biologically — yet we needed to make this concession to the MCM hypothesis in order to examine it more closely.

We first examined the MCM using highly idealized conditions. We caused all mutations to affect fitness in an equally deleterious way (each mutation, when in the homozygous form, reduced fitness by 0.001, relative to a reference genotype with a fitness of 1.0). We combined the fitness effects of such mutations additively within individuals. In this way we created a perfect correlation between genotypic fitness reduction and the individual's mutation count. We then applied zero environmental variation (heritability = 1.0), such that phenotypic fitness and genotypic fitness were identical. We then applied artificial truncation (wherein reproduction by a given individual depends exclusively on whether its phenotypic fitness is greater than an arbitrary fitness threshold).

Under these highly artificial conditions we found that MCM was indeed able to very effectively halt both mutation accumulation and fitness decline, as seen in Figure 1. However, using all the same parameters, but suspending sexual recombination (as would apply to any asexual species), completely abolished the MCM effect (Figure 1). Mutation accumulation and fitness decline were both perfectly linear without sexual recombination. Likewise, we found that using the original parameter settings and simply switching to probability selection essentially abolished the MCM effect (Figure 1), except as the population approached zero mean fitness (extinction). This extinction-related MCM effect must be seen as an artifact. As a population approaches a zero mean fitness, mutational load is so high that many individuals have a fitness of zero or less. These individuals are automatically and unconditionally removed from the population, forcing the population from probability selection into an artificially-induced form of truncation selection. However, in the natural world, a population would normally go extinct long before a large fraction of that population had zero biological functionality (for many reasons, including fertility decline and population collapse). Thus this type of MCM effect near the very end of our runs, whenever probability selection is in effect, must be viewed as an artifact of the simulation process which allows mean fitness to approach zero. Apart from this extinction-induced truncation phenomenon, we consistently see that mutation accumulation is essentially linear,

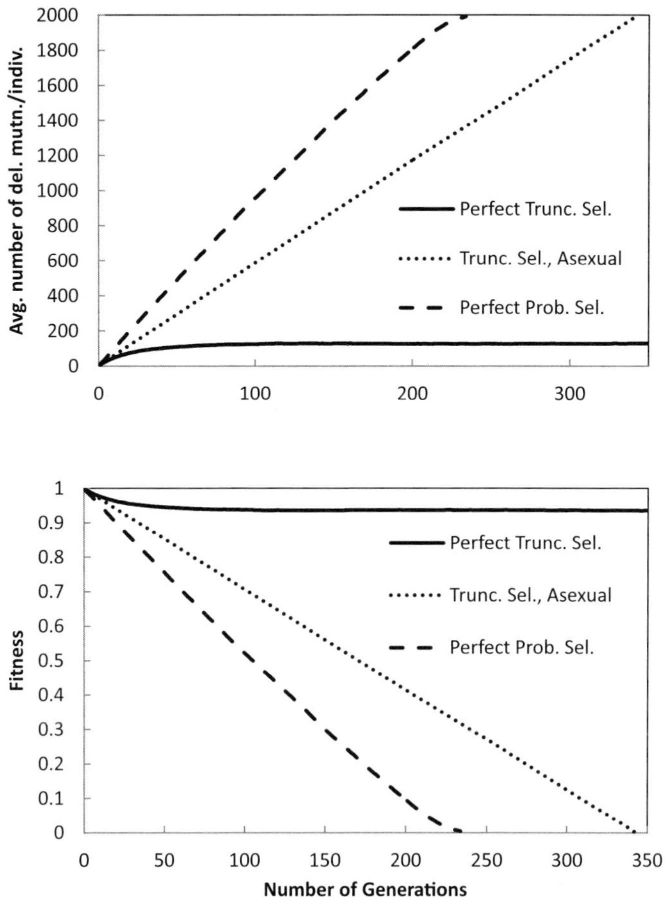

Fig. 1. Mean mutation accumulation per individual (top) and fitness history (bottom) for three experiments. Phenotypic fitness depended solely upon mutation count, that is, mutations all had the same effect (-0.001), and no environmental noise was added. Selection modes were: a) perfect truncation selection; b) perfect probability selection, and truncation selection without sexual recombination. Mutation count and fitness stabilized quickly when truncation selection was applied, due to the MCM effect. Either probability selection or asexual reproduction abolished the MCM effect.

even given idealized conditions, whenever probability selection is employed. In summary, Figure 1 shows us that the MCM can be effective, given equal mutation effects, zero environmental variance, and truncation selection. However, even with all mutation effects being equal, the MCM effect disappears whenever there is either asexual reproduction or probability selection.

We next examined the effect of partial truncation and environmental variance. We repeated the idealized experiment as described above with all mutations being equal, but instead employed partial truncation. We then did a series of runs where we studied the effect of environmental variance, and let the runs go longer (1000

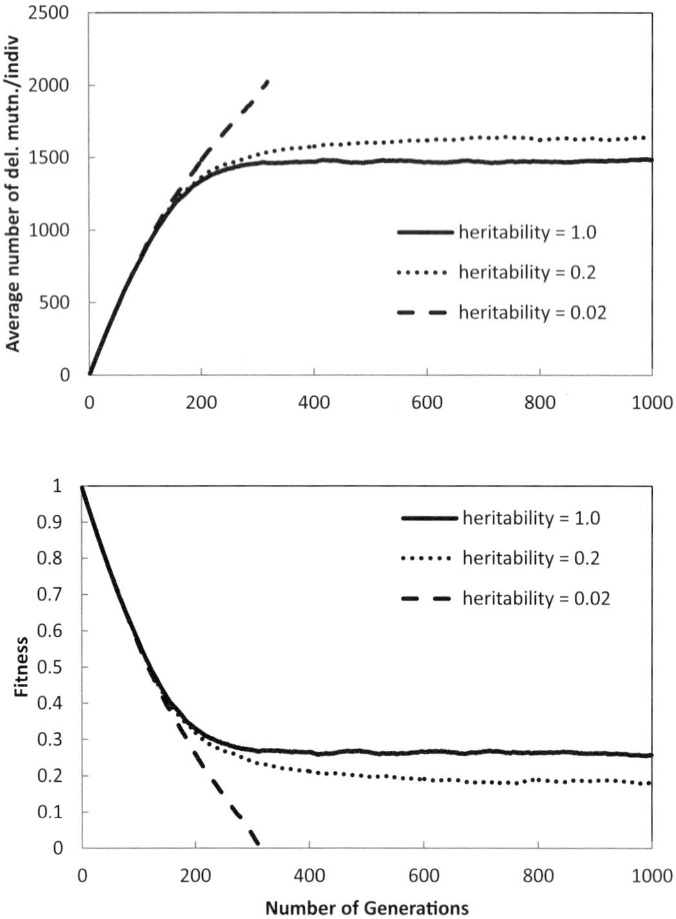

Fig. 2. Mean mutation accumulation per individual (top) and fitness history (bottom) for three experiments involving partial truncation (0.5) with varying amounts of environmental variance: zero environmental variance (heritability = 1.00), low environmental variance (heritability = 0.2), and high environmental variance (heritability = 0.02).

generations). Figure 2 shows that partial truncation (set at 0.5 – a selection mode halfway between perfect truncation and perfect probability selection), when combined with zero environmental variance, still produced a delayed, but still strong MCM effect. We then did experiments that added a low level of environmental variance and a high level of environmental variance. When we combine partial truncation with a low level of environmental noise (fitness heritability = 0.2), we saw that the MCM effect became somewhat weaker (Figure 2). When we combined partial truncation with a high level of environmental noise (fitness heritability = 0.02), we saw that the MCM effect was greatly reduced, becoming insufficient to prevent extinction under those settings (Figure 2).

Numerical simulations as described above, revealed evidence for a very significant MCM effect when mutation effects were perfectly uniform, and when selection was either full truncation or strong partial truncation. Addition of substantial environmental variation could greatly reduce the MCM, but did not entirely negate it. We therefore wished to examine how a moderate amount of variation in mutation effects might influence the efficacy of the MCM. Instead of using entirely uniform mutation fitness effects, we truncated our normal Weibull distribution of mutational effects so that the smallest mutational effect reduced fitness one part in 100,000 (3000-fold less than the Mendel default value). We then tested the four selection modes: full truncation, strong partial truncation (0.5), weak partial truncation (0.1), and probability selection. We let these experiments run 10,000 generations, introducing a modest amount of environmental variance (heritability = 0.2). The results of these experiments are shown in Figure 3. Given

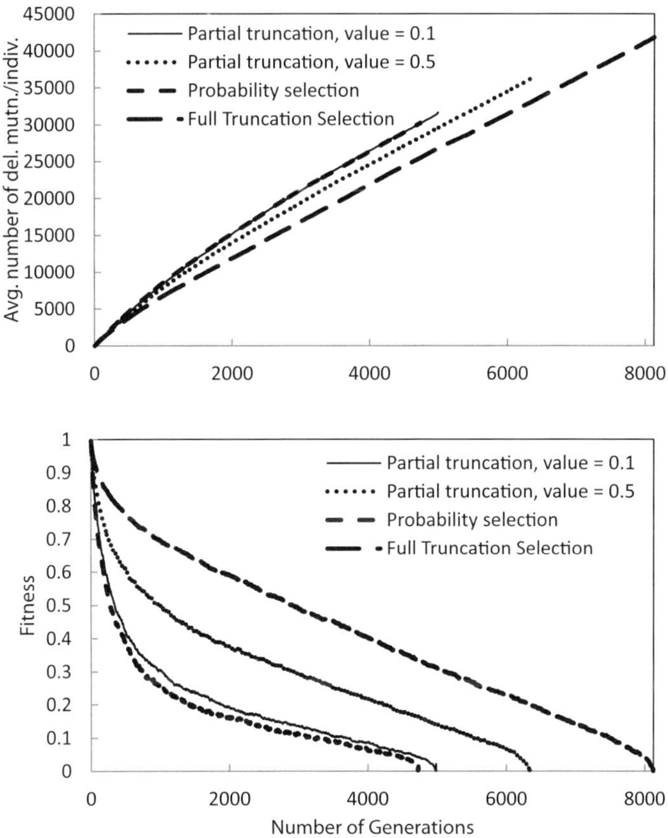

Fig. 3. Mean mutation accumulation per individual (top) and fitness history (bottom) for four experiments involving four modes of selection, given modest a amount of environmental variation (heritability = 0.2), and a relatively narrow range of fitness effects (lower limit = .00001). Selection modes were: full truncation, strong partial truncation selection (0.5), weak truncation selection (0.10), and probability selection.

these very favorable parameter settings, selection effectively removed all mutations with fitness effects of 0.001 or more (data not shown), and also removed most mutations with fitness effects between 0.0001 and 0.001. Therefore the accumulating mutations in these experiments were primarily in the range of 0.00001 to 0.0001 (varying across just over one order of magnitude). Despite this relatively narrow range of fitness effects, mutation accumulation eventually became essentially linear — regardless of whether selection mode was truncation, strong partial truncation (0.5), weak partial truncation (0.1), or probability selection. In the same way, fitness decline also became essentially linear regardless of selection mode, until population colapse occurred (mutational meltdown), as zero mean fitness was approached.

Discussion

Crow [19] recognized that if the deleterious mutation rate approached even one per generation, selective removal would fail and then de-evolution would logically result. Trying to escape this problem, he went back to the logic of Muller [1]. To quote Crow [19], "There is a way out, however. In stating his genetic death principle, Muller stated, 'For each mutation, then, a genetic death — except in so far as, by judicious choosing, several mutations may be picked off in the same victim.' Thus, natural selection… can indeed pick off several mutations at once…".

This is the essence of the mutation count mechanism — selecting away the highest mutation-count individuals by "judicious choosing", such that one death can remove more than one deleterious mutation. Our numerical simulations vividly illustrate the power of the MCM mechanism under ideal conditions (Figure 1). When all deleterious mutations have equal fitness effects, with no environmental variance, and with artificial truncation selection, mutation accumulation can be halted in very few generations.

Crow goes on to say "…such an efficient way of removal of mutations at small cost is strictly a consequence of sexual reproduction. An asexual species must either have a much lower mutation rate or suffer a large number of genetic deaths." Our numerical simulations also vividly confirm Crow's second assertion. Given the same idealized conditions as produced extremely effective halting of mutation accumulation, but excluding sexual recombination, the MCM effect vanishes completely (Figure 1). Genetic degeneration progresses like clockwork when we model asexual species, even given equal mutation effects, no environmental variance, and full truncation selection. Therefore the MCM mechanism does not appear to apply to dandelions, viruses, most bacteria, and innumerable other microbes. This means that the MCM mechanism is not generically applicable in the biological realm, and cannot be a generalized solution to the problem of mutation accumulation. The balance of this study has focused on populations having regular sexual recombination.

Even given normal sexual recombination combined with uniform mutation effects and zero environmental noise, the MCM effect essentially disappears given natural probability selection (Figure 1). It is widely understood that probability selection is what is generally happening in nature. Truncation selection is the type of artificial selection employed consciously by plant and animal breeders, and is not generally applicable to natural populations (truncation selection seems to primarily be invoked for natural populations only when the MCM is deemed desirable). However, it is significant to note that given uniform mutation effects and probability selection, as the population approaches zero mean fitness (extinction), we often observe clear evidence of the MCM effect, and this can slow or even stop mutation accumulation. This effect is weakly evident in Figure 2. But this special phenomenon actually helps prove the point, because what is happening as the population approaches extinction is that selection is forced from probability selection into a type of truncation selection. This actually helps demonstrate that some form of truncation is required to activate the MCM. In this particular case, as the population's mean fitness approaches zero, many individuals have a fitness of zero or less, and they are hence unconditionally removed from the population (truncation).

When selection regimes are employed that are intermediate between probability selection and truncation selection (partial truncation), with mutation effects still being equal and with no environmental variance, there is still a strong MCM effect — which can either slow or halt mutation accumulation (Figure 2). Low levels of environmental variation can interfere with the MCM effect under partial truncation, but cannot by itself negate it (Figure 2). However, higher levels of environmental variation can strongly interfere with the MCM effect (Figure 2), most especially in the case of full truncation selection (not shown).

Although it is instructive to model uniform mutation effects on fitness, we know that mutation fitness effects are never uniform, and are actually extremely variable in all living systems. Therefore we tested how effective the MCM might be, given a distribution of mutation effects which was intermediate between a totally uniform fitness effect and a realistic distribution for higher organisms. We did this by doing experiments using a Weibull distribution of mutation fitness effects having a higher than normal minimal fitness effect (.00001). This is 3,000 times greater than what we consider reasonable (i.e., the inverse of the functional genome size). In a large genome, there should be many mutation effects smaller than one is a million or even one in a billion. Even in free-living bacteria, deleterious mutation effects should minimally range down to .00001. We did a series of experiments using this more limited range of mutation effects. Given this distribution, the mutations that were accumulating only ranged from .001 to .00001 (just one to two orders of magnitude). We found that even given this relatively narrow range of accumulating fitness effects, mutation accumulation and fitness decline could not be halted, even under full truncation selection (Figure 3). Some non-linearity of mutation accumulation and fitness decline is evident early in these runs, but in all four experiments these rates eventually became very linear. Mutation accumulation and fitness decline then progressed at constant rates all the way to population collapse just prior to extinction, regardless of whether selection was full truncation, strong truncation, weak truncation, or probability selection. The selection mode merely affected the time to extinction (Figure 3).

We also experimented with an even narrower Weibull mutation distribution, with a lower limit of .001 (data not shown). When we combined this distribution with partial truncation selection (0.1), low environmental variance (heritability = 0.2), and a high mutation rate (10) the population went to extinction very rapidly, due to the high mean mutation effect. However if the mutation rate was reduced to 5, then there was sufficient time for the MCM mechanism to operate, and the population stabilized prior to extinction. This is hardly surprising when we consider that under those favorable conditions, the selection threshold was below .001, making the range of accumulation mutations extremely narrow (less than one order of magnitude). Because high-impact deleterious mutations (i.e., with fitness effects above .001) are rare, and because the few that do arise are rapidly removed from the population, the mutation accumulation problem is largely confined to low-impact mutations. To the extent that we can define conditions where there are no low-impact mutations, the mutation accumulation problem largely goes away. However, this is not realistic, especially for organisms with large functional genomes, where most mutations should have extremely subtle effects.

We believe that the lower limit of mutation effects for a given species can reasonably be approximated to be one over the functional genome size. In this light, a viroid might reasonably have mutation fitness effects that only range down to .001, and a typical virus might reasonably have fitness effects that only range down to .0001. Extremely small genomes of this type might reasonably be subject to the MCM — except for two problems. Firstly, most of these tiny genomes lack sexual recombination, and secondly such organisms should normally be subject to probability selection. Either of these is sufficient to negate the MCM effect. Indeed, even when we model the influenza virus (10,000 bp), which does have some limited recombination, the MCM effect is very weak. In such a case the mutation count increase is not initially strictly linear, yet mutation accumulation is not halted (data not shown). When we model genomes that would reflect any free-living organism (genomes of 10^6 bp or above), under all reasonable parameters settings, MCM very consistently fails and mutation accumulation is linear.

A possible objection to our methodology might involve the artificiality of defining certain nucleotides to be "mutant", since it might be argued from an evolutionary point of view that *all* nucleotides arose as mutations. This line of thinking would suggest that any hypothetical selection mechanism based upon "mutation count" is inherently contrived and artificial. This objection is reasonable; however it must be addressed to those who developed the MCM model in the first place. We are merely testing the viability of that concept. The MCM hypothesis obviously rests entirely on the idea that individuals within a population can actually have knowable and meaningful differences in their "mutation counts". On a practical level we have simulated this by assuming a genetically uniform population (as mighty arise after an extreme bottleneck), with all individuals initially having the same genotype and the same relative fitness of 1.0. This starting reference genotype then serves as our basis for defining all "new" mutations and for tracking each individual's subsequent "mutation count". All new mutations represent deviations from the starting reference genotype.

Careful numerical simulation reveals that the MCM hypothesis has very limited power to explain how deleterious mutation accumulation can be halted in natural populations. The mechanism works very well under highly unrealistic conditions, but fails when realistic parameters are applied. Previous numerical simulation studies have already clearly demonstrated that mean mutation count per individual consistently increases linearly over time [8, 9], given realistic parameter settings. Whenever there is a realistic distribution of mutation effects, even when all other relevant parameters are optimized, there is no stabilization of mutation count or fitness, indicating that meaningful selection against higher mutation-count individuals is not happening. We conclude that the MCM is not generally operational.

The primary reason MCM fails is because in real populations the distribution of deleterious mutational effects is never uniform, but must vary over many orders of magnitude. Deleterious mutation fitness effects should range from negative one (lethal), down to parts per million or even parts per billion. Therefore there must be a vanishingly small correlation between phenotypic fitness and actual mutation count. This means there can be no mechanism whereby natural selection can do any "judicious choosing" to remove individuals with slightly higher mutation counts, as required by Muller [21] and Crow [19, 22].

In this paper, we effectively falsify the general MCM hypothesis. In a companion paper [10], we falsify the synergistic epistasis hypothesis, which is a more elaborate model, but also employs the concept of focusing selection against high mutation-count individuals. These two hypotheses have been used for several decades, to try to dismiss the mutation accumulation problem. The falsification of both hypotheses leaves modern genetic theory without any credible mechanism that might halt genetic degeneration within natural populations. This strongly

Can Synergistic Epistasis Halt Mutation Accumulation? Results from Numerical Simulation

John R. Baumgardner[1*], Wesley H. Brewer[2] and John C. Sanford[3]

[1]*Department of Earth and Environmental Sciences, Ludwig Maxmilians University, Theresienstrasse 41, 80333 Munich, Germany.* [2]*Fluid Physics International, P.O. Box 4915, Mississippi State, MS 39762, USA.* [3]*Dept. Hort. Sci., NYSAES, Cornell University, Geneva, NY 14456, USA. (*Corresponding author: jrbaumgardner@cox.net).*

Abstract

The process of deleterious mutation accumulation is influenced by numerous biological factors, including the way in which the accumulating mutations interact with one another. The phenomenon of negative mutation-to-mutation interactions is known as synergistic epistasis (SE). It is widely believed that SE should enhance selective elimination of mutations and thereby diminish the problem of genetic degeneration. We apply numerical simulation to test this commonly expressed assertion.

We find that under biologically realistic conditions, synergistic epistasis exerts little to no discernible influence on mutation accumulation and genetic degeneration. When the synergistic effect is greatly exaggerated, mutation accumulation is not significantly affected, but genetic degeneration accelerates markedly. As the synergistic effect is exaggerated still more, degeneration becomes catastrophic and leads to rapid extinction. Even when conditions are optimized to enhance the SE effect, selection efficiency against deleterious mutation accumulation is not appreciably influenced.

We also evaluated SE using parameters that result in extreme and artificially high selection efficiency (truncation selection and perfect genotypic fitness heritability). Even under these conditions, synergistic epistasis causes accelerated degeneration and only minor reductions in the rate of mutation accumulation.

When we included the effect of linkage within chromosomal segments in our SE analyses, it made degeneration still worse and even interfered with mutation elimination. Our results therefore strongly suggest that commonly held perceptions concerning the role of synergistic epistasis in halting mutation accumulation are not correct.

Key words: mutation accumulation, synergistic epistasis, mutational meltdown, numerical simulation, Mendel's Accountant

Introduction

There is a significant body of literature indicating that direct selection against deleterious mutations is insufficient to halt mutation accumulation [1–5]. This has recently been validated using biologically realistic numerical simulations [6–9]. A primary reason for this result is that most deleterious mutations have extremely small effects on fitness and thus are invisible to selection [10–15].

Some have argued that this fundamental issue might be resolved if selection is not ultimately based directly upon the biological effects of individual mutations acting in isolation of one another, but instead is based largely upon interactions between mutations, interactions that act to compound the biological effects of the individual mutations. Such effect-enhancing interaction between deleterious mutations has been termed synergistic epistasis (SE). It is widely claimed that SE acts to slow deleterious mutation accumulation and thereby helps prevent genetic degeneration and mutational meltdown [16–29]. We will refer to this concept as the SE hypothesis.

The logic behind this hypothesis is somewhat counterintuitive. The reasoning is that, while the number of mutations per individual increases in roughly a linear manner, the number of potential mutation-mutation interactions increases in a non-linear fashion. The number of pair-wise interactions increases as the square of the mutation count, for example. Hence, if SE effects are significant, then at a certain point individuals who carry the most mutations might conceivably begin to display a significant reduction in fitness relative to the rest of the population. This, in turn, might increase selection against high mutation count individuals and thereby eliminate a larger total number of mutations from the population than would occur otherwise. Eventually, this intensifying selection against high mutation count individuals, if sufficiently strong, might stabilize the mutation count and thereby halt further genetic degeneration. This SE hypothesis is counterintuitive, because in most circumstances increasing the negative effects of deleterious mutations on fitness only serves to increase the rate of fitness decline and hasten mutational meltdown and extinction. For the SE hypothesis to be viable, the selection against high mutation count individuals must be sufficiently strong so that at some point it is able to counter the associated increased rate of fitness decline.

The circumstances under which selection, apart from any SE effects, can come to be based primarily upon mutation count, rather than the additive or multiplicative fitness effects of the individual's mutations, has been discussed by several investigators [17-20]. In a companion paper [9], we apply numerical simulation to test the efficacy of selection based upon mutation-count entirely apart from SE effects. In this paper we apply numerical simulation in a similar manner to evaluate whether or not SE has the ability to halt mutation accumulation.

Interactions among mutations within a genome are diverse in their impact. Any two mutations may act independently of each other (that is, have no interaction, which leads to the standard additive model), act multiplicatively (the multiplicative model), diminish each other's effect (antagonistic epistasis), or compound each other's effect (synergistic epistasis). Undoubtedly, all of these types of interactions operate in any sizeable genome. Therefore it is not reasonable to assume all mutation-mutation interactions in any genome are exclusively of a single type. Nevertheless, non-interaction should be the norm, with the other types of interaction being the exceptions. The only rationale for modeling a 100% multiplicative model or a model with SE contributions from 100% of the deleterious mutation interactions is to try to understand in which direction the exceptional interactions tend to pull the overall behavior away from the norm of additivity.

It is noteworthy that the main exceptions to the general rule of additivity pull in opposite directions. Both antagonistic epistasis and the multiplicative model drive population fitness in the direction opposite to that of synergistic epistasis. That is to say, as mutation count increases, both the antagonistic epistasis model and the multiplicative model cause fitness decline to slow down, while SE causes fitness to decline faster and faster. So when combined, the other types of interactions should cancel out the effects of the SE interactions in whole or in part, leaving what should closely approximate an additive model. Therefore, in a complex genome it would seem most realistic to assume the additive model, with interactions constituting a low level of "genetic noise" (which we would normally just refer to simply as "epistasis" or "general epistasis").

We therefore conclude that a genetic model in which all mutations interact in a synergistic manner is an artificial model, one that does not represent any real biological population. Moreover, such a model contradicts an extensive body of population genetics literature, which for nearly 90 years has been built on the assumption that most mutational effects combine either additively or multiplicatively (the latter effectively counteracting any generalized SE effect). The idea of genome-wide generic SE interaction is virtually never invoked, except as special pleading as a theoretical mechanism to halt mutation accumulation and degeneration. The present study uses numerical simulation to show that even if there were widespread and generic SE, it still could not halt mutation accumulation. Instead, what is seen is that as SE effects become stronger, there is more and more genetic degeneration, just as logic and common sense would suggest.

Methods

The program Mendel's Accountant [6], hereafter referred to as Mendel, is applied to study the effects of SE on mutation accumulation and genetic degeneration. This software uses realistic genetic accounting to study mutation accumulation [7–9].

There is enormous biological complexity inherent in the mutation/selection process when it is considered at the level of the whole genome and the whole population. It is not reasonable to assume that such complexity can be effectively captured by any tractable set of analytic equations. However, thanks to the computational capabilities now available, complex systems of this type can now be routinely analyzed using numerical simulation. Mendel, developed over the past five years, is a genetic accounting program which can actually do this. This software models and tracks a complete biological system, from individual mutations, to mutation-mutation interactions, to linkage blocks, to chromosomes, to genotypes, to phenotypes, to mating/recombination events, to sub-populations, to whole populations. Using Mendel, all the appropriate parameters are accounted for and are specified by the program user, and the computational processing is faithful to our understanding of how genetic systems operate.

The basic process underlying this numerical simulation is as follows. Mendel creates a population with specified biological characteristics. The individuals in this population are allowed to create gametes, mate, and generate offspring for a new generation. Each offspring inherits the mutations in the gametes from its two parents, including possible new mutations that arose in the germ line of the parents during their lifetime. Each new mutation has its own fitness effect and its own genome location involving a specific linkage block. Mendel then calculates the genotypic fitness of each offspring based upon the net effect of all the mutations it carries. Random environmental noise is next added to obtain a value for phenotypic fitness. Selection is then applied, based on phenotypic fitness, to determine which of the offspring will mate and reproduce to create the next generation. Although Mendel readily treats beneficial mutations, for the sake of clarity in this paper we include deleterious mutations only. We use Mendel's human default parameters, as might reflect a small human population, except as indicated. Apart from these exceptions, the default parameters in all our experiments are as follows: ploidy = diploid; reproduction = sexual; mating = random; linkage = dynamic; new mutations per individual = 10; beneficials = none; offspring per female = 4 (resulting in 50% selective elimination); population size = 1000; generations = 2000; haploid genome size = 3 billion; rate of high impact mutations (fitness impact of 0.1 or higher) = .001; gene expression = complete co-dominance; fitness heritability = 1.0; fertility decline with fitness decline = none; selection type = probability.

Modeling general epistasis

Mutational interactions are, by their very nature, unique and specific, so it is somewhat problematic to account for interactions in a generic manner. However, there is one generic aspect of nucleotide interactions which we can easily describe and model, namely, the phenomenon of general epistasis. General epistasis reflects the net effect of all types of mutation-mutation interaction. When there is genetic diversity within a sexual population and the segregating nucleotides recombine with each other every generation, many specific interactions in the parent are destroyed, and many new interactions are created in the progeny. These changing interactions generation to generation result in what is called epistasis. The overall effect of such epistasis is a type of non-heritable variation (noise), resulting in lower heritability and reduced selection efficiency. So the dominant effect of the ever-changing nucleotide interactions within the genome is generic epistasis, which hinders selection efficiency to a modest degree. In numerical simulations, the phenomenon of general epistasis can very reasonably be modeled simply by decreasing the genotypic heritability parameter by an appropriate amount.

Modeling additive interactions

While generic epistatic interaction as described above is significant, by far the most common relationship between any two given nucleotides should be *non-interaction* (or vanishingly small interaction). Like any two misspellings in a long text, any two nucleotides in a large genome will have a vanishingly small chance of having any meaningful direct interaction. When two letters are changed in a text, they generally need to be in the same word, or at least in the same sentence, to have any reasonable likelihood of interaction (wherein one affects the meaning of the other). In the same way, any two mutations are unlikely to interact significantly unless they are in the same gene, or at least in the same pathway. The vast majority of mutations should not significantly interact with one another.

The *non-interaction* of most mutations is the theoretical basis for the conventional additive model for combining the effects of mutations within an individual. The additive model assumes that as mutations accumulate, each new mutation affects fitness independently of the others. Under this model if an individual in a population has an initial fitness of 1.0, and we introduce two independent harmful mutations, with each reducing fitness by an increment of 0.1, the resulting fitness will be 0.8. If we then introduce a good mutation that increases fitness by an increment of 0.1, the new fitness will be 0.9. The mutational effects of all the mutations in a given individual are simply added. The additive model is commonly employed in population genetics because in a large genome it is only reasonable to assume that non-interaction is the rule and interaction is the exception. Mendel employs the additive model of mutation effect combination as its default.

Modeling multiplicative interactions

The most common alternative to the additive model is the multiplicative model. Under this model, as mutations accumulate, their mutational effects combine multiplicatively. This means that as deleterious mutations accumulate, they have less and less effect relative to the original fitness, while as beneficial mutations accumulate they have greater and greater effect. To draw an analogy, deleterious mutations act similarly to inflation eroding the value of a bank account, while beneficial mutations act as earned interest which is being compounded. This type of interaction is only reasonable where mutations act in a sequential manner, with one interaction building upon the effect of another, in series. This might plausibly occur when multiple mutations affect the same biochemical pathway. While some specific sets of mutations will doubtless interact multiplicatively, it is not reasonable to assume that all mutations would or could interact in this way. It is also not reasonable to use the multiplicative model as the primary method of combining mutational effects, because a purely multiplicative model can never reach a fitness of zero (i.e., extinction). In fact, under the strict multiplicative model, a small genome might have every nucleotide become mutated, with the genotype still retaining a positive fitness.

In the big picture, on the level the whole genome, the additive model should most generally be true, with the multiplicative model being applicable only to a limited number of special interactions. In other words, multiplicative interactions should only represent deviations from the norm of additive interaction. Mendel has been designed to allow any blend of additive and multiplicative interaction, ranging from 100% additive to 100% multiplicative. In our opinion, a fraction of 0.99 additive and 0.01 multiplicative interactions is reasonable, but this choice is left to the Mendel user.

By allowing any fraction of additive and multiplicative general interaction, and by adjusting heritability downward to allow for general epistatic noise, Mendel allows for the modeling of the primary mutation-mutation interactions.

Modeling synergistic epistasis

Mendel has also been designed, however, to handle the special type of reinforcing interaction between mutations known as synergistic epistasis. Like multiplicative interaction, SE interaction must be viewed as a deviation from the general rule of non-interaction (i.e., the additive model). SE interaction implies that as deleterious mutations accumulate, each additional mutation has a greater and greater effect on fitness. This is the exact antithesis of multiplicative interaction, wherein each deleterious mutation has less and less effect on fitness. Both multiplicative and SE interactions represent deviations from the additive model, but they pull in opposite directions. To the extent that multiplicative and SE interactions occur at a similar frequency, they should largely cancel each other. Viewing the genome as a whole, if 90% of all mutations combine additively, and 5% combine multiplicatively and 5% combine via SE, the result should be that the two types of deviation mostly cancel, yielding results nearly equivalent to a purely additive model. For most genetic simulations, a realistic and practical choice is simply to use the standard additive model.

Because SE has often been invoked as a hypothetical mechanism which might be able to halt mutation accumulation, we have included it as an option in Mendel. In doing so, we have endeavored to treat SE in as biologically realistic a manner as possible. Our implementation, however, involves a few assumptions which we shall now review.

First, we assume a reference genotype. From an evolutionary perspective all nucleotides have arisen by mutation, so viewed from that perspective, all nucleotides are "mutant". However, to treat SE in the normal sense of that term logically requires a reference genotype relative to which "mutations" may unambiguously be defined. The approach employed in Mendel is to assume a population with zero initial genetic variation, as might be approximated by a population after a severe bottleneck at a specific point in time. All subsequent mutations causing deviation from that starting genotype are tracked individually and contribute to the distinct set of mutations and hence to the mutation count of each member of the population in subsequent generations. This assumption of a reference genotype is inherent to Mendel's underlying formulation and does not apply in any special way to the treatment of SE. Note that when there is just one mutation in a genome, all the interactions involving that mutation are with non-mutant nucleotides, so 100% of that mutation's fitness effect is due to its interactions with non-mutant sites. Thus all solitary mutations have a non-epistatic effect on fitness that arises entirely from its interactions with non-mutant nucleotide sites.

As additional mutations accumulate, however, there are more and more potential mutation-mutation interactions. As the mutation count increases, the deleterious SE contribution to fitness increases at an accelerating rate, accelerating because the number of possible pair-wise interactions increases in proportion to the square of the number of mutations. A second assumption is that we restrict our SE treatment to these pair-wise interactions, that is, to interactions between pairs of individual mutations. A third is that we assume the strength of the SE effect on fitness is directly proportional to the non-epistatic fitness effects of each of the mutations in the pair. This means that if a mutation's effect on the non-mutant genome is small, then the SE contribution from its interactions with other mutations likewise is small.

We further assume that, in regard to SE interactions, it is proper to distinguish between linked mutation pairs, that is, those which reside within the same linkage block on a chromosome and those pairs which reside in separate linkage blocks. Linked mutations are inherited together. Not only are the non-epistatic fitness effects of all such mutations inherited together, but the SE effects of all their mutual interactions are as well. By contrast, genetic recombination progressively tends to scramble mutations that are not linked together. Hence, the SE contribution from non-linked mutations has a transient component. The SE effects arising from the non-linked interactions which change from one generation to the next act like a type of noise that interferes with the selection process. Therefore, realistic modeling of SE requires that linked and non-linked SE effects be treated separately. We therefore partition the SE effects on fitness into two parts, one involving interactions between deleterious mutations occurring in the same linkage block (linked interactions) and the other part involving interactions of deleterious mutations on different linkage blocks (non-linked interactions). SE effects from linked interactions are inherited, while part of those from non-linked interactions are transient and act, in effect, as a type of noise as far as the selection process is concerned.

Another major difference between linked and non-linked SE interactions is the relative magnitude of their effects. Intuitively, the strongest SE interactions should be within the same linkage block, even as two misspellings in an encyclopedia are likely to interact more strongly if they occur within the same chapter or paragraph or sentence. Two mutations are most likely to interact if they occur within the same protein-coding sequence or at least the same genic region. Therefore, the treatment in Mendel includes separate scaling factors for each of these two categories of SE effects. Normally, the scaling factor for linked interactions should be much larger (perhaps by a factor of 1000) than the one for non-linked interactions.

Since linked SE interactions are inherited perfectly, they must always make the degeneration problem worse. This is because the SE contributions act to reinforce the negative non-epistatic fitness effects of the mutations on each linkage block and, in effect, make the non-epistatic effects even more negative.

Let us now consider how Mendel actually treats the linked SE interactions. We assume the amplitude of the linked SE effect of each pair-wise interaction to be proportional to the product of non-epistatic fitness effects of the paired mutations. If a mutation's effect on the non-mutant genome is small, the SE contribution from its interactions with other mutations is likewise small. If we denote the number of mutations in a given linkage block by m, the number of pair-wise interactions each mutation has with the other mutations is m-1, and the total number of unique pair-wise interactions in the linkage block is $m(m-1)/2$. Mendel stores the fitness f (relative to unity, when no mutations are present) of each linkage block as well as the number m of mutations it carries.

Mendel computes the SE contribution to fitness whenever a new mutation is added to the linkage block. This contribution is proportional to the non-epistatic effect of the new mutation times the sum of the non-epistatic effects of each of the individual mutations already present on the block. When these SE contributions are accumulated, each of the $m(m-1)/2$ unique pair-wise interactions is accounted for. These contributions are scaled by a user-specified factor α. We also assume co-dominance for these SE interactions, which implies each haploid occurrence of a mutation gives 50% expression of the mutation's total non-epistatic value. This reduces the SE effect by a factor of 0.25. We note that, because mutations within a given linkage block are passed intact from one generation to the next, the SE effects arising from linked mutations are also passed intact from parent to offspring. Therefore, as we have already noted, the net result of including SE relative to linked deleterious mutations is always to increase the magnitude of their negative effect on fitness.

Mendel treats the non-linked SE interactions in a similar manner. Let M be the total number of mutations in the genome of a given member of the population and n be the number of equal-sized linkage blocks. The total number of unique pair-wise interactions between mutations is $M(M-1)/2$, the mean number of mutations per linkage block is M/n, and the approximate number of linked interactions is $n(M/n)[(M/n)-1]/2 = M(M-n)/2n$. With this approximation, the number of non-linked interactions becomes $(1-1/n)M^2/2$ and the ratio of the number of non-linked interactions to linked ones is $n-1/(1-n/M)$. With n typically 1000 or greater, as M becomes much greater than n, this ratio approaches n. In other words, as the total number of mutations becomes large relative to n, the number of non-linked mutations approaches n times the number of linked mutations.

Let us denote by F the overall genotypic fitness, apart from any SE effects, of a given member of the population. We assume the amplitude of the non-linked SE effect of each pair-wise interaction to be proportional to the product of non-epistatic fitness effects of the two mutations in each pair. The total non-linked SE fitness contribution is then nearly proportional to the sum of the non-epistatic fitness effects of all the individual mutations, $(1-F)$, but scaled to account for the portion of the mutations which are linked using the factor $(1-1/n)$, times the mean non-epistatic fitness effect of these mutations, $(1-F)/M$, times the number of unique pair-wise interactions, $(1-1/n)M/2$, that each non-linked mutation has with the others. This estimate has included the contributions from the self-interaction of each of the mutations, contributions that should not be included and which

Mendel omits. However, when the total number of mutations is large, the sum of these contributions is relatively small, in which case the estimate is reasonable accurate. We again assume co-dominance, which implies each haploid occurrence of a mutation gives 50% expression of the mutation's non-epistatic value. This reduces the overall contribution by a factor of 0.25. We scale this non-linked SE contribution with a user-specified input parameter β. As already mentioned, one expects that interaction between mutations within the same linkage block will, on average, have much greater SE effects than mutations which are more distant within the genome. Hence, a value for β much less than α is usually appropriate. The resulting approximate expression for the non-linked SE contribution to individual fitness is therefore $0.125\beta(1-F)^2(1-1/n)^2$. Mendel corrects this by subtracting away the sum of the self-interaction contributions.

We note that the negative SE contribution to fitness from all the non-linked interactions is proportional to $(1-F)^2$. Since the number of linkage blocks is typically 1000 or greater, the factor $(1-1/n)^2$ can usually be approximated as unity. The SE contribution from non-linked interactions is larger for individuals in the population with lower fitness and smaller for individuals with higher fitness. It therefore tends to accentuate the spread in fitness across the population and thus to enhance selection efficiency. Since fitness F tends to be dominated by the relatively few mutations in the high-impact tail of the fitness effect distribution, F is largely insensitive to mutation count. This non-linked SE contribution is therefore insensitive as well. Since the mean mutation fitness effect is directly proportional to $(1-F)$, the overall impact of this SE contribution from non-linked interactions is to increase the mean negative mutational fitness effect, just as is the case for the SE contribution from the linked interactions. Therefore, the net effect of SE for both linked and non-linked interactions should be a higher rate of fitness decline with time. There is nothing from a theoretical standpoint to suggest otherwise.

Finally in this section, let us estimate what a biologically reasonable value might be for the non-linked scaling factor β. The total SE fitness contribution in Mendel for non-linked mutations, assuming no linkage at all, is approximated by the expression $0.125\beta(1-F)2$, where F is the individual genotypic fitness. A plausibly hard upper bound on the magnitude of β might be the value that drives F to zero when, without SE, the fitness F of a given individual is 0.5. In this case, $\beta = 0.5/(0.125\times 0.5^2) = 16$. This means that, if the accumulated mutations in a given individual reduce the fitness of a given individual to 0.5 without SE, then with SE and $\beta = 16$, the fitness of this individual drops to zero. In our view, a biologically realistic value for the non-linked scaling factor β should therefore be no larger than 1.0 and more plausibly on the order of 0.1 or less.

Results

Preliminaries

We ran a number of experiments with Mendel's Accountant to ascertain a reasonable value for the linked scaling factor α relative to the non-linked factor β. We found that choosing α some 2000 times larger than β gave comparable SE contributions from linked relative to non-linked mutations. We considered cases with just under 2000 total linkage blocks for the diploid genome, or about 1000 for the haploid genome. This implies much larger linkage blocks and more linked mutations that observations would suggest for most organisms. Therefore, α should almost certainly be chosen larger than 2000 relative to β when the number of linkage blocks is increased if one wants the SE contribution from linked mutations to be comparable to that from linked mutations.

Large SE effects and modest selection pressure

We begin our exploration of the SE effects on fitness with SE scaling factors α and β that are large but with the selection pressure, controlled by fertility, relatively low. For a low level of selection pressure we chose a fertility of 1.1, which for a constant population size, implies that only 10% of the offspring in each generation do not reproduce. For SE scaling parameters we chose 10 for the non-linked mutation pairs and 2×10^4 for the linked mutation pairs, or 2000 times the non-linked scaling factor β. These parameter choices are about 100 times the maximum values we consider to be biologically realistic. What we found was that the effects on fitness after 2000 generations were too small to quantify, even though mean fitness due to normal mutation accumulation had decreased by 33%. Typically, we found that the mean number of accumulated mutations after 2000 generations was about 0.7% smaller with this level of SE relative to no SE. Despite the small effect on fitness, these values of 10 for β and 2×10^4 for α are likely still far higher than is realistic for most natural populations. Nevertheless, these experiments prompted us to explore what larger values for α and β were might reveal concerning SE behavior.

Let us consider cases with the same low selection intensity but with $\beta = 300$ and $\alpha = 6\times10^5$, both 30 times larger than before. Figure 1 displays the mutation accumulation and the population fitness histories for the following four cases: (1) no SE effects, (2) SE effects from non-linked interactions only, (3) SE effects from both linked and non-linked interactions, and (4) SE effects from both linked and non-linked interactions, but with both scaling factors doubled.

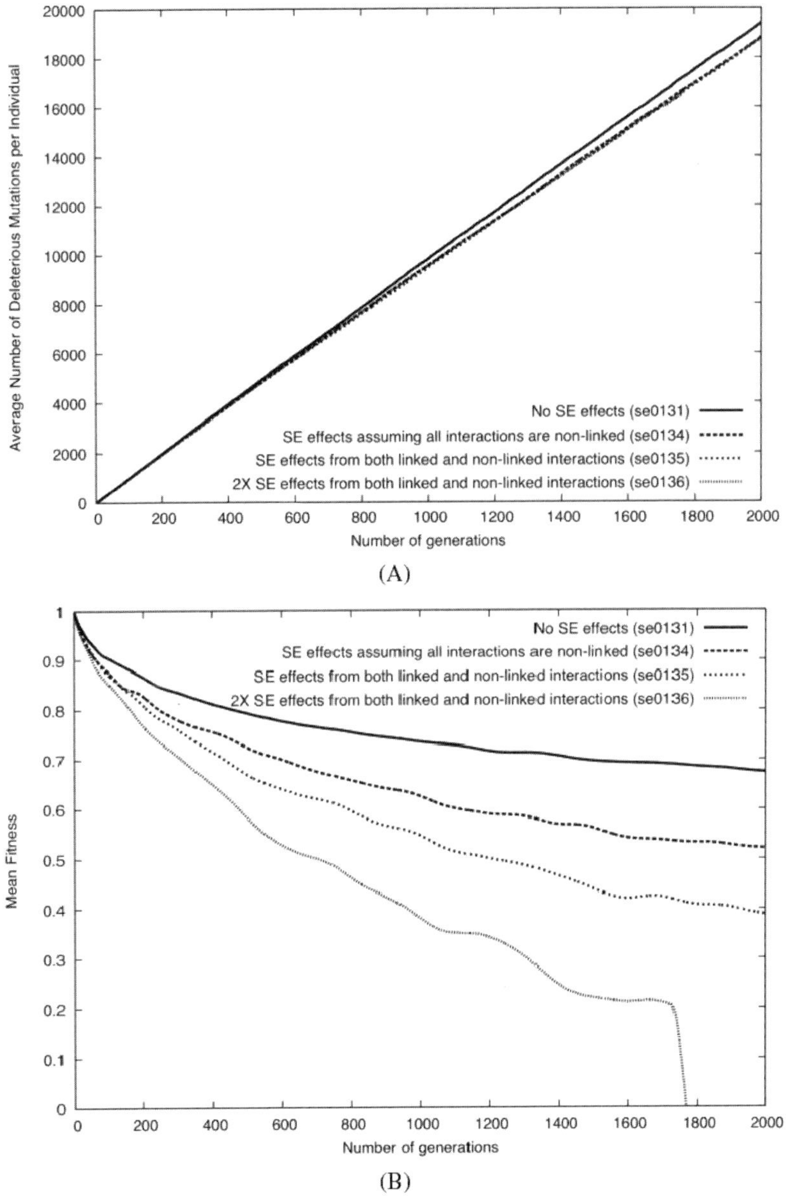

Fig. 1. Mutation accumulation (A) and the population fitness histories (B) for modest selection pressure and extreme SE interactions for four cases: (1) no SE effects, (2) SE effects from non-linked interactions only, (3) SE effects from both linked and non-linked interactions, and (4) SE effects from both linked and non-linked interactions, but with both scaling factors twice as large as in case (3). All cases apply probability selection, perfect genotypic heritability, and a fertility of 1.1, which implies 10% of the offspring in each generation do not reproduce in the next. The scaling factor for non-linked SE interactions in cases (2) and (3) is 3×10^2 and for linked interactions in case (3) is 6×10^5

Several features of these numerical experiments are readily apparent. First, the effects on mutation accumulation are relatively small given the large values of the SE scaling factors. With the mean mutation rate of 10 new mutations per offspring, if there were no selection, the average number of mutations per individual would be 20,000. The actual numbers of accumulated mutations per individual after 2000 generations for the first three cases are 19405, 18807, and 18763, respectively. The average number of accumulated mutations for case (3) is only 642 (3%) fewer than the case with no SE included, despite the large SE scaling factors. Also noteworthy is the fact that case (4) undergoes mutational meltdown at generation 1766 due to the strong deleterious SE effect on fitness.

These experiments show that it is possible, at least numerically, to make the SE effect sufficiently strong to drive a population to extinction. However, the scaling factors required for this to take place within 2000 generations are extreme.

Extreme SE effects and moderate selection pressure

In our next set of experiments we increase the selection pressure to a moderately high level. Instead of a fertility of 1.1, we choose a fertility of 2.0. This means that twice as many offspring are produced in each generation than are allowed to reproduce in the succeeding generation. That is, the selection process excludes half the offspring in each generation from reproducing in the next. For SE scaling factors we use 10^5 for non-linked interactions and 2×10^8 for linked interactions, and then examine a case with both scaling factors increased. Figure 2 displays the mutation accumulation and the population fitness histories for the following cases: (1) no SE effects, (2) SE effects assuming all interactions are non-linked, (3) SE effects from both linked and non-linked interactions, and (4) SE effects from both linked and non-linked interactions, but with scaling factors five times larger. The mean numbers of accumulated mutations after 2000 generations for the first three cases are 19570, 16510, and 16110, respectively. Cases (4) underwent mutational meltdown in generation 1960. We note that even with the SE effects exaggerated to this degree there is no hint that mutation accumulation can be halted, or even slowed to any significant degree, before mutational meltdown takes place.

Extremely exaggerated SE effects and extreme selection pressure

For this final set of cases we retain the fertility of 2.0, but instead of probability selection, we apply truncation selection. Truncation selection is artificial in that there is no randomness in the selection process. With a fertility of 2.0, each offspring

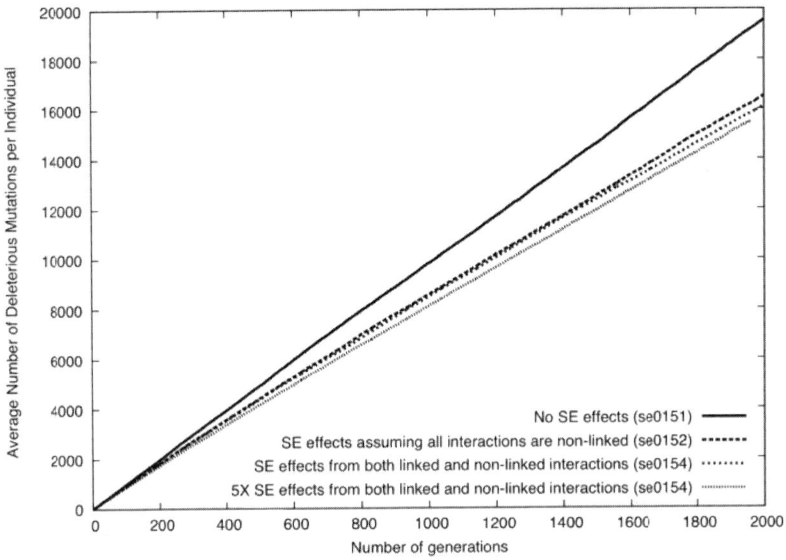

(A)

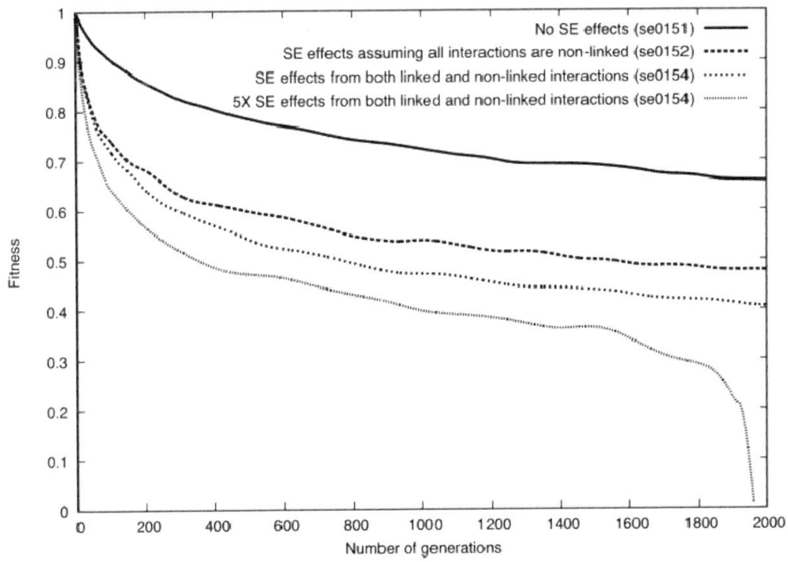

(B)

Fig. 2. Mutation accumulation (A) and the population fitness histories (B) for moderate selection pressure and extremely exaggerated SE interactions for cases: (1) no SE effects, (2) SE effects assuming all interactions are non-linked, (3) SE effects from both linked and non-linked interactions, and (4) SE effects from both linked and non-linked interactions, but with scaling factors 5 times larger. All cases apply probability selection, perfect genotypic heritability, and a fertility of 2.0, which implies half the offspring in each generation do not reproduce in the next. The scaling factor is 10^5 for non-linked SE interactions in cases (2) and (3) and 2×10^8 for linked interactions in case (3).

with fitness below the median value is selected away and does not reproduce in the succeeding generation, while each offspring with fitness above the median value does survive to reproduce. For SE scaling factors we use 5×10^5 for non-linked interactions and 10^9 for linked interactions, the same values that gave meltdown in the previous set of experiments. We also include a non-linked case with a scaling factor three times as large. Figure 3 displays the mutation accumulation and the population fitness histories for the following four cases: (1) no SE effects, (2) SE effects assuming all interactions are non-linked, (3) SE effects from both linked and non-linked interactions, and (4) SE effects assuming all interactions are non-linked with a scaling factor of 1.5×10^6. It is noteworthy that with truncation selection, fewer mutations accumulate for case (1) with no SE effects than for cases (2), (3), and (4) which include significant SE effects. The mean numbers of accumulated mutations after 2000 generations are 14388, 15480, 14510, and 14700 for these cases, respectively. In other words, instead of reducing mutation accumulation, SE actually *increases* the rate of mutation accumulation slightly in these experiments. This is almost certainly because SE increases the fitness variance considerably which makes the selection process less efficient. Also to be observed is that case (4) is in the process of mutational meltdown at generation 2000. These cases show persuasively that even with SE greatly exaggerated and selection efficiency also greatly exaggerated, SE fails to halt, or even slow, the accumulation of deleterious mutations.

Discussion

The importance of genic interactions

Like the letters in a text, nucleotides have meaning only within the context of other nucleotides, which is to say that nucleotides interact extensively. Such interaction between symbolic characters is the underlying basis for all language and all information systems. Functional genetic information is the basis of life and results from extensive networks of extremely specific, consistently positive, nucleotide-nucleotide interactions. Most mutations are deleterious because most represent disruptions of these networks of highly optimized sets of positive nucleotide-nucleotide interactions.

A given mutation's net biological effect arises from all of its actual interactions with other nucleotides within the genome. Each new mutation may have several or perhaps several dozen very specific significant interactions. A beneficial mutation is beneficial because it involves more positive total interactive effects than negative interactive effects. Most mutations are deleterious because, again, they disrupt

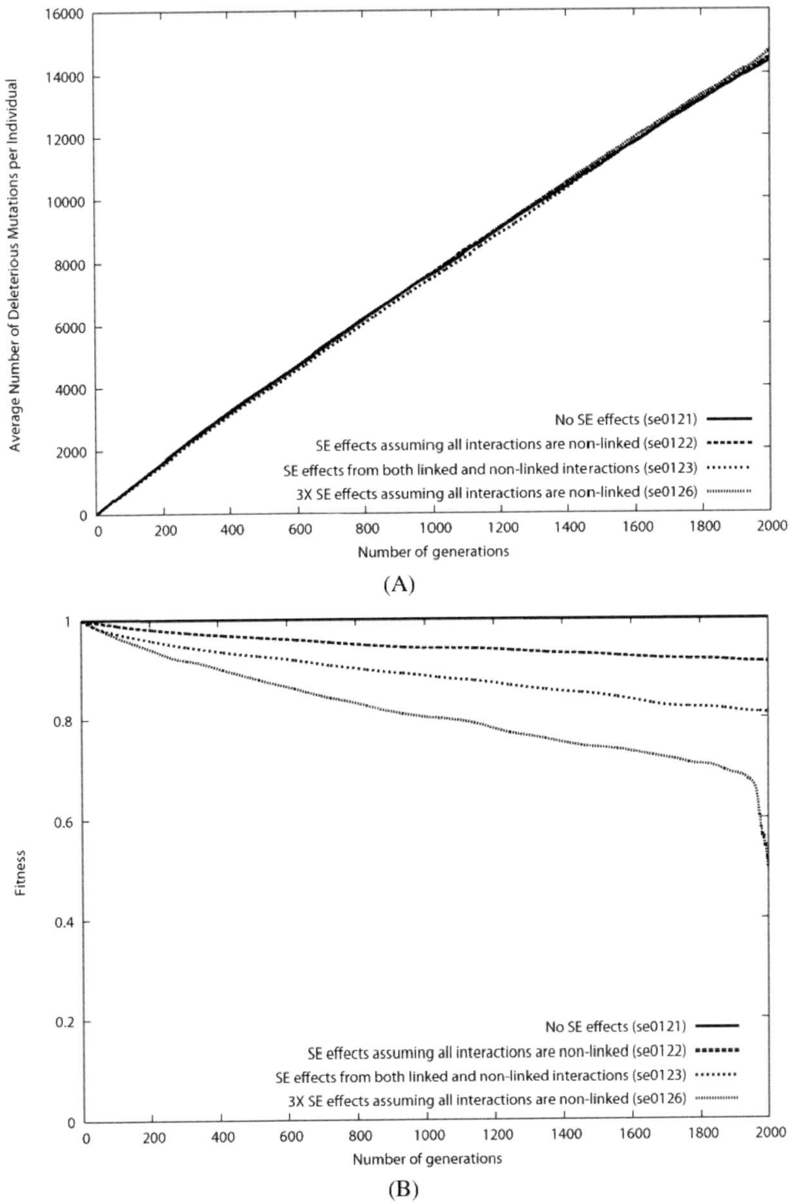

Fig. 3. Mutation accumulation (A) and the population fitness histories (B) for extremely exaggerated selection pressure and extremely exaggerated SE interactions for four cases: (1) no SE effects, (2) SE effects assuming all interactions are non-linked, (3) SE effects from both linked and non-linked interactions, and (4) SE effects assuming all interactions are non-linked, but with a scaling factor three times as large. All cases apply truncation selection, perfect genotypic heritability, and a fertility of 2.0, which implies 50% of the offspring in each generation do not reproduce in the next. The scaling factor for non-linked SE interactions for cases (2) and (3) is 5×10^5, for linked interactions in case (3) is 1×10^9, and for non-linked interactions in case (4) is 1.5×10^6.

or degrade existing highly specific positive nucleotide interactions that represent functional genetic specifications.

It is impossible to model all the possible interactions between nucleotides in a large genome. For example, the haploid genome of man has roughly 3 billion nucleotides. The number of potential pair-wise nucleotide interactions therefore is roughly 5×10^{18}. This is still an underestimate, because we are diploid and heterozygous at millions of sites, making the potential number of interactions even larger. Like widely spaced pairs of letters in a large book, the vast majority of nucleotide-nucleotide interactions surely have negligible effects. When there is a meaningful interaction, the biological effect can range from strongly negative to strongly positive. However, the vast majority of interactions that are not entirely neutral are surely still extremely subtle and nearly-neutral. We note that nearly-neutral interactions are beyond measurement, are not suited to empirical analysis, and therefore can be modeled only in a generic way.

The significance of SE

The primary reason that SE is of interest today is because it has been invoked as a mechanism that might possibly be able to halt mutation accumulation. This SE hypothesis, as we refer to it, has been embraced and advocated by several population geneticists, but it has never been demonstrated to work. In fact, the hypothesis is notably counterintuitive. In a non-selective setting, SE logically must accelerate genetic degeneration and lessen the time to extinction. This is because as deleterious mutations accumulate, SE guarantees that, on average, each new mutation must have a greater and greater deleterious effect.

However, it has been argued that, within a strongly selective setting, mutation accumulation might be halted if the SE effects were acute enough to activate what we refer to as the mutation-count mechanism (MCM). This mechanism requires selection to be strongly directed against those individuals within a population that have a higher mutation count than average. This conceivably might allow elimination of more mutations at less selective cost (that is, fewer individuals need be selected away). In a companion paper, we show that the MCM mechanism can operate only under certain highly artificial circumstances [9]. This special mechanism appears to be feasible only in sexually reproducing populations in which the range of mutational fitness effect variation is extremely narrow, the environmental variance is small, and truncation selection prevails. Arguably, these conditions never occur together in the natural world.

However, the MCM still might conceivably be activated, it has been argued, if extensive, strong, generic, non-linked SE interactions occur. Under such circumstances, fitness reduction from SE interactions might increase at an ever accelerating rate (while mutation count is increasing at a more or less constant rate), such that a *mutation-count threshold* arises. Above such a threshold, additional mutations might result in catastrophic fitness loss, triggering very strong truncation selection. If the SE effect were strong enough, mutation count might conceivably overwhelm the factors which otherwise would dominate (such as mutation rate, the mutation effect values themselves and their distribution, and environmental variance). At such a mutation-count threshold, truncation selection based primarily on mutation count might then potentially halt mutation accumulation and stop mutational degeneration completely. At the point of such a threshold, a newly arising small-effect mutation might have the same impact as a nearly lethal mutation (because both affect mutation count the same), even though in realty they might differ in their biological effects by orders of magnitude.

Is the SE mechanism described above even technically feasible? This study was designed to answer that question. If the SE effects are not actually strong enough to create the required level of truncation selection based on mutation count, then the very SE interactions conjectured to save the genome will instead more rapidly destroy it.

Testing the limits of SE

To probe the limits of how well the SE mechanism might conceivably work, we performed numerical experiments granting the SE hypothesis every possible advantage: 1) we allowed *all* mutation-mutation interactions to be SE interactions; 2) we included no interactions that were multiplicative or involved antagonistic or general epistasis; 3) we neglected the effects of linkage entirely; 4) we applied perfect truncation selection and perfect heritability; and 5) we allowed SE effects to assume extreme values, far beyond what is biologically realistic. Cases (2) and (4) of Figure 3 incorporate all of these generous concessions.

Are these concessions reasonable? No. It is not reasonable, for example, to make all mutations interact synergistically, because the vast majority of mutations should not interact with each other at all. In the big picture, non-interaction should be the norm, and simple additivity should describe how most mutations combine. Moreover, interactions that behave in a multiplicative manner as well as antagonistic epistatic interactions contribute to fitness in a manner opposite to that of SE. Further, it is not reasonable to neglect mutational linkage. Almost all SE interactions should be between mutation pairs that are tightly linked. Zero linkage is therefore a major concession benefiting the SE hypothesis. We make this concession simply because mutational linkage clearly neutralizes the mutation count mechanism [9]. When two mutations are linked, not only are the mutations inherited together but their SE effects are as well, and this results inexorably in accelerated fitness decline. Moreover, it is not reasonable to assume zero environmental noise (a heritability of 1) or to employ strict truncation selection. We make all these concessions only because in another paper we have already shown that the MCM is largely negated by low fitness heritability and probability selection [9].

Finally, although there should be a rational limit for how large each specific SE penalty should be relative to the basal, non-epistatic mutational fitness effect (as measured for a given mutation in an otherwise non-mutant genome), we allowed the amplitudes of the SE effects to become extreme. We showed earlier that the total SE fitness contribution in Mendel for non-linked mutations, assuming no linkage at all, is approximated by the expression $0.125\beta(1\text{-F})2$, where F is the genotypic fitness. We applied this formula to show that, if the accumulated mutations in a given individual reduce its fitness to 0.5 without SE, then with SE and a value for β of 16, the fitness of this individual drops to zero. We argued that a biologically realistic value for β should plausibly be on the order of 0.1 or less. In our numerical experiments we see a discernible SE effect only when we use unrealistically exaggerated non-linked SE scaling (300 and 600 in Figure 1, 10^5 and 5×10^5 in Figure 2, and 5×10^5 and 1.5×10^6 in Figure 3). In these experiments the scaling factor values for the SE contribution were orders of magnitude beyond a plausible upper limit. This represents a major concession to the SE model, yet, instead of activating a strong MCM, the large scaling values led consistently to accelerated genetic decline.

Cases (2) and (4) of Figure 3 incorporate all of these features that strongly favor the SE hypothesis. What we observe is that even with all these highly unrealistic concessions, the mutation count per individual *actually increases slightly*, rather than decreases, relative to the case of no SE. Even with exaggerated selection efficiency, both forms of SE cause starkly accelerated fitness decline relative to the default case of mutation non-interaction. We found that in order to see any noteworthy SE effect at all, the SE scaling factors must be larger than anything that seems biologically reasonable. Even when we do this, we do not observe the effects which are so widely ascribed to the SE mechanism (halting of mutation accumulation and stabilization of fitness). Instead we see the opposite. If SE has any effect at all, it consistently makes genetic degeneration worse. The larger the SE effect, the more rapid is the degeneration. This agrees with the logical expectation of what should happen when there is the on-going accumulation of increasingly severe mutational damage.

Modeling SE realistically

To model SE realistically, the net SE effect must be only a slight deviation from the standard additive model, most SE interactions must arise from mutations within the same linkage block, individual SE effects must have reasonable limits, there must be small fitness heritability, and selection must be characterized primarily by the probability model. These constraints all reflect biological reality as we understand it. Modeling SE in accord with any one of these five constraints *negates* the SE hypothesis. When we model SE under what we believe are the most realistic conditions, we consistently see no meaningful SE effect on either mutation accumulation or fitness decline. We feel this reflects biological reality; that is, generic SE effects are necessarily small, are strongly overshadowed by much more significant biological phenomena, and do not affect mutation accumulation in any significant way.

Pros and cons of the SE hypothesis

It might be argued that logically there should always be some selection against high mutation count individuals, so this should help slow mutation accumulation. In particular, the SE mechanism should create an increased penalty against the high mutation count individuals, strengthening the potential MCM. The problem with this line of reasoning is that, while higher mutation count will have some correlation with lower fitness, this correlation under natural conditions will be extremely weak. The major reason for this weak correlation is the large variation in the magnitude of mutation fitness effects. Some mutations have substantial effects, but most have small to vanishingly small effects. Individuals in a population with random mating should all have approximately the same number of mutations, due to averaging. Moreover, most mutations are nearly neutral. The primary reason some individuals display reduced fitness relative to the others is due to only a few substantial mutations and not because of some small difference in total mutation count. Realistic numerical simulation consistently confirms that this is true [this paper and 7–9].

Cases of genuine SE genetic interactions are well documented. Most involve the interactions of relatively large-impact mutations, usually within the same gene or samc pathway and affecting a single trait. These specific examples of SE should not be interpreted to imply, however, that SE effects arise from interactions from *every* pair of mutations throughout the genome. Naturally, high impact mutations can be expected to produce a few strong and measurable interactions, some of which will be synergistic. The interactions among such mutations, as well as the mutations themselves, are then highly selectable. For a simple trait whose character is determined by only few genes, each gene is highly significant relative to that trait. In a sense the "genome" for that trait is small, which makes every mutation in that limited system potentially significant. Because the "genome" is small, the likelihood that two mutations within it will display an SE interaction is larger that it would be otherwise. However, in a large functional genome with billions of nucleotides, which encode for thousands of traits, the likelihood that mutations in distant parts of the genome will have significant mutual SE interaction is tiny.

Experimental evidence of generic genome-wide SE in living populations has been inconclusive [30, 31]. The inferred absolute amplitudes of generic SE effects are small. These studies on the extent of generic SE in natural populations in no way support a conclusion that the SE mechanism acts to slow genomic degeneration. Our own analyses consistently show that regardless of the extent of generic SE in a genome, SE consistently accelerates degeneration and does almost nothing to slow mutation accumulation.

The SE hypothesis is that SE interactions cause truncation selection at a critical threshold, such that any further mutation (even the lowest impact mutation) acts essentially as if it were lethal. If SE stabilizes genomes and stops genomic degeneration in this way, then constant and intense selection must operate just below that threshold, such that any additional mutations will be severely detrimental. This means that the population stabilizes just a few mutations short of disaster (mutational meltdown). Another way of saying this is that the population is stabilized against mutational meltdown/extinction by maintaining itself on the verge of extinction. Ironically, in this state of extreme selective tension, an improvement in environmental conditions (e.g., good weather, fewer predators) could result in significantly relaxed selection, which could lead to mutation accumulation beyond the threshold, which could then lead to extinction in the more favorable environment. This seems more than counterintuitive. It is, in reality, entirely unreasonable. How could any population remain balanced on such a knife edge for millions, or even thousands, of generations?

Numerous mutation accumulation experiments have been performed involving a laboratory population of plants or animals placed in a state of relaxed selection for many generations. Such experiments cannot truly eliminate selection (there is always selection for embryo viability and fertility), but selection can be greatly reduced. Usually, the observed fitness decline is slow and gradual [32], consistent with very limited levels of SE. In the few cases where degeneration was more accelerated [27], it can readily be attributed to a few major interactions between a few high impact mutations (major mutations are naturally expected to have major interactions).

Genetic bottlenecks, often invoked in evolutionary scenarios, result in greatly reduced selection (because genetic drift overrides selection when population size is small). This also ought to result in mutation accumulation past the critical SE threshold, causing mutational meltdown and rapid extinction. Since such SE-induced meltdown is generally not thought to occur, this also seems to argue against the SE hypothesis.

Therefore, many lines of evidence, based upon both logic and biological data, argue strongly against the SE hypothesis. These evidences have now been validated by the numerical simulations carried out in this study. Our findings are consistent with the findings of Butcher [19], but apply to sexual as well as asexual species. While any one of these individual lines of evidence by itself might be insufficient to discredit the SE hypothesis, taken together they constitute an overwhelming case against the SE hypothesis, strong enough in our view to constitute falsification.

A very recent paper by Crow [33], forcefully argues against any significant role for epistasis in affecting selection efficiency. This would seem highly significant because the same author has for decades been a leading proponent for theoretical mechanisms that might resolve the mutational degeneration paradox, including the MCM and SE hypotheses. Crow now states, "My main objective here is to show that the breeders' practice of ignoring epistasis in quantitative selection is fully justified...In general, the smaller the effects, the more nearly additive they are. Experimental evidence for this is abundant...Multiple factors with individually small effects acting in a near-additive manner seem to be the rule... although there may be large dominance and epistatic components, selection acts only on the additive variance...For these reasons, one would expect that epistatic variance would have only a small effect on predicting the progress of selection...Any attempt to include epistatic terms in prediction formulae is likely to do more harm than good."

In summary, there appears to be neither theoretical nor observational support for the idea that a generic SE mechanism exists in nature capable of halting mutation accumulation or of stabilizing natural populations against mutational meltdown. Given that the SE hypothesis has so many glaring problems, one might ask how it ever became widely accepted. The SE hypothesis seems to have been proposed solely as a possible means for dealing with one of the as yet unsolved difficulties for the classic neo-Darwinian model. It appears to have become widely accepted only because no alternative mechanism could be identified that might conceivably stop deleterious mutation accumulation. We suggest that until a more credible mechanism can be discovered for halting deleterious mutation accumulation, the genetic degeneration problem should most honestly be described simply as a paradox that is yet to be explained.

Conclusions

1. Theoretical considerations show that SE should not be able to stop mutation accumulation. It has already been shown in our companion paper that with any realistic distribution of mutational fitness effects, the mutation count mechanism (MCM) does not operate and is of no avail in stopping deleterious mutation accumulation [9]. There is no theoretical basis for thinking that SE could stop mutation accumulation, even if it could activate the MCM effect. In this paper we show that for both linked and non-linked mutations, SE simply serves to *amplify* the fitness effect differences among mutations whenever the SE effect is directly related to the base, non-epistatic effect. In the case of linked mutations, the SE effects, like the linked mutations themselves, are inherited generation to generation, and therefore act simply as enhancements to the basal, non-epistatic mutational fitness effects. We show that the same is true of the non-linked SE interactions. Because of these enhancements to the basal mutation fitness effects, in both cases SE therefore logically can only serve to accelerate fitness decline and hasten mutational meltdown.
2. Consistent with simple logic, this paper's careful numerical simulations suggest that SE does nothing to halt mutation accumulation. In fact, even numerical experiments using truncation selection and perfect genotypic heritability show SE slightly *enhances* mutation accumulation. To the extent that SE has any noteworthy effect at all, it consistently accelerates degeneration. When realistic levels of linkage are included, this degeneration is accelerated even more.
3. If somehow these first two conclusions were not valid and the SE hypothesis were actually true, all species should mutate right up to the brink of their mutation-count threshold. Biological observations, however, do not support any type of mutation count threshold. In nature, if the SE hypothesis were true, any relaxation of selection pressure (a more favorable environment or a bottleneck episode) would be expected to cause rapid extinction. Likewise, lab mutation accumulation experiments, wherein selection is artificially relaxed, would be expected to result in rapid and catastrophic fitness meltdown. Neither result has ever been observed.
4. The SE hypothesis seems to have been proposed solely as a possible means for dealing with one of the as yet unsolved difficulties for the classic neo-Darwinian model. It appears to have become widely accepted only because no alternative mechanism has yet been identified that might conceivably stop deleterious mutation accumulation. The genetic degeneration problem remains unresolved.

***Addendum** – Since the finalization of this chapter, a significant new paper has been published. See: Sanford, J. & Nelson, C. (2012). The Next Step in Understanding Population Dynamics: Comprehensive Numerical Simulation, Studies in Population Genetics, in: M. Carmen Fusté (Ed.), ISBN: 978-953-51-0588-6, InTech, Available from: http://www.intechopen.com/books/studies-in-population-genetics/the-next-step-in-understanding-population-dynamics-comprehensive-numerical-simulation .*

References

1. Kondrashov AS (1995) Contamination of the genome by very slightly deleterious mutations: why have we not died 100 times over? J Theor Biol 175:583–594.
2. Lynch M, Conery J, Burger R (1995) Mutation accumulation and the extinction of small populations. Am Nat 146:489–518.
3. Lynch M, Conery J, Burger R (1995) Mutational meltdown in sexual populations. Evolution 49(6):1067–1080.
4. Higgins K, Lynch M (2001) Metapopulation extinction caused by mutation accumulation. Proc Natl Acad Sci USA 98:2928–2933.
5. Loewe L (2006) Quantifying the genomic decay paradox due to Muller's ratchet in human mitochondrial DNA. Genet Res 87:133–159.
6. Sanford J, Baumgardner J, Gibson P, Brewer W, ReMine W (2007) Mendel's Accountant: a biologically realistic forward–time population genetics program. Scalable Computing, Practice and Experience 8(2):147–165. http://www.scpe.org
7. Sanford J, Baumgardner J, Gibson P, Brewer W, ReMine W (2007) Using computer simulation to understand mutation accumulation dynamics and genetic load. In: Shi Y, van Albada GD, Dongarra J, Sloot PMA (eds) 7th International Conference on Computational Science, Beijing, China, May 27–30, 2007, Proceedings, Part II, LNCS 4488:386–392. Springer-Verlag, Berlin/Heidelberg.
8. Gibson P, Baumgardner JR, Brewer WH, Sanford JC (2013) Can purifying natural selection preserve biological information? In: Marks II RJ, Behe MJ, Dembski WA, Gordon B, Sanford JC (eds) Biological Information — New Perspectives. World Scientific, Singapore, pp. 232–263.
9. Brewer WH, Baumgardner JR, Gibson P, Sanford JC (2013) Using numerical simulation to test the "mutation–count" hypothesis. In: Marks II RJ, Behe MJ, Dembski WA, Gordon B, Sanford JC (eds) Biological Information — New Perspectives. World Scientific, Singapore, pp. 298–311.
10. Ohta T (1973) Slightly deleterious mutant substitutions in evolution. Nature 246:96–98.
11. Ohta T (1974) Mutational pressure as the main cause of molecular evolution and polymorphism. Nature 252:351–354.

12. Ohta T (1992) The nearly neutral theory of molecular evolution. Ann Rev Ecol Syst 23:263–286.
13. Ohta T (2002) Near-neutrality in evolution of genes and gene regulation. Proc Natl Acad Sci USA 99:16134–16137.
14. Kimura M (1979) Model of effectively neutral mutations in which selective constraint is incorporated. Proc Natl Acad Sci USA 76:3440–3444.
15. Kimura M (1983) Neutral Theory of Molecular Evolution. Cambridge University Press, New York.
16. Wolf JB, Brodie ED, III, Wade MJ (2000) Epistasis and the evolutionary process. Oxford University Press, Oxford.
17. Kondrashov AS (1988) Deleterious mutations and the evolution of sexual reproduction. Nature 336: 435–440.
18. Omholt SW, Plahte E, Oyehaug L, Xiang K (2000) Gene regulatory networks generating the phenomena of additivity, dominance and epistasis. Genetics 155:969–980.
19. Butcher D (1995) Muller's ratchet, epistasis and mutation effects. Genetics 141:431–437.
20. Sanjuán R, Elena SF (2006) Epistasis correlates to genomic complexity. Proc Natl Acad Sci USA 103: 14402–14405.
21. Bershtein S, Segal M, Bekerman R, Tokuriki N, Tawfik DS (2006) Robustness-epistasis link shapes the fitness landscape of a randomly drifting protein. Nature 444:929–932.
22. Wagner GP, Laubichler MD, Bagheri-Chaichian H (1998) Genetic measurement of theory of epistatic effects. Genetica 102–103:569–580.
23. Sanjuán R, Moya A, Elena SF (2004) The contribution of epistasis to the architecture of fitness in an RNA virus. Proc Natl Acad Sci USA 101:15376–15379.
24. Bonhoeffer S, Chappey C, Parkin NT, Whitcomb JM, Petropoulos CJ (2004) Evidence for positive epistasis in HIV-1. Science 306:1547–1550.
25. Whitlock MC, Bourguet D (2000) Factors affecting the genetic load in Drosophila: synergistic epistasis and correlations among fitness components. Evolution 54(5):1654–1660.
26. Crow JF, Kimura M (1979) Efficiency of truncation selection. Proc Natl Acad Sci USA 76:396–399.
27. Crow JF (1991) Professor Mukai: the man and his work. Japan J Gen 66:669–682.
28. Crow JF (1997) The high spontaneous mutation rate: a health risk? Proc Natl Acad Sci USA 94:8380–8386
29. Crow JF (2000) The origins, patterns and implications of human spontaneous mutation. Nat Rev 1:40–47.
30. Arjan J, deVisser GM, Hoekstra RF (1998) Synergistic epistasis between loci affecting fitness: evidence in plants and fungi. Gen Res 71:39–49.

31. Rosa JM, Camacho S, Garcia-Dorado A (2005) A measure of the within-chromosome synergistic epistasis for Drosophila viability. J Evol Biol 18:1130–1137.
32. Fry J (2004) On the rate and linearity of viability declines in Drosophila mutation-accumulation experiments: genomic mutation rates and synergistic epistasis revisited. Genetics 166:797–806.
33. Crow JF (2010) On epistasis: why it is unimportant in polygenic directional selection. Phil Trans R Soc B 365:1241–1244.

Computational Evolution Experiments Reveal a Net Loss of Genetic Information Despite Selection

Chase W. Nelson[1,*] and John C. Sanford[2]

*[1]Rainbow Technologies, Inc., 877 Marshall Rd., Waterloo, NY 13165, USA. [2]Department of Horticulture, NYSAES, Cornell University, Geneva, NY 14456, USA. (*Corresponding author: cwnelson88@gmail.com)*

Abstract

Computational evolution experiments using the population genetics simulation Mendel's Accountant have suggested that deleterious mutation accumulation may pose a threat to the long-term survival of many biological species. By contrast, experiments using the program Avida have suggested that purifying selection is extremely effective and that novel genetic information can arise via selection for high-impact beneficial mutations. The present study shows that these approaches yield seemingly contradictory results only because of disparate parameter settings. Both agree when similar settings are used, and both reveal a net loss of genetic information under biologically relevant conditions. Further, both approaches establish the existence of three potentially prohibitive barriers to the evolution of novel genetic information: (1) the selection threshold and resulting genetic decay; (2) the waiting time to beneficial mutation; and (3) the pressure of reductive evolution, i.e., the selective pressure to shrink the genome and disable unused functions. The adequacy of mutation and natural selection for producing and sustaining novel genetic information cannot be properly assessed without a careful study of these issues.

Key words: Avida, digital organisms, experimental evolution, genetic entropy, irreducible complexity, Mendel's Accountant, reductive evolution, selection threshold, waiting time to beneficial mutation

Introduction

Mathematical models and numerical simulation have long suggested that the accumulation of slightly deleterious mutations may pose a threat to the long-term survival of many biological species, including humans [1–4]. Computational evolution experiments with the forward-time population genetics simulation Mendel's Accountant have predicted a substantial fitness decline in the human species under biologically relevant conditions [5]. Moreover, experiments with biological organisms have raised similar concerns, revealing that the majority of adaptive mutations cause a loss of functionality [6–10]. Lethal mutagenesis may also play a key role in pathogen attenuation [11–13]. Recently, however, experiments using the digital genetics software Avida have suggested that purifying selection can be extremely effective and that novel genetic information can arise via selection for high-impact beneficial mutations [14]. Avida researchers have claimed a high degree of biological relevance for the program, using it to address numerous biological questions [15,16].

In this study, we investigate why Avida and Mendel's Accountant yield seemingly contradictory results. We find that most discrepancies are due to differences in default settings. Mendel's default settings implement values plausible for modeling the human species, while Avida's default settings have virtually no parallel in biological systems. Additionally, Avida introduces several un-biological mechanisms both for facilitating the development of novel genetic information and for preventing its loss. The most notable deviations from biological reality include the distribution of mutational fitness effects, the waiting time to high-impact beneficial mutation, and the selective neutrality of inert genomic material. When used with more realistic settings, Avida's results agree with other studies that reveal a net loss of genetic information under biologically realistic conditions. The results reported here suggest that three substantial barriers may prevent the evolution of genetic information by mutation and natural selection in biological organisms: (1) the selection threshold; (2) the waiting time to beneficial mutation; and (3) reductive evolution. Implications for theory and medicine are discussed.

Mendel's Accountant

Detailed descriptions of Mendel's Accountant (hereafter Mendel) are available elsewhere [17,18], and default settings are described in the Methods. Briefly, Mendel constitutes a numerical simulation that tracks mutations as they arise within the members of a model population. The user specifies parameters such as population size, genome size, mutation rate, and the proportion of beneficial mutations. Mutational fitness effects are represented by a Weibull distribution, with both deleterious and beneficial effects having lower and upper bounds. The largest deleterious fitness effect is –1.0 (lethal in most contexts), while the smallest effect is defined as the reciprocal of the functional genome size ($-1/G_e$), following the precedent of Kondrashov [2]. Beneficial mutations are limited by the same lower bound ($1/G_e$) and a user-defined upper bound (0.001 by default). Each mutation has its own fitness effect as well as its own location within an individual's genome, allowing the investigator to model linkage and recombination. To save computational resources, neutral mutations are not normally tracked. Instead, the mutation rate is scaled to exclude neutral mutations, such that the mutation rate defined by the user is the rate per effective genome, i.e., the rate of mutations affecting fitness.

The program periodically reports various statistics during an experiment, including the population's average fitness and the average number of deleterious and beneficial mutations per organism. The program is open source and is available online [19].

Avida

Avida differs from Mendel in that it represents genomes directly using machine code instructions, and generally requires more computer science knowledge for use and interpretation of results. Twenty-six genomic instructions are defined in the software, and each performs a specific computational task (e.g., adding two numbers). Individual genomes, called *digital organisms*, consist of about 100 instructions and undergo random mutation at a user-defined rate. Mutations may substitute, insert, or delete instructions at random. The Avidian organisms are themselves housed on a two dimensional grid. Replication is asexual, with daughter cells randomly replacing one of the eight surrounding neighbors. Because of this, replication rate determines fitness in Avida; any changes that allow an organism to copy its genome and replicate faster will allow it to replace other organisms, and its frequency in the population will increase.

Each organism in Avida has an associated *merit* value that determines its relative replication rate. This value reflects both genome size and the ability to perform one of nine computational functions (logic operations). Making merit proportional to genome size implements a scheme called *size neutrality* in which larger genomes are artificially given extra computational time. This removes the selective pressure to shrink genomes, making organisms with identical phenotypes but different genome sizes equivalent in fitness. Because of this, acquiring *merit bonuses* by performing any of the nine logic operations is the primary means by which organisms increase their replication rate in Avida. These functions arise when random mutations produce particular combinations of instructions that cause the functions to be executed. For example, the simplest logic operations, NAND and NOT, can occur when the instruction NAND arises in the correct combination with input-output and labeling instructions.

Considering its frequent application to biological questions, Avida's default range of beneficial mutational fitness effects is curiously high. The two simplest operations have a multiplicative merit bonus of 2, doubling an organism's fitness. Bonuses increase exponentially with the complexity of a function, and EQU (the most complex function in Avida) multiplies fitness by 32 (Table 1). For purposes of biological comparison, relative fitness may be defined as $w = 1 + s$, where s is the mutational fitness effect and w is the relative fitness of an organism expressing

Table 1. Default fitness bonuses for performing nine logic operations in Avida. Adapted from Lenski *et al.* [14].

Logic Operation	Computation	Number of NAND Operations Needed (n)	Default Multiplicative Bonus (2^n)	Default Fitness Effect ($w - 1$)
NOT	~A; ~B	1	2	1.0
NAND	~(A and B)	1	2	1.0
AND	A and B	2	4	3.0
ORNOT	(A or ~B); (~A or B)	2	4	3.0
OR	A or B	3	8	7.0
ANDNOT	(A and ~B); (~A and B)	3	8	7.0
NOR	~A and ~B	4	16	15.0
XOR	(A and ~B) or (~A and B)	4	16	15.0
EQU (XNOR)	(A and B) or (~A and ~B)	5	32	31.0

a particular function as compared to its function-free ancestor. Mutational fitness effects therefore range from 1.0 to 31.0 under Avida's default settings. The program is available online [20], and more detailed descriptions of the software are available elsewhere [21–23].

A previous study [24] has demonstrated that seven of the nine logic operations arise by mutation alone in Avida, without selection, reflecting their informational simplicity within the software environment. Under default settings lasting about 10,000 generations, an average of 8.6 (± 0.7) such functions successfully evolve (i.e., rise above a frequency of 50%), increasing fitness by an average of 20,000,000 fold. Increases of this magnitude are enabled by the large multiplicative fitness bonuses assigned to the logic operations ($2^2 \times 4^2 \times 8^2 \times 16^2 \times 32 = 33{,}554{,}432$; Table 1). Fitness increases observed in biological evolution experiments are negligible by comparison; e.g., in experiments with *E. coli*, fitness increased by only 75% after 20,000 generations [6]. Interestingly, the Avidian logic functions are prevented from reaching fixation by the relatively high mutation rate (approximately 0.85 mutations per genome per generation). Fitness eventually levels off, as only nine functions are available.

Although Avida's default mutational fitness effects range from 1.0 to 31.0, the user may specify other values. Using alternative values ranging from 0 to 1.0, Nelson and Sanford [24] used an empirical approach to demonstrate that Avidian populations experience a *selection threshold*, or a critical fitness effect

Methods

Experiments using Mendel's Accountant

All Mendel experiments used version 1.8.5. Random number seeds were chosen as integer values from 1 to 1,000. Experiments were performed using settings that: (1) approximate Avida's default settings, (2) employ Mendel's default settings, and (3) use Mendel settings more conducive to the occurrence of high-impact beneficial mutations. A full list of experimental settings appears in Table 2.

First, ten Mendel experiments were performed to approximate Avida's default results. The most notable changes to Mendel's default settings were a reduced genomic mutation rate of 0.01 (reflecting the size and selective neutrality of much of the ancestral Avidian genome), a proportion of 0.000023 mutations being beneficial, and uniform multiplicative beneficial fitness effects of 5.5. (Mendel does not lend itself to studying the large discrete fitness effects implemented in Avida, so uniform fitness effects were used.)

Next, ten experiments were performed under Mendel's default settings. Following this, twenty experiments were performed under settings more conducive to the occurrence and selection of high-impact beneficial mutations. The fraction of beneficial mutations was increased to 0.001, the maximum beneficial fitness effect increased to 0.5, heritability increased to 0.5, and experiment length increased to 1,000 generations.

Experiments using Avida

All Avida experiments used version 2.8.1. Random number seeds were chosen randomly as an integer value from 1 to 1,000,000,000. Two sets of experiments were performed, one in which various precursor functions were assigned neutral fitness bonuses, and one in which mechanisms preventing genome shrinkage were disabled.

For experiments in which functions were assigned neutral fitness bonuses, the number of neutral functions varied from zero to nine, with zero corresponding to Avida's default settings and nine corresponding to all functions (including EQU) having no fitness effect. Two sets of 20 replicates were performed, one in which functions were made neutral from simple-to-complex (beginning with NOT), and one in which functions were made neutral from complex-to-simple (beginning with XOR). Each replicate therefore consisted of 10 experiments, one for each combination of neutral functions. In all instances, EQU was the last function made neutral (all nine neutral functions). Default fitness bonuses were maintained for advantageous functions, and functions were made neutral by defining multiplicative

Table 2. Parameter settings used in experiments with Mendel's Accountant. A dash (–) indicates the use of default values.

Parameter Category	Parameter	Mendel Default Values (Expt 1)	Avida Approximation Values (Expt 2)	Altered Mutation Values (Expt 3)
Basic	New mutations per offspring	10	0.01	-
	Fraction of mutations beneficial	1.0×10^{-5}	2.3×10^{-5}	1.0×10^{-3}
	Offspring per female	6	4	-
	Population size (per tribe)	1000	3600	-
	Generations	500	10000	1000
Mutation	Functional genome size	3.0×10^{9}	100	-
	Fraction of mutations having a large effect	0.001	Not applicable	-
	Minimum deleterious mutation effect considered large	0.1	Not applicable	-
	Maximum beneficial effect per mutation	0.001	Not applicable	0.5
	Number of initial beneficial loci	0	-	-
	Fraction recessive	0	-	-
	Combine mutations in multiplicative manner	No	Yes	-
	Fraction multiplicative effect	Not applicable	1	-
	Consider all mutations equal	No	Yes	-
	Equal effect for each deleterious mutation	Not applicable	0.001	-
	Equal effect for each beneficial mutation	Not applicable	5.5	-
	Synergistic epistasis	No	-	-
	Allow back mutations	No	-	-
Selection	Random death	0	0.1	-
	Heritability	0.2	1	0.5
	Non-scaling noise	0		-
	Fertility declining with fitness	Yes	No	-
	Selection scheme	Unrestricted probability selection	-	-

(*Continued*)

Table 2. (*Continued*)

Population	Clonal reproduction	No	Yes	-
	Haploid	No	Yes	-
	Fraction self fertilization	0	-	-
	Initial heterozygous alleles	No	-	-
	Dynamic linkage	Yes	-	-
	Number of chromosome pairs	23	-	-
	Number of linkage subunits	989	-	-
	Dynamic population size	No	-	-
	Population substructure	No	-	-
	Bottleneck	No	-	-
Computation	Tracking threshold	1.0×10^{-5}	-	-
	Parallel processing	No	-	-
	Queuing system	PBS	-	-
	Simulation engine	Fortran	-	-

bonuses of 1.0 (fitness effect of 0) in the environment.cfg file (type=mult, value=1.0).

To examine the role of genome shrinkage in evolution, two sets of 30 replicates were performed, one each for genome sizes of 50 and 100. The default genome contained in default-classic.org was used for size 100, and genomes of size 50 were constructed by removing 50 of the unnecessary NOP-C instructions from the default genome. For each replicate, three alternative scenarios were compared: (1) size neutrality on (default; SNON); (2) size neutrality off (SNOFF); and (3) size neutrality off with mutations to the H-COPY instruction disabled (SNOFF NHC). To disable size neutrality, the avida.cfg file was altered to make base merit constant (BASE_MERIT_METHOD 0). To disable mutations to H-COPY, the instset-classic.cfg file was altered (h-copy 0). Mutations substituting the H-COPY instruction into the Avidian replication loop allow a doubling of the replication rate, and it was found that this process can circumvent the pressure to reduce genome size.

Results

Experiments using Mendel's Accountant

Under Mendel's default settings (Table 2), end-of-experiment fitness declined to an average of 0.76 (± 0.01) after 500 generations. Populations contained an average of 4,906.1 (± 34.3) deleterious mutations and 0.03 (± 0.04) beneficial mutations per genome. Figure 1(A) displays the fitness trajectory of a case study population under these conditions.

Under settings designed to approximate results obtained under Avida's default settings lasting 10,000 generations, fitness increased to an average of 35,730,000 (ranging up to 126,900,000) relative to the ancestral population. These results matched Avida very well, which produces an average fitness increase of approximately 19,749,130. Populations contained an average of 62.7 (± 5.2) deleterious mutations and 8.8 (± 0.9) beneficial mutations per genome. Figure 1(B) displays the fitness trajectory of a case study population under these conditions.

To explore evolution under conditions similar to the default settings but more favorable to beneficial mutation, the proportion of beneficial mutations was increased to 0.001, with a maximum effect of 0.5, heritability was increased to 0.5, and experiment length was increased to 1,000 generations. Under these conditions, end-of-experiment fitness decreased to an average of 0.8 (± 0.1), with an average of 9,739.3 (± 50.2) deleterious mutations and 14.8 (± 3.2) beneficial mutations per genome. Although no end-of-experiment fitnesses were above the ancestral fitness of 1.0, fitness did rise above 1.0 during the course of three (15%) of these experiments, with a maximum of 1.01. One of these cases is shown in Figure 1(C). Here, a high-impact beneficial mutation (fitness effect of approximately 0.2) occurred around generation 270 and rapidly moved to fixation. No other mutations (beneficial or deleterious) reached fixation over the 1,000 generations of this experiment. End-of-experiment fitness was 0.85.

Experiments using Avida

Experiments were conducted to determine how many functional precursors must be rewarded to enable the evolution of EQU in Avida. Results are summarized in Figure 2. EQU never evolved when seven or more precursor functions were neutral. It also never evolved with six neutral precursors under the complex-to-simple scenario, and evolved only once with six neutral functions under the simple-to-complex scenario. These findings expand the results of other studies, in which EQU never evolved when all simpler functions were neutral [14] and certain combinations of neutral functions involving NOR and XOR were found to hinder the evolution of EQU [37]. The evolution of XOR and EQU therefore requires selection for functional precursors, and at least two precursors must be rewarded for EQU to evolve. EQU is more likely to evolve when relatively complex operations are rewarded, because complex operations are less likely to arise without a selective advantage. Hitchhiking of neutral functions to high frequencies (> 50%) was common in these experiments.

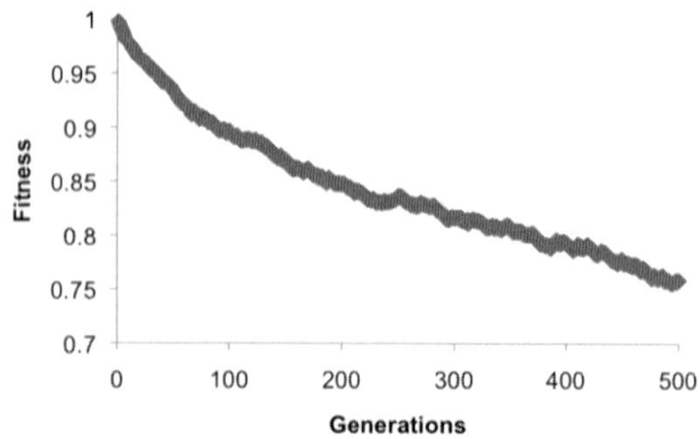

(A)

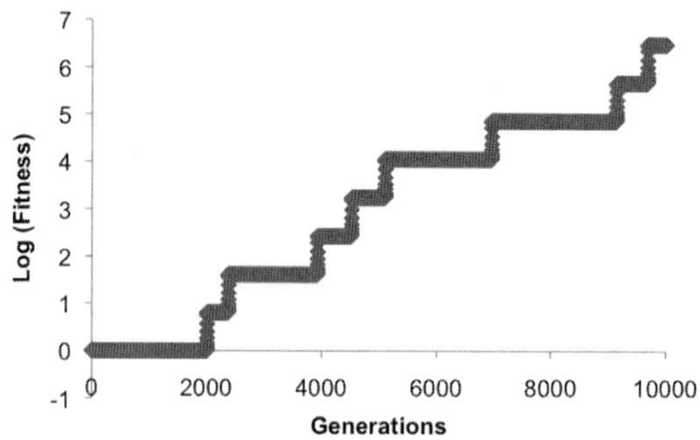

(B)

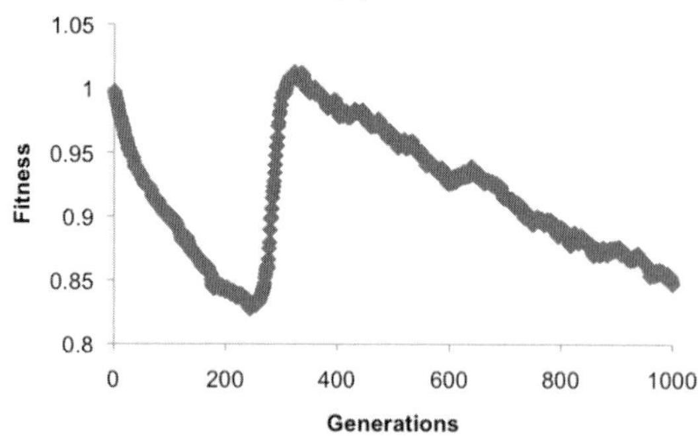

(C)

Fig. 1. Fitness trajectories of case study populations in Mendel's Accountant. Note that the axes differ. (A) Under Mendel's default conditions, fitness decayed to an end-of-experiment value of 0.76 as a result of the accumulation of approximately 4,897.2 deleterious mutations per individual. (B) Under conditions approximating Avida's default settings, fitness leaped in stages to an end-of-experiment value of 3,014,000 as a result of the spread of eight beneficial mutations with fitness effects of 5.5. Roughly 55.8 deleterious mutations were present per individual. Note that the y-axis is log base 10. (C) Under altered Mendel settings, fitness declined sharply, then leaped to 1.01 following the introduction of a high-impact beneficial mutation (fitness effect of approximately 0.2) around generation 270. This offset the adverse effects of approximately 2,643.4 deleterious mutations that had accumulated in the individual in which it occurred. End-of-experiment fitness was 0.85.

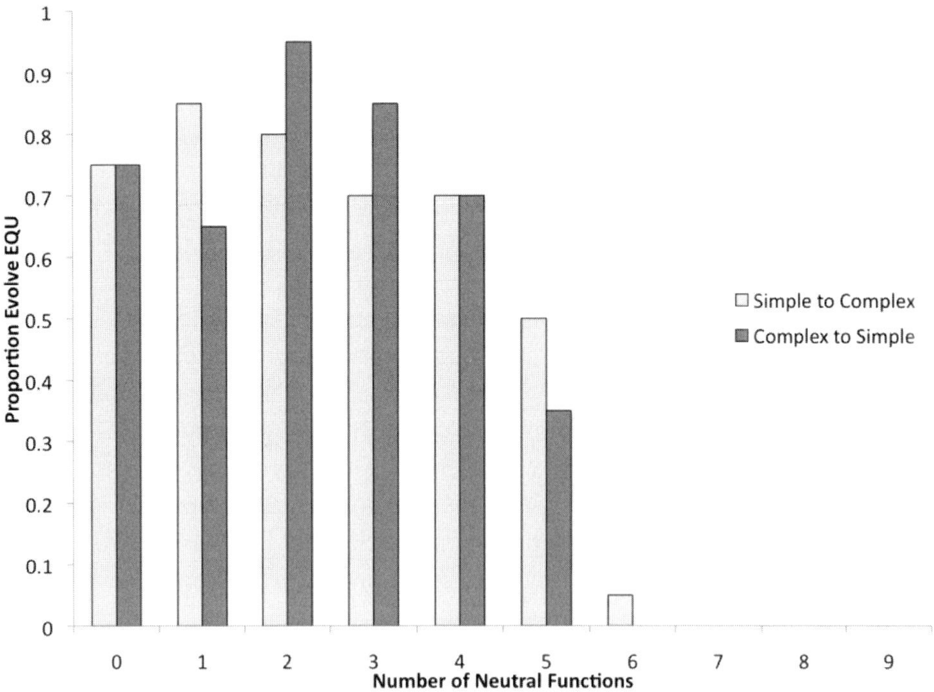

Fig. 2. The effects of selectively neutral precursors on the evolution of EQU. EQU never evolved when seven or more functions were assigned neutral fitness bonuses.

Avida experiments were also performed to examine the evolutionary consequences of selection acting on genome size (results not shown). Fewer functions evolved when size neutrality mechanisms were disabled, and this difference was more pronounced for organisms with smaller genomes. EQU evolved less often, and end-of-experiment fitnesses were lower for both genome sizes of 50 and 100. Though genome size tended to increase somewhat under default settings, this pattern was reversed when size neutrality was not enforced. Therefore, size neutrality artificially facilitates the evolution of complexity in Avida, presumably by maintaining inert genomic code that can be used as raw material for evolutionary innovation.

Discussion

Selection threshold and genetic entropy

A previous study [24] demonstrated that a fitness effect *selection threshold* exists in Avida. The selection threshold is defined as the mutational fitness effect at which natural selection and random genetic drift contribute equally to the fate of a mutation in the population. Practically, this is the fitness effect for which positive selection successfully captures half of the beneficial mutations that arise. In Avida, this occurs at a beneficial fitness effect of approximately 0.2 (of course, this is a lower estimate of the threshold value, as multiple mutations produce the same logic operations in each run). Moreover, zero new functions evolve when fitness effects are ≤ 0.075, and those that have previously evolved break down. Likewise, experiments with Mendel have estimated a selection threshold of approximately 10^{-4} to 10^{-3} under conditions typical of mammalian populations [5,34].

The selection threshold can act as a barrier preventing the evolutionary origin and maintenance of novel adaptive genetic information. Unless selection is able to "see" the fitness effects of mutations, they will drift in the population as if neutral. Because the majority of mutations are deleterious, the suspension of selection for low-impact mutations strongly favors the fixation of mutations that decrease fitness [32]. The net result is a phenomenon that Sanford [33] has termed *genetic entropy*. When this occurs, purifying selection is unable to counteract the accumulation of low-impact deleterious mutations. Even when rare beneficial mutations cause a selective sweep, they are linked to numerous deleterious mutations across many loci, such that the total number of functional loci decreases. Experiments with Mendel have confirmed that deleterious mutations accumulate in a linear fashion despite selection [5], consistent with biological studies (e.g., with *E. coli* [38]). It is worth emphasizing that the gradual fitness declines shown in Figures 1(A) and 1(C) occur despite the concurrent action of reasonably strong selection; in these cases, selection is simply unable to counteract the net adverse effects of new mutations.

Genetic entropy is not merely a theoretical concern. Numerous analyses have confirmed that the accumulation of slightly deleterious mutations can cause gradual fitness loss leading to extinction in asexual species [12,25,39–42], and similar processes are relevant to sexual species [1,43], including humans [2–4, 44–47]. Lethal mutagenesis of pathogens, due to elevated mutation rates and periodic bottlenecking upon infection, may also be applicable in novel medical approaches [11–13]. Novel means of genetic intervention to reduce mutation rates may be necessary to prevent the extinction of numerous species, though it is unclear whether this would be feasible.

High-impact beneficial mutations

The Mendel case study displayed in Figure 1(C) is an informative example of the effects of high-impact beneficial mutations. A single high-impact beneficial mutation (fitness effect of approximately 0.2) occurred around generation 270, offsetting the effects of approximately 2,643 deleterious mutations in the individual in which it arose. Beneficial mutations with large effects have certainly occurred in nature. For example, in the presence of an antibiotic, the fitness effect of any mutation conferring drug resistance is so large as to be mathematically undefined, as the ancestral fitness is rendered zero in that environment. Other examples of high-impact beneficial mutations have been reported in viruses in the presence of heat [48]. However, even though these mutations are beneficial in their respective environments, they work by damaging or eliminating genetic information [8], not producing it (see below).

This phenomenon highlights one disadvantage of Mendel's Accountant, namely, that it treats evolution merely as an accounting problem, in keeping with traditional population genetics. Evolution is seen as an exercise in fitness addition and subtraction, without any reference to the underlying genomic mechanisms or architecture. This favors progressive evolution, as it allows single beneficial mutations of large effect to compensate for large numbers of deleterious mutations. This phenomenon is made possible by the infinite allele model, and is precisely the process that Kimura invoked to explain the problem of very slightly deleterious mutation accumulation [27]. However, even though this model is clearly more conducive to progressive evolution, there are several reasons why it is not biologically realistic. Scenarios in which large numbers of deleterious mutations are regularly offset by relatively few high-impact beneficial mutations lead inevitably to shrinkage of the functional genome. If such beneficial mutations are the sole source of progressive evolution, the functional genome must shrink each time evolution takes a step forward (i.e., each selective sweep). This type of change is not sustainable and cannot constitute the sole source of progressive evolution. (For this reason, deleterious and beneficial mutations have heretofore been studied separately with Mendel, with high-impact beneficial mutations being studied as a special case [34].) Instead, plausible scenarios of progressive adaptive evolution must allow the deterministic elimination of most deleterious mutations through purifying selection. Additionally, the gradual accumulation of beneficial mutations through natural selection must have the potential to build every complex biological feature requiring explanation. This process requires qualities of linkage and functional integration that cannot be adequately represented with numerical simulation.

Distribution of mutational fitness effects

The mutational fitness effects implemented under Avida's default settings (1.0 – 31.0) are extremely rare or nonexistent in the biological realm (but see Bull *et al.* [48] on high-impact mutations in viruses). This renders published Avida results irrelevant to the great majority of biological mutations. Some readers may object that, while Avida's fitness effects are too large, those implemented in Mendel are too small. On the contrary, it is well established that (1) most mutations are deleterious, and (2) most mutations have very slight effects [28]. For example, a recent study of nonessential ribosomal genes in *Salmonella typhimurium* [49] examined a total of 126 single bp substitutions, revealing that 120 were weakly deleterious and 6 were neutral or nearly-neutral. Average deleterious selection coefficients were 0.0096 and 0.0131 for synonymous and nonsynonymous mutations, respectively. No significantly advantageous mutations were found, and no mutations caused a complete loss of function. In humans, most nonsynonymous mutations in protein coding regions have effects in the range of 10^{-3} to 10^{-1} [28]. Moreover, mutations in functional regions of the genome that are nonprotein-coding are likely to have even smaller effects. Viruses are somewhat exceptional for their high mutational sensitivity. Approximately 20 to 41% of viral mutations are lethal, while viable mutations have an average deleterious fitness effect of 0.10 to 0.13, and many mutations appear neutral [50,51]. However, viable mutations of small effect in viruses are still more abundant than those of large effect, and, as Lind *et al.* [49] have noted, it is possible that such experiments report large numbers of neutral mutations because of assays that lack sufficient sensitivity to detect low-impact mutations.

Junk DNA

A final concern is the existence of inert or "junk" DNA, i.e., genomic material for which mutation does not affect functionality. It does seem possible that many genomic sites play functional roles that are (at least partially) independent of sequence. Avida accounts for this by specifying *no-operation* instructions for 85% of the ancestral genome. Mendel also corrects for this possibility in two ways. First, Mendel models only the effective (functional) genome size, G_e, with 10% as the default. Second, to account for truly neutral mutations ($s = 0$), only the genomic rate of mutations affecting fitness, not the total rate, is used in default settings. Neutral mutations are thus excluded from the mutation rate. Mendel therefore uses a human mutation rate of 10 per genome per generation, rather than the actual mutation rate of approximately 50 – 100 [2,4,52–54], and a genome size of 3.0×10^8 (rather than 3.0×10^9). This genome size limits the magnitude of fitness effects to $1 / (3.0 \times 10^8) = 3.33 \times 10^{-9}$ and larger, allowing selection to act more effectively on mutations affecting fitness. These steps serve to account for neutral mutations and inert genomic material, to minimize the required computational resources, and to focus the use of Mendel on the effective (functional) genome (though the ability to track neutral mutations is currently being implemented).

The above considerations grant the common assumption that approximately 90% of the genome is indeed "junk." However, this has been subject to challenge for some time [55]. Importantly, the term "junk DNA" was first introduced by Ohno not as a result of experimentation, but rather as a theoretical necessity to avoid the evolutionary barrier of genetic entropy:

> ... there seems to be a strict upper limit for the number of gene loci which we can afford to keep in our genome. Consequently, only a fraction of our DNA appears to function as genes. ... the moment we acquire 10^5 gene loci, the overall deleterious mutation rate per generation becomes 1.0 which appears to represent an unbearably heavy genetic load. ... Even if allowance is made for the existence in multiplicates of certain genes, it is still concluded that, at the most, only 6% of our DNA base sequences is utilized as genes. ... More than 90% degeneracy contained within our genome should be kept in mind when we consider evolutional changes in genome sizes. ... it is not likely that these sequences came into being as a result of positive selection. Our view is that they are the remains of nature's experiments which failed [56].

This reasoning is common. For example, upon reporting a human mutation rate of 64 mutations per generation, Drake *et al.* [52] note that:

> It is hard to image [*sic*] that so many new deleterious mutations each generation is compatible with life, even with an efficient mechanism for mutation removal. Thus, the great majority of mutations in the noncoding DNA must be neutral.

Following the introduction of the junk DNA concept, many biologists quickly adopted the *selfish DNA* mechanism [57–59] to explain repetitive DNA [60], suggesting that "The search for other explanations may prove, if not intellectually sterile, ultimately futile" [58]. Others resisted this line of reasoning and suggested that repetitive DNA may function in gene regulation [61,62].

A full discussion of the functionality of nonprotein-coding DNA is beyond the scope of this study. However, it is worth noting that junk DNA assumptions have proven to be largely incorrect, while hypotheses suggesting functionality are being increasingly vindicated. Mattick has remarked that the junk DNA dogma may "be a classic story of orthodoxy derailing objective analysis of the facts, in this case for a quarter of a century... [it] may well go down as one of the biggest mistakes in the history of molecular biology" (quoted in reference [63]). A wide range of

Though it has been suggested that, counterintuitively, the waiting time to beneficial mutation does not increase exponentially with the number of necessary sites involved [81], the results reported here suggest otherwise. Further, Axe [74] has provided a detailed mathematical treatment of this evolutionary barrier by modeling a bacterial population of 10^9 individuals experiencing 1000 generations each year for all of Earth's history. Under these conditions, if intermediate states are neutral, adaptations involving at most six genomic sites can be expected to arise over the course of history; if intermediate states are maladaptive, adaptations involving at most two sites can be expected. (This hypothetical population strongly favors progressive evolution.) It follows that there is not enough time in Earth's history for mutation to generate any adaptive step involving > 6 genomic sites in any species. Several studies have alluded to these limitations. Orr has noted that "natural selection is essentially constrained to surveying those [sequences] that differ from wild-type by single-point mutations… Double mutations are too rare to be of much evolutionary significance" [82]. Similarly, the eventual stasis observed in long-term evolution experiments with *E. coli* has been explained thus: "Either further major improvements (with fitness increments of more than a few percent in this environment) do not exist or else they are evolutionarily inaccessible (e.g., adaptations requiring multiple genetic changes in which the intermediate states are unfit)" [83].

These concerns are usually discussed in terms of the waiting time to beneficial mutation, and generate spirited discussion in the literature [74,81,84–88]. However, although such calculations are usually interpreted to support the Darwinian mechanism of evolution, they are often incompatible with current theory. For example, Durrett and Schmidt [88] have calculated that the waiting time for a beneficial step involving only two sites, assuming a neutral intermediate, is roughly 100 million years in humans—yet humans are thought to have diverged from chimpanzees within the past 10 million years. Moreover, the challenge of generating the necessary adaptive mutations is complemented by the subsequent challenge of their fixation. This issue, classically known as the cost of substitution, is discussed elsewhere by ReMine [30,89].

The waiting time to beneficial mutation may alternatively be framed in terms of *irreducible complexity* [87,90]. The concept of irreducible complexity has had a great impact on the biological community, with numerous studies attempting to dismiss its importance. Avida has been used for this purpose [14–16,91]. Ironically, the program confirms that the problem is a reality by introducing what Dembski and Marks [92] have called *stair step active information* in order to evolve the EQU function, i.e., it provides information about the target (EQU) by rewarding the necessary building blocks, each of which can be feasibly constructed by mutation alone. This provides an easily scalable fitness landscape, in which successive steps are advantageous (Figure 3). Thus the EQU function can be built gradually from precursors of lower complexity, each of which is easy to generate through random mutation.

To justify the fitness scheme implemented in Avida, Lenski *et al.* have noted merely that this "is precisely what evolutionary theory requires" [14]. However, evidence suggests that paths to adaptive functions in biological organisms involve many genomic sites, with many of the intermediate states being maladaptive. For example, experiments with TEM-1 β-lactamase have shown that, for homologues of <~66% identity, intermediate protein sequences are typically non-functional when hybridized by random composite. This is the case even when only a fraction

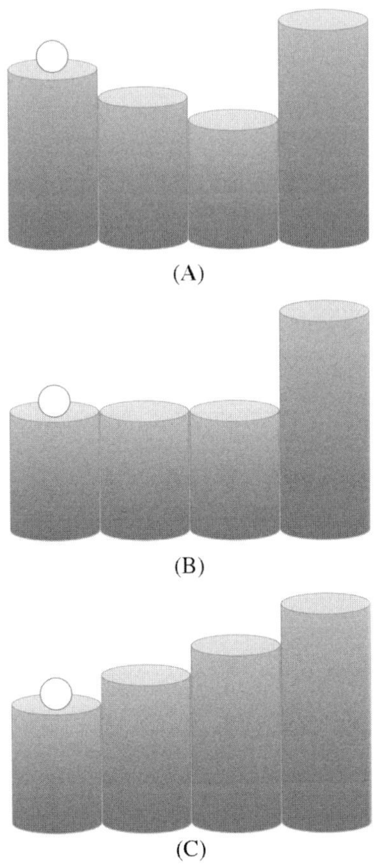

Fig. 3. Simple two-dimensional adaptive landscapes that become increasingly conducive to progressive evolution. The initial state is represented by the white ball. Natural selection can only promote intermediate states that increase fitness (steps "upward"). Shown are landscapes in which: (A) intermediate states are maladaptive; (B) intermediates states are neutral; and (C) intermediate states are beneficial.

of the total protein length is hybridized and sequences exhibit ~90% identity to wild type proteins [93]. These data suggest that contiguous stretches of co-optimized residues exist in biological proteins, and many intermediates between similar proteins may be nonfunctional. Moreover, most readily available adaptive changes are loss-of-function mutations [6,8,9]. These paths will be preferred by selection, as longer adaptive paths confer no advantage until distant targets are reached.

Avida demonstrates that the waiting time to beneficial mutation increases with the number of neutral intermediates, and that certain features cannot be expected to evolve unless simpler precursors are highly beneficial. While the problem of excessive waiting times does not make adaptive evolution formally impossible, it does render certain evolutionary scenarios implausible. Irreducible complexity means *complexity that is not reducible to parts that have a selective advantage on their own*, such that multiple coordinated changes are required without the help of selection. In other words, adaptations requiring multiple mutations are simply less likely, and the waiting time for their occurrence is greater. As Avida shows, this barrier can be prohibitive to progressive evolution. Unfortunately, computer simulations cannot provide a thorough understanding of the waiting times to adaptive steps in biology. As more is learned about the distribution of mutational fitness effects [28,94] and the genetic basis of adaptive change [8], the answers to these problems will become clearer.

Reductive evolution

Reductive evolution can entail an advantageous reduction in either genomic material or gene expression [9,95–97]. In both instances, organisms benefit from eliminating superfluous energy expenditure. The pressure to eliminate excess genomic material has been termed "compression selection" [97] and has been demonstrated in several biological systems. For example, in a classic serial transfer experiment with Qβ bacteriophage, replication rate increased by a factor of 15 and genome size decreased by 83%, with biological competency lost by the fifth transfer [95]. Some reductions in genome size have also been observed in evolution experiments with *E. coli* (e.g., reduced by 1.2% [38]). Although compression selection may not be strong in organisms for which the cost of maintaining and replicating DNA is a small fraction of the cell's total energy budget [26], it is clearly operational in some smaller systems.

More frequently, reductive evolution proceeds via the elimination of unnecessary gene expression. Gauger *et al.* [9] have shown that, because these types of mutations are relatively common [96], reductive evolutionary paths are usually taken even when short progressive paths are available. Long-term evolution experiments with *E. coli* have provided numerous examples of this process. One mutation that reduced *glmUS* expression by 10% was highly (~5%) beneficial [98], as was another mutation that reduced *spoT* expression [96]. Moreover, Cooper & Lenski [6] have reported that unused catabolic functions decayed as fitness increased in 12 experimental populations of *E. coli*, reducing diet breadth. One mutation, loss of the ability to use D-ribose, occurred in all 12 populations in the first 2,000 generations as a result of highly advantageous deletion mutations, increasing fitness by ~1.4% [7].

These studies indicate that the reduction of biological information can be highly advantageous. Recent reviews [8,10,99] have reported that the majority of studied adaptations involve the loss of traits and the reduction of genetic information. Whether a mutation is beneficial may depend critically upon the environment in which it arises (e.g., whether nutrients are available or antibiotics are present), meaning that the effect of a mutation on genetic information cannot be inferred from relative growth rates alone. Reductive changes are often (though not always) associated with fitness loss in other environments [100]. For example, the ability to transport (and therefore metabolize) citrate in oxic conditions evolved in one *E. coli* population after about 31,000 generations of experimental evolution [73]. However, the mutant is inferior on glucose, likely because it involves the alteration of a citrate transporter that normally operates only in anoxic conditions. Other decreases in channel constriction have also conferred advantages [100]. Similarly, Bull *et al.* [48] have reported high-impact beneficial mutations in the virus ϕX174 that increase fitness in an inhibitory, hot environment, but all of which reduce fitness at normal temperatures.

The Mendel software uses the classic infinite allele model, and so is not conducive to a straightforward study of the evolution of genome size. On the other hand, Avida is very tractable for this purpose. Importantly, the biological examples of adaptation discussed above involve reductive evolution, in contrast with adaptation under Avida's default settings, where novel functions arise and provide extreme advantages. It is somewhat surprising that Avidian populations achieve in only 10,000 generations what *E. coli* populations fail to glimpse in 50,000 generations. This occurs partly because artificial size neutrality mechanisms were introduced into the Avida software as a means of preventing the pressure of reductive evolution:

> The advantage gained by shrinking the code is so dramatic, however, that cells might even choose to shed sections of code that trigger moderate bonuses. Such a method certainly provides for very efficient optimization while discouraging the evolution of complex code by magnifying the barrier to neighboring local minima

> in the fitness landscape. ... Another possibility is to distribute CPU time in a manner proportional to the length of the code. This is the *size-neutral* scheme also used in tierra. The resulting fitness landscape is intuitively much smoother; strings that behave in the same way but differ in length of code are degenerate as far as their replication rate is concerned and far-lying regions in genotype-space can be accessed easily. Clearly this mechanism is much more conducive to the evolution of complexity… Note that enforcing size neutrality is strictly speaking un-biological, as it is known that self-replicating strings will shed all unnecessary instructions if given the opportunity. In avida, size neutrality is necessary in order to jump start the evolution of complexity [21].

Although the results reported here suggest that size neutrality is not strictly necessary for the evolution of complexity in Avida, it certainly improves success. Therefore Avida confirms that reductive evolution is also a potential barrier to the evolution of novel genetic information. Moreover, this barrier will be more prohibitive if new functions confer more realistic fitness bonuses.

The barrier that compression selection poses for progressive evolution is most extreme for small genomes. These results demonstrate that, when size neutrality is disabled, larger genomes evolve more logic operations than smaller genomes. This occurs because large genomes contain more superfluous material that may be used as raw material for evolutionary tinkering. If highly beneficial adaptations arise before prohibitive genome shrinkage occurs, the pressure to maintain highly beneficial functions can prevent further shrinking, which is only slightly adaptive. The large default rewards implemented in Avida dwarf the advantages gained by shrinking the genome, so evolved functions are retained once a minimal genome size is reached. This appears to be another case in which the waiting time to beneficial mutation is an important consideration, as innovations that require too much time may not arise before the extraneous genomic raw material is removed by selection.

Conclusions

This study used the evolutionary simulations Avida and Mendel's Accountant to examine three barriers to the production of genetic information by the neo-Darwinian mechanism of mutation and natural selection: (1) the selection threshold and resultant genetic entropy; (2) the waiting time to beneficial mutation, i.e., irreducible complexity; and (3) the pressure of reductive evolution, i.e., the pressure to shrink genomes and to disable unnecessary functions. The apparent disparity between the two programs results primarily from differences in default settings. When used with similar settings that reflect biological systems, both confirm that all three of the aforementioned barriers can prevent the progressive evolution of novel genetic information. Though neutral or even maladaptive changes (e.g., gene duplication) are often considered "complex features" [16,26,101,102], it is important to note that this is not synonymous with genetic information. Even adaptive changes typically eliminate genetic information within a genome [8,10].

The evolutionary barriers discussed in this report are not merely of theoretical importance. As Lynch [4] and others [2,3,44–47] have shown, the human species faces a potentially lethal threat from the accumulation of very slightly deleterious mutations. Additionally, the lethal mutagenesis of pathogen populations may be applicable in novel medical approaches to cure infection and thwart pandemics [11–13]. It may be the case that novel means of genetic intervention to reduce mutation will be necessary to prevent the extinction of numerous species, including our own.

While both Avida and Mendel demonstrate that neo-Darwinian evolution may be a theoretical possibility under certain conditions, both programs also suggest that it is not a plausible explanation of most biological information. Such computational approaches can provide informative predictions of the values that key parameters (e.g., the distribution of mutational fitness effects) must assume if neo-Darwinian theory is viable. However, biological studies will be necessary to determine the values that these parameters actually assume in nature.

Digital genetics pioneer Thomas Ray made the following point about computational evolutionary studies:

> To understand the biology of digital organisms requires a knowledge of the properties of machine instructions and machine language algorithms. ... there exists a complementary relationship between biological theory and the synthesis of life. Theory suggests how the synthesis can be achieved, while application of the theory in the synthesis is a test of the theory. If theory suggests that a certain factor will contribute to increasing diversity, then synthetic systems can be run with and without that factor. The process of synthesis becomes a test of the theory [103].

It would seem, then, that the "unbiological" [21] parameters required to make the neo-Darwinian mechanism succeed in computational experiments should call the biological theory into question. As science commentator David Berlinski has remarked, "Computer simulations of Darwinian evolution fail when they are honest and succeed only when they are not" [104]. As more is learned about the genetic basis of adaptive change and the distribution of mutational fitness effects, the severity of these concerns for theory and medicine will become clearer.

22. Ofria C, Wilke CO (2004) Avida: a software platform for research in computational evolutionary biology. Artificial Life 10:191–229.
23. The Avida Digital Life Platform [http://devolab.cse.msu.edu/software/avida/doc/]
24. Nelson CW, Sanford JC (2011) The effects of low-impact mutations in digital organisms. Theor Biol Med Model 8:9.
25. Muller HJ (1964) The relation of recombination to mutational advance. Mutat Res 1:2–9.
26. Lynch M (2007) The Origins of Genome Architecture. Sinauer Associates, Inc. Publishers, Sunderland, MA.
27. Kimura M (1983) The Neutral Theory of Molecular Evolution. Cambridge University Press, Cambridge.
28. Eyre-Walker A, Keightley PD (2007) The distribution of fitness effects of new mutations. Nat Rev Genet 8:610–618.
29. Nei M (2007) The new mutation theory of phenotypic evolution. Proc Natl Acad Sci USA 104:12235–12242.
30. ReMine WJ (1993) The Biotic Message. St. Paul Science, Saint Paul, MN.
31. Haldane JBS (1932) The Causes of Evolution. Harper, New York.
32. Lynch M (2007) The frailty of adaptive hypotheses for the origins of organismal complexity. Proc Natl Acad Sci USA, 104(Suppl 1):8597–8604.
33. Sanford JC (2005) Genetic entropy and the mystery of the genome. FMS Publications, Lima, NY.
34. Sanford JC, Baumgardner JR, Brewer WH (2013) Selection threshold severely constrains capture of beneficial mutations. In: Marks II RJ, Behe MJ, Dembski WA, Gordon B, Sanford JC (eds) Biological Information — New Perspectives. World Scientific, Singapore, pp. 264–297.
35. Baumgardner JR, Brewer WH, Sanford JC (2013) Can synergistic epistasis halt mutation accumulation? Results from numerical simulation. In: Marks II RJ, Behe MJ, Dembski WA, Gordon B, Sanford JC (eds) Biological Information — New Perspectives World Scientific, Singapore, pp. 312–337.
36. Brewer WH, Baumgardner JR, Gibson P, Sanford JC (2013) Using numerical simulation to test the "mutation-count" hypothesis. In: Marks II RJ, Behe MJ, Dembski WA, Gordon B, Sanford JC (eds) Biological Information — New Perspectives. World Scientific, Singapore, pp. 298–311.
37. Ewert W, Dembski WA, Marks II RJ (2009) Evolutionary synthesis of nand logic: dissecting a digital organism. In: Proceedings of the 2009 IEEE International Conference on Systems, Man, and Cybernetics, San Antonio, TX.
38. Barrick JE, Yu DS, Yoon SH, Jeong H, Oh TK, Schneider D, Lenski RE, Kim JF (2009) Genome evolution and adaptation in a long-term experiment with *Escherichia coli*. Nature 461:1243–1247.
39. Lynch M, Gabriel W (1990) Mutation load and the survival of small populations. Evolution 44:1725–1737.

40. Gabriel W, Lynch M, Bürger R (1993) Muller's ratchet and mutational meltdowns. Evolution 47:1744–1757.
41. Lynch M, Bürger R, Butcher D, Gabriel W (1993) The mutational meltdown in asexual populations. J Hered 84:339–344.
42. Butcher D (1995) Muller's ratchet, epistasis and mutation effects. Genetics 141: 431–437.
43. Lande R (1994) Risk of population extinction from fixation of new deleterious mutations. Evolution 48:1460–1469.
44. Muller HJ (1950) Our load of mutations. Am J Hum Genet 2:111–176.
45. Crow JF (1997) The high spontaneous mutation rate: is it a health risk? Proc Natl Acad Sci USA 94:8380–8386.
46. Eyre-Walker A, Keightley PD (1999) High genomic deleterious mutation rates in Hominids. Nature 397:344–347.
47. Eyre-Walker A, Woolfit M, Phelps T (2006) The distribution of fitness effects of new deleterious amino acid mutations in humans. Genetics 173:891–900.
48. Bull JJ, Badgett MR, Wichman HA (2000) Big-benefit mutations in a bacteriophage inhibited with heat. Mol Biol Evol 17:942–950.
49. Lind PA, Berg OG, Andersson DI (2010) Mutational robustness of ribosomal protein genes. Science 330:825–827.
50. Sanjuán R (2010) Mutational fitness effects in RNA and single-stranded DNA viruses: common patterns revealed by site-directed mutagenesis studies. Phil Trans R Soc B 365:1975–1982.
51. Peris JB, Davis P, Cuevas JM, Nebot MR, Sanjuán R (2010) Distribution of fitness effects caused by single-nucleotide substitutions in bacteriophage f1. Genetics 185:603–609.
52. Drake JW, Charlesworth B, Charlesworth D, Crow JF (1998) Rates of spontaneous mutation. Genetics 148:1667–1686.
53. Nachman MW, Crowell SL (2000) Estimate of the mutation rate per nucleotide in humans. Genetics 156:297–304.
54. Kondrashov AS (2002) Direct estimates of human per nucleotide mutation rates at 20 loci causing Mendelian diseases. Hum Mutat 21:12–27.
55. Nowak R (1994) Mining treasures from "junk DNA." Science 263:608–610.
56. Ohno S (1972) So much "junk" DNA in our genome. In: Smith HH (ed) Evol Genet Syst, Brookhaven Symp Biol 23:366–370.
57. Dawkins R (1976) The Selfish Gene. Oxford University Press, New York.
58. Doolittle WF, Sapienza C (1980) Selfish genes, the phenotype paradigm and genome evolution. Nature 284:601–603.
59. Orgel LE, Crick FHC (1980) Selfish DNA: the ultimate parasite. Nature 284: 601–603.
60. Britten RJ, Kohne DE (1968) Repeated sequences in DNA. Science 161:529–540.

61. Britten RJ, Davidson EH (1969) Gene regulation for higher cells: a theory. Science 165:349–357.
62. Davidson EH, Britten RJ (1979) Regulation of gene expression: possible roles of repetitive sequences. Science 204:1052–1059.
63. Gibbs WW (2003) The unseen genome: gems among the junk. Scientific American 289(5):46–53.
64. The ENCODE Project Consortium (2007) Identification and analysis of functional elements in 1% of the human genome by the ENCODE pilot project. Nature 447:799–816.
65. Kapranov P, Willingham AT, Gingeras TR (2007) Genome-wide transcription and the implications for genome organization. Nat Rev Genet 8:413–423.
66. Faulkner GJ, Kimura Y, Daub CO, Wani S, Plessy C, Irvine KM, Schroder K, Cloonan N, Steptoe AL, Lassmann T, Waki K, Hornig N, Arakawa T, Takahashi H, Kawai J, Forrest ARR, Suzuki H, Hayashizaki Y, Hume DA, Orlando V, Grimmond SM, Carninci P (2009) The regulated retrotransposon transcriptome of mammalian cells. Nat Genet 41:563–571.
67. van Bakel H, Nislow C, Blencowe BJ, Hughes TR (2010) Most "dark matter" transcripts are associated with known genes. PLoS Biol 8:e1000371.
68. Kapranov P, St. Laurent G, Raz T, Ozsolak F, Reynolds CP, Sorensen PH, Reaman G, Milos P, Arceci RJ, Thompson JF, Triche TJ (2010) The majority of total nuclear-encoded non-ribosomal RNA in a human cell is "dark matter" un-annotated RNA. BMC Biol 8:149.
69. Komar AA (2007) SNPs, silent by not invisible. Science 315:466–467.
70. Kimchi-Sarfaty C, Oh JM, Kim I-W, Sauna ZE, Calcagno AM, Ambudkar SV, Gottesman MM (2007) A "silent" polymorphism in the MDR1 gene changes substrate specificity. Science 315:525–528.
71. Seaman J (2013) DNA.EXE: A sequence comparison between the human genome and computer program code. In: Marks II RJ, Behe MJ, Dembski WA, Gordon B, Sanford JC (eds) Biological Information — New Perspectives. World Scientific, Singapore, pp. 385–401.
72. Montañez G, Marks II RJ, Fernandez J, Sanford JC (2013) Multiple overlapping genetic codes profoundly reduce the probability of beneficial mutation. In: Marks II RJ, Behe MJ, Dembski WA, Gordon B, Sanford JC (eds) Biological Information – New Perspectives. World Scientific, Singapore, pp. 139–167.
73. Blount ZD, Borland CZ, Lenski RE (2008) Historical contingency and the evolution of a key innovation in an experimental population of *Escherichia coli*. Proc Natl Acad Sci U S A 105:7899–7906.
74. Axe DD (2010) The limits of complex adaptation: an analysis based on a simple model of structured bacterial populations. BIO-Complexity 2010(4):1–10.
75. Axe DD (2004) Estimating the prevalence of protein sequences adopting functional enzyme folds. J Mol Biol 341:1295–1315.

76. Reidhaar-Olson JF, Sauer RT (1990) Functionally acceptable substitutions in two a-helical regions of l repressor. Proteins 7:306–316.
77. Axe DD (2010) The case against a Darwinian origin of protein folds. BIO-Complexity 2010(1):1–12.
78. Dembski WA (1998) The Design Inference: Eliminating Chance Through Small Probabilities. Cambridge University Press, New York.
79. Lloyd S (2002) Computational capacity of the universe. Phys Rev Lett 88:237901.
80. Abel DL (2009) The Universal Plausibility Metric (UPM) & Principle (UPP). Theor Biol Med Model 6:27.
81. Lynch M, Abegg A (2010) The rate of establishment of complex adaptations. Mol Biol Evol 27:1404–1414.
82. Orr HA (2003) The distribution of fitness effects among beneficial mutations. Genetics 163:1519–1526.
83. Lenski RE, Travisano M (1994) Dynamics of adaptation and diversification: a 10,000-generation experiment with bacterial populations. Proc Natl Acad Sci USA 91:6808–6814.
84. Behe MJ, Snoke DW (2004) Simulating evolution by gene duplication of protein features that require multiple amino acid residues. Protein Sci 13:2651–2664.
85. Lynch M (2005) Simple evolutionary pathways to complex proteins. Protein Sci 14:2217–2225.
86. Behe MJ, Snoke DW (2005) A response to Michael Lynch. Protein Sci 14: 2226–2227.
87. Behe MJ (2007) The Edge of Evolution. Free Press, New York.
88. Durrett R, Schmidt D (2008) Waiting for two mutations: with applications to regulatory sequence evolution and the limits of Darwinian evolution. Genetics 180:1501–1509.
89. ReMine WJ (2005) Cost theory and the cost of substitution — a clarification. TJ 19(1):113–125.
90. Behe MJ (1996) Darwin's Black Box. Free Press, New York.
91. Pennock RT (2004) DNA by design? In: Dembski WA, Ruse M (eds) Debating Design: From Darwin to DNA. Cambridge University Press, New York.
92. Dembski WA, Marks II RJ (2009) Conservation of information in search: measuring the cost of success. IEEE Trans Syst Man Cybern A Syst Hum 5:1051–1061.
93. Axe DD (2000) Extreme functional sensitivity to conservative amino acid changes on enzyme exteriors. J Mol Biol 301:585–595.
94. Keightley PD, Eyre-Walker A (2010) What can we learn about the distribution of fitness effects of new mutations from DNA sequence data? Phil Trans R Soc B 365:1187–1193.
95. Mills DR, Peterson RL, Spiegelman S (1967) An extracellular Darwinian experiment with a self-duplicating nucleic acid molecule. Biochemistry 58:217–224.

96. Cooper TF, Rozen DE, Lenski RE (2003) Parallel changes in gene expression after 20,000 generations of evolution in *Escherichia coli*. Proc Natl Acad Sci USA 100:1072–1077.
97. Ofria C, Adami C, Collier TC (2003) Selective pressures on genomes in molecular evolution. J Theor Biol 222:477–483.
98. Stanek MT, Cooper TF, Lenski RE (2009) Identification and dynamics of a beneficial mutation in a long-term evolution experiment with *Escherichia coli*. BMC Evol Biol 9:302.
99. Behe MJ (2013) Getting there first: an evolutionary rate advantage for adaptive loss-of-function mutations. In: Marks II RJ, Behe MJ, Dembski WA, Gordon B, Sanford JC (eds) Biological Information — New Perspectives. World Scientific, Singapore, pp. 450–473.
100. Elena SF, Lenski RE (2003) Evolutionary experiments with microorganisms: the dynamics and genetic bases of adaptation. Nat Rev Genet 4:457–469.
101. Lynch M, Conery JS (2003) The origins of genome complexity. Science 302: 1401–1404.
102. Gray MW, Lukes J, Archibald JM, Keeling PJ, Doolittle WF (2010) Irremediable complexity? Science 330:920–921.
103. Ray TS (1991) An approach to the synthesis of life. In: Langton CG, Taylor C, Farmer JD, Rasmussen S (eds) Artificial Life II, SFI Studies in the Sciences of Complexity, Vol. X. Addison-Wesley, Redwood City, CA.
104. Berlinski D (2008) The Devil's Delusion. Crown Forum, New York.

Information Loss: Potential for Accelerating Natural Genetic Attenuation of RNA Viruses

Wesley H. Brewer[1], Franzine D. Smith[2] and John C. Sanford[3]*

[1]*Fluid Physics International, P.O. Box 4915, Mississippi State, MS 39762, USA.*
[2]*Rainbow Technologies, Inc., Waterloo NY 13165, USA.* [3]*Dept. Hort. Sci., NYSAES, Cornell University, Geneva, NY 14456, USA. (*Corresponding author: jcs21@cornell.edu).*

Abstract

Loss of information is not always bad. In this paper, we investigate the potential for accelerating the genetic degeneration of RNA viruses as a means for slowing/containing pandemics. It has previously been shown that RNA viruses are vulnerable to *lethal mutagenesis* (the concept of inducing mutational degeneration in a given pathogen). This has led to the use of lethal mutagenesis as a clinical treatment for eradicating RNA virus from a given infected patient. The present study uses numerical simulation to explore the concept of accelerated mutagenesis as a way to enhance natural genetic attenuation of RNA viral strains at the epidemiological level. This concept is potentially relevant to improved management of pandemics, and may be applicable in certain instances where eradication of certain diseases is sought.

We propose that mutation accumulation is a major factor in the natural attenuation of pathogenic strains of RNA viruses, and that this may contribute to the disappearance of old pathogenic strains and natural cessation of pandemics. We use a numerical simulation program, Mendel's Accountant, to support this model and determine the primary factors that can enhance such degeneration. Our experiments suggest that natural genetic attenuation can be greatly enhanced by implementing three practices. (1) Strategic use of antiviral pharmaceuticals that increase RNA mutagenesis. (2) Improved hygiene to reduce inoculum levels and hence increase genetic bottlenecking. (3) Strategic use of broad-spectrum vaccines that induce partial immunity. In combination, these three practices should profoundly accelerate loss of biological information (attenuation) in RNA viruses.

Key words: mutation accumulation, lethal mutagenesis, error catastrophe, mutation meltdown, pandemic, Mendel's Accountant

Introduction

The concept of lethal mutagenesis has been put forward as a strategy for controlling pathogens [1, 2]. The idea of lethal mutagenesis is to enhance the mutation rate of the pathogen, thereby accelerating mutation accumulation and leading to mutational meltdown and extinction of the pathogen within a specific host individual. The concept of mutation accumulation in RNA viruses has been explored in biological experiments involving bacteriophage [3], tobacco etch virus [4], poliovirus [5], vesicular stomatitis virus [6, 5], and HIV [7–10]. All these researchers report rapid fitness declines of viral strains as deleterious mutations accumulate, often leading to the actual extinction of some strains. This strongly contradicts claims that RNA viruses are somehow robust against the accumulation of deleterious mutations [11–13]. Lethal mutagenesis is considered a potential antiviral therapy for infected patients and is also recognized as having relevance to management of pandemics [2].

RNA viruses are excellent candidates for genetic degeneration because they typically have an extraordinarily high mutation rate [14]. The higher mutation rate of RNA viruses is a consequence of the novel mechanisms required for RNA replication, which are especially prone to mutation, and the lack of effective repair enzymes for RNA replication. Even in RNA viruses with relatively small genomes, there appear to be as many as 0.1 to 1.0 new mutations per virus per replication cycle [15]. The mutation rate in RNA viruses is so high that it becomes difficult to speak of a given viral "strain", because any genotype quickly mutates into a complex of genotypes, such that any patient is soon infected with a "viral swarm". With such a high mutation rate, the large majority of viral genotypes in a patient must carry many deleterious mutations, and so will be inferior to the original infecting genotype. This implies the lack of a realistic mechanism to preserve a "standard genotype", and all RNA viral swarms should typically be on the verge of mutational meltdown.

When a virus is transmitted from one individual to the next, the first individual harbors a viral swarm. The second individual becomes infected by a random subset of that swarm (conceivably a single genotype). With this type of bottlenecking, the "best" viral genotypes within the first swarm have a small probability of being transmitted to the next host. This probability becomes especially small when infection arises from a single viral particle. Given a high mutation rate and regular bottlenecks, the operation of Muller's Ratchet becomes quite certain, which should result in a continuous ratchet-like mutational degeneration of the viral genome [6].

This type of genetic degeneration happens independently of specific virulence factors. A viral strain may have a few high-impact beneficial mutations that affect "virulence" (i.e., compatibility with a specific host), yet at the same time that same strain can be accumulating large numbers of low-impact mutations throughout its genome, which should systematically degrade function and reduce net fitness. Therefore such a strain can be undergoing genetic degeneration, even while it retains (or gains) favorable virulence factors.

In this light, it appears very likely that RNA viruses should have a strong. tendency to undergo what we will call "natural genetic attenuation". This should happen within the individual host organism as the mean fitness of the viral swarm continues to diminish with every replication cycle. This should happen even more dramatically as the viral swarm undergoes recurrent bottlenecking, as it passes from host individual to host individual. Such natural genetic attenuation should logically contribute to the transient nature of RNA viral infections within the individual, as well as the transient nature of pandemics caused by RNA viruses.

Historical evidence that RNA viruses undergo natural genetic attenuation

Dengue type-2 virus (DENV), a mosquito-borne, positive-sense, single-strand RNA virus, caused an epidemic in several Pacific Islands from 1971 to 1974. A recent paper [16] studied the epidemiological, clinical and biologic observations recorded during this time. The authors note that the time period, population dynamics and isolation of this epidemic gives a unique opportunity to study virus evolution minus many confounding factors. The initial outbreak of the disease, on Fiji and Tahiti, caused severe clinical symptoms, while the final outbreak on Tonga produced mild symptoms and near-silent transmission. Sequence and phylogenetic analysis showed that the outbreaks were genetically related and all due to a single introduction. Also these analyses placed the Tongan viral isolates in a single clade, with some unique site substitutions compared to viral isolates early in the epidemic. It is these deleterious genetic changes that Steel *et al.* [16] believe was responsible for the reduced epidemic severity on Tonga in 1973/1974.

Severe acute respiratory syndrome (SARS) caused by an animal-derived coronavirus appeared in the human population of Guangdong Province of China in late 2002. Sixty-one viral isolates from humans were sequenced from the early, middle and late phases of the outbreak in this region and were compared to animal derived viral sequences [17]. This epidemic was characterized by its sudden appearance, its extreme virulence, its rapid spread, and the rapid collapse of the pandemic after just two months [17]. This dramatic collapse cannot reasonably be attributed to human intervention. Given that SARS in man appears to have an inordinately high mutation rate of roughly 3 mutations per replication [18], and given that during this very short-term pandemic 291 mutations accumulated in the virus, it seems most reasonable to conclude that the outbreak ended prematurely because the virus underwent mutational degeneration and natural genetic attenuation.

Similarly, Ebola outbreaks have emerged explosively, initially being extremely virulent and extremely contagious, but very quickly they became self-contained apart from human intervention. While the Ebola virus appears to have an extremely wide host range, it has been almost impossible to find it in the natural fauna of the relevant regions [19]. This can most reasonably be explained by self-containment of the virus due to high mutation rates and natural genetic attenuation. Bowen *et al.* [20] cite the World Health Organization's report suggesting that such attenuation occurred after just 10–11 passages within the human population.

Influenza A virus causes respiratory infections in mammals and birds. In humans, this virus causes a yearly epidemic and an occasional pandemic. It appears that influenza strains are continuously going extinct at a high rate. The actual precursor strains of the H1N1 strain that caused the disastrous 1918 pandemic are unknown, and can be presumed to be extinct [i.e., 21, 22]. The H1N1 strain itself appears to have gone extinct in the mid-twentieth century, and apparently was inadvertently re-introduced from a researcher's lab freezer in 1977 [23, 24]. During the 2009 H1N1 pandemic, one of two original strains went extinct [25]. Given the global nature of influenza spread and distribution, it can very reasonably be asked - why does the previous year's strain of the flu routinely disappear so quickly? Why do most strains of influenza appear to routinely go extinct? The most reasonable answer would seem to be natural genetic attenuation due to mutation accumulation.

Methods and Results

We have conducted a series of numerical simulation experiments using the genetic accounting program Mendel's Accountant (Mendel). This program tracks mutation accumulation over time, as affected by the primary relevant variables such as mutation rate, distribution of mutational effects, selection pressure, and population size [26–31]. Although Mendel has traditionally been used to model higher organisms (e.g., diploid, sexually reproducing species), it has alternative parameter settings that allow us to model populations of organisms with small haploid genomes and which reproduce clonally.

In these experiments, we model a generic RNA virus similar to the influenza virus. We model only a single viral sub-strain, which becomes a viral swarm, which is then transmitted through a single lineage as it moves through a series of 100 individuals during a pandemic lasting 300 days. We model an RNA virus that employs RNA to RNA replication with a viral doubling time of one hour (24 replication cycles per day) [32, 33]. We assume that passage to a new host individual happens every 3 days [34], and that infection in the new host individual involves the transmission of either a low or high level of inoculum, depending on the model run (10 or 1000 viable viral particles randomly sampled from the viral swarm). Following each new host infection, the swarm is allowed to amplify in number until a specified steady-state population size is reached within the individual host. We use a maximal population size of 10,000 (in our experience, creating populations larger than this has minimal effect on selection efficiency and mutation accumulation, but consistently causes overflow of computer memory). We assume a functional genome size of 10,000 nucleotides, and we assume a starting baseline reference genotype, which we define as our "wild type" (having zero initial mutations by definition). We model 10% of all mutations as being perfectly neutral, with the remainder of mutations being 99% deleterious and 1% beneficial [35]. We model back-mutations based upon mutation rate and the fraction of nucleotides already mutated. We use the well-accepted Weibull distribution for mutation effects (a natural, exponential-type distribution [26]). In this type of distribution, low-impact mutations are much more abundant than high-impact mutations. The lowest impact mutation we model (excluding perfect neutrals) has a fitness effect which is the inverse of the genome size (such a mutation would reduce fitness by one part in 10,000 when arising in a genome of 10,000 functional nucleotides).

In order to be consistent with what is known about deleterious viral mutation distributions, we shape the mutation distribution such that there is a very substantial fraction of all mutations that have a large effect on fitness (10% of the deleterious mutations reduce fitness by 10% or more). We model beneficial mutations to have a similar distribution as deleterious mutations, but with a much narrower range (maximal fitness effect = 0.01). This upper limit excludes major virulence factor mutations, which are outside the scope of these experiments (we wish to study overall fitness, not singular host/pathogen compatibility factors). Viruses are recognized as having a much higher rate of lethal mutations than other organisms [35], and our Weibull distribution does not fully model this. However, since all viral particles with lethal mutations will fail to replicate, they are easily accounted for by simply adjusting the rate of "random deaths". Mutational effects are combined additively within a viral genotype [3].

Mutations were introduced into the viral population at rates ranging from 0.1 to 1.0 mutation per genome per replication [15, 25]. Viral replication was modeled as a simple asexual doubling every replication cycle, causing population size to double. After every replication, we eliminated the surplus population by applying natural selection (partial truncation selection) based upon phenotype, restoring the initial population size. When bottlenecking was modeled, every time a new host was infected the population size was reduced to either 10 or 1000 particles. The population was allowed to undergo rapid growth to restore population size. This was done by temporary partial relaxation of selection, such that roughly 50% of the surplus viral particles were not selected away but were allowed to contribute to population re-growth. As deleterious mutations accumulated to high levels, some viral particles had zero fitness and could not replicate. When there were not enough viable viral particles to repopulate the viral population after each selection cycle, the size of the viral population necessarily began to shrink each generation. If this continued, the viral population would shrink to zero, causing extinction of the viral swarm.

Our first experiment was a preliminary Mendel run using very conservative parameters. This was designed as a base-line for minimal genetic attenuation of our model RNA virus, as would occur during a 300-day pandemic. We used a low mutation rate of 0.1 mutations per virus particle per replication cycle. In every replication cycle the number of viral particles was allowed to double, and we modeled zero random death (zero percent of the viral particles were randomly lost). Phenotypic selection was applied (partial truncation), to eliminate all of the surplus population (50%), such that the initial population size was restored. In this first experiment, we did not model any population bottlenecks. The results of this experiment are summarized in Figure 1. we see that even using highly favorable assumptions and intense selection, the simulation failed to prevent mutation accumulation. After 7200 replication cycles, each virus accumulated an average of 235 deleterious, 74 neutral, and 9.4 beneficial mutations. There were 523 polymorphic mutant alleles segregating in the population, meaning that it was a very genetically diverse viral swarm. Although 7 beneficial mutations went to fixation within the swarm, these carried with them 180 deleterious mutations that also went to fixation. Fitness declined 16% in just 300 days. By the end of the experiment deleterious mutation count per virus was increasing at an essentially constant rate, and mean viral fitness was declining at nearly a constant rate. These results indicate the presence of strong forces working to attenuate any given strain, even when conditions for maintenance of the virus are optimal.

We then conducted a series of four simulations wherein we modeled the effect of factors that might accelerate natural genetic attenuation. Figure 2 summarizes the fitness decline seen in these experiments. In the first of these experiments, we introduced a realistic, but modest degree of random loss of viral particles (25%), as might be expected due to chance and various host defense mechanisms. Simultaneously, we introduced a very weak, and recurrent bottlenecking of population size (1000 viral particles/infection), corresponding to high inoculum levels during viral transmission to new host individuals. The result of this second experiment was a very slight acceleration in the rate of genetic attenuation compared to Figure 1 (final mean fitness was reduced from 0.84 to 0.82, see Figure 2).

In the second simulation, we tested the effect of increasing the random loss of viral particles as might arise, for example, due to host RNase activity, or as a result of antiviral pharmaceuticals, or as might arise due to partial immunity within the host. We eliminated 40% of all viral particles by random death, thus reducing the viral surplus population from 50% to 10%. This effectively reduces selection intensity. The result was another very slight acceleration of fitness decline (final mean fitness declined to 0.79, see Figure 2).

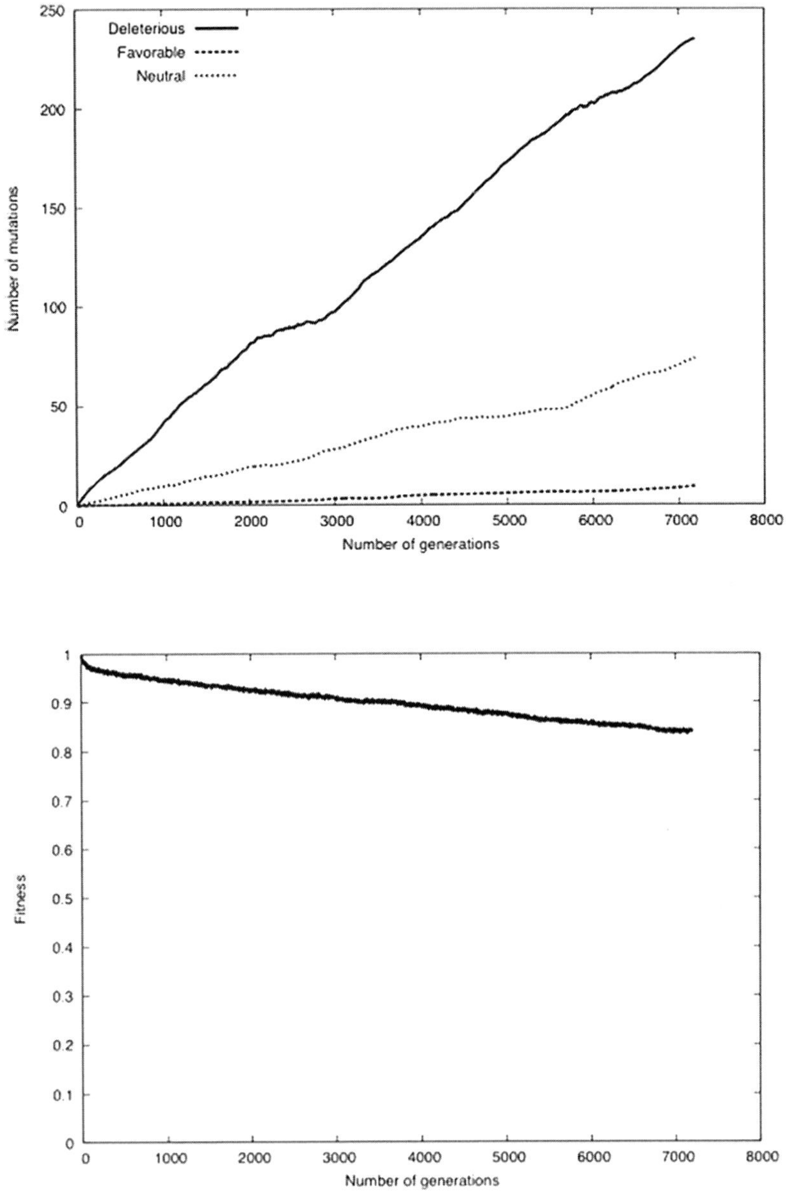

Fig.1. A preliminary numerical simulation experiment with parameters optimized for slowing genetic degeneration of a model RNA virus. Mutation rate = 0.1/genome/replication (89% of mutations deleterious, 10% neutral, and 1% beneficial). Partial truncation selection was employed (50% selective elimination, every replication cycle). No random death and no bottlenecking. Mean mutation count per virus over time (figure above), and fitness decline over time (figure below).

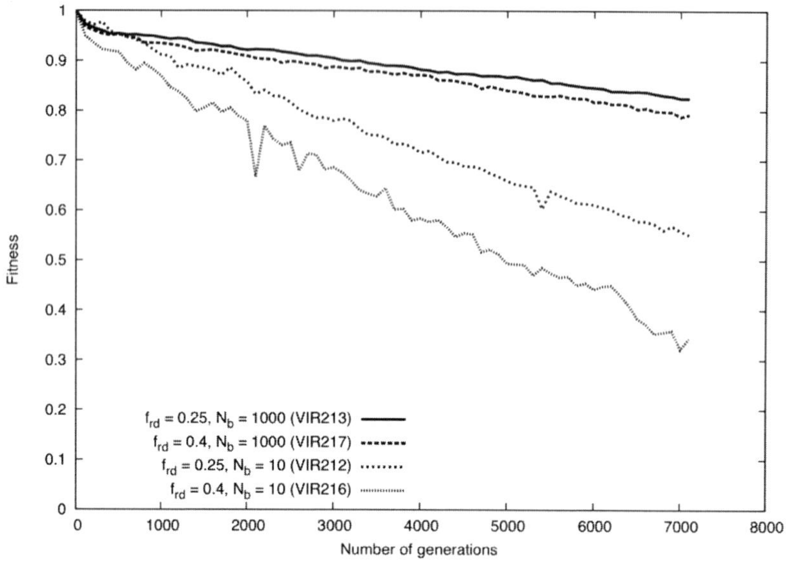

Fig. 2. Four simulations demonstraing the effect of bottlenecking and random death on fitness degradation during an RNA virus pandemic. VIR213 = minimal bottlenecking and modest random loss, VIR217 = more random loss, VIR212 = more severe bottlenecking, VIR216 = combination of more random loss and more severe bottlenecking. N_b is the population size during the bottleneck. F_{rd} is the fractional occurrence of random death.

In the third simulation, we tested the effect of much more severe bottlenecking, with just 10 viable viral particles per new infection. This might be clinically achieved either by use of antiviral pharmaceuticals or through better hygiene. We see that when we have strong bottlenecking, selection is significantly less effective and genetic attenuation is much faster. Fitness declined 45% in 300 days (final mean fitness was 0.55, see Figure 2). After 7200 replication cycles, each virus accumulated an average of 356 deleterious, 77 neutral, and 9.9 beneficial mutations. There were only 81 segregating polymorphic alleles in the population, reflecting the homogenizing effect of recurrent bottlenecking. Although 9 beneficial mutations went to fixation, along with them 338 deleterious mutations went to fixation.

In the fourth of these simulations, we modeled intensified bottlenecking combined with 40% random loss. The result was dramatically accelerated fitness decline (final mean fitness was 0.35, see Figure 2).

As evident from Figure 2, more severe bottlenecking and higher rates of random loss combine synergistically to greatly accelerate both fitness decline and genetic attenuation. More random death by itself had a very small effect, while the effect of bottlenecking by itself was more significant, but still fairly modest. However, a higher rate of random loss (hence lower viral titer) greatly amplified the bottleneck effect (because after a serious bottleneck, random loss increases the time needed for population size to recover, effectively extending the duration of each bottleneck).

We lastly conducted a series of four simulations wherein we examined the consequence of increasing mutation rate, as might be achieved by using a pharmaceutical such as Ribavirin. We used the most conservative settings shown in Figure 2 (weak bottlenecking and only a moderate rate of random loss), and we then examined the effect of increasing mutation rate from 0.1 to 0.2, 0.4, 0.6 and 0.8. The fitness decline resulting from an elevated mutation rate is shown in Figure 3.

As can be seen, even modest changes in the viral mutation rate had a substantial effect on viral fitness decline. This should not be surprising because it is known that RNA viruses are already near the edge of error catastrophe, due to mutation rates which are already very high. A mutation rate of 0.2 resulted in a final mean fitness of 0.57 (as opposed to a final fitness of 0.82 when the mutation rate was 0.1). A mutation rate of 0.4 caused strain extinction after 5,743 replications (239 days into the pandemic). A mutation rate of 0.6 caused strain extinction after 2,224 replications (93 days into the pandemic). A mutation rate of 0.8 caused strain extinction after 1,003 replications (42 days into the pandemic). As can be seen, even these very modest increases in mutation rate caused very rapid acceleration of fitness decline, due to the mutational meltdown phenomenon.

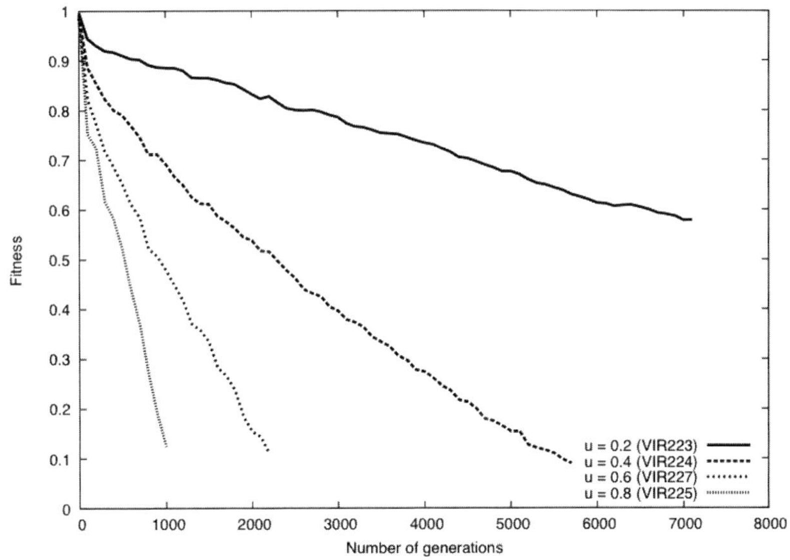

Fig. 3. Effect of mutation rate (u) on fitness degradation over time during an RNA virus pandemic. Four experiments showing that slight increases in mutation rate dramatically shorten pandemic duration. Mutations per virion per replication (u) were 0.2, 0.4, 0.6, and 0.8.

Discussion

Numerical simulations support our thesis that RNA viruses should be subject to natural genetic attenuation through mutation accumulation. Using conservative parameter settings for our model RNA virus, we observed continuous increases in the number of deleterious mutations per viral particle, and continuous genetic declines in viral fitness (Figure 1). Across a wide range of parameter settings, we have consistently observed that natural selection fails to remove a large fraction of the deleterious mutations, and that deleterious mutation count per viral particle increases linearly with time.

Our simulations indicate that genetic attenuation was accelerated by any of three factors (Figure 2), including: (1) increased rates of random death of virions (where there is significant loss of virions due either to poor assembly, degradation, or other host defenses); (2) more intense genetic bottlenecking; and (3) elevated mutation rates. Mutation rate had the greatest effect, and random death had the least effect. However, these three factors were most effective, by far, when acting in concert. When combined, these three factors caused very rapid genetic attenuation and would have clearly caused premature collapse of the model pandemic. How much fitness loss is required to stop a pandemic? This is unknown, but certainly fitness does not need to approach zero. The fitness loss in our most conservative case (16% decline, see Figure 1) may be sufficient in itself to explain the natural cessation of most pandemics.

We saw that even slight increases in the mutation rate had a profound effect on the rate of genetic attenuation. Just an 8-fold increase in the mutation rate was enough to cause rapid mutational meltdown and strain extinction after just 42 days (see Figure 3). This is consistent with Domingo *et al.* [5] who claim that even a 2.5-fold increase in mutation rate is sufficient to cause loss of infectivity of both poliovirus and vesicular stomatitis virus. Such elevated mutation rates can be readily achieved using certain pharmaceuticals [36–39, 9, 10]. Indeed, even if only half the infected people employed such antiviral medications, the use of such pharmaceuticals would be expected to increase the average mutation rate very significantly.

Use of mutation-enhancing pharmaceuticals should have the additional benefit of simultaneously reducing viral titers ("random loss"), and increasing the degree of bottlenecking. These benefits, along with an elevated mutation rate, should act synergistically in accelerating genetic attenuation. Better hygiene might also greatly increase bottlenecking, and thus significantly enhance genetic attenuation. Likewise, broad-spectrum vaccines, which help build more general immunity, along with other treatments that reduce viral titers, should also enhance attenuation. In some cases, these combined treatments might even be employed where a RNA virus has been targeted for eradication.

Our results suggest that, while lethal mutagenesis holds promise for treating individuals, a much more significant application may be on the epidemiological level. There appears to be a great potential for more effectively managing pandemics by increasing those factors described above. To the extent that we can significantly increase the mutation rate in RNA viruses, we can clearly accelerate natural genetic attenuation, and in many cases may be able to cause mutational meltdown of a given viral strain in a relatively short period of time. Deployment of better hygiene practices by itself should reduce inoculum levels, which should result in stronger bottlenecking and accelerated decline. Lastly, adding a higher rate of random elimination of viral particles, as might occur due to various factors favorable for viral elimination (i.e., partial immunity, fever, use of complimentary antiviral drugs, etc.), should further accelerate genetic attenuation. It is noteworthy that only 1% of poliovirus released from a host cell are able to complete a full cycle of replication [15]. These three factors (mutation rate, bottlenecking, and various mechanisms that reduce viral load) clearly combine synergistically to accelerate viral degeneration.

We acknowledge that our model virus may not precisely match any known RNA virus, but we feel it provides a reasonable approximation of a typical RNA virus. Our greatest reservation is that no one knows the precise shape of the distribution of mutation effects for a given virus. Our distribution of mutation effects may be skewed too far toward higher-impact mutations (the mean mutation effect in all these experiments, prior to selection, was 3.7% reduction in fitness). This may be causing unrealistically rapid genetic decline, resulting in over-estimation of the rate of fitness decline. Alternatively, our distribution of mutational fitness effects may be too skewed toward low-impact mutations, in which case the simulations would indicate unrealistically slow genetic decline, thereby resulting in under-estimating the rate of attenuation. However, we have consistently observed that when we shift the mutation effect distribution toward mutations with lower impact on fitness, the selection breakdown phenomenon is much more severe, such that a much higher proportion of deleterious mutations escape selection altogether. Shifting the distribution of mutation effects either up or down creates tradeoffs, resulting in only modest changes in the way the genetic damage accumulates. Therefore, we feel our model RNA virus is a useful approximation of how a real RNA virus should respond to the mutation/selection process.

We believe there is strong theoretical evidence that RNA viruses should systematically undergo natural attenuation, which in now supported by our numerical simulations. This raises the obvious question – if this is true, why have not all RNA viruses gone extinct? The most likely explanation seems to be that such viruses are preserved in natural reservoirs where they are more stable. The most obvious way for an RNA virus to be more genetically stable is to be in an environment where they have slower replication, and higher fidelity RNA replication. Since the host provides sub-units for the RNA replicase complex, the host should have very significant impact on both speed of replication and fidelity of replication, and therefore a specific host may foster much greater viral stability for a given virus. In the case of retroviruses, we know they can persist indefinitely in their DNA form (within the host genome). In this form they have very low mutation rates. Other viruses may lay dormant indefinitely in other states and in other types of natural reservoirs. For example, the H1N1 strain of influenza apparently went extinct for 20 years in the mid-twentieth century, but it is thought to have been resurrected from a researcher's freezer in 1977, and is once again circulating globally [23, 24]. The 1918 influenza virus strain (which gave rise to essentially all the current human and pig influenza strains) is assumed to have arisen from the natural reservoir of aquatic birds which harbor influenza viruses. However, there is no clear precursor for the 1918 strain in either bird hosts or other known hosts [i.e., 21, 22], so we can only say modern human/pig influenza emerged from an unknown natural reservoir around the turn of the twentieth century. There may be many ways that an RNA virus may be held in reserve for long periods of time in natural reservoirs.

Conclusions

Our findings are consistent with the idea that there are already very high rates of natural extinction among RNA viral strains, and that the vast majority of RNA viral strains die out naturally due to mutation accumulation. Such mutational degeneration should play a significant role in the natural progression of pandemics, with mutation accumulation causing the natural genetic attenuation of any given RNA viral strain. Our numerical simulations strongly indicate that such natural genetic attenuation can be enhanced during pandemics by: (a) employing strategic use of antiviral pharmaceuticals that increase RNA mutagenesis; (b) increasing genetic bottlenecking by reducing inoculum levels through improved hygiene and other means; and (c) strategic use of broad-spectrum vaccines that induce partial immunity and other means for reducing viral titers.

Addendum *— This study was purely theoretical, based upon biologically realistic numerical simulations. After this chapter was already accepted and finalized, an empirical analysis was initiated of actual mutation accumulation within the H1N1 Influenza viral genome since 1918. The results provided a remarkable validation of the present theoretical study. Within the human lineage, nearly every H1N1 strain that arose very quickly became extinct. All circulating human*

H1N1 strains went extinct in the mid-1950s, but the human H1N1 lineage was re-seeded into the human population in 1976, apparently from a researcher's freezer. The human lineage apparently again went extinct in 2009. During the entire history of H1N1 within man, mutations accumulated in a perfectly linear fashion – exactly as seen in this theoretical study. In the course of 90 years, almost 15% of the viral genome mutated, always at a very constant rate. Viral fitness, as reflected by associated human mortality rates, declined continuously and systematically from 1918 all the way to the apparent extinction of the human H1N1 strain in 2009. Because the publication of these proceedings was significantly delayed, the empirical study was published before the present theoretical study (which spawned the empirical study). See: Carter R.C. & Sanford, J.C. (2012). A new look at an old virus: patterns of mutation accumulation in the human H1N1 influenza virus since 1918. Theoretical Biology and Medical Modeling 9:42doi:10.1186/1742-4682-9-42.

Acknowledgments

The work was supported in part by FMS Foundation, Inc. and Rainbow Technologies, Inc.

References

1. Bull JJ, Sanjuan R, Wilke CO (2007). Theory of lethal mutagenesis for viruses. J.Virology, 81(6):2930–2939.
2. Martin G, Gandon S (2010) Lethal mutagenesis and evolutionary epidemiology. Phil Trans R Soc B 365:1953–1963.
3. De La Pena M, Santiago EF, Moya A (2000) Effect of deleterious mutation-accumulation on the fitness of RNA bacteriophage MS2. Evolution 54(2):686–691.
4. Iglesia F, Elena SF (2007). Fitness declines in tobacco etch virus upon serial bottleneck transfers. J Virology 81(10):4941–4947.
5. Domingo E, Ruiz-Jarabo CM, Sierra S, Arias A, Pariente N, Baranowski E, Escarmis C (2002) Emergence and selection of RNA virus variants: memory and extinction. Virus Res 82:39–44.
6. Duarte E, Clarke D, Moya A, Domingo D, Holland J (1992). Rapid fitness losses in mammalian RNA virus clones due to Muller's ratchet. Proc Natl Acad Sci USA 89:6015–6019.
7. Yuste E, *et al* (1999). Drastic fitness loss in Human Immunodeficiency Virus Type 1 upon serial bottleneck events. J Virology, 73(4):2745–2751.

8. Yuste E, *et al* (2000). Unusual distribution of mutations associated with serial bottleneck passages of Human Immunodeficiency Virus Type 1. J Virology 74(20):9546–9552.
9. Loeb LA, Essigmann JM, Kazazi F, Zhang J, Rose KD (1999) Lethal mutagenesis of HIV with mutagenic nucleoside analogs. Proc Natl Acad Sci USA 96:1492–1497.
10. Smith RA, Loeb LA, Preston BD (2005) Lethal mutagenesis of HIV. Virus Res 107:215–218.
11. Elena SF, Carrasco P, Daros JA, Sanjuan R (2006) Mechanisms of genetic robustness in RNA viruses. EMBO reports 7(2):168–173.
12. Sanjuan R, Cuevas JM, Furio V, Holmes EC, Moya A (2007) Selection for robustness in mutagenized RNA viruses. PLOS Genetics 3(6): 939–946.
13. Barr JN, Fearns R (2010) How RNA viruses maintain their genome integrity. J Gen Virol 91:1373–1387.
14. Drake JW (1993) Rates of spontaneous mutation among RNA viruses. Proc Natl Acad Sci USA 90:4171–4175.
15. Holmes EC (2010) The comparative genomics of viral emergence. Proc Natl Acad Sci USA 107(Suppl. 1):1742–1746.
16. Steel A, Gubler DJ, Bennett SN (2010) Natural attenuation of dengue virus type-2 after a series of island outbreaks: a retrospective phylogenetic study of events in the South Pacific three decades ago. Virology 405:505–512.
17. CSMEC (2004) Molecular evolution of the SARS coronavirus during the course of the SARS epidemic in China. Science 303 (5664): 1666–1669.
18. Lai MMC (2003) SARS Virus: the beginning of the unraveling of a new coronavirus. J Biomed Sci 10:664–675.
19. Holmes-Maybank K (2004). An Overview of Ebola. South Carolina Journal of Molecular Medicine (SCJMM) 1:15–21.
20. Bowen ETW, Platt GS, Lloyd G, Raymond RT, Simpson DIH (1980) A comparative study of strains of Ebola Virus isolated from Sudan and northern Zaire in 1976. J Med Virol 6:129–138.
21. Fanning TG, Slemons RD, Reid AH, Janczewski TA Dean J, Taubenberger JK (2002) 1917 Avian Influenza virus sequences suggest that the 1918 pandemic virus did not acquire its hemagglutinin directly from birds. J Virol 76(15):7860–7862.
22. Taubenberger JK, Morens DM (2006) 1918 Influenza: the mother of all pandemics. Emerging Infectious Diseases 12(1):15–22.
23. Palese P (2004) Influenza: old and new threats. Nature Med 10(12):S82–S87.
24. Oxford JS (2000) Influenza A pandemics of the 20th century with special reference to 1918: virology, pathology and epidemiology. Rev Med Virol 10:119–133.
25. Christman MC, Kedwaii A, Jianpeng X, Donis RO, Guoqing L (2011) Pandemic (H1N1) 2009 virus revisited: an evolutionary retrospective. Infect Genet Evol 11(5):803–811.

26. Sanford J, Baumgardner J, Gibson P, Brewer W, ReMine W (2007) Mendel's Accountant: a biologically realistic forward-time population genetics program. Scalable Computing, Practice and Experience 8(2):147–165. Available: http://scpe.info.uvt.ro/vols/vol08/no2/SCPE_8_2_02.pdf. Accessed 2 Apr 2009.
27. Sanford J, Baumgardner J, Gibson P, Brewer W, ReMine W (2007) Using computer simulation to understand mutation accumulation dynamics and genetic load. In 7th International Conference on Computational Science, Beijing, China, May 27–30, 2007, Proceedings, Part II, LNCS 4488. Edited by Shi Y, van Albada GD, Dongarra J, Sloot PMA, Berlin/Heidelberg: Springer-Verlag; 386–392
28. Gibson P, Baumgardner JR, Brewer WH, Sanford JC (2013) Can purifying natural selection preserve biological information? In: Marks II RJ, Behe MJ, Dembski WA, Gordon B, Sanford JC (eds) Biological Information — New Perspectives. World Scientific, Singapore, pp. 232–263.
29. Brewer W, Baumgardner J, Gibson P, Sanford J (2013) Using numerical simulation to test the "mutation-count" hypothesis. In: Marks II RJ, Behe MJ, Dembski WA, Gordon B, Sanford JC (eds) Biological Information — New Perspectives. World Scientific, Singapore, pp. 298–311.
30. Baumgardner J, Brewer W, Sanford J (2013) Can synergistic epistasis halt mutation accumulation? Results from numerical simulation. In: Marks II RJ, Behe MJ, Dembski WA, Gordon B, Sanford JC (eds) Biological Information — New Perspectives. World Scienitic, Singapore, pp. 312–337.
31. Sanford J, Baumgardner J, Brewer W (2013) Selection threshold severely constrains capture of beneficial mutations. In: Marks II RJ, Behe MJ, Dembski WA, Gordon B, Sanford JC (eds) Biological Information — New Perspectives. World Scientific, Singapore, pp. 264–297.
32. Horsfall FL (1954) On the reproduction of Influenza virus. Quantitative studies with procedures which enumerate infective and hemagglutinating virus particles. J Exp Med 100:135–161.
33. Horsfall FL (1955) Reproduction of Influenza viruses. Quantitative investigations with particle enumeration procedures on the dynamics of Influenza A and B virus Reproduction. J Exp Med 102(4):441–473.
34. Carrat F, Vergu E, Ferguson NM, Lamaitre M, Cauchemez S, Leach S, Valleron A-J (2008) Time lines of infection and disease in human Influenza: a review of volunteer challenge studies. Am J Epidemiol 167:775–785.
35. Sanjuan R, Moya A, Santiago EF (2004). The distribution of fitness effects caused by single-nucleotide substitutions in an RNA virus. Proc Natl Acad Sci USA 101(22):8396–8401.
36. Crotty S, Cameron CE, Andino R (2001) RNA virus error catastrophe: direct molecular test by using ribavirin. Proc Natl Acad Sci, USA 98(12):6895–6900.

37. Crotty S, Cameron CE, Andino R (2002) Ribavirin's antiviral mechanism of action: lethal mutagenesis? J Mol Med 80:86–95.
38. Crotty S, Andino R (2002) Implications of high RNA virus mutation rates: lethal mutagenesis and the antiviral drug ribavirin. Microbes and Infection 4:1301–1307.
39. Grande-Perez A, Gomez-Mariano G, Lowenstein PR, Domingo E (2005) Mutagenesis-induced, large fitness variations with an invariant arenavirus consensus genomic nucleotide sequence. J Virology 79(16):10451–10459.

DNA.EXE: A Sequence Comparison between the Human Genome and Computer Code

Josiah Seaman

University of Colorado, Anschutz Medical Campus, Computational Bioscience Program, Aurora, CO 80045–0511, USA. Josiah.Seaman@ucdenver.edu

Abstract

This study presents evidence that executable computer programs and human genomes contain similar patterns of repetitive code. When viewed with sequence visualization tools, these similarities are both striking and pervasive. The primary similarities are listed in order of scale: (1) homopolymers, (2) tandem repeats, (3) distributed repeats, (4) isochores, (5) and entire chromosome/file organization. Most strikingly, data visualization reveals that executable codes regularly make extensive use of tandem repeats which exhibit similar visual patterns as seen in higher genomes. In biology these tandem repeat patterns are normally attributed to replication errors, insertions, deletions, and substitutions. Similarly, on a larger scale, executable codes display regions with different ratios of 1's and 0's which parallel the isochore patterns within chromosomes, caused by local variation in the number of A/T vs. G/C. Further, blocks of data are stored at the beginning or end of a file, while the primary instructions occupy the middle of a file. This creates the same organizational patterns observed in human chromosome arms, where repetitive sequences are grouped near the telomeres and centromeres.

I propose that these similarities can be explained by universal constraints in efficient information encoding and execution. The genome may be viewed as the executable program that encodes life. Given the evidence that computer programs and genomes use many of the same patterns of organization, despite having very different context, it should be informative to explore the ways in which knowledge of computer architecture can be applied to biology and vice versa.

Key words: computer code, alu, tandem repeats, junk DNA, small RNA, biological computer, retrotransposon, programming, cybernetics, data visualization, data analysis

Introduction

The study of the human DNA sequence has been dominated by the study of protein coding genes. These protein coding regions, called exons, constitute a mere 1.2% of the human genome [1]. Exons use a very simple code called the codon code that can be expressed in terms of a single table of 64 values. Without the key of the codon code, exons would appear to be meaningless nonsense to us. Thankfully, the codon code is a (relatively) straightforward and known entity and with it we can predict the amino acid sequence of most genes in the nucleus. The codon code represents the first code in the human genome that we were able to decipher.

Those with even a passing understanding of human genetics should understand how incomplete this picture is. Exons using the codon code do not stand in isolation but are intertwined and dependent on numerous other genome elements which employ their own codes. Transcription of genes is regulated by promoters which use a transcription factor binding site code along with protein combinations. The newly formed pre-mRNA transcript contains elements called introns that use a third code, the splicing code, for determining how all of the exons are spliced together. All three of these codes are separate yet interdependent on each other to make the right protein product at the right time. Many other codes have been, and will continue to be discovered.

What is a code? A code is a precise mapping from a set of symbols to specified meanings, actions, and objects. We use codes for many purposes such as naming parts (e.g. A, B, and C) in an assembly manual. Human language itself is a type of code, though one that is much more elaborate and flexible than any other code. Cyphers used to conceal meaning for cryptography are sometimes called "codes" but cyphers are just one subset of codes. However, their use underscores a very important attribute of encoding: If one does not know the code in which something is written, it will appear to be meaningless nonsense.

In the previous century, the study of genomics was largely constricted to protein coding exons, which were already a formidable challenge to study. The other 98% of the human genome was dismissed as junk because it appeared to be meaningless nonsense [2, 3]. Slowly, exons' deadlock on genomics was loosened, first through gene expression analysis that showed the importance of promoters and enhancers, and then through the realization that alternative splicing was critical to understanding the complexity of the human body. The work of Barash *et al.* in 2009 was just the first step to cracking open the complexity of the splicing code [4]. There has been rapidly mounting evidence that various non-coding RNAs inside the cell serve useful functions and that there is a veritable zoo of RNA types. In 2007, the ENCODE project opened up the field by showing that over 90% of the human genome is transcribed [5, 6]. This forces us to conclude that either the cell wastes energy on extensive junk transcription, or, in keeping with the discovery of new RNA types, that the majority of the human genome is functional [7]. Transitioning from a 2–3% functional genome to >90% function is understandably unattractive to some, because it means that the majority of genetic research to date has only scratched the surface of all that the human genome actually encodes. An enormous task lies before us, as we endeavor to comprehend the many undiscovered functions of the multifaceted genome. To do this we need better tools and new approaches.

Methods

Skittle Genome Visualizer (Skittle for short) is a new sequence visualization tool suite [8]. This tool is especially sensitive for detection of any type of repeating pattern within sequence data. Although it was developed to analyze DNA sequences, it is very effective in analyzing repeat patterns in any type of sequence — including RNA, protein, written text, music, or computer code.

When DNA sequences from higher eukaryotes were examined using Skittle, extensive repetition was very clearly seen [8]. This might not seem surprising, as such repetition was previously known, and was presumed to be the result of numerous types of copying errors — generating a large amount of junk DNA.

During the course of browsing a number of chromosome sequence files using Skittle, a non-genome sequence file was opened accidentally. The program was directed to its own executable file: SkittleToo.exe. The same visualization that had been so successful in studying chromosome sequences was accidentally applied to computer machine code. Surprisingly, computer code revealed the same patterns and types of variations seen in the human genome. Yet none of these repeating patterns could be attributed to copying errors – every bit in the repeat patterns was there for a reason, and therefore the repeating patterns reflected the essential and inherent architecture of the computer code information system.

In order to make these visual comparisons easier, the executable computer code was converted to a base-4 symbol set of 'ACGT' (00 = A, 01 = C, 10 = G, 11 = T). In the figures, programs that have been converted to base-4 will have a .fa extension to indicate the change. For example, SkittleToo.exe becomes SkittleToo.fa when converted to the base-4 symbol set for comparison.

The main innovation in Skittle is to transform a sequence into an image by representing each letter with a color. This engages human visual recognition for structures and patterns instead of the target recognition that the brain uses when reading a sentence. In each of the figures (except Figure 2), the sequence is read from left to right starting at the upper right corner in the same way that one would read English text. Instead of displaying simple text, each letter is replaced with a colored square "pixel" that represents that letter. At Scale = 1 each pixel represents one letter or nucleotide. The visualization can "zoom out" by increasing the scale such that each pixel is the color average of multiple values (Figures 4B, 5, 6, 7). At Scale = 10 the color for one pixel is computed by taking the next 10 letters in the sequence, converting them into 10 colors and then averaging the colors together.

In addition to this, Skittle contains a suite of visualizations specialized for specific tasks (Figure 5, 7). For a more in depth explanation of the visualization methods and the pattern recognition algorithms used in this paper refer to "Skittle: a 2-Dimensional Genome Visualization Tool"[8].

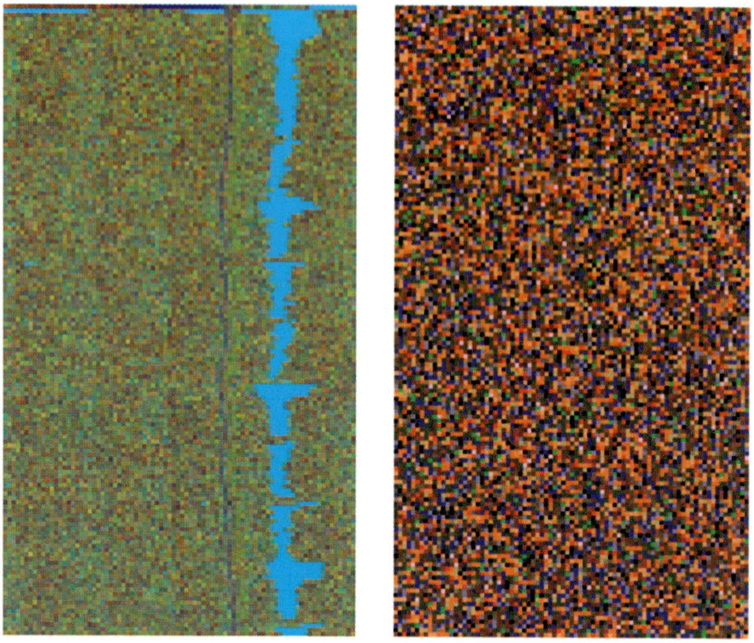

(A) (B)

Fig. 1. A) MP3 recording of Beethoven's Moonlight Sonata. MP3 files consist of thousands of copies of a repeat monomer. In this example, each line has a homo-polymer of variable length. In general, data files are a single repeat of one format in contrast to computer programs which contain many different types of information. B) Text of Moby Dick. The Nucleotide Display essentially looks like colored static with no changing color bias. Similarly, the figure shows no repeats. English prose actually shows almost no tandem repetition. The only detectable tandem repeat in the entire text of Moby Dick is a short song about the sea that repeats the chorus 3 times.

Results

Negative Controls

This study examines significant patterns of sequence repetition within genomes and within executable computer code. To determine how these patterns are distributed in other information formats a whole variety of file types were examined. Among the file types visualized in Skittle were: exe, dll, cab, zip, png, bmp, jpg, tif, mp3, wav, cod, and txt. Human text (cod, txt), such as books, had almost no discernible patterns (Figure 1B). On the other hand, data files (png, bmp, jpg, tif, mp3, wav) appeared as a single uniform tandem repeat (Figure 1A). Compressed regions of files (cab, zip) lack most visible patterns (data not shown). Executable programs (exe, dll) were the only kind of information examined that showed the same variation and diversity of repeat patterns that can be seen in eukaryotic genomes. Examples of the similarities between genomes and executable code were found at every scale.

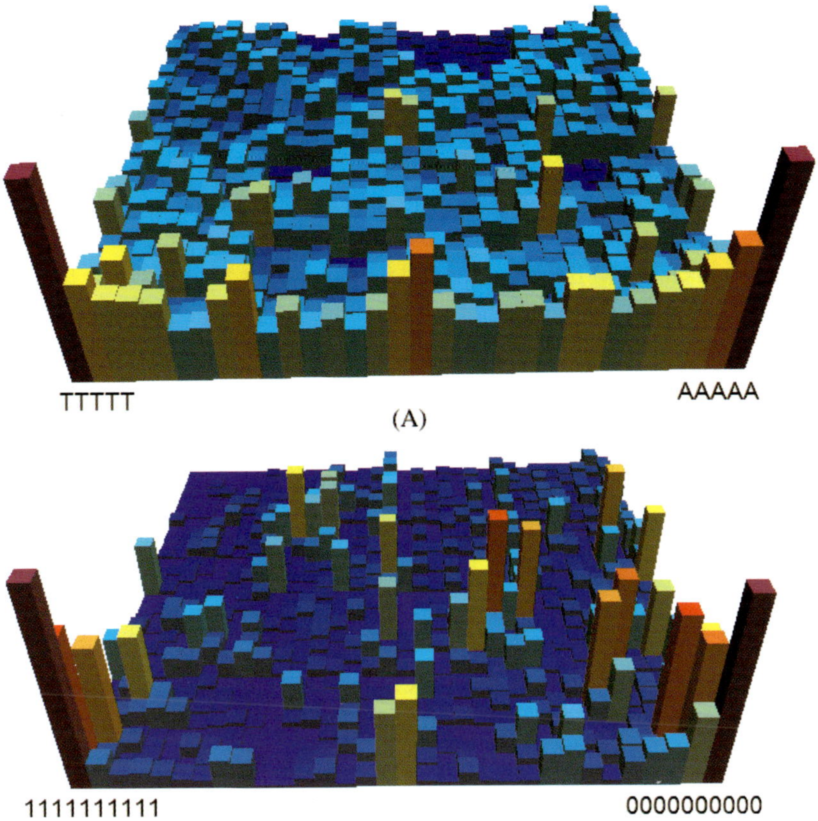

(B)

Fig. 2. These bar graphs show the abundance of all strings of length 5 (base-4). Strings are sorted by the CGR algorithm [9]. Coloring in this graph is based on abundance. Low abundance words are blue, medium are green to orange, with high abundance words being purple. A) All 5-mer strings in chromosome 21. In this graph, TTTTT is in the lower left, AAAAA is the lower right, CCCCC is the upper right, and GGGGG is the upper left. The poly-A and poly-T strings are by far the most frequent short strings in the genome. Also dominant in the lower center are two bars correspond to TAAAA and ATTTT respectively. B) All 5-mer (base-4) strings in the SkittleToo.fa computer code file. In this graph, 1111111111 is in the lower left, 0000000000 is the lower right. In computer programs the homopolymer-dominant pattern is just as strong as in the human genome. This computer code example contains the same pair of matching peaks (purple) as well as two smaller matching peaks in the bottom center, which correspond to TAAAA and ATTTT in the genome.

Results – From Small to Large

These results show five primary ways in which computer programs and chromosomes are similar. These five levels of similarity, arranged from smallest to largest, are: homopolymers, tandem repeats, distributed repeats, isochores (sequence bias), and whole chromosome/program structure. These five levels of organization are well known attributes of human chromosomes. Paradoxically, it was actually *easier* to find strong examples of these patterns in computer code than it was to find examples in DNA. Overall, computer code appears to be significantly more repetitive than the human genome.

Homopolymers — The human genome has an overabundance of strings that consist entirely of AAAAA or TTTTT (Figure 2). Since Adenine and Thymine bind less strongly than Guanine and Cytosine, these areas of the sequence are spots where the double helix can open more easily. Poly-A strings are found at the ends of mRNA strands as well. In computer code, 0 is often used as padding for a variety of reasons. Data files also contain homo-polymers in structured locations, but prose does not (Figure 1). A small positive number will have a long string of 0's at the beginning while a small negative number will have a long string of 1's. In both code and genomes they can be used as a dividing marker between different elements, similar to a space or paragraph break. Figure 2 demonstrates a strong similarity in the homopolymer patterns within the human genome and executable computer code.

Tandem repeats — The most striking visual patterns seen when higher genomes are visualized with Skittle are the tandem repeats (Figure 3). Tandem repeats in the human genome have long been considered junk DNA left over from replication errors. Tandem repeats are useful in forensics because of their anomalously high mutation rate, which can be up to one million times higher depending on the estimation technique. Both Weber and Brinkmann report mutation rates of at least 7×10^{-3} per locus per haploid per generation [10, 11], while the background mutation rate for the whole genome has recently been measured at 1.1×10^{-8} per position per haploid genome per generation [12]. Given that the patterns of variation visible within genomic repeats do not appear to be random (Figure 3) [8], it is reasonable to consider the possibility that such variation may not be the result of an entirely random mutation process.

In computer programs, tandem repeats are often used to store data in a structured format. When examined with Skittle, computer code shows remarkably similar tandem repeats as those seen in eukaryotic genomes. Also seen are the same types of internal structured variation as genomic tandem repeats (Figure 3).

In the genome we do not understand the function of tandem repeats, but in executable code, tandem repeats can be traced back to the original source code written by a programmer and to its function in the program. For example, visible tandem repeats can be mapped to data files (typically, columns of letters or numbers). Computer data consists of tokens that are often larger than one byte. When a token varies from the consensus, we observe straights columns or "covariance" (Figure 3B) and when a token has a variable length, we observe wavy columns or "indels" (Figure 4B).

Figures 3B and 4B show two specific examples of repeats in machine code, extracted from openofficeorg32.msi. This file is freely available, and is responsible for installing the Open Office software suite. When visualized in Skittle, the file is found to contain all the major features that biologists associate

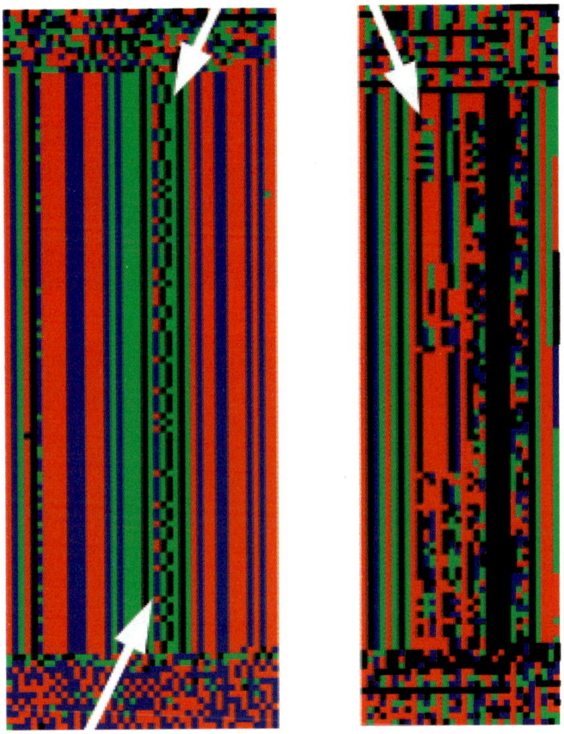

(A) (B)

Fig. 3. Structured variation in tandem repeats. Pixel coloring: 00 = A = Black, 01 = C = Red, 10 = G = Green, 11 = T = Blue. A) Chr Y: Start: 392,304 bp Length: 7,128 bp. This tandem repeat lies near the telomere of chromosome Y and was used in the first Skittle paper [8] as a strong example of nucleotide covariance. The arrows point to the substitutions in the 3 columns near the center, that show covariance. B) OpenOfficeorg32.msi: Start: 39,554,645 Length: 4,672 (x2) bits. This repeat shows a strong resemblance to Figure 3A, but it's found in computer code. The nucleotide covariance (arrow) is caused by replacing a token longer than 2 bits in the repeat, which involves simultaneously changing of a series of contiguous bits. Both repeats look like straight vertical bars because they contain no "insertions or deletions". Contrast this with the wavy, staggered appearance of Figure 4. Both repeats have columns that are highly variable, and columns that are entirely invariant.

with chromosomes. It contains isochores, segmental duplications, tandem repeats, distributed repeats, and sequence variations that have the same appearance as mutations. The lower sixth of the file consists of a segmental duplication with a length of 231,600 bytes per repeat monomer with 7 copies. The other half of the file is two larger segmental duplications of 295,852 bytes per repeat monomer.

Figure 4B shows a tandem repeat from openofficeorg32.msi visualized in Skittle. The text inside this particular example is in English, so we can see the content, while most computer code (in Hex) would be unreadable to the average person. A sample of the text in Figure 4B is shown in as follows.

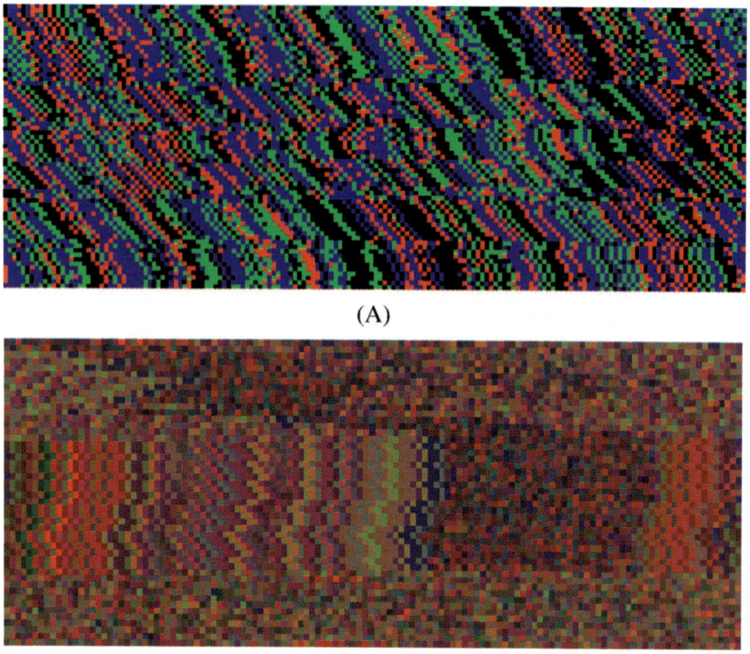

Fig. 4. Tandem repeats with indels. Pixel coloring: 00 = A = Black, 01 = C = Red, 10 = G = Green, 11 = T = Blue. A) Chr19: Start: 32,611,935 bp Length: 27,702 bp. Every human chromosome has a centromere, which is primarily a large tandem repeat. This alpha satellite repeat shows both substitutions and indels (wavy columns). B) OpenOfficeorg32.msi: Start: 212,581 Length: 8,001 (x2) bits Scale: 4 bp/pixel. This repeat was picked from computer code as an example that has both substitutions and indels (wavy columns). In this case the variable columns are concentrated together in the middle. These columns encode the unique ID that's written into the registry (see text). For clear visualization, each pixel is one byte. The color is an average of 4 "base pairs" per pixel or 8 bits per pixel. For a more detailed look at visualizing structured variation inside tandem repeats, including nucleotide covariance and indels please see "Skittle: a 2-Dimensional Genome Visualization Tool" [8].

…_REGISTRY_OPENOFFICEORG32{F0B285B1–7227-CDDC-6CEA-FA264CF46679}
(REGISTER_DOCX=1) AND (WRITE_REGISTRY…

…_REGISTRY_OPENOFFICEORG32{CF642EF8–3237–7F5A-6D31–18FFF1ACB2C1}
(REGISTER_DOT=1) AND (WRITE_REGISTRY=…

…_REGISTRY_OPENOFFICEORG32{98C7CE2B-EA3B-AB57–7F43–6987BCFF2C7E}
(REGISTER_DOTM=1) AND (WRITE_REGISTRY…

…_REGISTRY_OPENOFFICEORG32{C2E5C8BB-4D40–61A3–8D84–625E34119744}
(REGISTER_DOTX=1) AND (WRITE_REGISTRY…

…_REGISTRY_OPENOFFICEORG32{73E532F7-BAD7-F137–00C6–9188EA72701C}
(REGISTER_POT=1) AND (WRITE_REGISTRY=…

…_REGISTRY_OPENOFFICEORG32{837B2E93-F7D2–61BB-D711-E65E54F951AC}
(REGISTER_POTM=1) AND (WRITE_REGISTRY…

..._REGISTRY_OPENOFFICEORG32{1F1B63BC-3767–643B-9973–1A893B7488E5}
(REGISTER_POTX=1) AND (WRITE_REGISTRY...

..._REGISTRY_OPENOFFICEORG32{04A22EC4–39CD-4254-A2AD-22E5F170B043}
(REGISTER_PPS=1) AND (WRITE_REGISTRY=...

The repeated instruction "REGISTRY_OPENOFFICEORG32" provides context for the data. These are registry entries to be written during installation. The variable parts of this tandem repeat are the unique entries being entered e.g. "{F0B285B1–7227-CDDC-6CEA-FA264CF46679}". This setup means that each repeat entry can be read independently. The variable columns correspond to the unique entries while the duplicated text provides context. Also notice the next token lists various file types: e.g. "(REGISTER_DOCX=1)". DOCX, DOT, DOTM, PPS are all file types that Open Office can read. Most file types are 3 letters (TXT) but some file types are four letters long (DOCX). This token will vary length by one letter, creating the shifts referred to in biology as indels. Computer code also contains tandem repeats in the form of repetitive instructions.

Distributed repeats — Distributed repeats are sequences that are nearly identical and are found in many locations in a genome. The most common distributed repeats in the human genome are LINES and SINES, which have many functions, including regulation of transcription [14]. Similarly, computer programs have specific commands that are used frequently such as ADD, STORE, and LOAD. These common commands create a distributed repeat pattern. Distributed repeats are also the one repetitive pattern that can be observed in English prose. Words like "the" occur much more often than chance along with longer sentence fragments that are frequently reused. "For example", "the reality is", "a few", "a little", "about time", and "at this stage" are all used more often than would happen by chance. Figure 5A shows a given distributed repeat in the human genome, as mapped by Skittle, and Figure 5B shows a similar distributed repeat within SkittleToo.exe.

Isochore Patterns — In eukaryotic chromosomes, the sequence shows usage bias changes between G/C vs. A/T. This change in bias has been known since at least the 1970s, creating visible bands under a microscope when the chromosome is stained [15, 16]. These bands correlate with the presence of much larger DNA elements, and represent a basic division in eukaryotic sequences. G/C rich regions contain more genes, CpG islands, and are physically unpacked in the interphase nucleus. Unpacked regions are referred to as open chromatin because they are less dense [13]. A/T rich regions are associated with closed chromatin and lower levels of transcription.

Surprisingly, the isochore type of pattern is even stronger in computer code. We can observe the same kind of character usage bias, revealing both gradual changes, and sharp disjunctions. The change is due to the large-scale organization of types

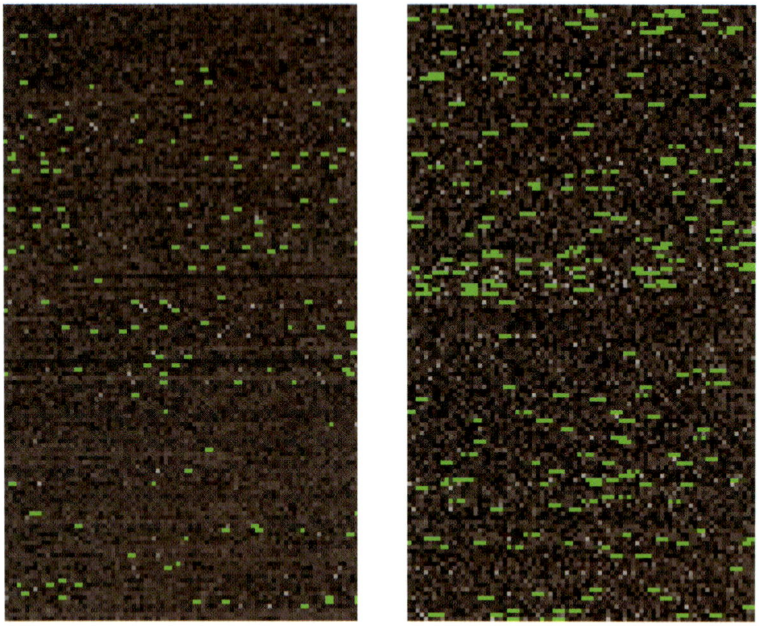

(A) (B)

Fig. 5. Distributed repeats are highlighted in green using the Skittle Sequence Highlighter similar to searching for text in a document. Gray pixels are non-matching sequence. A) Chr19 Start: 97,281 bp Length: 338,428 bp. Sequence: TGGGATTACAGGTGTGAGCCACCGCGCCCG at 80% similarity. The human genome is filled with distributed repeats, but their positioning is not entirely random. Some bands of the chromosome will have very few of a certain repeat sequence, while others will be very dense. These concentrated bands on the chromosome follow the isochore patterns (Figure 6). For example, Alu repeats are concentrated in GC rich regions along with genes [13]. B) SkittleToo.fa Sequence: AGAGCCACAAGCAAACAGAGCAC (0010001001010001000010010000000100100010010001) at 80% similarity. The banding pattern of repeats observed in human chromosomes is actually easier to see in computer code because the computer programs are more repetitive. Note the horizontal bands where many repeats are highlighted in green and the dark region just below it with no repeats highlighted.

of elements in code. Execution code and data are separated. Since data types are contiguous, the same encoding scheme (data format) will show up in discrete blocks. Elements in the program employ different codes and some codes are used in combination with each other. Each code will have a different bit distribution. For example, English letters occupy only 52 of the 256 possible values on the ASCII table. This means that by using Skittle one can visually differentiate English text from other types of codes. Isochores in programs are the result of large-scale organization and the use of multiple codes.

<u>Whole chromosome/program architecture</u> — Tandem repeats are much more common near the ends of chromosomes and near the centromere. Computer

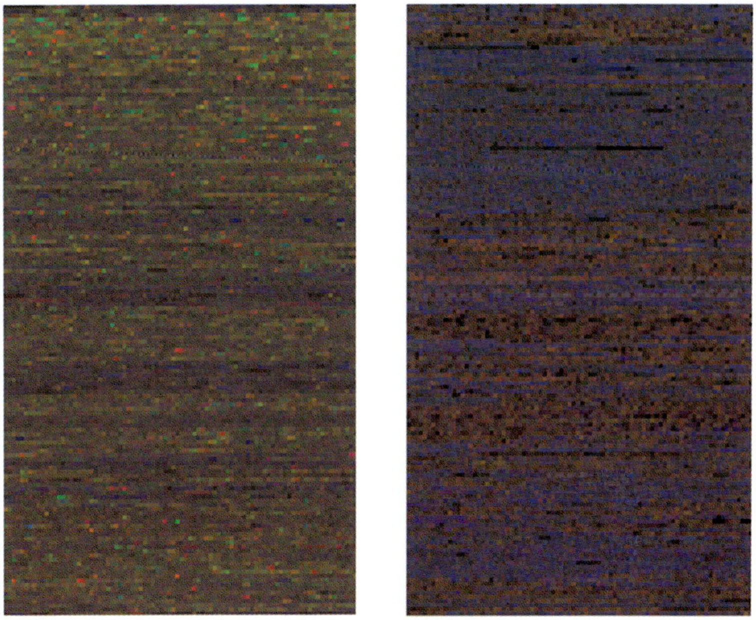

(A) (B)

Fig. 6. Examples of isochore-type structure seen at much larger scales. Pixel coloring: 00 = A = Black, 01 = C = Red, 10 = G = Green, 11 = T = Blue. This figure is a zoomed out view of a whole chromosome arm where each pixel is the color average of thousands of nucleotides (see Scale). A) Chr19: Start: 1 bp Length: 22,272,512 bp Scale: 1,061 bp/pixel. Using color averaging, the changing bias in GC content can be clearly seen on the short arm of chromosome 19. GC content has a high correlation with many other genome elements. B) SkittleToo.fa Start: 10,509,078 Length: 3,715,584 (x2) bits Scale: 177 bp/pixel. Isochore-type patterns can be clearly seen in computer code, even more clearly than the genome. Regions with many 1's in the code appear blue while the areas more rich in 0's have a dark reddish color. Even the variation in the size of the isochore bands is similar in both A and B, though the scale is different. Genomic tandem repeats appear at this scale as spots and streaks of bright color (usually green or red), and a similar pattern can be seen in the computer code in the small black horizontal lines that litter the image. (This image's contrast was increased for clarity in printing.)

programs also store large repetitive blocks of data at the end of programs. Many executable files actually contain a majority of repetitive sequences because of icons and graphics stored in the executable. Repeats are packed in at the end of computer programs because they are organized to make them easier to use. This organization extends to the RAM, where memory is allocated into the Stack and the Heap. The Stack contains the primary program and just a few local variables. The Heap contains the majority of the data and frequently changes size and composition during execution. Similarly in DNA, repetitive elements are packed towards the end of chromosome arms and tandem repeats show much higher mutation rates (often

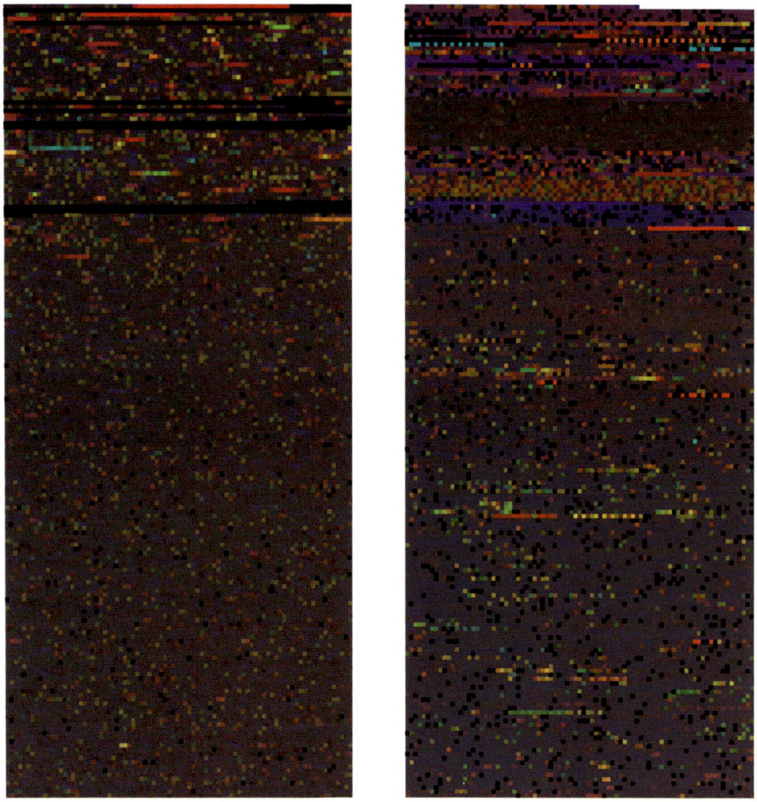

(A) (B)

Fig. 7. Skittle's Repeat Overview is used to highlight tandem repeats in bright colors to show the structure and distribution of a whole chromosome arm or an entire program. A) Chromosome X: Start: 1 bp Length: 8,670,000 bp Scale: 500 bp / pixel. Tandem repeats displayed in bright colors are primarily concentrated near the telomere in the top of this image. Large black lines are areas that were not sequenced, often because these regions are large tandem repeats. B) SkittleToo.fa Start: 26946779 Length: 21,501,600 (x2) bits Scale: 1,240 bp / pixel. The second half of the Skittle executable code can be seen here with repeats in bright colors and non-repetitive regions in dark blue. The image has been flipped vertically for comparison with Chr X. The less repetitive control code occupies the lower ⅔ of the image, while the repetitive icons, data tables, and other program resources are stored at the end of the file (displayed at top). In both sequences, repetitive structures are far more abundant near the ends.

seen as changes in size) than anywhere else in the genome [10, 11]. Like in computer programs, eukaryotic genomes tend to segregate highly repetitive sequences, which is further exemplified by recent research showing that chromosomes are organized in 3D space in the nucleus. For example, in the yeast genome, the centromeres of all the chromosomes can cluster together in one spot while the arms of the chromosome stretch out from that point. This is called a Rosette pattern [17].

The mammalian genome organization is noticeably more complex [18], but the pattern of aggregating similar sequences is still a major factor.

Discussion

The striking structural similarities between higher genomes and computer code strongly suggest they operate on similar principles, and that genomes and computer code may each instruct us on how to more fully understand the other. The results of this paper show that computers and cells use very similar encoding patterns despite the fact that the first code compilers could not have been designed to mimic the genome because the invention of compilers predates genome sequencing technology.

I propose that the simplest explanation for this similarity is that these findings represent convergent evolution driven by similar design constraints. Computers were not developed all at once. Instead, a number of different possibilities were tested. Similarly, compilers have gone through a series of revisions and optimizations. As an ongoing process of refinement, computer architecture is subject to many of the same constraints as biology, meaning that many of the optimal encoding patterns will be the same.

The comparison is half analogy and half reality. Obviously, there are major differences between the molecular computing of DNA and the electronic architecture of modern computers. Yet the first computer conceived by Alan Turing was an entirely mechanical apparatus moving along a tape — which has more resemblance to a polymerase on DNA than it does to modern computers [19]. Computer Science has shown that computation can take many forms, yet the fundamentals principles and constraints seem to remain the same.

The following are suggested as possible parallels between computers and biological systems:

Biology	Computer	Comments
DNA	**Hard Drive**	DNA is analogous to a hard drive because it serves as the canonical, non-volatile copy that is copied but not frequently edited.
RNA	**RAM**	RNA is analogous to the RAM in a computer because it acts as the active, working copy of the information that is edited, used, and then discarded.
Tandem Repeats	**Data blocks**	This explains the anomalously high mutation rates. Cells are purposefully storing inherited information in the DNA strand.

Continued

Continued

Polymerase + Ribosomes	**Processors**	The cell is a multi-processor system, with multiple parallel events occurring and being communicated through epigenetic modification and RNA. Variety in protein/RNA complexes are processors specialized for different tasks.
Cytoplasm Phenotype	**Output**	Most of the computation that goes into the decision process is never obvious to the user.
Nucleus	**Motherboard**	Computational center for the cell with hard drives integrated as closely as possible.
Nucleolus	**CPU**	Central area where most of the processors and memory is congregated for speed reasons.

Differences — While there are striking similarities between genomes and executable programs, there are also very important differences. These differences serve to highlight why the similarities are so informative: they reveal the underlying design constraints at work in both.

- DNA lacks large blocks of numbers sorted in ascending order.
- DNA does not have as many zero values as code in large blocks (such as padding), though the human genome does have a strong bias towards strings of A's or T's.
- Computers usually use a fixed word length, which shows up as a periodicity in the frequency graph. Exons in DNA show this same pattern because the codon code follows a fixed length look-up table, but there are many variable length elements as well.
- For structured variation in tandem repeats, computer code will often have zeroed out fields as part of covariance. A "zero value" has not been directly observed in biological sequences. With better token recognition, zero values could simply be skipped, in which case they would look like deletions.

These findings provide new tools to direct future research. In computer science, engineers use the attributes of an object to determine the type of object. This is called duck typing because it follows the phrase, "If it walks like a duck, and quacks like a duck, it's a duck". The comparison between computer programs and the human genome shows that elements in the genome share the same attributes with programming products. By applying duck typing, we get more than just a single hypothesis. We get a whole set of hypotheses about the function of any element in the genome that has similar properties to an element in a program. The starting hypothesis would be that the reason they look the same is because they are fulfilling the same types of function. This gives us a useful road map for designing biological experiments and predicting function.

Until recently, previous research has focused on genes and promoters, which constitute at most 3% of the genome. The rest of the genome has been a complete mystery. Despite some advances, the task ahead still remains daunting. The model of the genome as a computer system can act as a paradigm for exploring the whole genome, not just protein coding genes. Historically, Egyptian hieroglyphics were not deciphered until the discovery of the Rosetta stone. This was crucial because they found the same things written in Greek as was written in the indecipherable hieroglyphics. This allowed real translation between the two. It is possible that computer program architecture may be the Rosetta stone for unlocking the rest of the genome. Without a working model, the human genome appears to be indecipherable junk. But using a comparable architecture, we stand a real chance of deciphering the whole genome, starting with the basic components that make up all executable programs.

The rewards for such an endeavor are enormous. There are essentially two fields of science that can benefit from adapting the principles and knowledge of one into experiments and techniques in the other. Computer science can offer biologists information on system design, encapsulation, encoding, and abstraction. But biological systems are vastly more sophisticated than modern computers. Computer science can learn many lessons from biology about massively parallel architectures, self-assembling machines, overlapping codes, etc. With complete understanding of the mechanisms of biology, biologists might program an organism's metabolism for a specific task. With a clear understanding of the computational components of the cell, engineers might harness yeast as an all-purpose computer that could self-replicate, giving humans access to exponentially increasing computer power.

Conclusions

Executable code and genomes show striking similarities in the way information is structured, despite the fact that their physical mechanisms are completely different. I propose that this is because both kinds of code are subject to many of the same constraints dictated by information theory. Given the striking similarities between genomes and computer code, it will be fruitful to study the architecture of executable computer code, so that we might better understand how genomes function. We know that in computer code, there is no "junk code", and that all the structure we are seeing is functional. Therefore it is reasonable to expect that function underlies all the analogous structures seen within the genome.

Acknowledgments

The basic research that led to this discovery was supported by FMS Foundation.

References

1. Shapiro JA, von Sternberg R (2005) Why repetitive DNA is essential to genome function. Biol Rev 80:227–250.
2. Ohno S (1972) So much "junk" DNA in our genome. In: Smith HH (ed) Evol Genet Syst, Brookhaven Symp Biol 23:366–370.
3. Doolittle WF, Sapienza C (1980) Selfish genes, the phenotype paradigm and genome evolution. Nature 284:601–603.
4. Barash Y, et al (2009) Deciphering the splicing code. Nature 465:53–59.
5. Kapranov P, Willingham AT, Gingeras TR (2007) Genome-wide transcription and the implications for genome organization. Nat Rev Genet 8:413–423.
6. The ENCODE Project Consortium (2007) Identification and analysis of functional elements in 1% of the human genome by the ENCODE pilot project. Nature 447:799–816.
7. Pheasant M, Mattick JS (2007) Raising the estimate of functional human sequences. Genome Res 17:1245–1253.
8. Seaman JD, Sanford JC (2009) Skittle: a 2-Dimensional Genome Visualization Tool. BMC Bioinformatics 10:452.
9. Jeffrey HJ (1990) Chaos game representation of gene structure. Nucl Acids Res 18(8): 2163–2170.
10. Weber J, Wong C (1993) Mutation of human short tandem repeats. Hum Mol Genet 2(8): 1123–1128.
11. Brinkmann B, et al (1998) Mutation Rate in Human Microsatellites: Influence of the Structure and Length of the Tandem Repeat. AJHG 62 (6): 1408–1415.
12. Roach J, et al (2010) Analysis of Genetic Inheritance in a Family Quartet by Whole-Genome Sequencing. Science 328: 636–639.
13. Lieberman-Aiden E, et al (2009) Comprehensive Mapping of Long-Range Interactions Reveals Folding Principles of the Human Genome. Science 326: 289–293.
14. Faulkner GJ, et al (2009) The regulated retrotransposon transcriptome of mammalian cells. Nat Genet 41:563–571.
15. Macaya G, Thiery J-P, Bernardi G (1976) An approach to the organization of eukaryotic genomes at a macromolecular level. J Mol Biol 108:237–254.
16. Tykocinski ML, Max EE (1984) CG dinucleotide clusters in MHC genes and in 5' demethylated genes. Nuc Acids Res 12:4385–4396.

17. Duan Z, et al (2010) A three-dimensional model of the yeast genome. Nature Letters 465: 363–367.
18. Cremer T, et al (2001) Chromosome territories, nuclear architecture and gene regulation in mammalian cells. Nat Rev Genet 2: 292 – 301.
19. Turing, AM (1936) On Computable Numbers, with an Application to the Entscheidungs problem. Proceedings of the London Mathematical Society Ser. 2 Vol. 42: 230–65. 1937.

Biocybernetics and Biosemiosis

Donald Johnson

Retired Scientist and Professor (APU, U-MD, U-MN & U-WI), 38762 Fawn Springs Dr, Palm Desert, CA 92260, don@programmingoflife.info

Abstract

Biocybernetics is the study of life's hardware and software systems, which control the chemistry and physics of all of life's processes, including metabolism, manufacturing, control, and feedback. Unlike chemistry and physics, which are physical sciences, biology is an information science since what differentiates biology from complex organic chemistry is its information processing systems. Semiosis connects two independent worlds of signs and meaning by the conventional rules of a code. Many arbitrary coded symbol systems, with over 20 discovered in the past decade, play very important roles in communicating information between life's components. Life's networked computers and computer programs instantiated into DNA and RNA memory devices are discussed. A prescriptive algorithm can be implemented in either hardware or software. The "artificial genome" manufactured by Venter *et al.* demonstrates experimentally the reality of computer hardware and software in each cell.

Any serious origin-of-life or origin-of-species scenario must explain the origin of the required biological information. It is argued that each protein arises as the result of the execution of a genuine computer program. The creation of a functional protein via the mutation/selection paradigm lacks support from information science. Those who understand the reality of bioinformation, especially the prescriptive information of biocybernetics, will be able to incorporate that understanding into new models that will lead to a more complete understanding of life.

Key words: biosemiosis, biocybernetics, prescriptive information, DNA software, artificial genome, life's computers

Introduction

Biocybernetics is the study of life's hardware and software systems that use digital information processing to control, integrate, and maintain life's processes. While physics and chemistry are physical sciences whose interactions are wholly determined by physicality, biology is an information science since all of the defining characteristics of biology are controlled by life's information processing systems. Biology isn't just complicated chemistry, since it involves coded messages (semiotics) [1, 2] and coded algorithmic prescriptive instructions (instantiated computer programs) [3–5]. The vital nature of information in life has been downplayed by most materialists, since functional information has no feasible cause from physicality (though infeasible scenarios have been speculated). When addressed at all, the informational aspects of life are usually treated as metaphors or analogies, rather than realities.

Information is a non-material entity that can be instantiated into physicality for storage or communication. Information always involves contingency, such that it could have been different. If there is no contingency, the value is not informational, but instead is determined by natural law. Any property determined wholly by law is not informational. It is common to mistakenly view a physical property of an object as information. For example, the temperature of an object is a property totally determined by a number of physical laws related to mass, specific heat, energy flow, radiation, etc. Since temperature is determined by law, that property is not information, even though it can be transduced into functional information by use of a device known as a thermometer (which could be part of a thermostat for controlling the temperature). The temperature of an object could be information if it were contingent through appropriate choice of constraints. For example, a rock could convey binary information: hot = yes, cold = no, where the information rock is placed in either a bed of coals or a glacier to record the choice, before placing it in an insulated container for later examination. Obviously this stored information would be lost with time, as the rock nears ambient temperature, but a bit of RAM memory in your computer also requires a refresh to retain its information. The radiation from a star is totally determined by physicality, but a spectrophotometer could be used to produce information related to the star's temperature, composition, velocity, etc. The star's radiation has no contingency in that its properties cannot be otherwise give the initial constraints and the laws involved. The human measurements of that radiation, on the other hand, involve considerable contingency, and could even be incorrect if the instruments weren't properly calibrated. The weather is totally determined by law and initial constraints. Even a dark cloud with rain descending is not informational without an observer with the capability to ascertain meaning from the physical properties observed.

The broadest classification of information is that from information theory developed by Shannon [6], which requires nothing meaningful, except in the case of a coded information subset. Uncertainty is a better descriptive term since the Shannon "Information" of a pattern is inversely related to the pattern's probability. A random sequence has the highest possible uncertainty. A subset of the broadest category is functional information, which has meaning (such as in coded information). The most restrictive classification of information is prescriptive information, which is not only functional/meaningful, but is algorithmic (a recipe). Consider data typed into a word-processing program. Most such data is functional in that it has a purpose of communicating information to the ultimate reader of that information. If a monkey typed random data into the program, that complex data, produced by chance contingency, would have no purpose, but would have a very high Shannon uncertainty since that deals only with the probability of the data pattern, irrespective of any meaning. A computer program typed into the word processor is more than just meaningful, but is prescriptive in that it contains instructions to accomplish objectives based on data to be supplied during the execution of the program being typed. Prescriptive information may be a simple step-wise recipe, or may express the decisions to be made and the criteria for the different execution paths.

Some have questioned whether there are actual computer programs (instantiated algorithms) in life, or there is just a "resemblance" to computer-like characteristics. One of the most significant experimental confirmations of the reality that life is hardware/software was the announcement in 2010 of Venter's computer-generated artificial genome [7]. Venter stated "It certainly changed my views of definitions of life and how life works... Life is basically the result of an information process, a software process. Our genetic code is our software" [8]. Venter's team didn't "create life," but they put life synthesized pieces of the target DNA into yeast which assembled the target bacterium's genome. They didn't engineer specific instructions (algorithms), but rather combined DNA blocks that matched the target sequence. The assembled genome was transplanted into a different bacterium and 'booted up' to create a new synthetic version of the target. For this "proof of principle" instance, they synthesized a bacterium as close to the original genome as they could, using the original DNA as a "standard" for comparison, replacing the genome's application program set stored in the original organism's DNA memory with a genetically engineered application program set matching the target. The operating systems and the interacting computers in the cell whose genome was replaced remained intact and were able to function by using the replacement software. One of the things this research demonstrated is that (at least for the two bacteria involved) life uses common operating systems, programming languages, and devices (otherwise the programs for one machine wouldn't execute on another).

Since many believe it is important to differentiate hardware from software, perhaps it is beneficial to consider some related computer science principles. To be functional, both hardware and software are instantiations of algorithms, which are step-by-step solutions to problems. In the case of hardware, the algorithm could be developed using state-transition diagrams or a hardware description language before instantiation in an electronic circuit. Any hardware-generated control signals could be generated by software control. There is also an important distinction in computer science between architecture and organization.

Computer architecture refers to the machine characteristics visible to lowest-level user programming (the assembly language instruction set). Organization refers to the implementation of the architecture. For example, a CPU's control unit uses the fetched machine language instruction, along with other inputs, to generate the control signals needed to carry out the instruction. A control unit could be purely electronic, which is often done for the fastest computers. A control unit is usually implemented using a less-expensive control storage interpreter which uses the machine instruction to generate an address for reading the instruction's control signals from a control memory (microcode). The control storage can be permanent or writeable (allowing different machine architectures using the same hardware). The functionality is identical for any organization for the same architecture, and organization couldn't be ascertained from functionality. This is important for life's information because it may not be critical to identify what is software and what is hardware when analyzing functionality. Since hardware and software can't be differentiated based on functionality in electronic computers, there is no information science reason to expect differentiation would be possible based on functionality of biocybernetic systems. That differentiation may be important when ascertaining mechanisms, however.

Life's Computers

Most people tend to have a very narrow view as to what a computer can be. Realize that the first computer, Babbage's 1837 Analytic Engine (Fig.1), was totally mechanical, and yet "Turing complete" (could theoretically be programmed to compute anything possible to compute). Many architectures and organizations can be classified as "computers" since the necessary and sufficient requirements for a computer are: input (or embedded data), memory, an instantiated program, processing capability, and output. Note that the first electronic computer was not Turing complete (no branching capability) and couldn't be re-programmed, so those characteristics (as found in many biological computers) aren't required to be a "computer." There are many components of life that can thus be classified as computers or components of computers, so that the reality and variety of biological computers should not be surprising. For example, multiple proteins (including RNA polymerase) may form a computer to read the DNA memory/program to output the mRNA transcription. The "program" could be in the non-coding DNA, which could use the "gene" as data to transcribe. Some hypothesize that the transcription components are "merely" under control of a master computer, and are equivalent to a disk head assembly (perhaps with built-in control as found in a hard drive's read/write head assembly) [9]. Without being dogmatic,

Fig. 1. Babbage Mechanical Computer – Babbage's 1837 Analytic Engine, was totally mechanical, and yet "Turing complete".

that alternative approach wouldn't explain the fact that the replisome has higher priority than polymerase, causing a transcription in progress to abort [10] (why would a "master" control computer start such a transcription?). High-performance pipelined computers often use "optimistic scheduling" to start operations that won't ultimately complete, but this would seem to be a waste of energy for a process of life. Multiple networked interacting semi-autonomous computers seem to fit the observations better (at least based on what is currently known). Is the mRNA "simply" a coded digital message for the ribosome to process, or is mRNA a program to be interpreted by the ribosome? The later seems likely since mRNA can be generated by multiple means, each producing a protein as the output during the execution of the computationally-halting program (a requirement of a functional algorithm) when the mRNA program is interpreted by the ribosome (equivalent to a micro-programmed control unit interpreting a machine instruction sequence). Since the ribosome contains RNA memory, in addition to a multitude of proteins, and interprets the prescriptive program of its input mRNA, it seems likely that a ribosome is indeed a genuine specific-purpose computer (it has all necessary and sufficient requirements). It should also be noted that the epigenome, polypeptides (including proteins) and micro-RNAs of various lengths can serve as information-carrying structures and/or memories. The author has peer reviewed publications using serially-shared information within multiprocessor systems [11–13], and can attest to the importance of protocols for functionality when communicating such information.

When examining tRNA, a computer scientist quite naturally considers the purpose of its RNA memory structure. Does tRNA operate totally by "law," or might this be another computer? Although the total mechanism for attaching a particular amino acid so that it matches the codon on the opposite end of the tRNA is complex, and not fully understood, the presence of RNA, a memory structure, may indicate that multiple proteins can form a computer with the tRNA's instruction memory to select and attach the appropriate amino acid, and release it as output at the ribosome's request. Once again, the tRNA complex possesses the necessary and sufficient characteristics that define a computer. If a mechanism based on law can explain the functionality of tRNA, then perhaps its RNA memory simply serves as a separator, as opposed to being functional memory (which seems unlikely to the author). In any case, each protein is the result of the execution of a real computer program, ultimately instantiated in DNA for the protein's sequence. Venter's artificial genome experiment demonstrated that even mRNA generated by alternative mechanisms than direct transcription ultimately depends on the DNA memory.

Thus far, there has been no feasible mechanism proposed for writing computer programs by inanimate nature. There also has been no feasible mechanism for computer hardware being implemented from inanimacy. All known computer programs and hardware systems require formal solutions before a functional result is obtained. The prescriptive information incorporated in both life's hardware and software currently lacks any feasible explanation from chance and necessity. Scientific answers are needed, as no scenarios proposed so far are compatible with information science. The Origin-of-Life Prize (www.lifeorigin.info) highlights the major difficulties and "will be awarded for proposing a highly plausible mechanism for the spontaneous rise of genetic instructions in nature sufficient to give rise to life" [14]. OOL requires that each nucleotide of the RNA sequence be selected for potential function, as opposed to natural selection's favoring of existing functionality. Since natural selection depends on already existing protein structures of the phenotype, and each protein is a result of the genomic algorithm instantiated in the DNA, natural selection is not a mechanism for generation of new prescriptive information, but at most is a sorting procedure to weed out organisms that are less fit. What mutation/selection really says is that, "randomly changing a functional program can sometimes produce a modified program with improved functionality." Such a random net increase in non-trivial functionality has never been documented in computer science. For example, random changes in so-called "artificial life" programs use designed targets and fitness functions to steer results in desirable directions for functionality [15,16]. When irreducibly complex structures are considered, multiple programs would require simultaneous modification.

The polynucleotide sequence of DNA or RNA is an ideal information storage structure since each nucleotide has no dependence on preceding or following nucleotides, and can be arbitrarily set to the functional value desired from the four possible values. It should also be mentioned that within the DNA helix, only half of the nucleotides are informational since one strand is totally determined (and is redundant) by the other (informational) strand. Information requires contingency, and one strand has none. Note that other information for decoding overlapping genes and reverse transcription is not directly in the DNA sequence. The prescriptive information in a DNA sequence is a recipe or algorithm to accomplish a desired task. What complicates this is the fact that many nucleotides are components of multiple prescriptions, such as in overlapping genes or alternative splicing schemes. In those cases, the nucleotide has to be set so that it becomes a functional component of multiple algorithms. The algorithms can be those for protein generation or one of the numerous cellular controls. Sub-coded (codes within codes) information [17] and a second genetic code [18] characterizing alternative splicing have been discovered. Various transcribed RNAs are mixed and matched and spliced into mRNAs for specifying protein construction and other controls, sometimes joining messages that were separated by thousands of nucleotides. "For example, three neurexin genes can generate over 3,000 genetic messages that help control the wiring of the brain" [19]. Even "simple" prescription information lacks any feasible explanation using known science. Much more challenging are the explanations required for multiple and overlapping levels of prescriptive information.

Biosemiosis

"Physicality is the only reality" is a paradigm which encounters severe difficulties when confronted with biological coding systems (semiosis), and the associated formal operations which are required. What is a semiotic system? "A semiotic system is a system made of two independent worlds that are connected by the conventional rules of a code. A semiotic system ... is necessarily made of three distinct entities: signs, meanings and code" [20]. The best-known biological code is the codon-to-amino acid translation during protein construction which uses tRNAs to translate one codon from the 64-codon alphabet (a sign) into one amino acid in the 20 amino acid alphabet (meaning). There is no chemical or other deterministic link between the opposite ends of a tRNA that causes a particular amino acid to be associated with a particular codon. They are associated by an arbitrary rule determined by a code. Over 20 other semiotic codes have been discovered in life in the past decade, with each code having arbitrary rules agreed on by both sender and receiver of the coded information message, as described briefly below.

A coactivator code for coregulators may confer specificity to ubiquitous transcriptional regulatory factors, with wide-reaching implications. Cofactors use a variety of mechanisms to contribute to gene transcription activation and repression [21]. A protein destination code ensures delivery of the protein to the correct destination. "Proteins are the workhorses of the cell, but to get the most work out of them, they need to be in the right place. In neurons, for example, proteins needed at axons differ from those needed at dendrites, while in budding yeast cells, the daughter cell needs proteins the mother cell does not. In each case, one strategy for making sure a protein gets where it belongs is to shuttle its messenger RNA to the right spot before translating it. The destination for such an mRNA is encoded in a set of so-called "zipcode" elements, which loop out of the RNA string to link up with RNA-binding proteins. In yeast, these proteins join up with a myosin motor that taxis the complex to the encoded location" [22]. A code for resolving overlapping codes is needed to start transcription appropriately. "Genomes encode multiple signals, raising the question of how these different codes are organized along the linear genome sequence" [23]. The detailed coding "signals consist of both known and potentially novel codes, including position dependent secondary RNA structure, bacteria-specific depletion of transcription and translation initiation signals, and eukaryote-specific enrichment of microRNA target sites" [23].

The cytoskeleton anchoring code [1] determines the ultimate relationship between the cellular structures that the cytoskeleton is working on and the microtubule and microfilament components of the cytoskeleton. Every microtubule starts from a centrosome, with the other end growing or contracting in an exploratory "strategy" in a search for an anchor. There is a "dynamic instability" as monomers are added and taken away (if an anchor is not found), so the cytoskeleton can rapidly explore all of the cytoplasm's space, until a stable anchor code is found.

Barbieri lists 20 semiotic codes [20] from a variety of research papers, including adhesive code, sugar code, histone code, neural transcriptional codes, regulatory code in mammalian organogenesis, code of post translational modifications, nuclear receptors combinatorial code, transcription factors code, acetylation codes, estrogen receptor code, metabolic codes, rna codes, error-correcting codes, modular code of the cytoskeleton, lipid-based code in nuclear signaling, immune self-code, and signal transduction codes. In each case, the code provides an arbitrary translation between disjoint domains. This list only scratches the surface of all the codes that are still waiting to be discovered.

Since information is non-material, there have been no feasible scenarios for production of semiotic systems from physicality. Barbieri proposes "natural conventions" as the required codemaker that creates the required semiotic translation bridge between the sender and receiver [20]. Barbieri fails to present a feasible mechanism, but argues it "must have happened" since semiosis is a ubiquitous reality and is actually the mechanism he proposes for macroevolution. In his view, for something totally new to appear, a new organic code that had never existed before must come into being. Biological specificity (required for heredity and reproduction) was the result of the origin of the genetic code. Signal transduction codes allowed systems to produce their own signals, separating their internal space from the environment. The origin of the eukaryote nucleus was brought about by the origin of splicing codes [1,24]. The development of any coding system must account for information (especially transfer of information), in a manner compatible with information theory. The next paragraph provides the technical details, but the bottom line is that codes cannot evolve from simpler to more complex basic codes without violating an information theory theorem that has stood for over 60 years.

Given the probability vector, $\mathbf{p}_A$, of the elements of alphabet A in a source probability space [Ω, A, $\mathbf{p}_A$] and the probability vector, $\mathbf{p}_B$, of the elements of alphabet B in destination probability space [Ω, B, $\mathbf{p}_B$], then a unique mapping of the symbols of alphabet A onto the symbols of alphabet B is called a code [25]. Mutual entropy is a mathematical measure of the similarity between any two sequences one wishes to compare. Mutual entropy relates the input (x) and output (y) channels via: $\mathbf{I(B;A) = I(A;B) = H(x) - H(x|y)}$, where the conditional (x_i given y_i received) entropy is $\mathbf{H(x|y) = -\sum_{ij} p_j\, p(i|j) \log_2 p(i|j)}$, $\mathbf{p_j = \sum_i^n p_i p(j|i)}$ (which relates the probability vector, $\mathbf{p}$, elements to those of the conditional probability matrix, $\mathbf{P}$), and $\mathbf{H(x) = -\sum_{i=1}^{n} p_i \log_2(p_i)}$ is the information entropy. The Shannon Channel Capacity is also the maximum mutual entropy. For a transmitting system with fewer symbols in [Ω, A, $\mathbf{p}_A$] to pass information to [Ω, B, $\mathbf{p}_B$], the maximum mutual entropy would be exceeded. The channel capacity thus prohibits a simpler symbolic alphabet (e.g. a 2-nucleotide "codon") from evolving into an alphabet with more intrinsic symbols. Some have suggested it "must have happened," but have provided no falsification of Shannon Channel Capacity Theorem that has stood for over 60 years. Without such falsification, the original instantiation of any semiotic code would have an alphabet at least as symbolically complex as that currently used. Note that an alphabet symbol may consist of other symbols. For example, the American Standard Code for Information Interchange (ASCII) defines printable characters represented by seven bits, with a 7-bit group being one sign for the source alphabet.

The growing acknowledgment that the mutation/selection model of evolution is not sufficient to explain the origin of elaborate information processing systems seems to suggest that a major paradigm shift is imminent. The leading contender as a replacement is the "Extended Synthesis" [26], which is very flexible, incorporating essentially all proposed mechanisms for evolution, including concepts like "natural genetic engineering" [27, 28] for mutation selection and "natural conventions" [20] as semiotic code. Whatever the replacement will be, science needs to ensure that any scenarios are compatible with information science. The most difficult realities to accommodate will probably be the prescriptive information of biocybernetics and the arbitrary information translation codes of biosemiosis. There are numerous very specific problems that must explained if the the neo-Darwinian paradigm is to survive. Given all these challenges the defenders of the status quo must provide scientific answers to a series of extremely difficult questions, include the following [14].

1. *How did nature write the prescriptive programs needed to organize life-sustaining metabolism? Programs are shown by computer science to require a formal solution prior to implementation. How did inanimate nature formally solve these complex problems and write the programs? How did nature develop the operating systems and programming languages to implement the algorithms? How did nature develop Turing machines capable of computational halting? How did nature develop the arbitrary protocols for communication and coordination among the thousands (or millions) of computers in each cell?*

2. *How did nature develop multiple semiotic coding systems, including the redundant (surjective) codon-based coding system (for symbolic translation) that involves transcribing, communicating, and translating the symbolic triplet nucleotide block-codes into amino acids of the proteins? How did nature develop alternative generation of such messages using techniques such as overlapping genes, messages within messages, multi-level encryption, and consolidation of dispersed messages? A protein may obtain its consolidated prescriptive construction instructions from multiple genes and/or from the "junk" DNA, sometimes with over a million nucleotides separating the instructions to be combined.*

3. *How did nature defy computer science principles by avoiding software engineering's top-down approach required for complex programming systems? How did nature produce complex functional programs without planning, by randomly*

modifying existing algorithms? How did multiple such programs become simultaneously modified to result in the production of irreducibly complex structures?

These questions demand scientific answers that are compatible with information science. "It must have happened" is not science, but belief.

References

1. Barbieri M (2003) The Organic Codes. An introduction to semantic biology. Cambridge University Press, Cambridge, UK.
2. Barbieri M (editor of anthology) (2008) The Codes of Life: The Rules of Macroevolution.
3. Abel D (2008) The 'Cybernetic Cut': Progressing from Description to Prescription in Systems Theory. The Open Cybernetics and Systemics Journal 2: p252–262.
4. Abel D (2009) The GS (genetic selection) Principle. Frontiers in Bioscience 14: 2959–2969.
5. Abel D (2009) The Biosemiosis of Prescriptive Information. Semiotica, 1/4/09, p1–19.
6. Shannon (1948) A Mathematical Theory of Communication. Bell System Technical Journal: 27, July & October, p379–423 & 623–656.
7. Gibson D, et al (2010) Creation of a Bacterial Cell Controlled by a Chemically Synthesized Genome. Science 329(5987):52–56.
8. Venter (Craig), Interview, http://www.guardian.co.uk/science/video/2010/may/20/craig-venter-new-life-form
9. D'Onofrio D, An G (2010) A Comparative Approach for the Investigation of Biological Information Processing: an Examination of the Structure and Function of Computer Hard Drives and DNA. Theoretical Biology and Medical Modeling 7:3.
10. Rockefeller University (2010) Scientists crash test DNA's replication machinery. ScienceDaily, 3/1/10.
11. Johnson D, Lilja D, Riedl J, "A Circulating Active Barrier Synchronization Mechanism," International Conference on Parallel Processing I, 8/95, p202–209.
12. Johnson D, Lilja D, Riedl J, Anderson J, "Low-Cost, High-Performance Barrier Synchronization on Networks of Workstations," Jour Par & Distributed Computing, 2/97, p131–137.
13. Johnson D, Lilja D, Riedl J (2005) Circulating Shared-Registers for Multiprocessor Systems. Jour. of Systems Arch., 6/3/05, p152–168.
14. Johnson D (2010) Programming of Life, 2010, p83–84.
15. Truman R, "Evaluation of Neo-Darwinian Theory Using the Avida Platform," PCID 3.1.1,11/04.

16. Gilder G (2006). Evolution and Me. National Review, 7/17/06.
17. Cannarozzi (G.), N. Schraudolph, M. Faty, P. von Rohr, M. Friberg, A. Roth, P. Gonnet, G. Gonnet, & Y. Barral, "Role for Codon Order in Translation Dynamics," Cell 141, 4/16/10, p355–367.
18. Barash Y, Calarco J, Gao W, Pan Q, Wang X, Shai O, Blencowe B, B, Frey B (2010) Deciphering the Splicing Code. Nature 465 (7294):53–9.
19. Frey (J.), quoted in "Researchers Crack 'Splicing Code,' Solve a Mystery Underlying Biological Complexity," 5/5/10, www.physorg.com/news192282850.html
20. Barbieri M (2008) Biosemiotics: a new understanding of life. Naturwissenschaften 95:577–599.
21. Gamble M, Freedman L (2002) A coactivator code for transcription. TRENDS in Biochemical Sciences 27 (4):165–167.
22. Robinson R (2011) A two-step process gets mRNA loaded and ready to go. Public Library of Science: Biology, 9(4): e1001047. doi:10.1371/journal.pbio.1001047.
23. Itzkovitz S, Hodis E, Segal E (2010) Overlapping codes within protein-coding sequences. Genome Res 20:1582–1589.
24. Barbieri M (1998) The Organic Codes. The Basic Mechanism of Macroevolution," Rivista di Biologia-Biology Forum 91:481–514.
25. Yockey H (2005) Information Theory, Evolution, and the Origin of Life. Cambridge University Press, Cambridge.
26. Pigliucci M, Müller G (2010) Elements of an extended evolutionary synthesis. In: Pigliucci M, Müller G (eds) Evolution—the Extended Synthesis, The MIT Press, Cambridge, MA.
27. Witzany (Günther), Natural Genetic Engineering and Natural Genome Editing (Proceedings), 12/09, 276 pgs.
28. Shapiro J (2010) Mobile DNA and Evolution in the 21st Century. Mobile DNA 1:4.

Section Three — Theoretical Molecular Biology: Introductory Comments

Michael J. Behe — Section Chairman

Biological information must be expressed to be consequential. In the past half century, science has discovered that expression often takes the form of sophisticated molecular machinery. Information resides in the very shape of the machinery itself, as well as in the instructions to build the machinery, to regulate it, to allow separate systems to communicate with it, and more. In all these cases the information must be physically instantiated to be effective. This section focuses on systems that are known, or speculated, to instantiate information, and how they may be affected by evolutionary forces.

The chapter by **Macosko and Smelser** reviews the evidence that the genetic code used by nearly all life forms on earth is far from a "frozen accident" — that is, far from an arbitrary assignment of codons to amino acids that got locked into place because changing it would have been beyond the reach of Darwinian processes, as was speculated by early investigators. Rather, close analysis of the code has shown it to be better than the vast majority of possible codes in resisting the occurrence of deleterious mutations, in resisting their impact if they occur, and more. The conjecture that the code was optimized by Darwinian selection runs headlong into the profound difficulty that a change in the code used by an organism would affect all proteins coded for by its genome. Almost certainly such a change would negatively impact the functioning of many proteins, and be resisted by the very selection that is posited to shepherd the code to greater efficiency. Macosko and Smelser argue that the hypothesis of purposeful intelligent design better fits the data, and can lead to new insights into this basic feature of life.

The chapter by **Dent** seeks to discern how widely-separated molecules obtain the information with which to find their targets. Deeming the standard explanation of a random Brownian search to be inadequate in many cases, Dent hypothesizes that there exist coherent oscillator structures within chromosomes and proteins with a narrow range of resonant frequencies. Such oscillations are thought to attract biomolecules to one another with great specificity. In experiments using ultra-high-frequency Doppler vibrometry, live onion cells and fish eggs were scanned for the presence of vibrations in the gigahertz range, predicted of DNA. Although such signals were not detected in the present study, they may not in fact reach the cell surface, but be confined to the nucleus. Further work is planned to investigate this possibility.

An Ode to the Code: Evidence for Fine-Tuning in the Standard Codon Table

Jed C. Macosko* and Amanda M. Smelser

*Department of Physics, Wake Forest University, Winston-Salem, NC 27109, USA. (*Corresponding author: macoskjc@wfu.edu)*

Abstract

The Standard Codon Table (SCT) records the correlation observed in nature between the complete set of 64 trinucleotide codons and the 20 amino acids plus 3 nonsense (i.e. stop or termination) signals. This table was called a frozen accident by Francis Crick, yet current evidence points to optimization that minimizes harmful effects of mutations and mistranslations while maximizing the encoding of multiple messages into a single sequence. For example, a recent article with the running title "The best of all possible codes?" concluded that "evidence is clear" for the optimized nature of the SCT, and another study found that difficult-to-encode secondary signals are minimized in the SCT. Additionally, the initiating amino acid methionine has been found to minimize the nascent peptide chain's barrier to exit the ribosome. Moreover, the symmetry in the SCT between 4-fold-synonymous and <4-fold synonymous codons has been explained in terms of minimizing mistranslation. In this paper, the hypothesis that the finely tuned optimization of the SCT originates in external intelligence is compared to the hypothesis that its fine tuning is due to the adaptive selection of earlier codes. It is concluded that, in the absence of metaphysical biases against this hypothesis, external intelligence better explains the origin of the SCT. Additionally, this hypothesis prompts lines of inquiry that, 50 years ago, would have accelerated the discovery of the now-known features of the SCT and that, today, can lead to new discoveries.

Key words: genetic code, origin of life, adaptive code, error minimizing code, stereochemical origin, frozen accident, amino acid biosynthesis, coevolution, family non-family symmetry

Introduction

In 1976, Francis Crick and coauthors wrote, "The origin of protein synthesis is a notoriously difficult problem" [1]. Proteins are synthesized based on information contained in mRNA, according to an easily-represented map between RNA trinucleotides and protein building blocks [2]. This map describes the flow of information from mRNA to protein in nearly every organism and is usually called "the genetic code".

Here, the map (Figure 1) is called the Standard Codon Table (SCT) to distinguish it both from the *physical machinery* (Figure 2) that enables this flow of

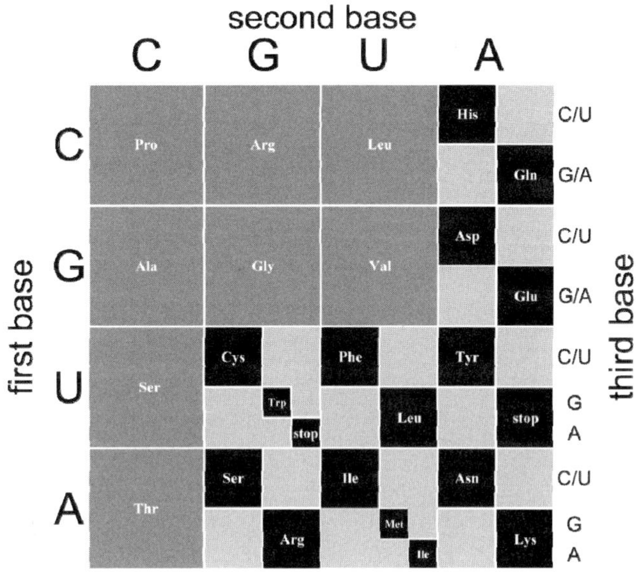

Fig. 1. The Standard Codon Table (SCT) arranged to highlight the family/split-box symmetry. In gray are eight "family" amino acids, specified by four codons each for a total of 32 codons. In black are the other 32 codons: the three stop codons and the codons for the 12 "split-box" amino acids that are coded by three or less codons each. Three amino acids — serine, arginine and leucine — use both family and split-box codons. For purposes of tRNA comparison, the tRNAs that recognize the grey ser, arg, and leu codons are considered family tRNAs and those that recognize the black ser, arg, and leu codons are considered split-box tRNAs.

information and from *additional codes* of secondary signals. These so-called "sub-codes" or "second-layer codes", and the coding machinery itself, are integral parts of the true genetic code, i.e. the full code that starts with the genetic information in DNA and ends with the protein and RNA machines that keep organisms alive [3].

The evolutionary origin of the protein synthesis scheme shown in Figure 2 is what Crick considered a "difficult problem" [1]. There are two parts of this problem: first, how the general coding scheme (Figure 2) originated, and second, how the specific correspondence between trinucleotides and amino acids, i.e the SCT (Figure 1), came about. These two parts are interrelated, but it is helpful at first to consider them separately.

Theories of the Origin of the Standard Codon Table

Currently there are four theories that, alone or in combination, address the origin of the SCT (see review: [4]). First, there is the **frozen accident model**, which takes its name from Crick's suggestion that the SCT was a frozen accident [2]. In other

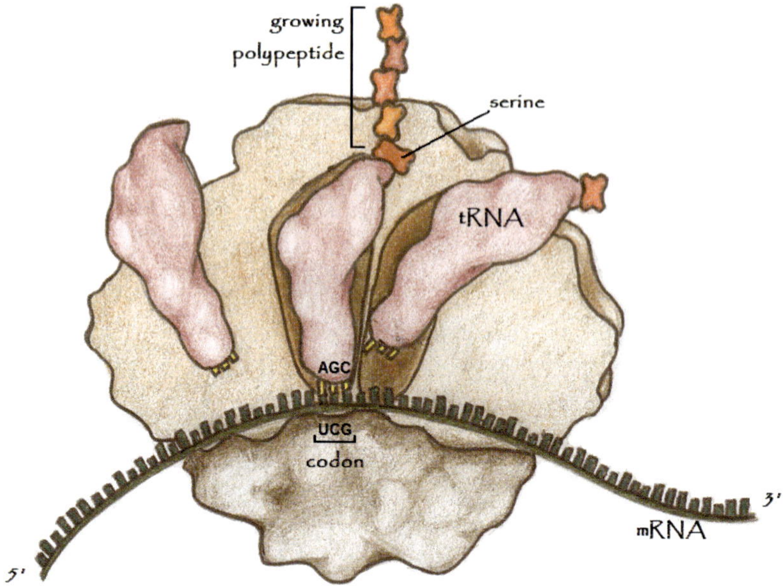

Fig. 2. The tRNAs, shown inside a ribosome, are key pieces of the physical machinery that actualizes the information flow from the mRNA to the polypeptide (protein) chain. This flow follows the SCT; for example, the mRNA letters UCG are recognized by the tRNA that has CGA (as read in the 5' to 3' direction) in its anticodon loop and that carries serine at its opposite end. This example of UCG=serine is shown in Figure 1 (see the grey box labeled "serine" at the intersection of "first base U" and "second base C").

words, neither the mechanism that led to the general coding scheme (Figure 2), nor any other mechanism, dictated the pattern in the SCT (Figure 1). It was purely an accident; the SCT could have ended up with any arbitrary structure. Thus, the current structure does not reveal any information about a past mechanism.

The other three theories all assume that the SCT was not an accident but was formed by a mechanism. By examining the nature of the SCT, one can learn about the mechanism that formed it. The first of these theories is the **error minimization model**. In this model, the SCT was formed by a mechanism that primarily minimized the negative impact of DNA mutations, of mRNA mistranscriptions, and of protein chain mistranslations [5]. Thus, the arrangement of amino acids in Figure 1 is not accidental. For example, once a guanine (G) base in the first codon position and an adenine (A) base in the second position came to represent one of the negatively charged amino acids, then both negatively charged amino acids became encoded with the sequence GAN (where N is any base) so that a mutation in the third position would simply exchange one negatively charged amino acid for another.

Another theory proposes that the origin of the SCT is linked to, or coevolved with, primordial **amino acid biosynthesis** [6]. Several of the 20 amino acids shown in Figure 1 are synthesized in living cells starting from other amino acids. For example, the negatively charged amino acid, aspartic acid, is known to be a precursor for methionine, threonine, isoleucine, and lysine [7]. These four amino acids are encoded by ANA and ANG codons (see Figure 1), which some take as evidence in favor of this theory [8].

The final theory depends on **stereochemical interactions** between amino acids and their respective trinucleotide codons (Figure 1) or anticodons. This model was popular immediately after the elucidation of the SCT, since it postulated a simple mechanism for the origin of the codon assignments: each codon (or anticodon) had a physical affinity for its respective amino acid, and not for other amino acids [9]. Thus, had this theory proved true, the assignments shown in Figure 1 would have been biochemically predestined by virtue of stereochemical interactions. As it is, the evidence is limited with respect to statistically significant interactions between the codons or anticodons and their respective amino acids. Of the 20 amino acids, only seven (phenylalanine, isoleucine, leucine, histidine, arginine, tyrosine, and tryptophan) show such interactions, and the preference for codon versus anticodon involvement appears random [10].

Of the four theories, error minimization and amino acid biosynthesis are currently favored, though some claim these mechanisms are minor influences compared to the overall frozen accident nature of the SCT [11].

It is important to remember that these four SCT origin theories do *not* explain the origin of the machinery (e.g. Figure 2) that is responsible for converting mRNA information into amino acid sequences. Theories for the origin of the coding machinery are abundant and are generally viewed as extremely speculative (e.g. [12] and reviewer comments). As such, this paper does not address these theories but focuses on just the origin of the codon assignments themselves.

In the next section, we present four studies that describe SCT features that are optimal and are orthogonal, i.e. the optimality of one would not necessarily lead to the optimality of the others. These features are 1) similar amino acids are coded by similar codons thus minimizing the impact of errors, 2) the family/non-family symmetry minimizes mistranslations while maximizing tRNA usage efficiency, 3) the stop codons are related to commonly occurring amino acids in a way that optimizes second-layer codes, and 4) methionine is an optimal initiating amino acid due to its minimized energy for exiting the ribosome.

Orthogonally Optimized Features of the Standard Codon Table

Previous studies [5, 13–16] have compared the optimality of the SCT to those of alternative codon tables in terms of how they mitigate genetic errors by ensuring that similar amino acids are coded with similar codons (see the "error minimization" theory above). One of these studies in 2000 by Freeland *et al.* determined the most optimized code, given different values of two parameters [15]. The first parameter was the relative likelihood of transitions — A:G or thymine(T):cytosine(C) exchanges — and transversions — A or G exchanging with T or C. The second parameter was the relative impact of mutation as modulated by the power to which the error equation is raised. For most of the intermediate values of these two parameters, the real SCT was the single most optimized codon table — the "best of all possible codes" as this paper's running title suggested. Interestingly, this 100% optimization of the SCT was demonstrated within a restricted set of codon tables. The restricted set reflected the amino acid biosynthesis theory described above. Thus, this paper blended the two favored mechanisms for the origin of the SCT — error minimization and biosynthesis — and quantified a level of optimization that was near or at the global maximum.

Freeland *et al.*'s landmark study tacitly assumes that an optimized code imparts to its owner a selectable advantage over organisms that have not-as-optimized codes. Recent work by Geiler-Samerotte *et al.* helps to answer the question, "What selective effect would a more optimal code have?" [17]. These authors compared the fitness of mutant yeasts expressing a gratuitous protein that misfolded to varying extents. When the protein mostly misfolded and was present at high levels (47,000 copies out of ~40 million total protein molecules per cell, or ~0.1%) the selective disadvantage was 3.2%. Ideally, a selectable disadvantage might be purged from a population when the disadvantage exceeds the inverse of population size, which in yeast is ~10^7 (i.e. 0.00001% when inverted). The authors extrapolate from 47,000 copies to just one misfolded molecule per cell and predict a fitness disadvantage of 0.00008%, that is to say, 8 times greater than the selection threshold. Thus, relative to less optimal codes, any code that results in one less misfolded protein molecule per cell, or even per ~8 cells, can produce a selective advantage. How many less misfolded molecules arise thanks to a "best of all possible" code or a "one in a million" code is still an open question that awaits a direct experimental link between mistranslation rates and misfolding probability.[1]

[1] Interestingly, the Geiler-Samerotte et al. paper nearly provides this experimental link. They state that "random PCR mutagenesis' was performed to generate mutants of the gratuitous protein. 10 mutations out of 238 amino acids were found to cause misfolding. These mutations were: N23I, E32K, G40V, M78V, K101E, I123V, D155G, V163A, Q183H, and S208P. If we assume that these were the complete set of single amino-acid changes that resulted in perceptible misfolding, then the probability that a wrong amino acid causes perceptable misfolding is 10 out of 4522 (i.e. the 238 amino acids multiplied by the 19 possible wrong amino acids at each position). In their study, "perceptible" misfolding appears to be 10%. Thus, for a typical mistranslation rate of 10^{-4} per codon, ~500 codons per protein, and 4×10^7 total proteins per cell, there are >4400 misfolded proteins per

While Freeland *et al.* reported on how the SCT minimizes the impact of errors, another study found an SCT feature that avoids errors in the first place. In 2001, Lim and Curran modeled the specificity of correct codon-anticodon duplex formation during translation [18]. One of the propositions of their model is that, for ribosomes to reject an incorrect duplex, the incorrect duplex must have at least one uncompensated hydrogen bond. This criteria for rejection is problematic when duplexes have a pair of pyrimidines — U (uracil, the RNA equivalent of T) or C — in the wobble position (i.e. third position in codon, first position in anticodon). Pyrimidine bases are smaller than the G and A purine bases and, if they are in the wobble position, they allow certain mismatches in the *second* position to form non-Watson-Crick pairs thereby compensating their missing hydrogen bonds. These mismatches in the second position then fail to be properly rejected and result in a mistranslation event.

This problem of failed rejection nicely explains why 32 codons in the SCT are in "split boxes", and the other 32 are in "family boxes", i.e. the so called family/non-family symmetry of the SCT (see Figure 1). This explanation begins with the observation that the failed rejection problem can be solved by modifying an anticodon's pyrimidine in the wobble position such that it can no longer form a pyrimidine pair. If pyrimidines are modified in this way, then a single anticodon that could have recognized four codons can now only recognize two codons. In other words, there will now need to be one tRNA for the third position pyrimidines, U and C, and another tRNA for the third position purines, A and G.

Lim and Curran's explanation continues with another observation. If each tRNA could recognize four codons apiece, there only would need to be 16 tRNAs for 64 codons. However, these 16 tRNAs could then only encode 16 amino acids. Life requires 20 amino acids and one termination signal, therefore at least some tRNAs must recognize less than four codons (see Figure 3). Conveniently, Lim and Curran showed that there is already a set of tRNAs that must recognize less than four codons — those that are modified to avoid the failed rejection problem.

The choice of which codon boxes in the SCT should be "split" is thus predetermined by the same stereochemistry that determines which mismatches in the second position fall prey to the failed rejection problem. The codons that are susceptible to failed rejection are those with N_1A_2, U_1 or A_1U_2, and U_1 or A_1G_2 — i.e. exactly the split boxes of Figure 1. The symmetry that is observed in the SCT

cell. Which means that if a genetic code caused a ~0.02% increase in "wrong" amino acids relative to a different genetic code, it would result in one additional misfolded protein and would therefore, by Geiler-Samerotte *et al.*'s argument, experience negative selection. For comparison, a completely randomized code increases "wrong" amino acids >100 times more, relative to the universal code, than this factor of 0.02%.

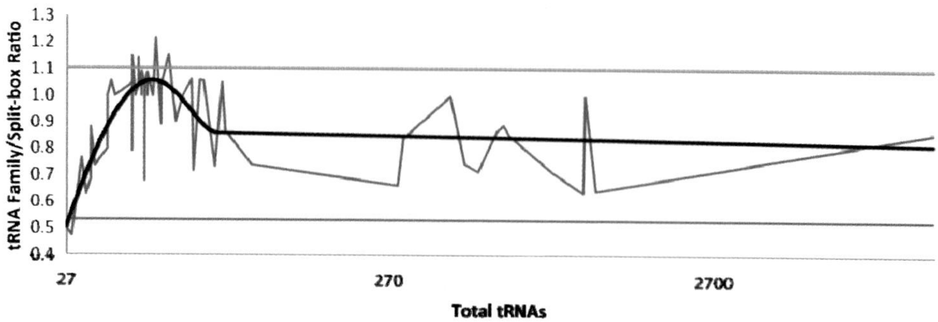

Fig. 3. Family/split-box ratio as a function of total tRNA count (shown in blue, fit with a black line). If each codon had one tRNA, the total tRNA count would be 61 (the three stop codons do not require tRNA) and the family tRNAs to split-box tRNAs ratio would be 32/29 (=1.1, green line). If each amino acid used only one tRNA, the total tRNAs count would be 23 (not 20, since we are double counting arg, leu and ser, as described in the text) and the ratio would be 8/15 (=0.53, purple line). The actual ratio, below 75 total tRNAs, starts at an absolute minimum of 9/18 and climbs to an average that is slightly below 1.1 before settling into an average of 0.85 for organisms with >75 tRNAs (linear fit). The fact that the ratio is below 1.1 for most organisms indicates that tRNA usage is economized via the mechanism described by Lim and Curran (see text).

is not an accident, it is precisely the symmetry one would expect if the SCT was optimized to avoid translation errors, in particular the failed rejection errors due to unmodified pyrimidines in the wobble position.

Itzkovitz and Alon in 2007 described a third remarkable orthogonal advantage of the SCT: the assignments of UAA, UAG, and UGA as stop codons [19]. High frequency codons, such as those coding for aspartic or glutamic acid, can frequently form stop codons if the reading frame shifts. Consequently, translation of a frame-shift error is halted more quickly on average in the real genetic code than in 99.3% of alternative codes, thus saving the cell significant expense. Correlated with this advantage is the SCT's nearly optimal ability to contain secondary signal sequences within the protein-coding sequence, for example, those that encode regulatory and structural protein binding, and splicing sites.

The reason for the correlation between these two advantages is quite simple. Secondary signal sequences are likely to contain all trinucleotide combinations, including UAA, UAG, or UGA, but if any of these three combinations appear as in-frame codons in the protein-coding sequence they will be read as stop codons during translation. However, since, as noted above, UAA, UAG and UGA are frame-shifts of common codons, it is more probable that they can be successfully embedded in the protein-coding sequence. In other words, the first advantage of the SCT (translation of frame shifted sequences stops sooner) leads to the second advantage (secondary signals are embedded more successfully) and vice versa.

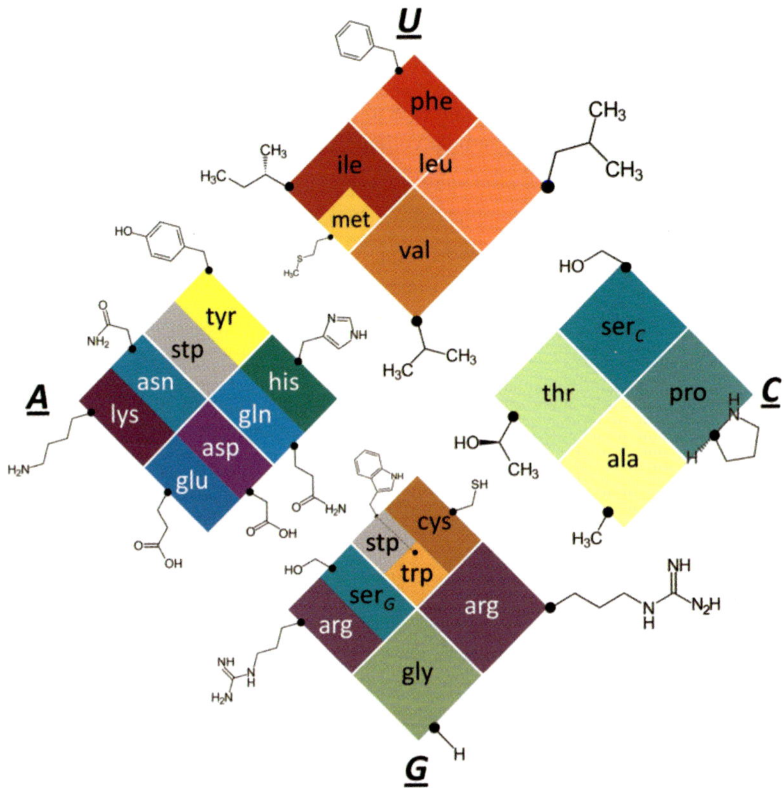

Fig. 4. A new format for displaying the SCT. This version of the new format shows the structure of the 20 amino acid side chains. To identify which trinucleotide codons match which amino acids, follow four steps: 1) Find the quadrant that matches the 2nd base (U=north, G=south, C=east, A=west); 2) Find the square within this quadrant that corresponds to the 1st base (U=north, etc.); 3) Go to the corner of this square that corresponds to the 3rd base (U=north, etc.); 4) Read off amino acid. For example, the AUG codon stands for methionine and has its: 1) second base in the U (north) quadrant 2) first base in the A (west) square 3) third base in the G (south) corner. This new format is useful for showing different patterns in the SCT (see next figure). The rainbow color scheme used here is: most red for most hydrophobic, most blue for most hydrophilic, and grey for the three stop codons. Note, the "family" serine region of the SCT is labeled Ser_C and the "split-box" serine region is labeled Ser_G. Serine is the only amino acid that has codons on the SCT that are not contiguous, i.e. they cannot be connected by single mutations. To go from a Ser_C to a Ser_G codon requires at least two simultaneous mutations.

The fourth orthogonal feature of the SCT is its use of methionine as the initiating amino acid. In 2011 Lim, Curran and Garber devised a novel theory explaining interactions between biomolecules in solution.[2] The lowest barrier to interaction

[2]Lim VI, Curan JF, Garber MB (2012) Hydration shells of molecules in molecular association. A mechanism for biomolecular recognition. J Theo Bio 301:42.

Table 1. Summary of four orthogonally optimized features of the SCT.

Name	Evidence	Extent of optimization
Error impact minimization	Similar amino acids encoded by similar codons	Best possible codes, with restrictions[1]
Error occurrence minimization	Family/ split box symmetry, computer simulation	Specifies symmetry of code[2]
Secondary signal maximization	Stop codons frame shift to common codons	Stop codons vis-a vis common codons
Exit barrier minimization	Initiating methionine has lowest exit barrier	specifies the initiating amino acids

[1]The three restrictions are that all possible codes must have 1) the synonymous codon groupings of the SCT, 2) the stop codons of the SCT, and 3) must not be allowed to change the SCT's groupings of biosynthetically related amino acids.

[2]Placing 32 codons into four-fold synonymous groupings and the other, symmetry-related 32 codons into two-fold synonymous groupings reduces the number of possible codes from 21^{64} ($\sim 10^{84}$) to $21^{8}\times 21^{16}$ ($\sim 10^{31}$) or a 10^{53}-fold optimization.

results from hydrophobic molecules that present one another with the smallest surface area. A quick inspection of Figure 4 shows that lysine and methionine are the longest, unbranched amino acid residues. Of these two, only methionine is also hydrophobic. Indeed when Lim *et al.* calculated which residue had the lowest interaction barrier, methionine was by far the most optimal.

Besides these four orthogonal features (summarized in Table 1), there are additional SCT features that appear to be orthogonally optimized — three that will be given here as examples. First, the SCT uses fewer codons for rarer and more energetically costly amino acids, thus conserving cellular resources, particularly in mitochondria [20]. Second, it has been shown that frame shifts of the coding and non-coding strands of genes (i.e. protein coding DNA) are more likely to translate into folded proteins than frame shifts of non-genes. In other words, the SCT facilitates the encoding of several proteins in a single region of DNA up to a maximum of six: three on one strand and three on the complementary strand [21]. This high compression of protein data occurs naturally in some viruses that, due the small volume of their capsids, must encode their protein data in their DNA genome as efficiently as possible [22]. Third, the SCT ensures that more common amino acids are less prone to change due to a single base mutation relative to less common ones. This keeps the total number of amino acid changes lower. Interestingly, alternate codon tables that ensure this effect on *both* strands of the DNA are extremely rare, and again the SCT is "one in a million" in this respect [23].

These three additional features are reminders that there are undoubtedly more optimal aspects of the SCT that are awaiting discovery. In the next section, two theories for the origin of optimality in the SCT will be compared. The first theory depends on the adaptive selection of earlier codes. The second theory depends on the influence of external intelligence. These theories will be evaluated based on whether they plausibly explain the origin of the SCT's optimality in the absence of metaphysical biases. They will also be evaluated based on whether they are conducive to future discoveries of SCT features.

The Origin of Optimality in the Standard Codon Table

The first section of this paper outlined the four theories for the origin of the SCT: frozen accident, error minimization, biosynthesis, and stereochemistry. The second section examined orthogonally optimal features of the code, without specifying models for their origin. In this section, origins are again discussed, but only the origin of the *optimality* of the SCT is considered. Since frozen accident, biosynthesis and stereochemistry are not optimizing mechanisms and produce optimal features only as a collateral effect, they will not be discussed in this section; rather, the error minimization theory will be examined in more detail and compared to the hypothesis that an external intelligence is responsible for the observed optimal features.

Table 1 lists four orthogonally optimal features and the extent of optimization in the SCT due to each one. At first glance, it may seem that one feature — error impact minimization — completely determines any and all optimization in the SCT, since using the error impact criterion alone the SCT was shown to be the most optimal of all possible codes [15]. However, there are three important restrictions placed on the possible codes to which the SCT is compared. First, these other codes must match the SCT in terms of synonymous codons, i.e. the other codes will have the same grey and black boxes shown in Figure 1, but with different amino acids in each box. Second, the other codes must match the SCT in terms of their stop codons, i.e. they all use UAA, UAG, and UGA as stop codons. Third, to construct an alternate code, amino acids cannot swap their positions in Figure 1 with *all* others but only biosynthetically related ones. The four groups of related amino acids used to construct the alternate codes were: 1) Phe, Ser, Tyr, Cys, Trp; 2) Leu, Pro, His, Gln, Arg; 3) Ile, Met, Tyr, Asn, Lys; and 4) Val, Ala, Asp, Glu, Gly.

The SCT is the best of all possible codes within a *specific subset* of possible codes. If one of the three restrictions is relaxed, the SCT is no longer the best of all. For example, the prior work of Freeland *et al.* did not include the third restriction; as a result they found one alternative codon table out of one million attempts that outperformed the SCT in terms of error impact minimization [13]. Interestingly, the other two restrictions are at least partially set by optimal features discussed earlier (Table 1). Error occurrence minimization [18] partially sets the first restriction—matching synonymous codon boxes — and secondary signal maximization [19] roughly sets the second restriction — UAA, UAG, and UGA stop codons. With two of three restrictions in place, to a first approximation the SCT appears to be at least a "one in a million" code.

The question at this point is: What is the mechanism for the SCT's optimization? It is useful to consider three hypotheses — law, chance, and intelligence [24]. In other words, is the optimization best explained by a predictable, law-like process, by random chance, or by intelligent causation? To distinguish between these choices, it is useful to evaluate them sequentially, beginning with law-like processes. If no law-like processes explain the effect, the probability that chance processes should be considered. Finally, if chance is ruled out based on low probabilities relative to the available time and opportunities, then intelligent causation is by default the best explanation for the effect.

Is there a law that can explain the SCT optimization? Several papers have considered this possibility [4, 11, 25]. For example, if there were primordial organisms that all used different codon tables and if these organisms competed such that only the most fit lineage survived, then by the law-like process of natural selection this lineage would become the last universal common ancestor (LUCA) and its codon table would become the standard for all of life.

Competition between separate lineages with different codes is deemed more likely than a changing code over time within a lineage, where each changed code would need to be backward compatible to the genetic messages of the previous code [2]. Yet despite being more likely, many publications have argued that the laws of competition between lineages cannot explain the SCT's optimization [6, 10, 16, 26–30]. The problem is that if the SCT is "one in a million" there must be a million competing genetic codes in the population of primordial organisms. This problem becomes worse when the optimization of the SCT approaches the "best of all possible codes". In that case, the population of competing codes would need to approach 10^{84} — a ludicrous population size, considering that 10^{84} carbon atoms are a trillion, trillion, trillion times more massive than the earth.

Is chance, then, a reasonable explanation for the SCT's optimization? In 2007 Eugene Koonin invoked the chance hypothesis to explain the complexity of a "translation-replication" system, which would include the SCT, translation components such as shown in Figure 2, and a host of other translation and replication machines [12]. How could a chance occurrence possibly explain even more complexity and optimization than the SCT alone? Koonin's answer is that, if our universe is but one of many in an infinite multiverse, "emergence of highly complex systems by chance is inevitable".

Koonin was criticized by Eric Bapteste in the open access reviewers' comments that accompanied this paper for using a metaphysical argument that "could open a huge door to the tenants of intelligent design". An appeal to an infinite multiverse, which has never been nor can ever be observed, is a poor way to rescue the chance hypothesis from overwhelmingly low probabilities. Better to rule out the chance hypothesis and proceed to the next hypothesis, for even if the particular intelligence responsible for a low probability effect is not known, the general pattern of intelligence producing finely-tuned, optimized effects is well-known and well-studied.

Design is not controversial, but a designer is. All scientists admit that aspects of the universe — and biological systems in particular — conform to various designs that achieve various functions. Yet most scientists reject the possibility that an external intelligence, i.e. a designer, is responsible for the observed design.

There is a persistent, pervasive bias against the design hypothesis, which ensures that even if law and chance fail to explain a biological effect (e.g. the optimization of the SCT), external intelligence will never be considered as an option. However, once this bias is removed, the external intelligence hypothesis becomes the best working hypothesis. Therefore, it should be considered the most viable explanation until a natural mechanism can be found that explains the degree of SCT optimization, or until new data show that the current assessment of optimization is grossly overestimated.

A lingering question is: Why this bias against external intelligence? Possibly, scientists worry that explaining some natural effects via an intelligent force will encourage *all* effects to be explained in this way, thereby dooming the scientific method. This is a reasonable concern. The final section of this paper, therefore, examines the benefits of using external intelligence as a working hypothesis in the specific case of SCT optimization.

Using the Hypothesis of External Intelligence to Guide Discovery

Before the discovery of the SCT in the early 1960's, many researchers assumed that the code would be optimal in some respect. For example, the "diamond" code proposed by George Gamow in 1954 was optimal in its information storage [31]. A chain of N amino acids could be coded by a chain of N+2 mRNA letters, whereas, in the real SCT, N amino acids are specified by 3N mRNA letters. Another pre-SCT code, proposed by Crick, Griffith and Orgel in 1957, was "comma free" and optimal for avoiding frame shifts [32]. Still other codes had interesting mechanisms for automatically correcting errors in translation [33].

With the discovery of the real SCT (see Figures 4 and 5 for a format that is slightly different than Figure 1), two features were immediately recognized: the SCT lacked the host of “nonsense” codons that were required in the comma free codes, and the SCT assigned similar codons to similar amino acids [34–36]. The first feature implied that the physical machinery of the genetic code (e.g. Figure 2) had to be vastly more complex — or more of a random accident — than originally envisioned. The second feature revealed a new type of optimization that was not anticipated, and, surprisingly, was not readily accepted as an optimization. The majority of publications for 30 years seemed intent on explaining *away* this optimization and interpreting the lack of nonsense codons as evidence of randomness rather than complexity (see [37]) of the biosynthetic SCT origin theory via codon expansion, also called “codon capture”, where biosynthetically related amino acids capture the codons of amino acids that are already being used in the SCT [38]. In this theory, physiochemical similarities, not biosynthetic pathways, determined how similar codons were assigned to groups of amino acids.

Would SCT research have taken a different tack if external intelligence was considered as its possible source? Would it have taken over 30 years to demonstrate that the obvious pattern of similar amino acids in similar codons confers an impressive level of error impact minimization?

Would other features — secondary signal encoding and error occurrence minimization — have been discovered earlier?

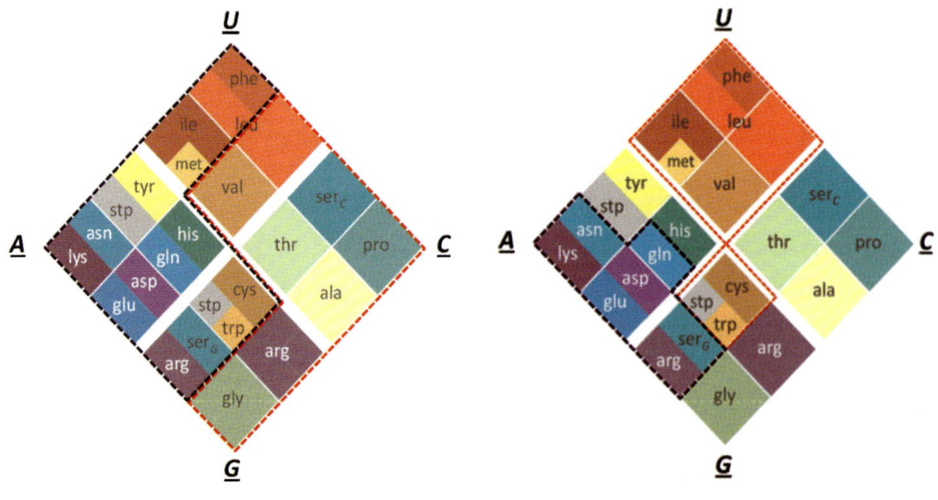

Fig. 5. Example patterns in the standard codon table. Left: Family/split-box symmetry. Right: Hydrophobic (red dashed lines) and hydrophilic (blue dashed lines) amino acids; rankings are the average of five commonly used indices.

At least two papers in the late 1960s suggested that the observed pattern was real optimization and not an artifact of biosynthesis or codon expansion [37, 39]. However, only one of these took an experimental approach and actually tested the SCT against other possible codes, showing that it was more optimal than a random code [37]. This study from 1969 was only cited three times in the 1970s, but gained citations as interest in the optimization of the SCT grew in the late 1980s and into the 1990s. By the time Freeland and Hurst published their "one in a million" paper in 1998, discussion of error impact minimization in the SCT was in full swing.

It is impossible to state unequivocally that optimized features in the SCT would have been discovered and discussed more rapidly in the absence of a bias against external intelligence. However, it is instructive to look at an example from archeology, where external intelligence — i.e. human intelligence — is assumed to account for many features. The Rosetta Stone's discovery in 1799 sparked widespread global interest [40]. Copies were circulated to museums, and each new observation that brought scholars closer to cracking the hieroglyphs was heralded across Europe.

Contrast this scene with the discovery of the SCT. Certainly there was widespread interest, though perhaps shorter lived; an article published three years after the SCT's discovery bore the title "The Genetic Code after the excitement" [41].

The main difference was that the features in the SCT that we now know to be highly optimized were noticed immediately but explained away. Would the discovery today of an intergalactic Rosetta Stone, with the potential to decipher an extraterrestrial language be explained away as an artifact? Certainly not. The bias for or against external intelligence makes all the difference.

There are more features of the SCT that merit examination. Does the proximity in the SCT of biosynthetically related amino acids merely reflect its historical evolution or could this, too, be an optimized feature? Is it significant that the SCT's stop codons would have the weakest codon-anticodon interactions? These and other features will surely be investigated, but the speed at which they will be studied would accelerate if researchers considered the SCT a possible product of external intelligence, with optimized, carefully-engineered features awaiting discovery.

Conclusion

The SCT is by no means the most complex piece of the biological world. On the contrary, its relative simplicity is the reason it has been examined in this paper. Since it is an arrangement of 20 amino acids (and the signal for "stop polymerizing amino acids") with known properties onto 64 trinucleotides with known properties, it is an ideal test case to examine orthogonal optimized features and to apply the filter of law, chance, and intelligence. If the optimization of the SCT lies between "one in a million" and "the best of all possible codes" as is likely to be the case, the law and chance hypotheses are increasingly untenable and external intelligence becomes the most promising working hypothesis. As new orthogonally optimized features are discovered, the explanatory divide between law and chance on one hand and intelligence on the other becomes more pronounced.

References

1. Crick FHC, Brenner S et al (1976) A speculation on the origin of protein synthesis. Origins of Life and Evolution of Biospheres 7(4): 389–397.
2. Crick FH (1968) The origin of the genetic code. J Mol Biol 38(3): 367–379.
3. Berleant D, White M et al (2009). The genetic code — more than just a table. Cell Biochem Biophys 55(2): 107–116.
4. Knight RD, Freeland SJ et al (1999) Selection, history and chemistry: the three faces of the genetic code. Trends Biochem Sci 24(6): 241–247.
5. Freeland SJ, Wu T et al (2003) The case for an error minimizing standard genetic code. Orig Life Evol Biosph 33(4–5): 457–477.
6. Wong JT (2005) Coevolution theory of the genetic code at age thirty. Bioessays 27(4): 416–425.
7. Lehninger AL, Nelson DL et al (2000) Lehninger principles of biochemistry. Worth Publishers, New York.
8. Wong JT (2007) Question 6: coevolution theory of the genetic code: a proven theory. Orig Life Evol Biosph 37(4–5): 403–408.
9. Woese CR, Dugre DH, et al (1966) On the fundamental nature and evolution of the genetic code. Cold Spring Harb Symp Quant Biol 31: 723–736.
10. Yarus M, Caporaso JG et al (2005) Origins of the genetic code: the escaped triplet theory. Annu Rev Biochem 74: 179–198.
11. Wolf YI, Koonin EV (2007) On the origin of the translation system and the genetic code in the RNA world by means of natural selection, exaptation and subfunctionalization. Biol Direct 2: 14.
12. Koonin EV (2007) The cosmological model of eternal inflation and the transition from chance to biological evolution in the history of life. Biol Direct 2:15.
13. Freeland SJ, Hurst LD (1998) The genetic code is one in a million. J Mol Evol 47(3): 238–248.
14. Freeland SJ, Knight RD et al (2000). Measuring adaptation within the genetic code. Trends Biochem Sci 25(2): 44–45.

15. Freeland SJ, Knight RD et al (2000) Early fixation of an optimal genetic code. Mol Biol Evol 17(4): 511–518.
16. Ronneberg TA, Landweber LF et al (2000) Testing a biosynthesis theory of the genetic code: fact or artifact? Proc Natl Acad Sci USA 97(25): 13690–13695.
17. Geiler-Samerotte KA, et al (2011) Misfolded proteins impose a dosage-dependent fitness cost and trigger a cytosolic unfolded protein response in yeast. Proc Natl Acad Sci USA 108(2): 680–5.
18. Lim VI, Curran JF (2001) Analysis of codon: anticodon interactions within the ribosome provides new insights into codon reading and the genetic code structure. RNA 7(7): 942–957.
19. Itzkovitz S, Alon U (2007) The genetic code is nearly optimal for allowing additional information within protein-coding sequences. Genome Res 17(4): 405- 412.
20. Swire J, Judson OP, et al (2005) Mitochondrial genetic codes evolve to match amino acid requirements of proteins. J Mol Evol 60(1): 128–139.
21. Chirico N, Vianelli A et al (2010) Why genes overlap in viruses. Proc Biol Sci 277(1701): 3809–3817.
22. Barrell BG, Air GM et al (1976) Overlapping genes bacteriophage phiX174. Nature 264(5581): 34–41.
23. Bubak M, van Albada G et al (2008). Optimization of asymmetric mutational pressure and selection pressure around the universal genetic code. In: Computational Science-ICCS 2008. Springer, Berlin/ Heidelberg.
24. Dembski WA (1998) The design inference: eliminating chance through small probabilities. Cambridge University Press, Cambridge/New York.
25. Koonin EV, Novozhilov AS (2008) Origin and evolution of the genetic code: The universal enigma. IUBMB Life.
26. Kobayashi K, Furuta H et al (1989) [RNA world, a key step in the origin of life: inspection from the view point of chemical evolution]. Tanpakushitsu Kakusan Koso 34(2): 124–137.
27. Lazcano A, Miller SL (1996) The origin and early evolution of life: prebiotic chemistry, the pre-RNA world, and time. Cell 85(6): 793–798.
28. Soll D, RajBhandary UL (2006) The genetic code-thawing the ‘frozen accident’. J Biosci 31(4): 459–463.
29. Stoltzfus A, Yampolsky LY (2007) Amino acid exchangeability and the adaptive code hypothesis. J Mol Evol 65(4): 456–462.
30. Carter CW Jr (2008) Whence the genetic code? Thawing the ‘frozen accident’. Heredity 100(4): 339–340.
31. Gamow G (1954) Possible relation between deoxyribonucleic acid and protein structures. Nature 173(4398): 318.
32. Crick FHC, Griffith JS et al (1957) CODES /WITHOUT COMMAS. Proc Natl Acad Sci USA 43(5): 416–421.

33. Golomb S, Davey J et al (1963). Synchronization. Comm Sys, IEEE Trans 11(4): 481–491.
34. Matthaei H, Nirenberg MW (1961) The dependence of cell-free protein synthesis in E. coli upon RNA prepared from ribosomes. Biochem Biophys Res Commun 4: 404–408.
35. Nirenberg MW, Matthaei JH (1961) The dependence of cell-free protein synthesis in E. coli upon naturally occurring or synthetic polyribonucleotides. Proc Natl Acad Sci USA 47: 1588–1602.
36. Nirenberg M, Leder P (1964) RNA codewords and protein synthesis. The effect of trinucleotides upon the binding of sRNA to ribosomes. Science 145: 1399–1407.
37. Alff-Steinberger C (1969). The genetic code and error transmission. Proc Natl Acad Sci USA 64(2): 584–591.
38. Wu H-L, Bagby S (2005) Evolution of the genetic triplet code via two types of doublet codons. J Mol Evol 61(1): 54–64.
39. Goldberg AL, Wittes RE (1966) Genetic Code: aspects of organization. Science 153(3734): 420–424.
40. Parkinson RB (1999) Cracking the code: the Rosetta Stone and the art of decipherment. British Museum, London.
41. Sadgopal A (1968) The genetic code after the excitement. Adv Genet 14: 325–404.

A New Model of Intracellular Communication Based on Coherent, High-Frequency Vibrations in Biomolecules

L. Dent

Pepperdine University/NASC, 24255 Pacific Coast Highway, Malibu, CA 90263–4321, USA. laurieanne.dent@pepperdine.edu

Abstract

Chemistry has been the ruling paradigm for understanding the communication network that integrates a living cell. However, biochemistry alone is insufficient to explain how widely-separated biomolecules locate and move toward one another with accuracy and speed. We propose a new model wherein cytoplasmic motion is vibrationally-directed due to a community of oscillating biomolecules. DNA vibrations have been predicted in the 2-GHz range, thus we used high-frequency laser-Doppler vibrometry to test the hypothesis that resonance-driven molecular motion would be detectable as picometer surface displacements in live onion epidermal cells and fish eggs but would be absent in dead cells. Although, no surface vibrations were detected under these conditions, we discuss implications for the vibrational model of intracellular communication and suggest future experiments.

Key words: cellular communication networks, intracellular signaling, DNA vibrations, biomolecular resonance, biological oscillators

Introduction

Cells are constantly processing information from their external and internal environments in order to function properly. Information about the status of energy sources, cell-specific functions, and the condition of the genome must be communicated continuously. To give a few examples: a) the presence of a carbon source, lactose, in the environment induces expression of bacterial *lac* operon genes, efficiently optimizing the biochemistry of the cell for lactose utilization; b) reciprocal signaling between adjacent cells expressing *wingless* and *hedgehog* genes maintains precise segmental boundaries in the developing fruit fly; and c) the p53 DNA repair pathway is activated when chromosomal damage has been detected. These astonishingly complex cellular communication systems, constituted of biochemical pathways and signaling cascades and involving interactions between myriads of biomolecules, have been described only in small part by biochemists and molecular biologists.

So complicated are subcellular processes, that in recent years computer animators have been employed to help us visualize these sophisticated molecular machines and processes at work. When watching an animation of, say, gene expression, we see biomolecules "flying around", apparently guided to their targets; protein transcription factors glide to their specific DNA-binding sequences, mRNA transcripts seem to be directed through nuclear pores toward ribosomes for translation, tRNAs land on a ribosome and perfectly align with mRNA to make the anticodon-codon hybrid. What we cannot appreciate from these animations are the blinding speeds at which these processes occur. DNA replication occurs at a rate of 50 nucleotides per second in humans, and *Escherichia coli* bacteria can add 40 amino acids per second during protein synthesis. Animators have greatly reduced the speed of these systems so the motion can be apprehended by the human eye. How, then, do these biomolecules locate each other with such accuracy and rapidity?

Classically, Brownian motion has been invoked as the mechanism in cellular biochemistry wherein two biomolecules perform a "random walk" through the cell, and then by chance, collide at just the right orientation to allow a chemical reaction [reviewed in 1]. This may happen in cases where enzymes and substrates are at high density and close proximity — such as in a typical bacterium, where each soluble enzyme contacts every other enzyme and substrate once every second — ,but it is not sufficient to explain cases where large molecules must find each other, starting from relatively great distances. Illustrations of this problem would include the precise synapsis of homologous chromosomes during meiosis I, or the *trans*-acting factors (proteins that initiate transcription) which must locate a specific DNA sequence on a specific chromosome amid billions of base pairs during gene regulation. Brownian motion simply does not appear to be adequate to overcome the "crowded-cell problem" in terms of the need for vast numbers of macromolecules to find their distant targets quickly. These well-established biochemical models of cellular communication do not adequately consider the localization and transport information that is required for all the component parts to find each other and react in assembly line fashion. For example, forty years ago, it was shown that the targeting of the *lac* repressor to its DNA-binding site occurred up to 1,000 times faster than the predictions of diffusion and random collision [2]. This finding spawned what has become a very large research effort in structural and molecular biology focused on discovery of protein-nucleic acid target-search mechanisms. These studies have yielded hypothetical "one-dimensional diffusion" mechanisms of protein hopping, sliding, and intersegmental transfer. In their excellent review of experiments investigating these mechanisms, Gorman and Green [3] conclude,

> Importantly, none of the published studies where one-dimensional diffusion was visualized used DNA substrates that actually contained specific target sites for the

proteins being studied, so diffusion and target binding still remain to be seen in the same reaction trajectory

Thus, a chronic challenge in diffusion and macromolecular crowding research is the difficulty of interpreting data from experimental or theoretical studies that strain to approximate the complex intracellular environment [1, 4, 5].

While the role of cytoskeletal trafficking of biomolecules within the cytoplasmic [*Drosophila* axis patterning: 6, 7, 8; signal transduction pathways in yeast: 9; hyphal polarity in fungi:10; review: 11] and nucleoplasmic [plant chromatin remodeling: 12; interphase chromosome repositioning: 13, 14; Cajal bodies and U2 snRNA gene:15; nuclear rearrangement and transcription enhancement: 16, 17] compartments is an integral principle, cytoskeletal mechanisms cannot account for all instances of biomolecular transport. Curiously, filamentous actin (the cytoplasmic type) is not found in the nucleus whereas many actin-binding proteins (ABPs) are. The ABPs of the nucleus accomplish chromatin remodeling via nucleosome and histone interactions. When actin is translocated to the nucleus, it appears to be facilitated by cofilin [18]. These beg the question of how actin and non-filamentous actin-related proteins themselves find their specific nuclear targets. An as yet undiscovered mechanism must be at work in marshaling distant biomolecules involved in coordinated cellular functions.

To address the problem of how biomolecules might find each other apart from simple Brownian motion, we have developed the following hypothetical model. We propose that the molecular motion in the cytoplasm is not truly random, but is vibrationally-directed and coherent due to a community of oscillator structures within chromosomes and proteins, within a narrow distribution of resonant frequencies. We predict that specific nucleotide sequences will vibrate at characteristic resonances, and that these are closely matched to the inherent oscillation frequencies of the α-helices in functionally-linked proteins (e.g., transcription factors). Such harmonic interaction might facilitate the mutual identification and attraction of protein-DNA binding. These vibrationally-coupled "communication channels" may then synchronize the resonant motifs within other biomolecules, perhaps establishing oscillations across a family of harmonic frequencies, with the DNA molecule vibrating at the fundamental frequency; and in so doing, attract biomolecules to one another with great specificity while providing an essential cell "lubricant" to free cellular molecules from the "stickiness" associated with the cytoskeleton and crowded cytoplasm so that molecules can find each other with greater rapidity. This novel hypothesis may help us to understand how molecules might interact from a distance, and if correct, would reveal an entirely new level of biological information.

Vibrations in DNA molecules and proteins have been known for more than twenty years. Vibrational modes in DNA and proteins have been predicted theoretically [19–23] and measured experimentally using Raman spectroscopy techniques [24–28]. One of the principal investigators, studying theoretical models of DNA vibrations in the microwave range, was K. C. Chou who predicted an ultra-high frequency vibrational mode in DNA, around 2 GHz [23].

In this light, we felt our model was potential useful and should be tested. Since DNA vibrations in the gigahertz range have been predicted [23] and eukaryotic nuclei have a high DNA content, it seemed reasonable to begin looking for ultra-high frequency vibrations in the vicinity of a cell nucleus. These collective vibrations may be transmitted to the cell surface and detectable as ultra-high frequency displacements. In this paper, we present preliminary experiments aimed at detecting such coherent molecular motion within living cells, which should be absent from dead cells, in onion cells and fish eggs using ultra-high frequency laser-Doppler vibrometry.

Materials and Methods

Cellular material

We prepared plant and animal cell specimens in order to investigate the possible presence of high frequency vibrations at the cell surface. Onion epidermal cells were selected as representative plant cells because of the following attractive features: size, proximity of nucleus to cell surface, and ease of preparation. Cells are large (approximate length = 100 μm) and form a flat monolayer which can be teased easily from an onion scale by use of fine forceps. Also, the nucleus is relatively large (approximately 10-μm diameter), larger than some eukaryotic cell types. Owing to the size of the nucleus, it easily visible under low magnification and lies near the cell surface which consists of a plasma membrane covered by a cell wall. Live cells were obtained from freshly-harvested green onions. We determined whether cells were living by observing cytoplasmic streaming under light microscopy.

Animal cells were obtained from freshly killed female jacksmelt fish which had been caught the same day by fishermen at Newport Beach, California. We manually expelled roe from two gravid females, and hundreds of unfertilized fish eggs were available for immediate analysis. Eggs were spread into a single layer in a plastic petri dish and assumed to be viable based on the rapid collection protocol.

Laser-Doppler vibrometry

A 0.5-cm^2 section of live, green onion epidermis was excised, and then flattened onto a dry microscope slide with the waxy surface facing up. The specimen was positioned on the stage of the vibration-isolated workstation of a Micro System Analyzer (MSA-500-TPM2–20-D, Polytec, Inc., Irvine, California, USA) which combined microscopy with scanning laser-Doppler vibrometry for detection of surface vibrational signals across a large bandwidth of frequencies.

First, a living onion cell was located under 50X magnification via the live video stream capabilities of the MSA Optical Unit, and then a 1-μm laser spot was focused over the nucleus. We acquired cell surface displacement data over two frequency ranges, 0–20 kHz and 30 kHz–24 MHz. For the first range, the surface velocity was measured and then converted to displacement using the Polytec vibrometer software. For the latter frequency range, the surface displacement was measured directly. Data acquisition as well as conversion of the raw data into the frequency domain, using a Fast Fourier Transform (FFT), was performed within the Polytec vibrometer software. We utilized an extremely broadband approach because we hypothesized that the supercoiling of DNA molecules may lower the functional frequencies, although the work of Chou and his colleagues predicted DNA vibrations in the ultra-high frequency range. In probing the cell surface for the presence of vibratory signals, both single-point measurement and scanning routines were used. In the latter, the optical unit was programmed to analyze several points across a two-dimensional array of the cell surface above the nucleus, and then the beam collected velocity/displacement measurements at each point according to this pre-programmed routine. Off-line data analyses were performed in MATLAB (v. 7.10.0.499, R2010a, The MathWorks, Inc.).

For comparison with live cells, we continued to collect measurements from cells that showed no cytoplasmic streaming after having been probed for several minutes with the beam; these were presumed to be dead due to damage from the laser, although no defects could be seen in the vicinity of the laser spot. Also, we took measurements on other varieties of onions, red and white, that did not show cytoplasmic streaming; however, we could not ascertain whether the cells were dormant or dead.

On the basis of Chou's theoretical modeling of DNA vibrational modes in the gigahertz range, onion cells were also examined for the presence of ultra-high frequency surface vibratory signals, up to 1.2 GHz, using the UHF-120 Ultra High Frequency Vibrometer (Polytec, Inc., Irvine, CA, USA). As in the case of the MSA-500 for the frequency range of 30 kHz to 24 MHz, the UHF-120 also measures surface displacement directly. Experimental protocols and data analysis similar to those used with the MSA-500 system were carried out for ultra-high frequencies.

We repeated these tests on fish eggs, employing the same measurement protocols described above. Here, the chief difficulty was determining the health of the cells; we did not have an assay for live versus dead animal cells.

Both the MSA-500 and the UHF-120 are laser Doppler vibrometers, which are precision non-contact optical transducers used for detecting vibration velocity and displacement at a fixed location. The technology is based on the Doppler Effect, sensing the frequency shift of the back scattered laser light. The surface velocity is determined using the following relationship:

$$v = \frac{f_D * \lambda}{2}$$

v is the surface velocity of the object at the location of the laser spot, f_D is the Doppler shift in frequency and λ is the wavelength of the laser light.

Results

We tested the prediction that ultra-high frequency vibrations emanating from the nucleus of live onion epidermal cells and fish eggs would be detectable at the cell surface, whereas in dead cells, no surface vibrations would be present.

Cell surface vibrations are not detectable in onion cells

Highly-sensitive laser-Doppler vibrometry capable of detecting frequencies up to 1.2 GHz with picometer displacement resolution did not reveal a passive vibratory signal at the surface of an onion epidermal cell (Fig. 1a-c). The only peaks present in the lower frequency bandwidth, 0–40 kHz (Fig. 1a), were noise components from the electronic circuitry. For example, peaks below 2.5 kHz included 60 Hz plus harmonics associated with the power grid; also, the two peaks around 24 and 25 kHz (arrow) were characteristic laser resonances from its power supply. We are confident that any nucleus-originating vibratory signal propagated to the cell surface would have been detectable with this instrumentation. It is generally accepted that plasma membranes are 7–10 nm-thick, so with a broadband noise floor in the order of a few picometers, any peaks of biological origin would have been apparent. Similarly, Figure 1b shows no cell surface vibrations across a frequency bandwidth of 30 kHz to 20 MHz. Here, the noise floor encompasses about 350 pm. The tailing off of signal magnitude observed at the upper ends of the frequency spectra is a function of filter roll-off (Fig. 1a and b, asterisks). Finally, no ultra-high frequency vibrations up to 1.2 GHz were present at the cell surface (Fig. 1c); the noise floor was about 20 pm. Resonances of biological origin characteristically produce broad peaks or "humps" in a magnitude-frequency plot.

Here, only "lines" are present, spikes of energy at a single frequency, typical of laser resonances (coherent light) and electronic artifacts.

Measurements collected from epidermal cells in other varieties of onions, red and white, showed a similar lack of cell surface vibrations (data not shown).

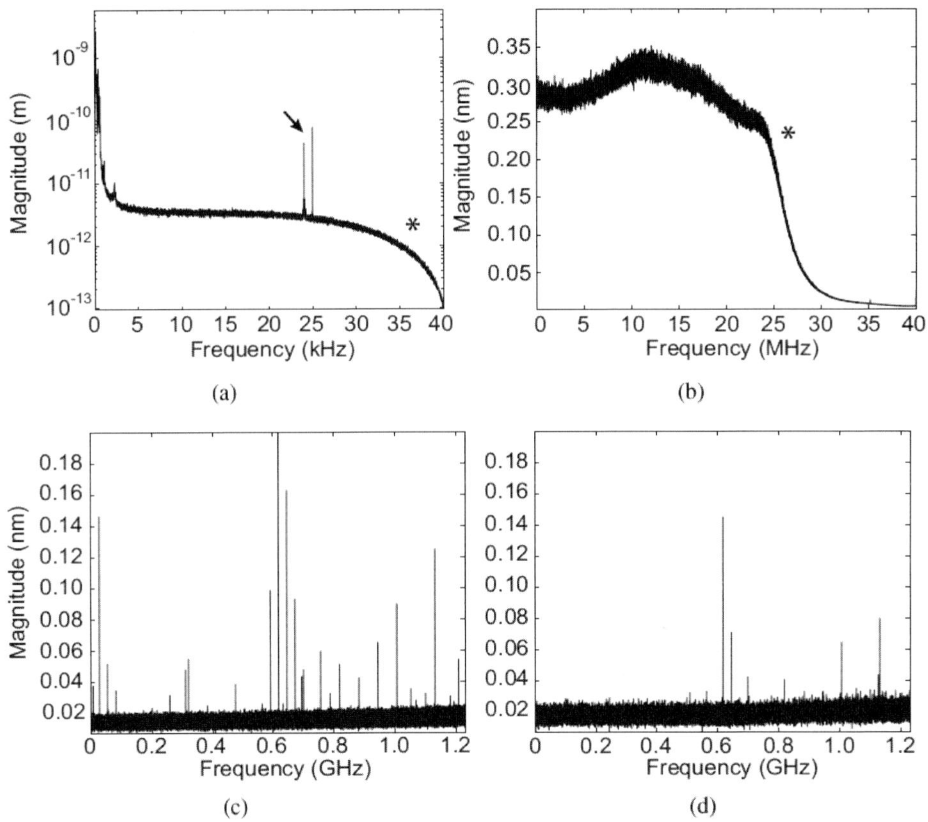

Fig. 1. Broadband frequency analyses performed by laser-Doppler vibrometry show no cell surface vibrations in onion epidermal cells or fish eggs. (a) Surface displacement magnitude as a function of lower frequencies, 0–40 kHz, is shown for a region over the nucleus of a live green onion. Low-frequency peaks, <2.5 kHz, are non-biological and represent electronic noise (e.g., fundamental and harmonics of 60 Hz contribution from electrical power grid). The log-transformed y-axis reveals two prominent peaks around 24–25 kHz (arrow) known to represent the laser power supply. The upper bandwidth limit of the decoder is evident in the filter roll-off response at higher frequencies (asterisk). (b) The same specimen in 'a', analyzed for higher frequency vibrations across a range of 30 kHz–20 MHz, produces no signal above the noise; and then, again, (c) for ultra-high frequencies, up to 1.2 GHz, where peaks present are not of biological origin. (d) Also, no peaks of biological origin are present in a fish egg analyzed for ultra-high frequency surface vibrations. It is important to keep in mind that all the vibration signals represented in the plots (a) through (d) are well below 1 nm, which is very small.

Ultra-high frequency cell surface vibrations are not detectable in fish eggs

We used ultra-high frequency laser-Doppler vibrometry to probe for cell plasma membrane vibrations on unfertilized fish eggs for comparison with onion cells which have a cell wall. No resonance peaks of biological origin were present in frequency analyses up to 1.2 GHz (Fig. 1d; similar to experimental parameters of Fig. 1c); only artifactual resonance lines appeared above the approximately 15-pm noise floor. We conservatively suggest that these data were collected from viable oocytes, since only two hours had lapsed between collection of fish at the pier and analysis in the laboratory.

Discussion

In this study, we are seeking to test one hypothesis that arises from a new model of a communication network that may integrate a living cell. Our model proposes that nuclear-originating, broadband vibrational frequencies elicit sympathetic vibrations in functionally-related biomolecules and order the molecular motion of the cytoplasm. To our knowledge, this study is the first test of the vibrational model of intracellular communication. We tested the specific prediction that cell surface vibrations will be present as ultra-high frequencies due to the propagation of coherent molecular motion, especially emanating from the nucleus; furthermore, we predicted that vibrations would be present in live cells but absent in dead cells. Although we were unable to detect vibrations on the surface of living cells across a broad frequency range using the highest-precision, most-sensitive instrumentation available, we propose the following causes may have prevented signal detection, including (i) constraints of cell architecture, (ii) heat damage during laser measurements, and (iii) the limitation of detecting frequencies above 1.2 GHz.

Plant cells are surrounded by a rigid cell wall. This thick, inflexible structure may have damped any high frequency vibrations that may have propagated to the perimeter of the cell. However, animal cells do not have a cell wall, and we were not able to detect any surface vibrations on the fish eggs. It is possible that nucleus-originating vibrations are not reaching the cell's plasma membrane. The filtering properties of the nuclear envelope in response to compressional waves are not known. Its double-thick phospholipid bilayer, separated by a space of 20–40 nm, may be sufficient to restrict intranuclear vibrations to the nuclear compartment. Even if some vibratory energy were transmitted to the cytoplasm, the filtered signal may attenuate rapidly, especially as frequency increases. One might imagine that the intranuclear environment — a relatively small space packed with nucleic acids and other biomolecules — requires a communication system which utilizes ultra-high frequency carrier signals for high-energy, short-range signaling. If, therefore, DNA is oscillating at frequencies in the gigahertz range, it may not be possible, due to the short reach of the vibrational energy, to detect the signal at the cell surface, whether plant or animal.

We also considered that the cells may have sustained heat damage from the laser beam during scans, denaturing DNA and proteins and, thus, disrupting normal cell functions. For this reason, we were careful to check for cytoplasmic streaming after each point scan of onion epidermis to confirm the health of the cell. However, following one extended multi-point scanning routine, no cytoplasmic streaming was observed, and we inferred the cell had sustained heat damage from the laser measurement. The fish eggs, without the additional protection of a cell wall, may be even more liable to laser damage. In both laser-Doppler vibrometry systems, the MSA-500 and the UHF-120, mechanisms are in place to minimize exposure of the specimen to the laser; the laser power can be attenuated manually (MSA-500), or a built-in gating function dims the laser when measurements are not being taken (UHF-120).

Detecting out-of-plane vibrations of 1.2 GHz is at the limit of cutting-edge laser-Doppler vibrometry, however, it is unlikely to be "good enough." On the basis of the predictions of Zhang and Chou [23], though, 2-GHz vibrational modes in DNA would, indeed, be beyond the detection capabilities of the instruments used in these experiments.

Of these possible explanations for the absence of cell surface vibrations in our experiments, I believe the most likely is the compartmental organization of the cell. Suppose the mechanism of intracellular communication includes an intranuclear communication network consisting of ultra-high frequencies generated by DNA. One might expect the vibratory frequency to be predictably related to DNA nucleotide sequence; possibly, prominent resonances could develop most easily across sequences of tandem DNA repeats. It is interesting that more than 50% of the human genome consists of repetitive DNA. Much of this repetitive DNA is located in the centromeres of chromosomes (which facilitate proper segregation of replicated chromosomes during mitosis). Centromeric DNA is characterized by repetitive, simple, non-coding sequences called "satellite" DNA; one type in particular, α-satellite DNA, consists of a 171 bp-long repeating unit, and thousands of tandem arrays may stretch over one million bases of a chromosome. The higher-order molecular structure of centromeric DNA has been difficult to study, however it may be similar to non-centromeric DNA structure which has been compared to a "solenoid" — 160 bp of DNA wrapped twice around a histone octamer core (nucleosome) which is further coiled into a superhelix that contains six nucleosomes per turn [29]. (The assumption of higher-order supercoiling in centromeric DNA motivated our search for cell surface vibrations in the megahertz frequency range — larger coils should affect downward frequency modulations.) It may be that the three-dimensional architecture of a centromere fundamentally based on tandem repeats of DNA generates a standing oscillation which could act as a vibratory signal, possibly ultra-high frequency, to other biomolecules within the nuclear compartment.

It would seem, then, that the operation of the vibrational model may be limited to the intranuclear compartment. How might other biomolecules in the nucleus receive this signal? The first three-dimensional structure of a biomacromolecule to be solved was the α-helix in proteins in 1948 by Linus Pauling who later won a Nobel Prize for his discovery. These α-helices may function as resonance structures within proteins, something like an antenna. Notably, many proteins that interact with specific DNA sequences have multiple α-helix domains; for example, the α-helices of the p53 tetramer (modulates the cell cycle by controlling expression of DNA repair proteins) are closely associated with the DNA helix in predicted three-dimentional models of the complex. It may be that the α-helices of DNA-binding proteins have characteristic resonances that are related fundamentally or by harmonics to the resonance of its specific DNA sequence. These functionally-related biomolecules may oscillate within a narrow bandwidth such that spontaneous sympathetic vibration occurs, generating directed, rapid movement between protein and target DNA sequence. In the relatively small, "noisy" nuclear compartment, densely populated with nucleic acids and proteins, an ultra-high frequency, high-intensity, but short-range signaling network, shielded from the rest of the cell by a double-thick phospholipid bilayer nuclear envelope, may constitute ideal conditions for rapid, high-precision intranuclear communication.

Molecular vibration as a mechanism for carrying information via biomolecules is not entirely without precedent. Recently, behavioral studies in fruit flies (*Drosophila melanogaster*) showed that the animals could discriminate between isotopes of the same odorant [30]. The researchers were interested in testing the mechanism of odorant-receptor recognition, which is not understood. Traditionally, odor recognition has been attributed to a biochemical mechanism where binding affinity depends on a "lock-and-key" fit between odorant and receptor. In this study, flies were trained to choose between deuterated and nondeuterated odorants that would have had the same molecular shape but would have differed in vibrational modes due to differences in mass numbers of the atomic nuclei. Although this study of odorant recognition strongly suggests that vibrational differences in molecules carry different information detectable by the animal's nervous system, it does not go as far as our hypothesis which suggests that different vibrational modes in biomolecules give rise to directed motion in a medium.

The theoretical grounds for resonant mechanical vibrations giving rise to directed motion may be found in the asymmetry of biomolecules. Asymmetrical objects that are enveloped by moving fluids experience differential pressures at different points on the object, resulting in motion of the object down the pressure gradient. A classic example is an airfoil that experiences lift due to lower fluid pressure on the more-curved upper surface where fluid is flowing faster relative to the flat lower surface. Asymmetrical biomolecules like DNA and proteins vibrating in resonance may create regions of low pressure by displacing more fluid on one side, between them where the molecules can "fall together", sliding down a steep pressure gradient, perhaps something like the nodal patterns of Chladni plates that change as a function of resonant frequency. An interesting study by Baldwin and colleagues [31] showed that DNA molecules aggregate *in vitro* in a sequence-specific fashion. They constituted a mixture of two types of double-helical DNA molecules with similar nucleotide composition and length but differing in nucleotide sequence, labeled with green or red fluorescent dye, and then used confocal imaging to quantify the fluorescence to indirectly measure the segregation of DNA. They observed that in a protein-free, electrolytic environment, DNA molecules with similar sequences aggregated, while DNA molecules with dissimilar sequences segregated as evidenced by significant color separation. The mechanism they proposed was based on the ability of double helices to remain in register because of the sequence-dependent pitch of juxtaposed DNA molecules. In an important DNA–DNA interaction like homologous chromosome synapsis, the result of this study is powerfully suggestive because it rules out mechanisms of Brownian motion and cytoskeletal transport.

If our vibration-based model is limited by the nuclear envelope to intranuclear communication, what form of energy might carry the information that integrates the *entire* cell? For example, how might a specific transcription factor manufactured by ribosomes in the cytoplasm get "called up" for translocation to the nuclear compartment? It is known that many cellular processes are affected by electromagnetic fields [32–35]. Thus, another component of our model, not addressed by the preliminary experiments of this paper, proposes there is a cellular ("global") positioning system based on electromagnetism that establishes a three-dimensional coordinate system across the cell. In addition to providing spatial coordinates, it may be that DNA or another nucleus-associated biomolecule, generates timing information, a kind of "clock frequency" (as in a computer), that provides the fundamental frequency and harmonizes cellular components and biomolecules via families of harmonic frequencies. More locally, functionally-related biomolecules may have intrinsically-oscillating electromagnetic resonances that vibrate sympathetically and generate a local field within which directed motion may occur. We have begun to explore the theoretical basis for resonance between biomolecules that must locate a specific target within a broad search area — transcription factors and their specific DNA-binding sequences. Specifically, we are in the process of planning computer simulation-based investigations of the model's prediction that molecules with similar vibrational signatures may attract at a distance, allowing directed molecular motion in the cell.

While this study provided no evidence for our hypothesis that living, nucleated cells have a vibration that may originate in the nucleus and cause coherent cytoplasmic motion, we hope to find a suitable cell model and the right experimental and computational approaches to continue testing the vibrational model of intracellular communication. Developing an *in vitro* system where resonance correlations between DNA-binding proteins and DNA can be studied may be the next logical step. Raman spectroscopy techniques may reveal frequency patterns in functional families of biomolecules. I strongly believe the attraction-at-a-distance mechanism is based on a resonance principle, but whether the resonance may be mechanical or electromagnetic or a combination of both—will have to wait for future experiments.

Addendum

Other researchers are also seeking evidence for a resonance principle at work in directing important cellular events. They have predicted from theoretical models that electrical fields arising from synchronized oscillations within centrosomes, microtubules, and chromatin drive centrosome movements and homologous chromosome synapsis during mitosis and meiosis (Zhao and Zhan, 2012a); and that "chromatin oscillation cluster" formation may coordinate the efficient transcription of genes across the genome (Zhao and Zhan, 2012b). See references Zhao Y, Zhan Q (2012a) Electrical fields generated by synchronized oscillations of microtubules, centrosomes and chromosomes regulate the dynamics of mitosis and meiosis. Theor Biol Med Model 9:26.; Zhao Y, Zhan Q (2012b) Electrical oscillation and coupling of chromatin regulate chromosome packaging and transcription in eukaryotic cells. Theor Biol Med Model 9:27.

Acknowledgments

The author would like to thank Tom English and Stan Sholar for invaluable collaborations throughout this project. These experiments could not have been carried out without the provision of specimens by Jim Pamplin. The author is grateful to the engineers at Polytec, Jerome Eichenberger and Anthony Brock, for their expertise in the use of the laser-Doppler vibrometry systems for data acquisition and analyses; to Mr. Eichenberger, in addition, for critical discussions and manuscript comments regarding technical aspects of the project. The author also thanks two anonymous reviewers for their critiques that greatly strengthened the paper. Finally, the author acknowledges John C. Sanford for the original idea and conception of the model. This work was funded by Rainbow Technologies, Inc.

References

1. Dix JA, Verkman AS (2008) Crowding effects on diffusion in solutions and cells. Annu Rev Biophys 37:247–263.
2. Riggs AD, Bourgeois S, Cohn M (1970) The *lac* repressor-operator interaction. 3. Kinetic studies. J Mol Biol 53:401–417.
3. Gorman J, Greene EC (2008) Visualizing one-dimensional diffusion of proteins along DNA. Nat Struct Mol Biol 15:768–774.
4. Ellis J, Minton AP (2003) Join the crowd. Nature 425:27–28.
5. Hall D, Hoshino M (2010) Effects of macromolecular crowding on intracellular diffusion from a single particle perspective. Biophys Rev 2:39–53.
6. Theurkauf WE, Hazelrigg TI (1998) In vivo analyses of cytoplasmic transport and cytoskeletal organization during *Drosophila* oogenesis: characterization of a multi-step anterior localization pathway. Development 125:3655–3666.
7. Byeong-Jik C, Koppetsch BS, Theurkauf WE (2001) In vivo analysis of *Drosophila* bicoid mRNA localization reveals a novel microtubule-dependent axis specification pathway. Cell 106:35–46.
8. Kugler JM, Lasko P (2009) Localization, anchoring and translational control of oskar, gurken, bicoid and nanos mRNA during *Drosophila* oogenesis. Fly 3:15–28.
9. Forgacs G, Yook SH, Janmey PA, Jeong H, Burd CG (2004) Role of cytoskeleton in signaling networks. J Cell Sci 117:2769–2775.
10. Zarnack K, Feldbrügge M (2010) Microtubule-dependent mRNA transport in fungi. Eukaryot Cell 9:982–990.
11. Mammoto T, Ingber DE (2010) Mechanical control of tissue and organ development. Development 137:1407–1420.
12. Meagher RB, Deal RB, Kandasamy MK, McKinney EC (2005) Nuclear actin-related proteins as epigenetic regulators of development. Plant Physiol 139:1576–1585.
13. Chuang CH, Carpenter AE, Fuchsova B, Johnson T, Lanerolle P, Belmont AS (2006) Long-range directional movement of an interphase chromosome site. Curr Biol 16:825–831.
14. Kumaran RI, Rajika T, Spector DL (2008) Chromatin dynamics and gene positioning. Cell 132:929–934.

15. Dundr M, Ospina JK, Sung MH, John S, Upender M, Ried T, Hager GL, Matera AG (2007) Actin-dependent intranuclear repositioning of an active gene locus in vivo. J Cell Biol 179:1095–1103.
16. Hu Q, Kwon YS, Nunez E, Cardamone MD, Hutt KR, Ohgi KA, Garcia-Bassets I, Rose DW, Glass CK, Rosenfeld MG, Fu XD (2008) Enhancing nuclear receptor-induced transcription requires nuclear motor and LSD1-dependent gene networking in interchromatin granules. Proc Natl Acad Sci USA 105:10199–19204.
17. Castano E, Philimonenko VV, Kahle M, Fukalova J, Kalendova A, Yildirim S, Dzijak R, Dingova-Krasna H, Hozak P (2010) Actin complexes in the cell nucleus: new stones in an old field. Histochem and Cell Biol 133:607–626.
18. McDonald D, Carrero G, Andrin C, de Vries G, Hendzel MJ (2006) Nucleoplasmic beta-actin exists in a dynamic equilibrium between low-mobility polymeric species and rapidly diffusing populations. J Cell Biol 172:541–552.
19. Chou KC (1983) Identification of low-frequency modes in protein molecules. Biochem J 215:465–469.
20. Chou KC (1984) Low-frequency vibrations of DNA molecules. Biochem J 221: 27–31.
21. Chou KC (1988) Low-frequency collective motion in biomacromolecules and its biological functions. Biophys Chem 30:3–48.
22. Chou KC, Maggiora GM, Mao B (1989) Quasi-continuum models of twist-like and accordion-like low-frequency motions in DNA. Biophys J 56:295–305.
23. Zhang CT, Chou KC (1996) Beat motion in DNA double helix and a mechanism of energy exchange its two strands with microwave frequency. Chem Phys 206: 271–277.
24. Painter PC, Mosher LE, Rhoads C (1981) Low-frequency modes in the Raman spectrum of DNA. Biopolymers 20:243–247.
25. Urabe H, Tominaga Y (1982) Low-lying collective modes of DNA double helix by Raman spectroscopy. Biopolymers 21:2477–2481.
26. Urabe H, Tominaga Y, Kubota K (1983) Experimental evidence of collective vibrations in DNA double helix Raman spectroscopy. J Chem Phys 78:5937–5939.
27. Lindsay SM, Powell JW, Rupprecht A (1984) Observation of low-lying Raman bands in DNA by tandem interferometry. Phys Rev Lett 53:1853–1855.
28. Lamba OP, Wang AHJ, Thomas Jr GJ (1989) Low-frequency dynamics and Raman scattering of crystals, of B-, A-, and Z-DNA, and fibers of C-DNA. Biopolymers 28:667–678.
29. Hartwell LH, Hood L, Goldberg ML, Reynolds AE, Silver LM, Veres RC (2008) Genetics: From Genes to Genomes, 3rd edn. McGraw-Hill Companies, Inc., New York.
30. Franco MI, Turin L, Mershin A, Skoulakis EMC (2011) Molecular vibration-sensing component in Drosophila melanogaster olfaction. Proc Natl Acad Sci USA 108: 3797–3802.

31. Baldwin GS, Brooks NJ, Robson RE, Wynveen A, Goldar A, Leikin S, Seddon JM, Kornyshev AA (2008) DNA double helices recognize mutual sequence homology in a protein free environment. J Phys Chem B 112–1060–1064.
32. Shi R, Borgens RB (1995) Three-dimensional gradients of voltage during development of the nervous system as invisible coordinates for the establishment of embryonic pattern. Dev Dynam 202:101–114.
33. Song B, Zhao M, Forrester JV, McCaig CD (2002) Electrical cues regulate the orientation and frequency of cell division and the rate of wound healing in vivo. Proc Natl Acad Sci USA 99:13577–13582.
34. Blank M, Goodman R (2004) Initial interactions in electromagnetic field-induced biosynthesis. J Cell Physiol 199:359–363.
35. Cifra M, Fields JZ, Farhadi A (2011) Electromagnetic cellular interactions. Prog Biophys Mol Biol 105:223–246.

Getting There First: An Evolutionary Rate Advantage for Adaptive Loss-of-Function Mutations

Michael J. Behe[1]

[1]*Department of Biological Sciences, Lehigh University, Bethlehem, PA 18015, USA. mjb1@lehigh.edu*

Abstract

Over the course of evolution organisms have adapted to their environments by mutating to gain new functions or to lose pre-existing ones. Because adaptation can occur by either of these modes, it is of basic interest to assess under what, if any, evolutionary circumstances one of them may predominate. Since mutation occurs at the molecular level, one must look there to discern if an adaptation involves gain- or loss-of-function. Here I present a simple, deterministic model for the occurrence and spread of adaptive gain-of-function versus loss-of-function mutations, and compare the results to laboratory evolution experiments and studies of evolution in nature. The results demonstrate that loss-of-function mutations generally have an intrinsic evolutionary rate advantage over gain-of-function mutations, but that the advantage depends radically on population size, ratio of selection coefficients of competing adaptive mutations, and ratio of the mutation rates to the adaptive states.

Key words: gain-of-function mutation, loss-of-function mutation, rate of fixation

1. Introduction

In *On the Origin of Species* Charles Darwin emphasized that natural selection is relentless, continuously monitoring each organism for its fitness, selecting those with an advantage and weeding out the disadvantaged [1]. However, as Darwin also knew, an organism's advantage in a particular set of circumstances did not have to involve the gain of a new ability, such as the power to fly or swim. Indeed, it could involve the loss of those abilities. Flightless birds had adapted to their habitats partially by abandoning such a faculty. Some organisms went even further. Darwin described some barnacles in which the male was reduced to a transparent sac, with little but a reproductive system remaining [2]. By specializing in this way, the barnacles and their descendants presumably gained an adaptive advantage over competitors.

In the nineteenth century Darwin and his contemporaries could identify mutations only through their phenotypic effects. However, with the progress of biology especially in the last half-century, contemporary science can now characterize mutations also by their molecular effects to the genetic material of a species. In order to understand the roles of loss-of-function (LOF) versus gain-of-function (GOF) mutations, one must keep phenotypic versus molecular changes separate. An altered, visually observable phenotype may be due to any of a number of disparate underlying molecular mutations. For example, a mutant mouse that is 50% larger than its litter mates might have had the gene for a repressor protein that switches off production of growth hormone deleted. At the molecular level, that would be an LOF mutation, since a functional molecular feature was deleted, even though the increased size of the mouse may strike the casual observer as a gain-of-function. On the other hand, a large mutant mouse might be due to the formation of a new promoter site for a transcription factor near a gene involved in growth, which would be a GOF mutation, since a new functional molecular feature (the promoter site) was produced. In this paper I will consider LOF and GOF mutations as affecting functional molecular features such as genes and regulatory elements, no matter what their possible phenotypic effects may be.

2. The model

Consider a population of organisms that comes under a new selective pressure. To respond to the pressure two different adaptive mutations are postulated to be potentially available: one which results in the gain of a molecular function, and another which results in the loss of one. What factors might affect the probabilities of either kind of mutation becoming fixed in the population in competition with the other? One factor of immediate importance is the rate of appearance of the adaptive mutations. It is very often possible to eliminate a molecular function by a variety of mutations. GOF mutations, on the other hand, are generally much more specific, sometimes being produced in only one way.

As an illustration, consider several mutations to human genes that give a measure of resistance to malaria. The best known such mutation is the sickle cell gene in which, by means of a single A→T transversion, the codon for a glutamic acid residue in the sixth position of the β-chain globin gene is converted to a codon for valine [3]. This can be considered a GOF mutation, because the hemoglobin gains a self-association site on its surface, allowing the individual proteins upon deoxygenation to aggregate into microtubular-like structures [4]. By an as-yet-unknown mechanism, the polymerization negatively affects the growth of the malarial parasite (which spends part of its life cycle in the red blood cell) [5, 6]. Another mutation which confers a measure of resistance to malaria is deficiency of glucose-6-phosphate dehydrogenase (G6PD), in which a mutant gene produces little or no functional enzyme [7]. For reasons that are unclear, this interferes with parasite viability. Population genetic studies have shown that hundreds of separate mutations have led to deficiency of wild-type G6PD in populations at risk for malaria. On the other hand, the mutation producing the sickle gene is thought to have arisen de novo only a few times in the last 10,000 years, or perhaps only once [8].

The reason for the disparity in the number of *de novo* mutations is straightforward. To secure a sickle mutation a particular nucleotide of the β-globin gene must be substituted. Since the nucleotide mutation rate of humans is on the order of 10^{-8} substitutions/ generation, that is also the de novo rate of appearance of the sickle gene [9]. On the other hand, there are many ways to produce a nonfunctional protein such as malaria-resistant G6PD. For example, during replication the insertion of a nucleotide anywhere within the coding sequence results in a frame-shift and likely an inactive polypeptide. Deletion of a nucleotide in the coding region will have the same affect, as will alteration of a codon from sense to nonsense. Longer insertions and deletions will frequently have the same effect. Missense mutations, although likely not completely inactivating the protein, will often make the protein less stable or less functional. Thus, considered as a class, the mutation rate from a functional to a nonfunctional gene may be several orders of magnitude greater than the basic nucleotide mutation rate. (Indeed, the adaptation rate of *E. coli*, whose generational nucleotide mutation rate is 50-fold lower than that of humans, has recently been measured as 10^{-5})[10]. For the two classes of mutations, in this paper I explore the effect of this factor on the evolutionary rate of spread of adaptive mutations as a function of population size, mutation rate, selection coefficient, ratio of selection coefficients of the competing adaptive mutations, and ratio of mutation rates to the adaptive state.

Calculations were performed using *Mathematica* [34].

3. Results

3.1 *Relatively small population sizes*

In this section I consider small population sizes ($N_e << 1/v$), where N_e is the effective population size and v is the mutation rate per generation. Unless otherwise stated, organisms are assumed to be haploid (because most laboratory evolution experiments have been done with haploids), and the model is developed accordingly. The resulting equations can be applied to diploid organisms by replacing N_e by $2N_e$.

In order for an adaptive mutation to become fixed in a population of relatively small size two separate processes must occur, each with its own time scale: (1) if the mutation does not yet exist in the population when the selective pressure begins, then the expected waiting time to the appearance of the selected mutation is $t_{w1} = 1/(2N_e vs)$, where s is the selection coefficient; (2) once the selected mutation appears, the time for it to fix in the population is $t_{fx1} = (2 \ln N_e)/s$ [11].

If one is comparing two distinct mutations in the same population that are responsive to the same selective pressure, however, both the rates of mutation to the adaptive state and the selection coefficients may differ. For the second mutation, the expected waiting time to the appearance of the selected mutation may be written as $t_{w2} = 1/(2N_e vsr_v r_s)$, where r_v is the ratio of the mutation rates to the adaptive state and r_s is the ratio of the selection coefficients for the two cases. The expected time for the second mutation to spread to fixation in the population can be written $t_{fx2} = (2 \ln N_e)/r_s s$. Considering the case of a GOF versus LOF mutation, if we take v to be the nucleotide mutation rate, then in general r_v will range from 1 to ~1000 for an LOF mutation. r_s can take any positive value (both selection coefficients are positive because both the GOF and the LOF mutations are postulated to be adaptive).

A useful metric for comparing the prospects of fixation for the GOF versus LOF mutations is $r_{D/fx}$, which is defined as the expected time to appearance of an adaptive GOF mutation minus that for an adaptive LOF mutation, divided by the time for the LOF mutant to spread to fixation in the population. If the difference in the expected waiting times between the selected GOF versus LOF mutations is greater than the time required for the LOF mutation to spread, then the LOF mutation will have already fixed in the population before the expected appearance of the selected GOF mutation. The expected difference in waiting time to appearance of the selected mutations is

$$t_D = t_{wG} - t_{wL} = \frac{1}{2N_e vs} - \frac{1}{2N_e vsr_v r_s} = \frac{1}{2N_e vs}\left(1 - \frac{1}{r_v r_s}\right) \tag{1}$$

The ratio of the time difference t_D to the time for the LOF mutation to spread to fixation in the population, t_{fxL}, is

$$r_{D/fx} = \frac{t_D}{t_{fxL}} = \frac{\frac{1}{2N_e vs}\left(1 - \frac{1}{r_v r_s}\right)}{\frac{2}{r_s s}\ln N_e} = \frac{1}{4N_e v \ln N_e}\left(r_s - \frac{1}{r_v}\right) \tag{2}$$

Thus whenever $r_{D/fx} > 1$, the LOF mutation is expected to fix in the population before the selected GOF mutation appears. Figure 1 illustrates this situation. Two curves are plotted for the appearance and subsequent spread of an LOF and a GOF mutation in a population of 10^6 organisms. The selection coefficient for the GOF is 0.1 and for the LOF is 0.01; thus r_s is 0.1. The basic nucleotide mutation rate is taken to be 10^{-9}, and r_v, the ratio of the mutation rate to the adaptive state for the LOF vs GOF mutation, is set at 100. The expected waiting time to the appearance of the selected LOF mutation under these circumstances is 500 generations, while for the GOF mutation the time is 5,000 generations. On average the GOF mutation would take 276 generations to fix in the population; the LOF mutation would require 2763 generations. Figure 1 shows that under such circumstances the selected LOF mutation would be expected to fix in the population before the selected GOF mutation appeared. Equation 2 determines the ratio $r_{D/fx}$ for this situation to be 1.62.

If $r_s = 1/r_v$, then equation 2 evaluates to zero, which means there is no expected difference t_D in the waiting time to the appearance of the selected LOF versus GOF mutations — the rate advantage of the LOF mutation is exactly offset by the relative weakness of its selection coefficient. If $r_s < 1/r_v$, then $r_{D/fx}$ will be negative, which means that there is less time to the appearance of the selected GOF mutation than to the LOF mutation — the rate disadvantage of the GOF mutation is more than offset by the relative strength of its selection coefficient.

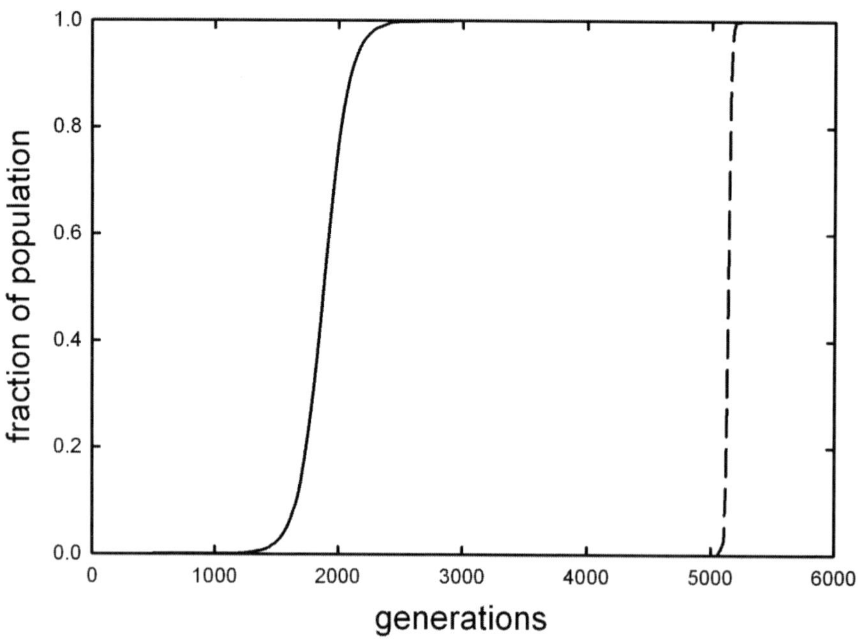

Fig. 1. Time in generations to occurrence and spread of an adaptive LOF mutation versus GOF mutation. The LOF mutant (———) has a selection coefficient 0.1-times that of the GOF mutant (— — —), but a mutation rate to the adaptive state 100-times that of the GOF mutant. The effective population size N_e is set at 10^6. The GOF mutation rate v is 10^{-9} per generation and the GOF selection coefficient $s = 0.1$.

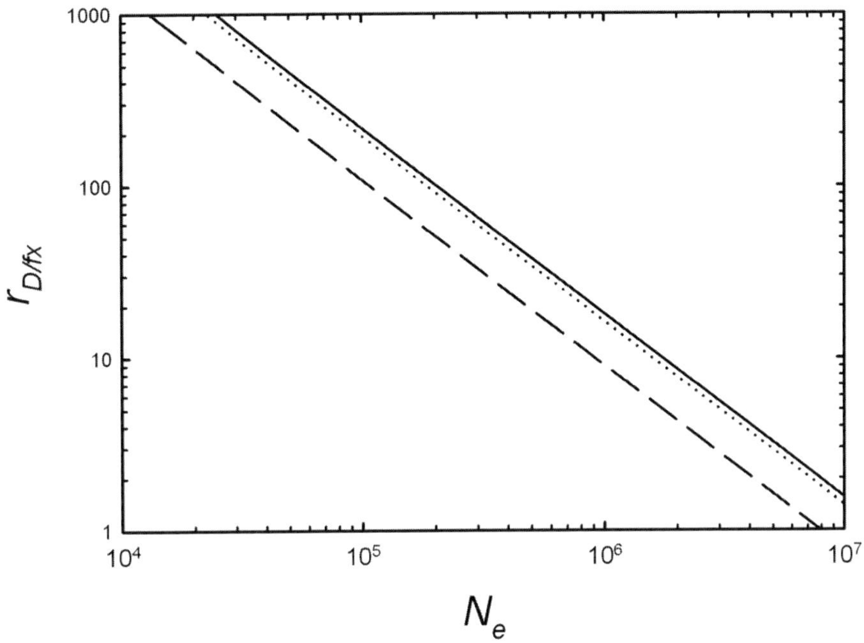

Fig. 2. The ratio $r_{D/fx}$ versus effective population size N_e. $r_{D/fx}$ is the time to appearance of an adaptive GOF mutation minus that for an adaptive LOF mutation, divided by the time for the LOF mutant to spread to fixation in the population. In this figure the LOF and GOF selection coefficients are equal. The mutation rate v is 10^{-9} per generation. (———) $r_v = 1000$; (⋯⋯⋯) $r_v = 10$; (— — —) $r_v = 2$.

Figure 2 plots the value of $r_{D/fx}$ versus the effective population size N_e for several values of r_v, with r_s held constant at one. As can be seen, the value of $r_{D/fx}$ is largely insensitive to changes in r_v, the ratio of the mutation rates to the adaptive state. Decreasing r_v 100-fold from 1000 to 10 leaves the value of $r_{D/fx}$ little changed. In all of these circumstances (except where $r_v = 2$ at effective population sizes very near 10^7) the ratio of the time for the LOF mutation to spread to the difference in the expected waiting time to the selected GOF versus LOF mutations, $r_{D/fx}$, is well above one.

Figure 3 examines the relationship between the value of $r_{D/fx}$ versus the effective population size N_e for several values of r_s, with r_v held constant at 1000, its likely maximum for a typical gene. In this case $r_{D/fx}$ depends linearly on the ratio of the selection coefficients: at any population size in the range, a decrease of a factor of 10 in r_s decreases $r_{D/fx}$ by approximately the same factor. (The magnitude of s, the selection coefficient itself, which is absent from equation 2, does not affect the results.) Thus, when r_s is 0.01 (that is, when the selection coefficient for the LOF mutation is only 1% of that of the GOF mutation), $r_{D/fx}$ decreases to a value of one at a population size of about 1.5×10^5, versus a population size of 1.5×10^7 when r_s is one.

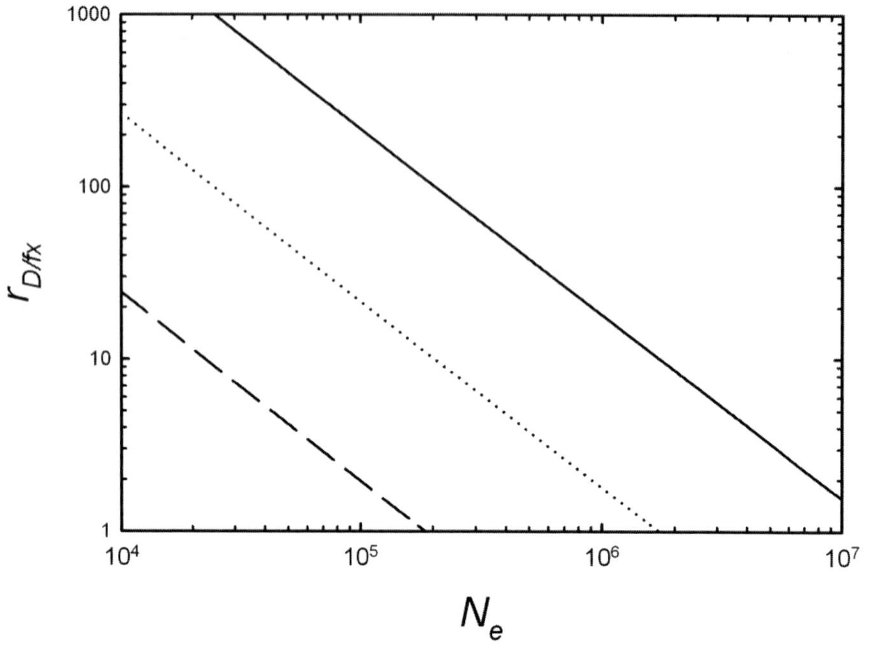

Fig. 3. The ratio $r_{D/fx}$ versus effective population size N_e. $r_{D/fx}$ is the time to appearance of an adaptive GOF mutation minus that for an adaptive LOF mutation, divided by the time for the LOF mutant to spread to fixation in the population. In this figure the rate of mutation to the adaptive state of the LOF mutant is 1000-times that of the GOF mutant. The mutation rate v is 10^{-9} per generation. (———) $r_s = 1$; (············) $r_s = 0.1$; (— — —) $r_s = 0.01$.

Figure 4 shows the dependence of $r_{D/fx}$ on r_v and r_s at a fixed value of $N_e = 10^6$. As can be seen $r_{D/fx}$ is essentially independent of r_v over a wide range, but is linearly dependent on r_s. The pronounced curvature for both plots at lower values on the x-axis reflects the approach of the factor (r_s - $1/r_v$) to zero.

3.2 *Relatively large population sizes*

In this section I consider relatively large population sizes ($N_e \geq 1/v$). As population size increases, the expected waiting time to the appearance of either or both selected mutations can shrink to much less than the expected time for the mutations to spread in the population. In fact, one or both mutations may be present continuously in the population at a low percentage as a neutral or detrimental allele before the new selective pressure makes it adaptive. Thus in this population size range a different metric is required to follow the relative advantage of LOF versus GOF mutations.

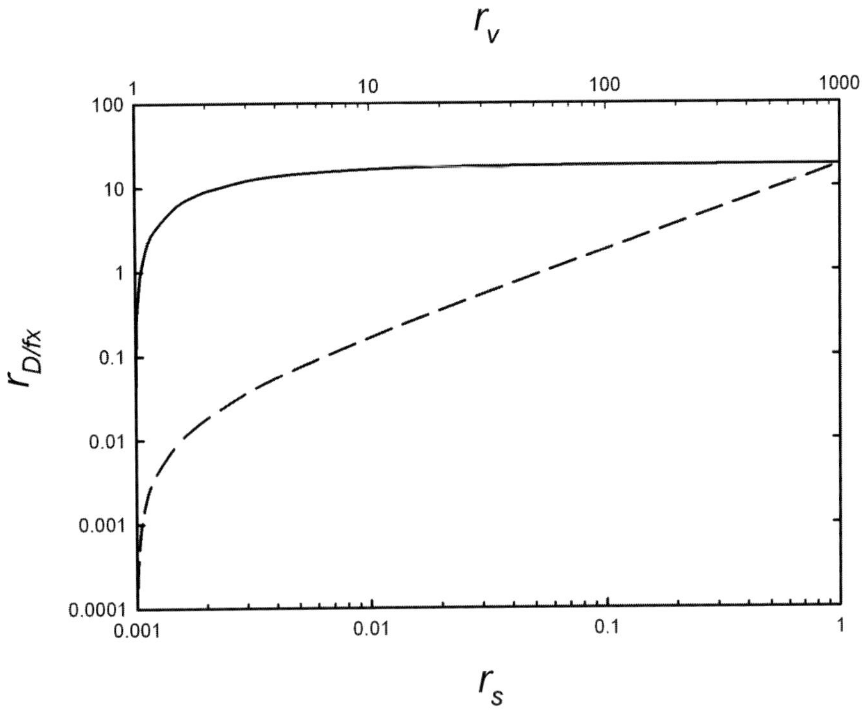

Fig. 4. The ratio $r_{D/fx}$ versus r_s and r_v. $r_{D/fx}$ is the time to appearance of an adaptive GOF mutation minus that for an adaptive LOF mutation, divided by the time for the LOF mutant to spread to fixation in the population. r_s is the ratio of the LOF to GOF selection coefficients. r_v is the ratio of the rate of LOF to GOF mutation to the adaptive state. The effective population size N_e is set at 10^6 and the GOF mutation rate v is 10^{-9} per generation. (———) r_s is set at 1 and r_v ranges from 1 to 1000; (— — —) r_v is set at 1000 and r_s ranges from 0.001 to 1.

A useful measure is the ratio of the fractions of LOF to GOF mutations in the population when the sum of those fractions first increases to 1.0. The time t in generations required to increase the frequency of a selected mutation from a value of q_0 to q_t can be calculated from [11]:

$$q_t = \frac{1}{1+\left(\frac{1-q_0}{q_0}\right)e^{-st}}$$

Thus (ignoring double mutants) the number of generations required for the fractions of an LOF and GOF mutation to sum to one can be calculated from:

$$\frac{1}{1+\left(\frac{1-q_{0G}}{q_{0G}}\right)e^{-st}}+\frac{1}{1+\left(\frac{1-q_{0L}}{q_{0L}}\right)e^{-r_s st}}=1 \quad (3)$$

The initial fraction q_0 when a selected mutation begins to increase in a haploid population is at a minimum $1/N_e$. However, for population sizes greater than the inverse of the mutation rate, numerous mutants are expected to be present in the initial population. For example, if the mutation rate v is 10^{-9} and the population size is 10^{12}, then there will be 10^3 mutants produced in the first generation. So the initial fraction q_{0G} is at least $\frac{1}{N_e}+\frac{N_e v}{N_e}=\frac{1}{N_e}+v$, and q_{0L} is at least $\frac{1}{N_e}+r_v v$.

The time t in equation (3) is the time required for the selected mutation to spread. Thus if we are counting generations from the first application of the selective pressure, then the expected waiting time for the selected mutation must be accounted for. As mentioned previously, for a haploid GOF mutation this is $t_{wG} = 1/(2N_e vs)$ and for an LOF mutation $t_{wL} = 1/(2N_e vsr_v r_s)$. Equation (3) can then be re-written as:

$$\frac{1}{1+\left(\frac{1-q_{0G}}{q_{0G}}\right)e^{-s(t_{fx}-t_{wG})}}+\frac{1}{1+\left(\frac{1-q_{0L}}{q_{0L}}\right)e^{-r_s s(t_{fx}-t_{wL})}}=1 \tag{4}$$

where $(t_{fx} - t_w)$ is the time for the mutations to spread to a sum fraction of 1.0 after the waiting time for at least one kind of selected mutation to first appear in the population. Given N_e, v, s, r_v, and r_s, equation 4 can be solved for t_{fx} and the value used to determine r_{fx}, which is the fraction of adaptive LOF mutations divided by the fraction of adaptive GOF mutations in the population when the two fractions first sum to one:

$$r_{fx}=\frac{1+\left(\frac{1-q_{0G}}{q_{0G}}\right)e^{-s(t_{fx}-t_{wG})}}{1+\left(\frac{1-q_{0L}}{q_{0L}}\right)e^{-r_s s(t_{fx}-t_{wL})}} \tag{5}$$

Figure 5 plots r_{fx} from equation 5 at $r_s = 1$ and $r_v = 1000$ for population sizes N_e ranging from 10^7 to 10^{10}. It is seen that at lower values of N_e, r_{fx} increases very rapidly. Indeed, at population sizes of 10^7 or less, r_{fx} is greater than N_e, reflecting the fact that less than one GOF mutant is expected to be present in the population when the LOF mutant has fixed. As N_e increases, r_{fx} approaches a constant value of approximately 31.6. Thus when the population initially consists entirely of LOF and GOF mutants and $N_e \geq 10^9$, under the circumstances described in Figure 5 LOF mutants will represent about 97% of the population.

Figure 6 plots r_{fx} as a function of r_v for population sizes from $10^{6.5}$ to 10^{12}, with $r_s = 1$. At the smallest population sizes the fixation ratio is extremely sensitive to the ratio of mutation rates. As N_e increases, however, and it becomes more likely

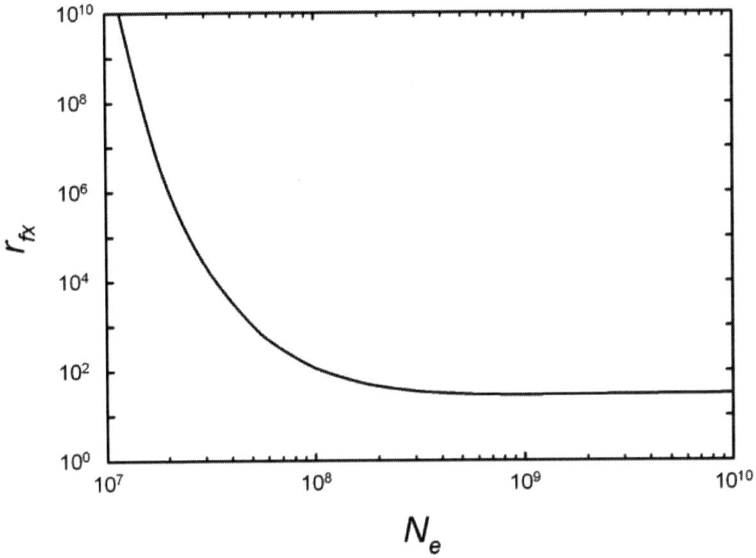

Fig. 5. The ratio r_{fx} versus the effective population size N_e. r_{fx} is the fraction of adaptive LOF mutations divided by the fraction of adaptive GOF mutations in the population when the two fractions first sum to one. $r_s = 1$; $r_v = 1000$; $s = 0.1$; $v = 10^{-9}$ per generation.

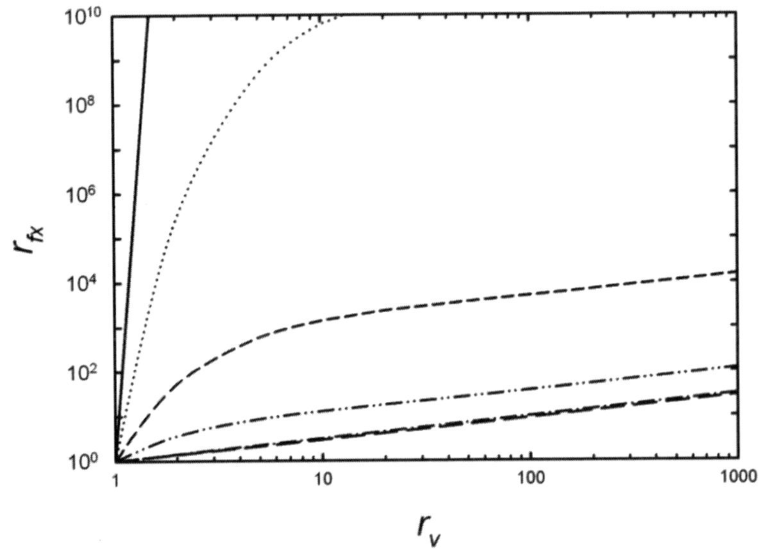

Fig. 6. The ratio r_{fx} versus the ratio r_v. r_{fx} is the fraction of adaptive LOF mutations divided by the fraction of adaptive GOF mutations in the population when the two fractions first sum to one. r_v is the ratio of the rate of LOF to GOF mutation to the adaptive state. r_s is set at 1; v is 10^{-9}. (———) $N_e = 10^{6.5}$; (…………) $N_e = 10^7$; (- - - - -) $N_e = 10^{7.5}$; (– ·· – ·· – ·· –) $N_e = 10^8$; (— — —) $N_e = 10^9$; (– · – · – · –) $N_e = 10^{12}$.

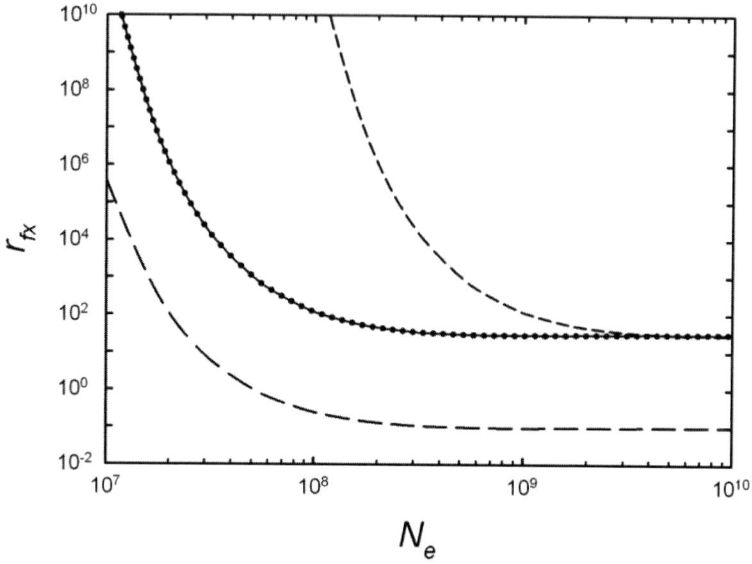

Fig. 7. The ratio r_{fx} versus the effective population size N_e. r_{fx} is the fraction of adaptive LOF mutations divided by the fraction of adaptive GOF mutations in the population when the two fractions first sum to one. For all curves r_v is set to 1000. (———) $s = 0.1$, $r_s = 1$, $v = 10^{-9}$; (············) $s = 0.0001$, $r_s = 1$, $v = 10^{-9}$; (— — —) $s = 0.1$, $r_s = 0.5$, $v = 10^{-9}$; (- - - - -) $s = 0.1$, $r_s = 1$, $v = 10^{-10}$.

that the mutants are present in the population from the first generation, the sensitivity decreases. As seen in the figure, the plots of r_{fx} versus r_v for values of $N_e \geq 1/v$ are essentially superimposable, and closely approximate the relationship $r_{fx} = \sqrt{r_v}$.

Figure 7 plots r_{fx} versus N_e for several variables. The solid curve reproduces the values from Figure 5 of $s = 0.1$ and $r_s = 1$. Coinciding with the solid curve is a dotted curve for which $s = 0.0001$, demonstrating the insensitivity of the curve to changes in the selection coefficient itself. The long-dashed curve uses the same parameters as the solid curve except that the value of r_s has been decreased to 0.5. As can be seen, this decreases the value of r_{fx} by several orders of magnitude, so that at large population sizes the value is below one, and the GOF mutation predominates at fixation, despite the initial 1,000-fold advantage of the LOF mutation rate. The short-dashed curve uses the same parameters as the solid curve except that the value of v has been decreased from 10^{-9} to 10^{-10}. As can be seen, this has the effect of simply moving the curve an order of magnitude to the right on the population axis.

Figure 8 plots r_{fx} versus r_v at three values of r_s with $N_e \gg 1/v$. As seen, modestly varying the ratio of the selection coefficients displaces the curve considerably along the r_{fx} axis and slightly alters its slope. Figure 9 compares r_v and r_s versus

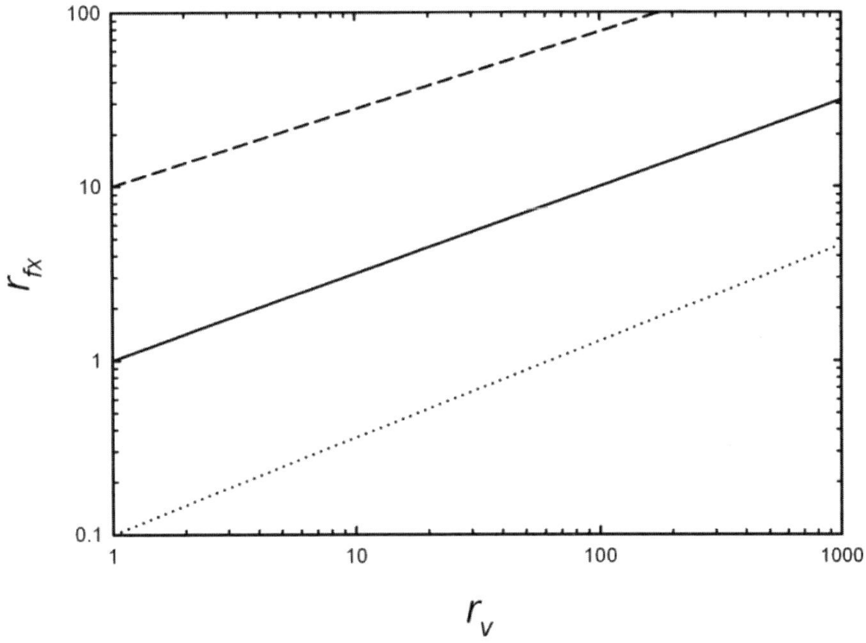

Fig. 8. The ratio r_{fx} versus the ratio r_v. r_{fx} is the fraction of adaptive LOF mutations divided by the fraction of adaptive GOF mutations in the population when the two fractions first sum to one. r_v is the ratio of the rate of LOF to GOF mutation to the adaptive state. For all curves, N_e is set at 10^{12} and v is 10^{-9}. (———) $r_s = 1$; (············) $r_s = 0.8$; (- - - - -) $r_s = 1.25$.

r_{fx}, showing the relative sensitivity of the fixation ratio to those parameters at large N_e. Figure 9 plots values for r_s including from one to 100; that is, for situations in which the selection coefficient of the LOF mutation is greater than or equal to that of the GOF mutation. r_{fx} is greater than one and increases rapidly in this region. In general, whenever $r_s \geq 1$ and $r_v > 1$, r_{fx} will be greater than one at any population size. That is, the LOF mutation will always be the majority of the population when the entire population is initially comprised of LOF and GOF mutations.

4. Discussion

4.1 *LOF versus GOF adaptive mutations*

Organisms can adapt to their environment either by acquiring new abilities or by abandoning old ones. This can be observed in such examples as legless snakes and sightless cave fish. Science has learned especially in the last fifty years that altered, observable phenotypes are the manifestation of changes to the genetic endowment

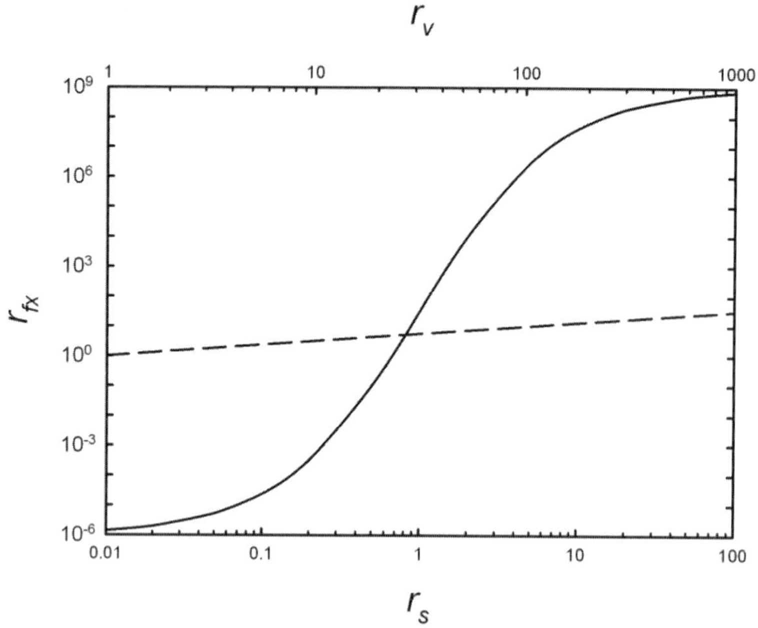

Fig. 9. The ratio r_{fx} versus r_s and versus r_v. r_{fx} is the fraction of adaptive LOF mutations divided by the fraction of adaptive GOF mutations in the population when the two fractions first sum to one. r_s is the ratio of the LOF to GOF selection coefficients. r_v is the ratio of the rate of LOF to GOF mutation to the adaptive state. For both curves, N_e is set to 10^{12} and v is 10^{-9}. (———) r_{fx} versus r_s, r_v is set at 1000; (— — —) r_{fx} versus r_v, r_s is set at 1.

of an organism. It has also learned that there is not a necessary correlation between loss or gain of an ability at the phenotypic level and loss or gain of a functional genetic element at the molecular level. In other words, what strikes an observer as a phenotypic gain of function may be caused by either a molecular loss or gain of function. The same holds for a phenotypic loss of function: it may be the result of a genetic gain or loss of function. Because organisms can adapt by either molecular GOF mutations or LOF mutations it is of basic interest to determine which, if either, kind of mutation will dominate under various circumstances.

Research over the past fifty years has shown that many genetic elements consist of multiple nucleotides. Protein coding regions can be thousands of nucleotides in length; RNA genes can be hundreds of nucleotides; regulatory elements and processing signals can be several nucleotides to dozens of nucleotides long. A substantial portion of possible mutations in these elements will result in the diminution or loss of their function. Thus, as a class, LOF mutations for a particular genetic element will occur at a rate from several times to several-orders-of-magnitude times the basic nucleotide substitution rate.

That is not the case for GOF mutations. Consider two examples: First, a transcription factor binding site that is 10 nucleotides in length, and a second DNA sequence which has 9 of 10 nucleotides that are necessary to form a second regulatory site. Suppose that in response to a new selective pressure an adaptive effect could be secured either by mutating the first site so that it lost its function or by mutating the single mismatching residue of the second site so that it gained function. The LOF mutation would on average appear at 10-times the nucleotide substitution rate, simply because there are multiple ways to break the functioning element. The GOF mutation, however, would appear at even less than the basic rate of nucleotide substitution (because for a currently-nonfunctional, potential genetic element there it is possible that one of the "correct" nucleotides in the sequence will mutate before the "incorrect" one [12]). Second, consider a recently duplicated gene which could provide an adaptive effect in response to a new selective pressure if a certain nucleotide in the gene were altered (allowing the duplicate gene product to, say, diverge productively in activity from the parent gene product). Suppose, however, that an adaptive effect could also be had by reducing or eliminating the activity of another, separate gene. Because of the many ways in which a gene can be altered to lose function, the LOF mutation would have a rate several orders of magnitude greater than that of the GOF mutation for the duplicated gene.

There can be cases in which a GOF mutation may appear at several times the nucleotide substitution rate. I discussed earlier the sickle mutation, in which a single particular nucleotide in the β-globin gene must be changed. Yet in other cases of GOF, there can be several possible nucleotides to change, each of which will suffice. For example, Couñago *et al.* [13] replaced the essential gene for adenylate kinase in *Geobacillus stearothermophilus* — a moderate thermophile — with that of *Bacillus subtilis*, a mesophile, which they then grew in a turbidostat at increasing temperatures. Over the course of 1500 generations they isolated six thermostable mutants of the enzyme — one single point mutant and five double point mutants derived from the single mutant. Thus in this circumstance the enzyme could gain the function of being active in a hostile environment by altering any of six positions. Nonetheless, the number of ways to break a functional element will almost always be much greater than the number of ways to construct one, so that in almost all cases r_v would be expected to be greater than one.

4.2 *Effect of disparity in adaptive rate*

In this chapter I investigate the effect of the disparity in rate of mutation to an adaptive state for LOF and GOF mutations as a function of several parameters.

The model presented here is a simple, deterministic one, which does not consider the probabilistic nature of changes in allele frequencies [11]. Because of its simplicity, the general behavior of the investigated model is visible with considerable clarity and the issue of the evolutionary rate advantage of adaptive LOF mutations is highlighted.

The behavior at relatively small population sizes is governed by equation 2, which accounts for the two separate phases of fixation of a new mutation: the expected waiting time to the appearance of the selected mutation, and the time taken for the mutation to spread within the population. An interesting aspect of the equation is that it does not contain the selection coefficient s; that is, the ratio of the selection coefficients r_s influences the competition between the two mutations rather than the absolute value of either or both selection coefficients. (This also is the case at relatively large population sizes, as shown by Figure 7.) Whenever equation 2 evaluates to $r_{D/fx} > 1$, then the LOF mutation is expected to fix in the population before a selected GOF mutation appears. Thus, as illustrated in Figure 1, an LOF mutation whose selection coefficient is ten-fold weaker than an adaptive GOF mutation can outrace it to fixation, due to its greater rate of mutation to an adaptive state.

Figures 2 and 3 show that this effect exerts substantial influence at relatively low population sizes. For a population size of $< 10^7$, if $r_s \geq 1$ and $r_v > 1$, then an LOF mutant is always expected to fix in the population before a selected GOF mutant appears. Because of an increasing disparity in waiting times, at population sizes $<<10^7$ an LOF mutant may be fixed in the population first even if its selective advantage is considerably less than that of a GOF mutant. For example, for a population size of 10^5, an LOF adaptive mutation will become fixed first at $r_v \geq 1$ even if its selection coefficient is only one-hundredth that of a GOF adaptive mutation; that is, if $r_s \geq 0.01$. At smaller population sizes, the advantage for the LOF mutation increases linearly with $1/N_e$.

If an LOF mutation with a smaller selection coefficient is first fixed in a population, what scenario is most likely to occur after the GOF mutation eventually appears? The answer to that question is likely to depend sharply on the specific genetic elements involved. One possible scenario is that the GOF mutation also spreads to fixation, and the LOF mutation remains fixed. A second possibility is that, depending on the physical nature of the mutation, the LOF mutation may be repaired by subsequent mutation after the GOF mutation spreads in the population. If it cannot be repaired, it may be replaced by horizontal gene transfer or by having its function taken over by another genetic element, or the organism may adapt in other ways to its loss. Penman *et al.* [14] recently demonstrated that the outcome in competition between the sickle mutation (which is highly protective against malaria) and various thalassemic disorders (which are less protective) is quite difficult to predict because of epistatic effects unrelated to their anti-malarial activities. Thus the future course of the evolution of a system after initial fixation of an LOF mutation might be considerably more complex than a linear succession of mutations with increasing selective value.

For $v = 10^{-9}$, at population sizes $N_e > 10^8$ an LOF mutation is no longer expected to fix in the population before a selectable GOF mutation appears, even if $r_s = 1$, because the larger population sizes produce both types of mutations within the time it would take for the LOF mutation to spread in the population. Nonetheless, even though the metric $r_{D/fx}$ decreases below one in this range, in many cases the LOF mutation will become the dominant mutation in the population. In order to assess the advantages of LOF versus GOF mutations in this population range, a new metric, r_{fx}, was introduced in equation 5. r_{fx} is the ratio of LOF to GOF mutants when their fraction of the population first sums to one.

Figure 6 shows that LOF mutations always possess a rate advantage over GOF mutations if the respective selection coefficients are equal; that is, if $r_s = 1$. Under these circumstances at large population sizes ($N_e \geq 1/v$), $r_{fx} \approx \sqrt{r_v}$, and the ratio of LOF to GOF mutations when their fraction first sums to one will range from 1.41 to 31.6 for values of r_v ranging from 2 to 1000. Thus the LOF mutant will comprise from 59% of the population to 97% of the population. If at this point the mutants then drift neutrally in the population (because it is postulated that neither has a selective advantage over the other), the LOF mutant is expected to become fixed with a probability equal to its population fraction [15].

Under what circumstances would two selection coefficients be expected to be equal? If two mutations both met the new selective pressure without causing deleterious pleiotropic effects, then their selection coefficients would be expected to be the same. Thus whenever such a situation presents itself, the LOF mutation would have an advantage.

If the selection coefficients are not equal, how likely is it that a GOF mutation will have a value of s greater than that of an LOF mutation, or vice-versa? The answer to that question is not known, but both LOF and GOF mutations can have significant selection coefficients. The selection coefficient for LOF mutations of the *rpoS* gene of *E. coli* has been measured at 0.217, a substantial value [16]. The selection coefficient for the GOF sickle mutation has been estimated as 0.05 to 0.18, again a large value [17]. If in general there is no overall correlation between adaptive GOF versus LOF mutations and the magnitude of the selection coefficient, then the intrinsic rate advantage enjoyed by LOF mutations will bias long-term evolution in that direction.

4.3 Comparison to laboratory evolution experiments

Over the past forty years many laboratories have conducted evolution experiments, observing adaptation of micro-organisms to varying environmental conditions, and in many cases identifying the molecular changes that comprised the adaptive mutation [4]. How do the results obtained in this chapter bear on the interpretation of those experiments?

Comparison to experiments where $N_e < 1/v$: The most extensive laboratory evolution experiment to date has been performed under the direction of Richard Lenski at Michigan State University [18]. Starting in the early 1990s, Lenski and colleagues began growing 10 ml cultures of *E. coli*, which undergo six to seven doublings per day. Each day they transferred 1% of the culture to fresh medium. Over the years the cultures have undergone more than 50,000 generations. All adaptive mutations identified to date appear to be LOF ones [4]. The single most beneficial mutation was the destruction of the *rbs* operon by insertion sequences. The value of the selective coefficient for this was approximately 0.02 [19]. Other identified LOF mutations include ones in the *pykF, nadR, pbpA-rodA, hokB/sokB, malT*, and *topA* genes. A number of other adaptive genes have been identified to date, but the natures of the mutations, whether LOF or GOF, have not yet been reported [20].

The rate of nucleotide mutations per generation of *E. coli* is ~5×10^{-10} [21]. The effective population size N_e of Lenski's [18] cultures of *E. coli* is ~2×10^7, which is the harmonic mean between the initial population of the day's culture ($5{\times}10^6$) and the final population of the day ($5{\times}10^8$) if the population is assumed to double in discrete generations [11]. Substituting these numbers into equation 2 shows that $r_{D/fx}$ would be 1.47 — greater than one — if r_s were one and r_v were 100. An LOF mutation would thus be expected to be fixed in the population before a GOF mutation appeared if their selection coefficients were equal. How great of a selective advantage must a GOF mutation have to outcompete an LOF mutation under these circumstances? Using equation 2 it is seen that if r_s is 0.68, then $r_{D/fx}$ falls slightly below one. In other words, a GOF mutation would have to have a selection coefficient about 50% greater than an LOF mutation in these circumstances in order to at least appear in the population before the LOF mutation were fixed.

To find out how much greater the selection coefficient must be to actually outcompete the LOF mutation, we must use equations 4 and 5 to calculate r_{fx}. Assuming r_s were 0.68, there would be approximately one GOF allele in the population per ~$2{\times}10^7$ LOF alleles. In order to overcome the LOF rate advantage, however, r_s would have to fall to ~0.25. In other words, if the selection coefficient of the GOF mutation were approximately four times that of the LOF mutations, then the GOF mutation would be slightly more than half the population. In order to dominate the population by ~90% r_s would have to be ~0.2; that is, the selection coefficient of the GOF mutation would have to be about five-fold that of the LOF mutation. Since no GOF mutations have yet been seen, we can tentatively conclude that there are no GOF mutations available whose selective value is five-fold greater than the least-adaptive LOF mutations seen in this series of experiments. (Lenski's group recently reported a very adaptive Cit+ phenotype, which apparently required both LOF and gene duplication mutations [22]). If an LOF mutation appeared within the first 25,000 generations, it would require a minimum selection coefficient of 0.00076 to spread to fixation in the next 25,000 generations. To outcompete it, a GOF mutation would require a minimum selection coefficient of five-times this value, i.e. ~0.0038. Thus it can be concluded that there are no GOF mutations available under the circumstances of the experiment whose selection coefficients exceed that number.

The question might be asked, what if a potential GOF mutation with a sufficiently strong selection coefficient existed, but simply failed to arise during the term of the experiment? That is always a possibility, but an unlikely one. Given the scale of the Lenski experiment [20], with an effective population size of 2×10^7 over 50,000 generations and a nucleotide mutation rate of $\sim5\times10^{-10}$, each nucleotide is expected to be substituted 500-fold over the course of the experiment. Deletions, additions, and other kinds of mutations would similarly be expected to occur multiple times. There were many redundant opportunities for all simple mutations to arise (the Cit^+ phenotype apparently needed several mutations to arise). Thus we can be confident that if a particular mutation, or kind of mutation, was not observed, then it is very unlikely to have the necessary selection coefficient.

Comparison to experiments where $N_e > 1/v$: As seen in Figures 7–9, at $N_e \geq 1/v$, r_{fx} is much more sensitive to r_s than at smaller population sizes. Just a slight advantage in the selection coefficient for a GOF mutation is sufficient to offset a 1,000-fold advantage in the rate of LOF mutation. This great sensitivity can be used to infer whether such a GOF mutation is available under particular environmental circumstances. That is, if a certain selective pressure is applied, one or more LOF mutations are observed, and $N_e \geq 1/v$, then the failure to observe a GOF mutation would imply that no GOF mutation is available within a single mutational step that had a somewhat greater selection coefficient than the LOF mutations(s). Conversely, if a GOF mutation were observed but no LOF mutation, we could deduce that no LOF mutation was available that had a selection coefficient greater than or equal to the GOF mutation.

As mentioned earlier, Couñago *et al.* [13] replaced the essential gene for adenylate kinase in *Geobacillus stearothermophilus* — a moderate thermophile — with that of

Bacillus subtilis, a mesophile, which they then grew in a turbidostat at increasing temperatures. Over the course of 1500 generations they isolated six thermostable mutants of the enzyme — one single point mutant and five double point mutants derived from the single mutant, which can all be classified as GOF mutations. The mutation rate of *G. stearothermophilus* can be estimated by using a value for the mutation rate of approximately 0.003 per genome per generation for DNA-based microbes, which yields a value of about 5×10^{-10} mutations per generation [21]. Since the authors maintained a continuous culture, the population was not subject to the large changes in size seen in Lenski's experiments, so the effective population of microbes per generation in the turbidostat was $\sim5\times10^{10}$. In other words, in the Couñago *et al.* [13] experiment, $N_e > 1/v$. Inserting these values into equations 4 and 5 shows that if a potentially adaptive LOF mutation were available with the same selection coefficient as a GOF mutation, then it would dominate the population with an r_{fx} of 9.9; in other words, it would comprise ~91% of the population. Thus it can be concluded that, despite the frequency of adaptive LOF mutations in Lenski's work, no LOF mutation with an $r_s \geq 1$ compared to the observed GOF mutations was available in the experiment conducted by Couñago *et al.* [13]. The likely reason for the disparate results is the differing experimental regimens. Lenski did not put strong constraints on the direction for *E. coli* to evolve, but Couñago *et al.* [13] replaced an essential gene with a substitute optimized for a different growth temperature before applying selective pressure, which they termed a "weak link" method. Furthermore, Couñago *et al.* [13] used an N_e that was more than three orders of magnitude greater than Lenski's group. The activity of the thermophilic adenylate kinase activity had to be replaced or compensated for. Apparently, the fastest way available to do so at high N_e was by GOF point mutations to the mesophilic substitute gene.

4.3.1 *Comparison to experiments where two selective routes were potentially available*

An interesting conceptual blend of the Lenski [18] and Couñago *et al.* [13] approaches was recently published by Gauger *et al.* [23]. This group mutated two amino acid residues of a plasmid-borne trpA gene of *E. coli*, transfected a Trp^- bacterial strain with the plasmid, and grew it in a tryptophan-limiting medium. One of the mutations (E49V) alone completely inactivates the gene product; the other mutation (D60N), when present alone, allows weak Trp^+ activity and supports growth in Trp^- media when the plasmid-borne gene is overexpressed. The authors expected cells containing the double mutant plasmid to take a short, selected route to full Trp^+ activity when grown in tryptophan-limiting medium by first reverting the inactivating mutation at position 49 (allowing the resumption of weak Trp^+ activity) and then reverting the second mutation at position 60 to regain full activity. However, almost all mutants recovered after sustained growth had not taken even the first step on that expected pathway. Rather, the expression of the *trpA* gene was decreased either by deletion, insertion of an IS element, or by various point mutations, apparently saving the cell the energy of overproducing the protein.

The *E. coli* point mutation rate is 5×10^{-10}. Gauger *et al.* [23] grew liquid cultures to an effective population size N_e of ~0.6×10^7 cells per generation. Substituting these numbers into equation 2 shows that $r_{D/fx}$ would be 5.3 — greater than one — if r_s were one and r_v were 100. That is, if the selective advantage the cell received from shutting down overexpression of the plasmid-borne gene were equal to the selective advantage it would receive from taking the first GOF mutational step to partial Trp^+ activity, the LOF mutation would be expected to easily be fixed in the population well before a GOF mutation appeared. For one partial-revertant to be expected to appear before the LOF mutant fixed under the conditions of the experiment r_s would have to be about 0.2. That is, the selection coefficient for the GOF mutation would have to be approximately five-fold greater than that of the LOF mutation. Equations 4 and 5 can be used to show that for the GOF mutant to be expected to dominate the population to >90%, the GOF selection coefficient would have to be about 12.5-times that for the LOF mutation. Apparently, regaining merely limited Trp^+ activity did not have 12.5-times the selective value of the decrease in expression of the plasmid gene caused by the LOF mutations. Thus, under the conditions of the experiment, the selective pathway back to full Trp^+ activity is blocked at the first step. Interestingly, if cells transfected with either singly-mutated plasmid (E49V or D60N) were grown in liquid culture, Trp^+ revertants quickly took over the culture, indicating the selection coefficient for full-reversion was greater than 12.5-times the selection coefficient for saving the cell the energy of overproducing the protein [23].

4.4 *Comparison to short-term evolution in the wild*

A possible objection to results from laboratory evolution experiments is that they are artificial. The organisms are housed in special environments and not exposed to the rigor and variety of challenges they would encounter in nature. Thus the many advantageous LOF mutations observed in experimental work may not reflect what happens in nature, since presumably the great majority of an organism's genes are required in the wild, and therefore few if any adaptive LOF mutations are available in nature.

While that may turn out to be the case, and more data will be required to come to a definitive conclusion, an increasing number of results from nature appear to ratify the importance of adaptive LOF mutations in the wild. One class of such LOF mutations which I have mentioned previously includes genes that help adapt humans to the presence of malaria [4]. Other important human adaptive mutations are also LOF mutations: immunity to HIV due to a deletion variant of CCR-5 [24]; and resistance to tuberculosis by a deletion variant of SLC11A1 [25]. Development of lactose tolerance in adult humans [26] also seems a good candidate for an adaptive LOF mutation, perhaps by loss of a repressor binding site, although that has not yet been confirmed. In a recent survey of multiple human genomes it has been determined that for humans, "On average, each person is found to carry approximately 250 to 300 loss-of-function variants in annotated genes...," over 1% of the total number of human genes [27].

A second example of LOF mutation in nature is seen in the evolution of the plague bacterium *Yersinia pestis*. A plausible evolutionary scenario to explain its great virulence is that it serially acquired several plasmids which conferred on it the ability to be transferred between mammalian hosts by flea bite [28, 29]. After the acquisition of these plasmids (which are GOF events), the *Y. pestis* genome lost several hundred genes, apparently because they were no longer necessary for its new life cycle [29, 30]. Thus, after several GOF events, the plague bacterium adjusted to its new environment by much more numerous and rapid LOF adaptive mutations.

Nadeau and Jiggins [31] have recently reviewed genomic studies of adaptation in natural populations and note that "Many of the well-studied examples of adaptive evolution have involved trait loss, such as the loss of bony structures in freshwater stickleback populations and the reduction of pigmentation and eyes in cavefish." Although, as mentioned earlier in this chapter, there is not a necessary correlation between phenotypic trait loss and adaptive LOF mutations, in the cases mentioned by Nadeau and Jiggins [31] they coincide. Loss of pelvic spines in freshwater sticklebacks has been traced to deletion of a Pitx1 enhancer [32]. Eye reduction in cavefish apparently involves multiple genes [33]. Of those that have been identified three involve decreased expression of the gene (*γ-M crystallin*, *rhodopsin*, and *αA crystallin*). One gene, *hsp90α*, has increased expression, and it appears to be involved in promoting apoptosis.

5. Conclusion

Organisms have adapted over evolutionary history both by gaining and losing functions. Therefore it is of basic interest to determine if one or the other dominates during particular circumstances. Until the past few decades, however, the molecular events underlying these processes were obscure. In recent decades science has in some cases gained the ability to determine whether the events behind a phenotypic adaptation involve an adaptive GOF mutation or an adaptive LOF mutation [4].

Both experimental laboratory work over the past few decades and recent genomic studies of adaptation in natural populations attest to the importance, even dominance, of LOF mutations in short term evolutionary episodes. The work presented in this paper helps show why this should be the case. Functional genetic elements such as genes and regulatory regions are built of multiple nucleotides, and a substantial fraction of mutations to these elements will cause them to lose their function. Thus the LOF mutation rate can be orders of magnitude greater than the nucleotide substitution rate. On the other hand, GOF mutations tend to be quite specific. So the rate for adaptive GOF mutations tends to be equal or very similar to the nucleotide mutation rate. As shown here, for some population size regions and for some values for the ratio of selection coefficients, the greater rate of mutation to the adaptive state for LOF versus GOF gives adaptive LOF mutations an intrinsic edge over adaptive GOF mutations.

In retrospect, the result is straightforward. Yet it also seems somewhat surprising because, as Nadeau and Jiggins [31] write, "there clearly are complex structures that are gained during evolution ... and we currently know little about how this process takes place." It may be hoped that understanding how organisms survive in the short term by adaptive LOF mutations will be a step toward understanding how complex structures are built over the long term.

References

1. Darwin C (1859) The Origin of Species. Bantam Books, New York.
2. Stott R (2003) Darwin and the Barnacle. W. W. Norton, New York.
3. Chang JC, Kan YW (1981) Antenatal diagnosis of sickle cell anaemia by direct analysis of the sickle mutation. Lancet 2:1127–1129.
4. Behe MJ (2010) Experimental evolution, loss-of-function mutations, and "The First Rule of Adaptive Evolution". Quart Rev Biol 85:419–445.
5. Friedman MJ (1978) Erythrocytic mechanism of sickle cell resistance to malaria. Proc Natl Acad Sci USA 75:1994–1997.
6. Verra F, Simpore J, Warimwe GM, Tetteh KK, Howard T, Osier FH, Bancone G, Avellino P, Blot I, Fegan G, Bull PC, Williams TN, Conway DJ, Marsh K, Modiano D (2007) Haemoglobin C and S role in acquired immunity against *Plasmodium falciparum* malaria. PloS ONE 2:e978.

7. Ruwende C, Khoo SC, Snow RW, Yates SN, Kwiatkowski D, Gupta S, Warn P, Allsopp CE, Gilbert SC, Peschu N, Newbold CI, Greenwood BM, Marsh K, Hill AVS (1995) Natural selection of hemi- and heterozygotes for G6PD deficiency in Africa by resistance to severe malaria. Nature 376:246–249.
8. Cavalli-Sforza LL, Menozzi P, Piazza A (1994) The History and Geography of Human Genes. Princeton University Press, Princeton, NJ.
9. Lynch M (2010) Rate, molecular spectrum, and consequences of human mutation. Proc Natl Acad Sci USA 107:961–968.
10. Perfeito L, Fernandes L, Mota C, Gordo I (2007) Adaptive mutations in bacteria: high rate and small effects. Science 317:813–815.
11. Li WH (1997) Molecular Evolution. Sinauer Associates, Sunderland, Mass.
12. Lynch M, Conery JS (2000) The evolutionary fate and consequences of duplicate genes. Science 290:1151–1155.
13. Couñago R, Chen S, Shamoo Y (2006) In vivo molecular evolution reveals biophysical origins of organismal fitness. Mol Cell 22:441–449.
14. Penman BS, Pybus OG, Weatherall DJ, Gupta S (2009) Epistatic interactions between genetic disorders of hemoglobin can explain why the sickle-cell gene is uncommon in the Mediterranean. Proc Natl Acad Sci USA 106:21242–21246.
15. Kimura M (1983) The Neutral Theory of Molecular Evolution. Cambridge University Press, Cambridge.
16. Ferenci T (2008) The spread of a beneficial mutation in experimental bacterial populations: the influence of the environment and genotype on the fixation of *rpoS* mutations. Heredity 100: 446–452.
17. Li WH (1975) The first arrival time and mean age of a deleterious mutant gene in a finite population. Am J Hum Gen 27:274–286.
18. Lenski RE (2004) Phenotypic and genomic evolution during a 20,000-generation experiment with the bacterium *Escherichia coli*. Plant Breeding Rev 24:225–265.
19. Cooper VS, Schneider D, Blot M, Lenski RE (2001) Mechanisms causing rapid and parallel losses of ribose catabolism in evolving populations of *Escherichia coli* B. J Bacteriology 183: 2834–2841.
20. Barrick JE, Yu DS, Yoon SH, Jeong H, Oh TK, Schneider D, Lenski RE, Kim JF (2009) Genome evolution and adaptation in a long-term experiment with *Escherichia coli*. Nature 461: 1219–1221.
21. Drake JW, Charlesworth B, Charlesworth D, Crow JF (1998) Rates of spontaneous mutation. Genetics 148:1667–1686.
22. Blount ZD, Barrick JE, Davidson CJ, Lenski RE (2012) Genomic analysis of a key innovation in an experimental *Escherichia coli* population. Nature 489: 7899-7906.
23. Gauger AK, Ebnet S, Fahey PF, Seelke R (2010) Reductive evolution can prevent populations from taking simple adaptive paths to high fitness. BIO-Complexity 2:1–9.

24. Samson M, Libert F, Doranz BJ, Rucker J, Liesnard C, Farber CM, Saragosti S, Lapoumeroulie C, Cognaux J, Forceille C, Muyldermans G, Verhofstede C, Burtonboy G, Georges M, Imai T, Rana S, Yi Y, Smyth RJ, Collman RG, Doms RW, Vassart G, Parmentier M (1996) Resistance to HIV-1 infection in caucasian individuals bearing mutant alleles of the CCR-5 chemokine receptor gene. Nature 382:722–725.
25. Barnes I, Duda A, Pybus OG, Thomas MG (2010) Ancient urbanization predicts genetic resistance to tuberculosis. Evolution doi:10.1111/j.1558–5646.2010. 01132.x
26. Tishkoff SA, Reed FA, Ranciaro A, Voight BF, Babbitt CC, Silverman JS, Powell K, Mortensen HM, Hirbo JB, Osman M, Ibrahim M, Omar SA, Lema G, Nyambo TB, Ghori J, Bumpstead S, Pritchard JK, Wray GA, Deloukas P (2007) Convergent adaptation of human lactase persistence in Africa and Europe. Nat Genetics 39:31–40.
27. The 1000 Genomes Project Consortium (2010) A map of human genome variation from population-scale sequencing. Nature 467:1061–1073.
28. Wren BW (2003) The yersiniae–a model genus to study the rapid evolution of bacterial pathogens. Nat Rev Microbiol 1:55–64.
29. Carniel E (2003) Evolution of pathogenic Yersinia: Some lights in the dark. In: Skurnik M, Bengoechea JA, Granfors K (eds) The Genus Yersinia: Entering the Functional Genomic Era, Kluwer Academic/Plenum, New York.
30. Chain PS, Hu P, Malfatti SA, Radnedge L, Larimer F, Vergez LM, Worsham P, Chu MC, Andersen GL (2006) Complete genome sequence of *Yersinia pestis* strains Antiqua and Nepal516: evidence of gene reduction in an emerging pathogen. J Bacteriol 188:4453–4463.
31. Nadeau NJ, Jiggins CD (2010) A golden age for evolutionary genetics? Genomic studies of adaptation in natural populations. Trends Gen 26:484–492.
32. Chan YF, Marks ME, Jones FC, Villarreal G Jr., Shapiro MD, Brady SD, Southwick AM, Absher DM , Grimwood J, Schmutz J, Myers RM, Petrov D, Jónsson B, Schluter D, Bell MA, Kingsley DM (2010) Adaptive evolution of pelvic reduction in sticklebacks by recurrent deletion of a Pitx1 enhancer. Science 327:302–305.
33. Jeffery WR (2009) Regressive evolution in *Astyanax* cavefish. Ann Rev Genetics 43:25–47.
34. Wolfram S (2003) *The Mathematica Book*. Wolfram Media, Champaign, IL.

arrays and spatially localized targets already present in the oocyte; those polarizations and localizations, in turn, derive from prior asymmetries inherent in the ovary [8–14].

Spatial information also precedes and directs the GRNs in embryos of the nematode *Caenorhabditis elegans*. The sperm centrosome first establishes an anterior-posterior axis by initiating cytoskeletal changes that produce a polarized distribution of zygotic proteins. These in turn lead to asymmetrical cell divisions and subsequent differentiation [15–17].

In ascidian oocytes, the cortex (the cell membrane plus underlying cytoplasmic and cytoskeletal elements) already contains spatially localized morphogenetic determinants that specify the primary axis of the embryo. Upon fertilization, the sperm centrosome induces cytoskeletal changes that reorganize those determinants and establish the second (dorsal-ventral) axis [18,19].

Oocytes of the frog *Xenopus laevis* also have a primary axis before the sperm enters. The sperm establishes a second axis by aligning a microtubule array in the zygote that directs morphogenetic determinants to the future dorsal side of the embryo [20–22].

In all of these cases, spatial coordinates are established in the embryo before zygotic GRNs become active. Such coordinates provide biological information by specifying domains in the embryo that later differentiate by means of GRNs in progressively finer detail. Spatial information can be mediated by polarized cytoskeletal arrays, which in some embryos are reorganized by the sperm upon fertilization. Other spatial information is mediated by cortical or membrane patterns. The remainder of this paper focuses on the latter.

Endogenous Electric Fields

One way membranes can provide spatial information is by generating electric fields. Indeed, all living cells produce electric fields by transporting ions across their membranes. The sodium-potassium pump utilizes energy from ATP to move three sodium ions out of the cell while taking in only two potassium ions [23]. With each cycle of the pump the interior of the cell thus acquires a net negative charge equivalent to one electron. So the inside of every living cell is electrically negative with respect to its external environment, and the voltage across the membrane — the "membrane potential" — ranges from about 50 to 200 mV DC (average ~70 mV). This produces a steady endogenous electric field in the 10–100 mV/mm range [24].

Multicellular organisms, and their organs, are covered by an epithelium — a single layer of cells laterally connected by tight junctions that block the flow of ions. Epithelia are polarized, in the sense that the ion channels on the side facing away from the organ or organism are different from the ion channels on the side facing the organ or organism. The result is a "transepithelial potential" that (unlike the transmembrane potential of individual cells) is usually negative on the outside of the organ or organism and positive on the inside. The transepithelial potential typically ranges from 15 to 60 mV [24].

Xenopus laevis embryos generate endogenous electric fields from the single cell stage through at least the neurula stage [25–27]. In the embryos of chicks (*Gallus gallus*) and mice (*Mus musculus*), large ionic currents pass through the primitive streak, a furrow through which cells move into the interior as they differentiate into tissues and organs [28,29].

In 1995, Riyi Shi and Richard Borgens proposed that endogenous electric fields could "both polarize the early vertebrate embryo and serve as cues for morphogenesis and pattern." If this were true, they wrote, "at least five corollaries must be satisfied: (1) embryonic cells must be responsive to extracellular voltages within the range of magnitudes measured within embryos, (2) disturbance of these endogenous gradients of voltage by imposed voltages in the physiological range should result in developmental arrest or abnormality, (3) this disturbance should be most profound at the embryonic stages when endogenous fields are present within the embryo, (4) since the internal voltages are spatially polarized during development, the form of teratological change in the embryo produced by an artificially imposed field should be predictable based on its orientation relative to the embryo's orientation, and (5) any technique that will reduce or eliminate an endogenous voltage gradient should lead to developmental arrest or retardation. All five of these requirements have been met" [30].

For example, applied electric fields of physiological strength can induce and guide cell migration *in vitro* [31–39]. Furthermore, targeted disruption of endogenous electric fields disrupts normal development in ways that suggest the fields are controlling morphogenesis [40–43]. There is also evidence that direct currents in the physiological range can affect gene expression [44,45].

(Note that this has nothing to with the controversy surrounding the alleged effects of environmental electromagnetic fields — whether extremely low frequency or microwave frequency. The endogenous electric fields that concern us here are steady, not oscillating.)

Since the topology of an endogenous electric field would depend on the spatial arrangement of ion channels in the membrane or epithelium, such a field could be one way that membrane patterns provide spatial coordinates for embryo development. Another way that membrane patterns could affect development is through intracellular signaling.

Membrane Proteins and Intracellular Signaling

Networks of intracellular signaling molecules regulate a cell's morphology, physiology. They also interface with GRNs to regulate gene expression, and they mediate a cell's response to extracellular signals such as hormones and growth factors.

Membrane proteins are key nodes in such networks. Many intracellular signals originate with them, and their spatial localization is often essential to their proper functioning. Some of the more important membrane-bound signaling molecules are the Ras proteins (so called because they were originally found in cells transformed by Rat sarcoma viruses) [46].

Ras proteins are localized mostly on the inner face of the plasma membrane, though they also occur in inner membranes such as the Golgi apparatus [47]. They come in many forms: in humans alone, the Ras superfamily includes more than 150 different members [48]. Distinct Ras isoforms have distinct functions [49], including the regulation of ion channels [50], cell migration [51], and cytoskeletal remodeling [52]. Proper Ras functioning is essential to mammalian development, and its disruption has been linked to cancer [53].

Ras proteins are organized in the membrane into spatially segregated "nanoclusters," each containing several proteins [54–56]. The spatial localization of Ras proteins in nanoclusters is essential for generating and regulating spatially distinct intracellular signaling circuits [57,58]. In 2008, Angus Harding and John Hancock wrote that those circuits "integrate and process signals to operate as switches, oscillators, logic gates, memory modules and many other types of control system. These complex processing capabilities enable cells to respond appropriately to the myriad of external cues that direct growth and development." Harding and Hancock identified "common design principles that highlight how the spatial organization of signal transduction circuits can be used as a fundamental control mechanism to modulate system outputs *in vivo*" [59].

For example, Ras nanoclusters operate as analog-digital-analog converters. Ras is either non-activated (off) or activated (on); it responds to an external signaling molecule such as epidermal growth factor by switching on; the concentration of the external signaling molecule determines how many Ras molecules are activated; and the number of activated Ras molecules determines the downstream concentration of an intracellular molecule that interacts with other signaling networks and regulates gene expression. The spatial organization of Ras molecules in nanoclusters is essential to reduce noise and produce high fidelity signal transmission across the membrane [60–62].

So spatial organization is essential to the proper functioning of membrane proteins, and those proteins can generate intracellular signals that regulate gene expression. The gene regulatory networks described by Davidson and Carroll are related to DNA information at one end and spatial information at the other. Neither source of information can be discounted.

The Sugar Code

Cell–cell interactions — including those in developing embryos — depend on carbohydrates localized on the surface of each cell. Sugars can be attached either to lipids (glycolipids) or to membrane proteins (glycoproteins). Carbohydrate-binding proteins (lectins) mediate their interactions. Because sugars can be covalently linked in a variety of ways (unlike amino acids in a protein, which are all linked by identical peptide bonds), the diversity of side chains on glycolipids and glycoproteins is enormous.

In 1985 Ronald Schnaar wrote, "There appears to be a code on the surface of each cell that specifies its function and directs its interactions with other cells, a code in some ways comparable to the genetic code carried on the DNA molecules *inside* each cell." The "letters" of the cell surface code to which Schnaar was referring are sugar molecules. A few monosaccharide building blocks can produce the enormous diversity of "words" needed to identify the many different kinds of cells in a complex organism, Schnaar explained, because "each building block can assume several different positions. It is as if an *A* could serve as four different letters, depending on whether it was standing upright, turned upside down, or laid on either of its sides. In fact, seven simple sugars can be rearranged to form hundreds of thousands of unique words, most of which have no more than five letters. (This alphabet is even more efficient than the genetic code: the four nucleic acids that constitute DNA — guanine, adenine, thymine, and cytosine — can be connected only front to back, like roller coaster cars.) So, not only are sugars in the right place to serve as the alphabet for the cell-surface code, they have the requisite structural flexibility too." Schnaar concluded, "It may be that as much control over the cell's fate, and as much of the language of life's unfolding, reside on the cell's surface as in its nucleus" [63].

Hans-Joachim Gabius has called this the "sugar code." According to Gabius, sugars provide a "high-density coding system" that is "essential to allow cells to communicate efficiently and swiftly through complex surface interactions." This is because "all the structural requirements for forming a wide array of signals with a system of minimal size are met by oligomers of carbohydrates. These molecules surpass amino acids and nucleotides by far in information-storing capacity and serve as ligands in biorecognition processes for the transfer of information" [64,65]. In 2009, Lopez and Schnaar provided evidence that membrane patterns in cells of the immune system and the nervous system depend in part on lateral interactions among their constituent glycolipids [66].

So the sugar code carries essential biological information in addition to that carried by DNA sequences. It is not known whether the sugar code can be directly inherited, but there is evidence that other cell surface patterns are heritable independently of DNA sequences.

Some Membrane Patterns Can Be Inherited apart from the DNA

In single-celled protozoa, changes in cilia patterns in the cortex can be inherited apart from changes in the DNA. In 1965, Beisson and Sonneborn induced one member of a conjugating pair of *Paramecium aurelia* to transfer to its partner a section of cortex that had been surgically inverted 180º relative to the surrounding cortex. The DNA was unchanged. Ciliates with artificially inverted rows have been stably maintained for thousands of generations [67,68].

In 1977, Ng and Frankel reported similar results with *Tetrahymena pyriformis* and concluded, "The cell as an architect thus not only makes use of the genomic information to produce the appropriate building blocks, but, in addition, also arranges the building blocks according to the blueprint as defined in the preexisting architecture" [69]. Frankel called this extra-genic blueprint the "corticotype" [70]. Similar results have been reported in *Tetrahymena* by Nanney and in *Stylonychia* by Grimes [71,72]. Clearly, cortical patterns in ciliates can serve as their own templates when they replicate.

There is also evidence that some cellular patterns in multicellular organisms are heritable apart from the DNA. In 1977, Albrecht-Buehler reported that after mitoses in cultured 3T3 mouse fibroblast cells, about 40% of daughter cells contained mirror symmetrical actin-bundle patterns and performed directional changes in a mirror symmetrical way. He concluded that the "organizations of daughter 3T3 cells form mirror images of each other" at the time of mitosis [73].

In 1979, Solomon observed that about 60% of sister pairs in cultured neuroblastoma cells displayed analogous morphologies. He concluded that "determinants of biologically functional shape can be dictated to some extent by the cells themselves. Such a program of information can be heritable through mitosis," though "we do not know, of course, how or in what structures this information is stored" [74]. In 1981, Solomon found additional circumstantial evidence for endogenous determinants of morphology, and he concluded, "It is possible that detailed cell morphology is specified by structures which nucleate the assembly of the cytoskeletal fibers that underlie that morphology," though "an alternative model is that the endogenous determinants of neuroblastoma morphology may reside at the cell surface" [75].

In 1990, Locke reported paired patterns in caterpillar epidermis cells that "imply that a part of the epigenetic sequence leading to the formation of the pattern has replicated [and been] inherited by daughter cells. It is not just genetic material that is inherited but part of a cell in a particular state. Inheritance is somatic, in the sense that it is part of the operation of an epigenetic determinant that has been inherited." According to Locke, the problem with such inheritance is that it "requires more than number and kind of molecule. The duplication of pattern involves relative position and orientation," factors that "cannot be specified only by a base sequence." Locke concluded, "The observations suggest that while the detailed arrangement of cell components may be variable and not under direct genetic control, some patterns result from epigenetic determinants that replicate and are inherited from one mitosis to the next" [76]. The following year, Locke and his colleagues published "further evidence for the operation of transiently heritable factors as determinants for cell pattern" [77], and in 2007 an international team of biologists reported that similar mirror-symmetric divisions are essential for proper neural tube development in zebrafish embryos [78].

As Solomon pointed out, such symmetrical divisions may be due to the inheritance of cytoskeletal patterns, or membrane patterns, or both. In the case of membrane patterns, proteins from the cell interior are incorporated during membrane growth only if they match the existing matrix. George Palade wrote in 1983 that membranes "recognize and incorporate like components, grow by expansion in two dimensions, and eventually divide into two sets of descendant membranes, one for each daughter cell. These sets are qualitatively identical" [79].

Robert Poyton has proposed a detailed hypothesis to explain how this process might work. According to Poyton, the units of epigenetic spatial memory in membranes are hetero-oligomeric membrane proteins, of which there are many kinds. These proteins are localized on membrane surfaces in quasistable "unit areas." When phospholipids are incorporated into the membrane in preparation for replication, the hetero-oligomers dissociate into their subunits. Then newly synthesized subunits in the cytoplasm associate with the corresponding older subunits to form hybrid hetero-oligomers that are chemically identical to the originals. Thus membrane replication — like DNA replication — is semi-conservative. Poyton wrote, "It is the preexisting spatial memory encoded in a membrane that brings new proteins to its surface... Realizing that genetic memory is one-dimensional, along a DNA molecule, whereas spatial memory is likely to be two-dimensional, along membrane surfaces, and three-dimensional within the cellular interior, it is probable that spatial memory is more complicated and diverse than genetic memory" [80].

Some recently published work is consistent with several aspects of Poyton's hypothesis. First, empirical and theoretical studies indicate that the interaction of membrane proteins — in particular, the stability of homo- and hetero-dimers — is affected by the extent of their dilution in lipid bilayers [81,82]. Second, as prions demonstrate, proteins can serve as templates for their own self-replication [83–85]. Third, experiments show that membrane proteins selectively recruit other proteins to Ras nanoclusters and adjust their orientation to maintain intracellular signaling [86–89].

Implications for Modern Evolutionary Theory

Clearly, the biological information needed for embryogenesis exceeds the information encoded in DNA sequences. RNAs and proteins encoded by DNA form gene regulatory networks that are essential for development, but those networks must be localized in spatial domains for the embryo to differentiate into various cell types and organs, and those domains must be spatially ordered with respect to each other for the organism to develop its proper morphology.

Two features of cells and embryos that provide spatial cues are the membrane and the cytoskeleton. Both are composed of subunits that are encoded in DNA, but their two- or three-dimensional patterns are not determined by those subunits, just as the structure of a house is not determined by its bricks.

The arrangement of proteins and carbohydrates in a membrane is analogous to a two-dimensional code that specifies many aspects of a cell's morphology and physiology, as well as its interactions with other cells. Indeed, several membrane codes can be distinguished: the pattern of ion channels in the epithelium of an embryo generates an endogenous electric field that provides a three-dimensional coordinate system to guide migrating cells; the pattern of membrane-bound proteins such as those in the Ras family spatially organizes intracellular signaling and mediates responses to extracellular signals; and the complex pattern of carbohydrates on a cell surface is essential for cell–cell interactions.

Membrane patterns in ciliates are known to be heritable independently of the information in DNA sequences, and there is evidence that some cytoskeletal and membrane patterns in the cells of multicellular organisms can also be inherited apart from the DNA. Taken together, the data suggest that embryo development is not controlled by DNA alone, and thus that DNA mutations are not sufficient to provide raw materials for evolution.

In 1983, John Maynard Smith defended the gene-centered view of development and evolution and asserted that the DNA-independent inheritance of cortical patterns in ciliates constituted "the only significant experimental threat" to that view [90]. It now appears that ciliates are not the only example of non-genic developmental information and DNA-independent inheritance.

One could speculate that accidental changes in membrane patterns — analogous to accidental mutations in DNA — could provide the missing raw materials for evolution. Yet two- and three-dimensional information-carrying patterns are likely to entail more specified complexity than the one-dimensional information in DNA sequences, making beneficial "mutations" in such patterns much less probable than beneficial mutations in DNA. At the very least, calculations of the time required for evolution will now have to take into account these higher dimensions of biological information.

Acknowledgments

The author gratefully acknowledges the financial support of the Discovery Institute.

References

1. Webster G, Goodwin BC (1996) Form and Transformation: Generative and Relational Principles in Biology. Cambridge University Press, Cambridge.
2. Harold FM (2001) The Way of the Cell: Molecules, Organisms and the Order of Life. Oxford University Press, Oxford.
3. Müller GB, Newman SA (2003) Origination of Organismal Form: Beyond the Gene in Developmental and Evolutionary Biology. MIT Press; Cambridge, MA.
4. Davidson EH (2010) Emerging properties of animal gene regulatory networks. Nature 468: 911–920.
5. Carroll SB (2008) Evo Devo and an expanding evolutionary synthesis: a genetic theory of morphological evolution. Cell 134: 25–36.
6. Davidson EH (2006) The Regulatory Genome: Gene Regulatory Networks in Development and Evolution. Academic Press, Amsterdam, pp. 90, 94.
7. Carroll SB (2005) Endless Forms Most Beautiful: The New Science of Evo Devo. W. W. Norton, New York, p. 116.
8. Frohnhöfer HG, Nüsslein-Volhard C (1986) Organization of anterior pattern in the *Drosophila* embryo by the maternal gene *bicoid*. Nature 324: 120–125.
9. Lehmann R, Nüsslein-Volhard C (1991) The maternal gene *nanos* has a central role in posterior pattern formation of the *Drosophila* embryo. Development 112: 679–691.

10. Theurkauf WE, Hazelrigg, TI (1998) In vivo analyses of cytoplasmic transport and cytoskeletal organization during *Drosophila* oogenesis: characterization of a multi-step anterior localization pathway. Development 125: 3655–3666.
11. Forrest KM, Gavis ER (2003) Live imaging of endogenous rna reveals a diffusion and entrapment mechanism for *nanos* mRNA localization in *Drosophila.* Current Biology 13: 1159–1168.
12. Mhlanga MM, *et al* (2009) In vivo colocalisation of oskar mRNA and trans-acting proteins revealed by quantitative imaging of the *Drosophila* oocyte. PLoS One 4: e6241.
13. Irion U, St Johnston D (2007) Bicoid RNA localization requires specific binding of an endosomal sorting complex. Nature 445: 554–558.
14. Roth S, Lynch JA (2009) Symmetry breaking during *Drosophila* oogenesis. Cold Spring Harbor Perspectives in Biology 1: a001891.
15. Cowan CR, Hyman AH (2004) Asymmetric cell division in *C. elegans*: cortical polarity and spindle positioning. Annual Review of Cell and Developmental Biology 20: 427–453.
16. Gönczy P, Rose LS (2005) Asymmetric cell division and axis formation in the embryo. WormBook (October 15): 1–20.
17. Tsai M-C, Ahringer J (2007) Microtubules are involved in anterior-posterior axis formation in *C. elegans* embryos. Journal of Cell Biology 179: 397–402.
18. Sardet S, Dru P, Prodon F (2005) Maternal determinants and mRNAs in the cortex of ascidian oocytes, zygotes and embryos. Biology of the Cell 97: 35–49.
19. Sardet S, *et al* (2007) From oocyte to 16-cell stage: cytoplasmic and cortical reorganizations that pattern the ascidian embryo. Developmental Dynamics 236: 1716–1731.
20. Larabell CA, *et al* (1996) Confocal microscopy analysis of living *Xenopus* eggs and the mechanism of cortical rotation. Development 122: 1281–1289.
21. Rowning BA, *et al* (1997) Microtubule-mediated transport of organelles and localization of beta-catenin to the future dorsal side of *Xenopus* eggs. Proc Natl Acad Sci USA 94: 1224–1229.
22. Weaver C, Kimelman D (2004) Move it or lose it: axis specification in *Xenopus*. Development 131: 3491–3499.
23. Skou JC (1989) The identification of the sodium-pump as the membrane-bound Na^+/K^+-ATPase: a commentary. Biochimica et Biophysica Acta 1000: 435–438.
24. Nuccitelli R (2006) Endogenous Electric Fields in Animals, Chapter 2. In: Handbook of Biological Effects of Electromagnetic Fields, 3rd edn. Taylor & Francis; Boca Raton, FL.
25. Kline D, Robinson KR, Nuccitelli R (1983) Ion currents and membrane domains in the cleaving *Xenopus* egg. Journal of Cell Biology 97: 1753–1761.
26. Metcalf MEM, Shi R, Borgens RB (1994) Endogenous ionic currents and voltages in amphibian embryos. Journal of Experimental Zoology 268: 307–322.

27. Robinson KR, Stump RF (1984) Self-generated electrical currents through *Xenopus* neurulae. Journal of Physiology 352: 339–352.
28. Jaffe LF, Stern CD (1979) Strong electrical currents leave the primitive streak of chick embryos. Science 206: 569–571.
29. Winkel GK, Nuccitelli R (1989) Large ionic currents leave the primitive streak of the 7.5-Day mouse embryo. Biological Bulletin Supplement 176: 110–117.
30. Shi R, Borgens RB (1995) Three-dimensional gradients of voltage during development of the nervous system as invisible coordinates for the establishment of embryonic pattern. Developmental Dynamics 202: 101–114.
31. Hinkle L, McCaig CD, Robinson KR (1981) The direction of growth of differentiating neurons and myoblasts from frog embryos in an applied electric field. Journal of Physiology 314: 121–135.
32. Stump RF, Robinson KR (1983) *Xenopus* neural crest cell migration in an applied electrical field. Journal of Cell Biology 97: 1226–1233.
33. Cooper MS, Keller RE (1984) Perpendicular orientation and directional migration of amphibian neural crest cells in dc electrical fields. Proc Natl Acad Sci USA 81: 160–164.
34. Erickson CA, Nuccitelli R (1984) Embryonic fibroblast motility and orientation can be influenced by physiological electric fields. Journal of Cell Biology 98: 296–307.
35. Gruler H, Nuccitelli R (1991) Neural crest cell galvanotaxis: new data and a novel approach to the analysis of both galvanotaxis and chemotaxis. Cell Motility and the Cytoskeleton 19: 121–133.
36. Nishimura KY, Isseroff RR, Nuccitelli R (1996) Human keratinocytes migrate to the negative pole in direct current electric fields comparable to those measured in mammalian wounds. Journal of Cell Science 109: 199–207.
37 Farboud B, *et al* (2000) DC electric fields induce rapid directional migration in cultured human corneal epithelial cells. Experimental Eye Research 70: 667–673.
38. McCaig CD, *et al* (2005) Controlling Cell Behavior Electrically: Current Views and Future Potential. Physiological Reviews 85: 943–978.
39. Rajnicek AM, Foubister LE, McCaig CD (2006) Growth cone steering by a physiological electric field requires dynamic microtubules, microfilaments and Rac-mediated filopodial asymmetry. Journal of Cell Science 119: 1736–1745.
40. Hotary KB, Robinson KR (1992) Evidence of a role for endogenous electric fields in chick embryo development. Development 114: 985–996.
41. Metcalf MEM, Borgens RB (1994) Weak applied voltages interfere with amphibian morphogenesis and pattern. Journal of Experimental Zoology 268: 323–338.
42. Hotary KB, Robinson KR (1994) Endogenous electrical currents and voltage gradients in *Xenopus* embryos and the consequences of their disruption. Developmental Biology 166: 789–800.
43. Levin M (2003) Bioelectromagnetics in morphogenesis. Bioelectromagnetics 24: 295–315.

44. Jennings J, Chen D, Feldman D (2008) Transcriptional response of dermal fibroblasts in direct current electric fields. Bioelectromagnetics 29: 394–405.
45. Morokuma J, *et al* (2008) Modulation of potassium channel function confers a hyper-proliferative invasive phenotype on embryonic stem cells. Proc Natl Acad Sci USA 105: 16608–16613.
46. Willingham MC, *et al* (1980) Localization of the src gene product of the Harvey strain of MSV to plasma membrane of transformed cells by electron microscopic immuno-cytochemistry. Cell 19: 1005–1014.
47. Walker SA, Lockyer PJ (2004) Visualizing Ras signalling in real-time. Journal of Cell Science 117: 2879–2886.
48. Wennerberg K, Rossman KL, Der CJ (2005) The Ras superfamily at a glance. Journal of Cell Science 118: 843–846.
49. Omerovic J, Laude AJ, Prior IA (2007) Ras proteins: paradigms for compartmentalised and isoform-specific signaling. Cellular and Molecular Life Sciences 64: 2575–2589.
50. Pochynyuk O, Stockand JD, Staruschenko A (2007) Ion channel regulation by Ras, Rho, and Rab small GTPases. Experimental Biology and Medicine 232: 1258–1265.
51. Charest PG, Firtel RA (2007) Big roles for small GTPases in the control of directed cell movement. Biochemical Journal 401: 377–390.
52. Correll RN, *et al* (2008) The RGK family of GTP-binding proteins: regulators of voltage-dependent calcium channels and cytoskeleton remodeling. Cellular Signalling 20: 292–300.
53. Ramjaun AR, Downward J (2007) Ras and phosphoinositide 3-kinase: partners in development and tumorigenesis. Cell Cycle 6: 2902–2905.
54. Plowman SJ, *et al* (2005) H-ras, K-ras, and inner plasma membrane raft proteins operate in nanoclusters with differential dependence on the actin cytoskeleton. Proc Natl Acad Sci USA 102: 15500–15505.
55. Abankwa D, Gorfe AA, Hancock JF (2007) Ras nanoclusters: molecular structure and assembly. Seminars in Cell & Developmental Biology 18: 599–607.
56. Goswami D, *et al* (2008) Nanoclusters of GPI-anchored proteins are formed by cortical actin-driven activity. Cell 135: 1085–1097.
57. Abankwa D, *et al* (2008) A novel switch region regulates H-ras membrane orientation and signal output. EMBO Journal 27: 727–735.
58. Henis YI, Hancock JF, Prior IA (2009) Ras acylation, compartmentalization and signaling nanoclusters (Review). Molecular Membrane Biology 26: 80–92.
59. Harding AS, Hancock JF (2008) Using plasma membrane nanoclusters to build better signaling circuits. Trends in Cell Biology 18: 364–371.
60. Tian T, *et al* (2007) Plasma membrane nanoswitches generate high-fidelity Ras signal transduction. Nature Cell Biology 9: 905–914.
61. Harding AS, Hancock JF (2008) Ras nanoclusters: combining digital and analog signaling. Cell Cycle 7: 127–134.

62. Gurry T, Kahramanoğullari O, Endres RG (2009) Biophysical mechanism for ras-nanocluster formation and signaling in plasma membrane. PLoS One 4: e6148.
63. Schnaar RL (1985) The Membrane is the message. The Sciences (May-June): 34–40.
64. Gabius H-J (2000) Biological information transfer beyond the genetic code: the sugar code. Naturwissenschaften 87: 108–121.
65. Gabius H-J, *et al* (2004) Chemical biology of the sugar code. Chembiochem 5: 740–764.
66. Lopez PHH, Schnaar RL (2009) Gangliosides in cell recognition and membrane protein regulation. Current Opinion in Structural Biology 19: 549–557.
67. Beisson J, Sonneborn TM (1965) Cytoplasmic Inheritance of the Organization of the Cell Cortex in *Paramecium Aurelia*. Proc Natl Acad Sci USA 53: 275–282.
68. Grimes GW, Aufderheide KJ (1991) *Cellular Aspects of Pattern Formation: The Problem of Assembly*, Monographs in Developmental Biology Vol. 22 (Karger, Basel), pp. 22–26.
69. Ng SF, Frankel J (1977) 180º rotation of ciliary rows and its morphogenetic implications in *Tetrahymena pyriformis*. Proc Natl Acad Sci USA 74: 1115–1118.
70. Frankel J (1980) Propagation of Cortical Differences in *Tetrahymena*. Genetics 94: 607–623.
71. Nanney DL (1968) Cortical Patterns in Cellular Morphogenesis. Science 160: 496–502.
72. Grimes GW (1982) Pattern Determination in Hypotrich Ciliates. American Zoologist 22: 35–46.
73. Albrecht-Buehler G (1977) Daughter 3T3 cells: are they mirror images of each other? Journal of Cell Biology 72: 595–603.
74. Solomon S (1979) Detailed Neurite Morphologies of Sister Neuroblastoma Cells Are Related. Cell 16: 165–169.
75. Solomon S (1981) Specification of cell morphology by endogenous determinants. Journal of Cell Biology 90: 547–553.
76. Locke M (1990) Is there somatic inheritance of intracellular patterns? Journal of Cell Science 96: 563–567.
77. Delhanty P, Leung H, Locke M (1991) Paired cytoskeletal patterns in an epithelium of siamese twin cells. European Journal of Cell Biology 56: 443–450.
78. Tawk M, *et al* (2007) A mirror-symmetric cell division that orchestrates neuroepithelial morphogenesis. Nature 446: 797–800.
79. Palade GE (1983) Membrane Biogenesis: An Overview. Methods in Enzymology 96: xxix-lv.
80. Poyton RO (1983) Memory and Membranes: The Expression of Genetic and Spatial Memory During the Assembly of Organelle Macrocompartments. Modern Cell Biology 2: 15–72.

81. Hong H, *et al* (2010) Method to measure strong protein-protein interactions in lipid bilayers using a steric trap. Proc Natl Acad Sci USA 107: 19802–19807.
82. Psachoulia E, Marshall DP, Sansom MS (2010) Molecular dynamics simulations of the dimerization of transmembrane alpha-helices. Accounts of Chemical Research 43: 388–396.
83. Pezza JA, Serio TR (2007) Prion propagation: the role of protein dynamics. Prion 1: 36–43.
84. Krammer C, Schätzl HM, Vorberg I (2009) Prion-like propagation of cytosolic protein aggregates: insights from cell culture models. Prion 3: 206–212.
85. Sindi SS, Serio TR (2009) Prion dynamics and the quest for the genetic determinant in protein-only inheritance. Current Opinion in Microbiology 12: 623–630.
86. Plowman SJ, *et al* (2008) Electrostatic Interactions Positively Regulate K-Ras Nanocluster Formation and Function. Molecular and Cellular Biology 28: 4377–4385.
87. Shalom-Feuerstein R, *et al* (2008) K-ras nanoclustering is subverted by overexpression of the scaffold protein galectin-3. Cancer Research 68: 6608–6616.
88. Abankwa D, *et al* (2010) Ras membrane orientation and nanodomain localization generate isoform diversity. Proc Natl Acad Sci USA 107: 1130–1135.
89. Ariotti N, *et al* (2010) Epidermal growth factor receptor activation remodels the plasma membrane lipid environment to induce nanocluster formation. Molecular and Cellular Biology 30: 3795–3804.
90. Maynard Smith J (1983) Evolution and Development, pp. 33–46 in Goodwin BC, Holder N & Wylie CC (eds), *Development and Evolution*, The Sixth Symposium of the British Society for Developmental Biology (Cambridge University Press, Cambridge).

Explaining Metabolic Innovation: Neo-Darwinism *versus* Design

Douglas D. Axe and Ann K. Gauger

Biologic Institute, 16310 NE 80th Street, Redmond, WA 98052, USA.
daxe@biologicinstitute.org, agauger@biologicinstitute.org

Abstract

Like all life, bacterial life depends on a complex, integrated network of precise metabolic processes. These processes are carried out by more than a thousand enzymes — genetically encoded proteins with information-rich three-dimensional structures that catalyze specific chemical reactions. Can neo-Darwinian theory explain the origin of this network of enzymes that orchestrates metabolic complexity? Building on previous experimental and theoretical work, we argue here that it cannot. But instead of merely listing the theory's shortcomings, we attempt to construct a full and coherent picture of *how* it has failed to explain metabolic innovation, from the level of single enzymes all the way up to the network of enzymatic pathways that composes metabolism as a whole. Then, from this critical synthesis we identify six key principles of a new theory of biological innovation. Although these principles only hint at the substance of the new theory, they show clearly that it will be strikingly unlike neo-Darwinism. Whereas the old theory focuses on the simple material processes of mutation and selection in the hope that these can drive innovation, the new one focuses on innovation itself — on the concepts that guide effective designs. Consequently, the new theory will look more like the systematic concepts of an engineering discipline than a set of causal laws.

Key words: metabolic complexity, innovation, pathway evolution, complex adaptation, enzyme recruitment, cost of gene expression, causal circularity, design principles, critique of neo-Darwinism

Introduction

Life exhibits extraordinary functional complexity on many scales, from the molecular to the organismal and on up to whole ecosystems. Near the bottom of this arrangement is *metabolic* complexity, which refers to the intricate networks of coordinated chemical reactions that undergird all biological phenomena. Even the very simplest organisms, bacteria, are highly complex in this respect, which makes metabolic complexity a universal hallmark of life. Its universality also makes it a *benchmark* for assessing theories of biological origins. That is, any theory claiming to explain the origin of biological complexity in general must tackle the particular challenge of explaining metabolic complexity.

How well has the dominant theory, neo-Darwinism, met this challenge? The structure of metabolism itself suggests that this should be assessed in a hierarchical way. At the lower level the question is how well the theory explains the origin of new functions for single enzymes, while at the higher level it is how well it explains the origin of the more complex metabolic functions that emerge when enzyme functions are combined to form metabolic pathways, and the integrated networks of pathways that constitute metabolism as a whole. Notice that natural selection relates more directly to the higher level, in that this is where phenotypic traits are manifested, whereas mutation relates more directly to the lower level, in that individual mutations typically alter single genes, and therefore single enzymes. The perennial challenge for neo-Darwinism has been to explain how mutation and selection, two disparate phenomena operating at different levels, can combine to produce such spectacular functional innovations at both levels.

The hope has always been that explaining evolutionary innovation at the level of single genes would eventually simplify the task of explaining innovation at the level of complete pathways. That reductionistic hope seems to be fading. Even at the level of single genes, explaining innovation is growing harder, not easier, as more and more distinct protein structures are discovered. The count of fundamentally distinct structures, or *folds*[1] as they are known, now stands at about 2,000, with more being added every year.

The extraordinary difficulty that neo-Darwinism encounters with single-gene innovations requiring a new protein fold has recently been described in detail [1]. That raises an obvious question. If the Darwinian mechanism cannot reliably explain innovation at the level of a single protein fold, what *can* it explain? This prompted us to investigate the more modest case of enzymatic innovation within a fold family, which we regard as metabolic innovation on the smallest scale possible.[2] With that aim, we attempted to modify one particular bacterial enzyme so as to make it perform the function of another that closely resembles it [2]. Although we were ultimately unable to achieve this functional conversion, extensive testing of the kinds of amino-acid substitutions that ought to promote it

[1]Proteins have three-dimensional folded structures that determine their function. Those with secondary structural elements (alpha helices and beta strands) in the same order and similar spatial arrangement are said to have a *common* fold, or in other words, to be members of the same fold family. Proteins with fundamentally *distinct* folds differ in the arrangements of secondary structural elements and/or in their order.

[2]Although adaptations can certainly occur on a smaller scale, 'innovation' refers to the first-time appearance of a genuinely new function, not the adjustment of an existing function.

demonstrated that success would, for our test case, require many more specific changes than the Darwinian mechanism can accomplish, even over billions of years.

It would be tempting to disregard that result if there were a body of contrary evidence. Instead, as we have discussed [2], our result is just one contribution to a consistent picture based on numerous studies (see below). No one denies the possibility of converting enzymes to new functions, but it seems that anyone attempting it with the assumption that it can be done with just a few nucleotide changes is in for a surprise.

Where to go from here is a matter of perspective. Darwin's theory certainly will not benefit from ignoring or denying the severity of the problems that have beset it. Once that is conceded, the most important question is whether the theory needs to be remedied or replaced. Among the things that will be needed to answer that question is a full picture of what has gone wrong with the standard evolutionary account. In other words, it will be increasingly helpful to go beyond a mere catalog of inexplicable facts to something more like a *synthesis* of the whole problem. We use the word 'helpful' here because a synthesis of this kind should, we think, be the start of something much more positive than the dismantling of an old theory. It should instead be seen as an opportunity to gain key insights for constructing a new theory by building a clear understanding of how the old theory went wrong.

With that in mind, we here take a step toward such a synthesis by describing briefly the general aspects of metabolic innovation that most profoundly challenge the current neo-Darwinian model. The aspects are logically separable, which allows them to be examined as distinct topics, but their effects are highly interconnected. We will show this by developing a synthesis of the whole problem in a progressive way as each aspect is considered. Based on this critical synthesis we then offer the beginnings of a *positive* synthesis — a set of principles that hint at a new theory of innovation. The ultimate aim, of course, is to develop a theoretical framework from which to understand all biological innovation. Metabolic innovation will admittedly be only a small part of that big picture, but its relative simplicity makes it a promising *first* part for getting the whole project underway.

As should now be obvious, this paper is written primarily for readers who are willing to at least consider the possibility that Darwin's theory might be fundamentally deficient as an explanation for innovation in the history of life. We recognize that a great many talented biologists may not place themselves in that category, but we think the time is right for the evidential case against the standard Darwinian model to be presented in order to begin a serious discussion of the alternatives.

Problem 1: Offsetting the cost of gene expression

The most widely accepted explanation for the origin of new enzymes is gene duplication and recruitment [3, 4]. This process involves duplication of an existing gene, followed by divergent evolution of one of the copies to a new function. For this process to work, though, the diverging duplicate must continue to be transcribed and translated. But these processes of gene expression carry a resource cost [5–8]. Consequently, a duplicate gene undergoing divergent evolution will only confer a net benefit if that cost is more than offset by its positive biological contribution. In many cases this makes cost reduction by deletion or inactivation of the duplicate gene much more likely than innovation as an adaptive response. Several recent papers have demonstrated this by finding that cells reduce expression of nonessential or duplicate genes, or completely inactivate them, in competitive environments [8–12]. When under continuous selection for metabolic efficiency, such as when growing under nutrient-limiting conditions, cells that reduce the total cost of gene expression by inactivating or deleting unneeded genes have a significant fitness advantage and can quickly overtake the population [8, 10].

In judging the degree to which the cost of gene expression impedes metabolic innovation, it is particularly important to distinguish natural selection from laboratory selection. Reported experimental conversions of two enzymes to *o*-succinylbenzoate synthase (OSBS) activity illustrate this point. Working with an *Escherichia coli* (*E. coli*) strain in which the chromosomal gene encoding OSBS was deleted, Schmidt and coworkers identified single mutations that enable two other genes to replace this missing function well enough for selection *in vivo* under specified laboratory conditions. Among those conditions, though, was high-level expression of the replacing gene,[3] which was needed in order to compensate for the very low activity of the converted function (0.0004% or 0.06% of wild-type activity based on k_{cat}/K_m, depending on the source gene [13]). Even with the boosted expression, though, the converted genes fell well short of fully restoring growth [13]. So while the enzyme conversions reported in that study provide useful information, it should not be assumed that they would succeed in nature.

Considering that newly evolved functions are likely to be extremely weak, it should be expected that they would need amplified expression in order to be of any use. But if so, the expression cost might easily outweigh any functional benefit. Natural genes, of course, escape this dilemma by having extremely high catalytic proficiencies and by minimizing expression costs through regulated expression (turning expression off when it is not needed).

[3]Achieved with an induced *tac* promoter on a multi-copy plasmid [13]. For vector details, see http://www.ncbi.nlm.nih.gov/entrez/viewer.fcgi?db=nuccore&id=45614.

First obstacle: *Because gene expression is costly, it cannot be assumed that weakly converted enzyme functions isolated by laboratory selection would provide net selective benefit in wild populations.*

Problem 2: Winning the fixation lottery

Bacteria reproduce rapidly enough to exhaust any pool of nutrients, no matter how large, in a short time frame. This means that local extinction (by starvation) figures much more prominently in the dynamics of bacterial populations than it does for higher organisms. Many bacterial cells alive now, for example, will manage to have billions of descendants alive a year from now. But for each of these cells destined for success, billions in the current population are destined to expire in that time frame, leaving no descendants. Thus, losers in the game of bacterial procreation vastly outnumber winners.

The overall effect of these frequent local extinctions or near-extinctions is a dramatic reduction in genetic variability, which means a dramatic increase in the time required for rare genotypic variants to become *fixed* (i.e., to become the new wild-type). In population genetics, the parameter that characterizes this phenomenon is the *effective population size*, N_e. Roughly speaking, N_e is the size of the subpopulation in each generation that will influence the genetic makeup of future generations. So the smaller N_e is relative to the true population size, N, the more rare winners are in the propagation lottery.

The estimated value of N_e for wild bacterial populations is 10^9 [14, 15], roughly eleven orders of magnitude lower than estimates of N [16]. Consequently, particular beneficial mutations have to appear on the order of 10^{11} times before they have any reasonable likelihood of being fixed. And because that likelihood scales with the coefficient of selection, s [17], which is commonly assumed to have a small fractional value, something like 10^{12} or more appearances may be needed in order for fixation to become probable. In a population of 10^{20} organisms that passes through 10^3 generations per year [18], this does not prevent fixation of common mutations. A beneficial mutation that occurs once in 10^9 cells, for example, will appear 10^{11} times per generation, which means that a cell line destined to carry this mutation to fixation will probably be present within roughly 10 generations. But the situation changes for rare mutations or rare combinations of mutations. At an incidence rate of one new carrier in the population per generation, some 10^{12} generations (~10^9 years) would be required for fixation to become likely, even though the genotype in question exists somewhere in the population most of the time.

Second obstacle: *Beneficial mutations appearing less than about once per generation in a global bacterial population may remain unfixed for a billion years or more.*

Problem 3: Complex adaptation — Combining rare genetic events

From here on it will become increasingly apparent that each of the problems we describe is compounded by the others. If new enzyme functions can evolve by consecutive adaptive mutations,[4] each known to occur spontaneously with reasonable frequency, then Problem 2 would be of no consequence. The difficulty arises from the fact that they typically appear *not* to be achievable in this way.

As mentioned in the introduction, when we attempted to convert an enzyme to perform a new function, we found it to be surprisingly difficult [2]. The starting point was an enzyme we designated Kbl_2 (2-amino-3-ketobutyrate CoA ligase), and the target function was that of $BioF_2$ (8-amino-7-oxononanoate synthase). The structures of Kbl_2 and $BioF_2$ are so similar (Fig. 1) that the enzymes are commonly assumed to be close evolutionary relatives. However, after extensive testing of mutations that were carefully chosen for their potential to achieve the desired conversion, we found success to be elusive. Still, we were able to deduce from our results that the shortest path to conversion would involve seven or more mutations. That is, at least seven mutations would be required before any level of the new function would be achieved. The true number is probably much higher, considering that we introduced many more than seven substitutions without success. But seven is high enough to cause a severe problem. Mathematical analysis shows that even this seemingly modest number of mutations places the conversion *well* beyond what neo-Darwinian evolution can explain (Fig. 2) [2, 21].

There is an understandable tendency for defenders of a theory, when faced with challenging evidence like this, to marshal as much opposing evidence as can be found. Indeed, if there were a solid body of evidence showing that genuine conversions of enzyme function usually *are* achievable with one or two nucleotide substitutions, we would conclude that the case we examined happened to be exceptionally problematic. But the result of our study is actually quite consistent with the whole body of work on functional conversions in enzymes, even as others have summarized it. For example, two well-known contributors to the field, John Gerlt and Patricia Babbit, recently gave this sobering assessment of the field:

> Interchanging reactions catalyzed by members of mechanistically diverse superfamilies might be envisioned as "easy" exercises in (re)design: if Nature did it, why can't we? ...Anecdotally, many attempts at interchanging activities in mechanistically diverse superfamilies have since been attempted, but few successes have been realized [22].

[4]Adaptive mutations are those that increase the fitness of the organism that carries them, meaning that the organism can grow and reproduce faster than its neighbors. Most mutations are neutral or deleterious.

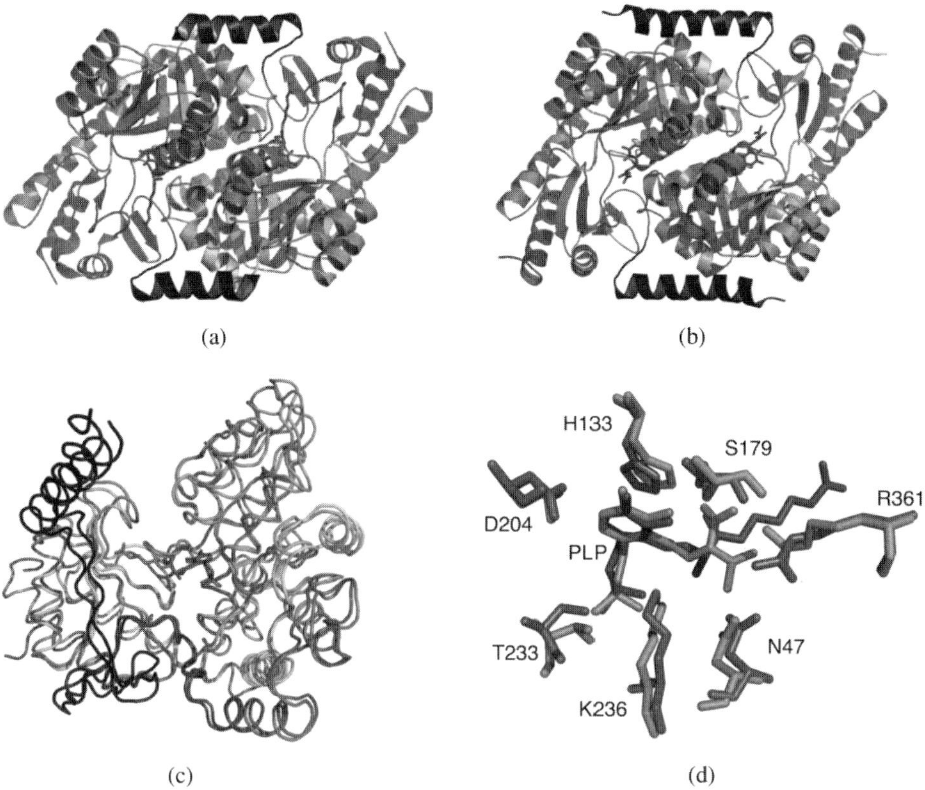

Fig. 1. Structural similarity of BioF and Kbl. a) Dimeric enzymes $BioF_2$ (left; 1DJ9 [17]) and Kbl_2 (right; 1FC4 [18]) viewed along axes of symmetry. Active sites are at the monomer interfaces. b) Aligned backbones of BioF and Kbl monomers. c) Identical side chains in the $BioF_2$ (dark) and Kbl_2 (light) active sites, labeled according to BioF positions. PLP-external aldimines are shown in the center of the active sites. This figure was originally published as Fig. 5 in reference 2.

Similarly, Philip Romero and Frances Arnold drew the conclusion that many researchers (including us) have reached:

> Some functions, however, simply cannot be reached through a series of small uphill steps and instead require longer jumps that include mutations that would be neutral or even deleterious when made individually. Examples of functions that might require multiple simultaneous mutations include the appearance of a new catalytic activity… [23]

Apart from neo-Darwinian expectations, perhaps the difficulty of enzyme conversion should not have been a surprise. The information content of an enzyme is quite large. Its one-dimensional protein sequence bears a complex causal relationship to

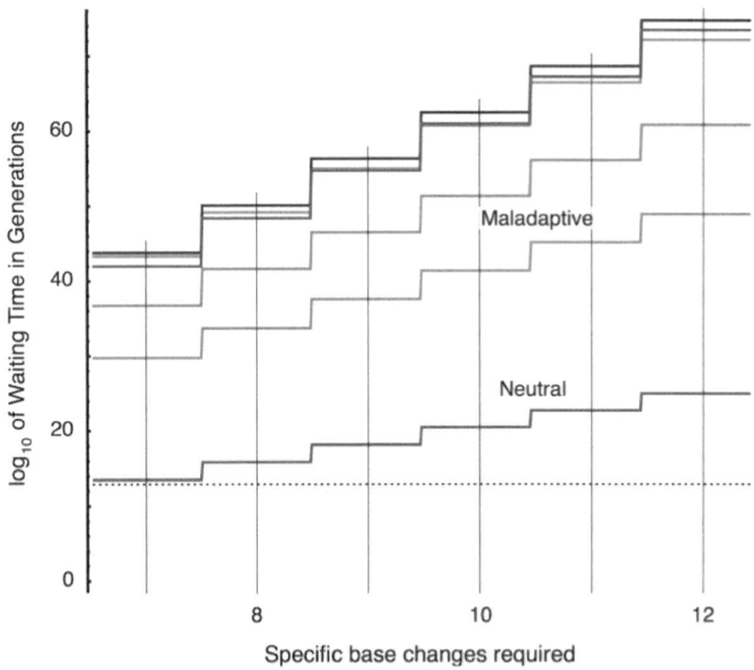

Fig. 2. Expected waiting times for an enzyme conversion requiring from seven to twelve specific base changes. The assumed starting point is a population lacking a duplicate version of the gene to be converted. As discussed (Problem 1), cells in which a duplicate appears are disadvantaged by the cost of expressing a raw duplicate. Shown are the predicted times for a 1% fitness cost ($\bar{s} = -0.01$; top line), a 0.1% fitness cost ($\bar{s} = -0.001$; second from top), and a 0.01% fitness cost ($\bar{s} = -0.0001$; third from top), and no cost (bottom line). Other parameter values are as listed in Table 1 of [19]. The dashed line marks the boundary between feasible waiting times (below) and waiting times that exceed the age of life on earth (above), assuming 10^3 generations per year. This figure was originally published as Fig. 11 of reference 2.

its three-dimensional folded structure *and* to its dynamic behavior as an enzyme. Its activity depends upon many distinct and context-dependent interactions that enable it to form a stable folded structure and to carry out its chemistry. Converting an enzyme to a new catalytic activity is therefore likely to require the simultaneous reconfiguration of many amino-acid interactions, so any step-wise process of enzymatic conversion almost inevitably will involve non-functional intermediates.

In the end, two things seem inescapable. One is that enzymatic innovations requiring more than two specific mutations in a spare gene (provided by a duplication event) are implausible in neo-Darwinian terms [21]. The other is that once this limitation is taken into account, most reported experimental conversions of enzyme function are beyond the reach of neo-Darwinian processes under natural conditions.

Third obstacle: *Adaptations requiring duplication and modification of an existing gene should not be presumed feasible if they require more than two specific base substitutions, which seems to exclude most functional conversions.*

Problem 4: The complexity of metabolic pathways

The severe challenge to the Darwinian model posed by the first three problems becomes exponentially more severe when we recognize that the relevant scale of genetic innovation is not a single new enzyme function, but rather the coordinated sequence of enzymatic steps needed to produce a new phenotypic trait. Our reported attempt to change Kbl_2 into a $BioF_2$-like enzyme in *E. coli* illustrates this point [2]. To make selection of successful mutants possible, one of us (AG) engineered a strain that lacks the gene encoding $BioF_2$. Without that gene the engineered strain is unable to make biotin, an essential cofactor for fatty acid biosynthesis and other carboxylation reactions [24–26]. This makes growth impossible unless either functional conversion is achieved (which never happened) or biotin is supplied as a nutrient (which is how we maintained the strain). This suited our experimental objectives well, but it is important to recognize that our engineered strain is *wholly* unrealistic as a natural evolutionary context for the origin of $BioF_2$.

The complete metabolic pathway for biotin synthesis (Fig. 3) shows why this is so. $BioF_2$ is just one of four enzymes that are exclusively dedicated to biotin production. This means that any proposed explanation of the origin of biotin production as a phenotypic trait must account for innovation on a considerably larger scale than the already problematic scale of a single functional conversion. The full impact of this becomes evident when we realize that quadrupling the scale of a complex adaptation increases the evolutionary difficulty not merely by a *factor* of four, but rather by a *power* of four [21].[5]

The biotin example illustrates the problem of pathway complexity nicely, but is it typical or exceptional for metabolic pathways to depend on four dedicated enzymes? To answer this we need to examine the whole metabolic picture. When we do this, we see that the biotin pathway is unexceptional in its complexity. According to EcoCyc, a comprehensive database of metabolic information on *E. coli*, this common bacterium uses 1,467 enzymes to carry out the functions of 281 metabolic pathways.[6] That amounts to just over five enzymes per pathway, on

[5]More precisely, it increases the required probabilistic resources (opportunities for success) by a power of four, which would increase the waiting time in generations by *more* than a power of four (assuming each generation provides multiple opportunities for success).

[6]See http://ecocyc.org/ECOLI/organism-summary?object=ECOLI.

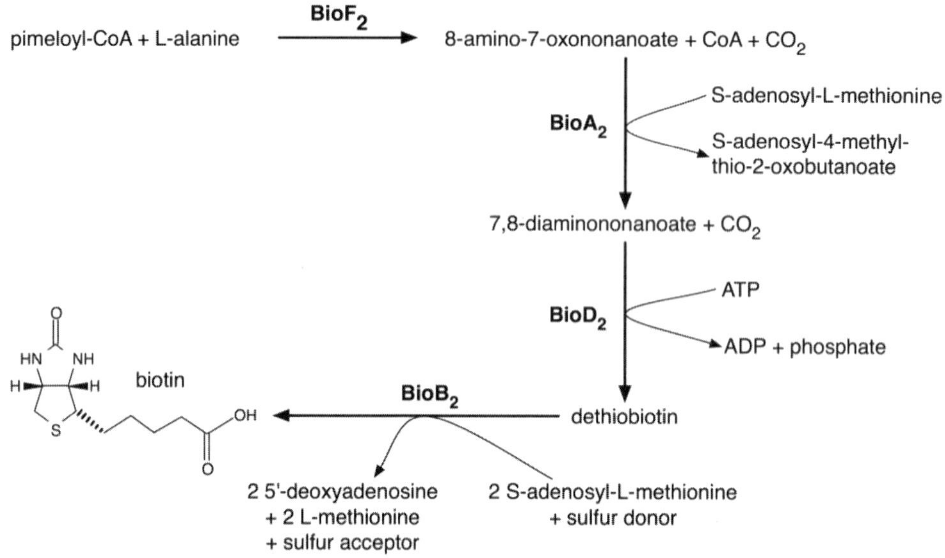

Fig. 3. The dedicated pathway for microbial biotin biosynthesis.

average. Similarly, in 2001 Teichmann *et al.* reported 581 proteins used in 106 small-molecule metabolic pathways in *E. coli* [27]. Although the definition of "pathway" is somewhat imprecise, these figures give us at least a rough picture of the complexity of metabolic processes in terms of enzymatic steps, and from that picture we deduce that most of the innovations that brought new metabolic traits did indeed involve multiple enzymatic innovations.

This poses a severe challenge for neo-Darwinism. Mechanisms that have been proposed in attempts to meet this challenge, such as retrograde evolution [28], or serial duplication and recruitment [29] do not match the actual distribution of protein domains across and within pathways [30]. Rather, most pathways employ several different protein folds, which, as we discuss next, raises another problem.

> ***Fourth obstacle:*** *Accounts of metabolic innovation must recognize that beneficial metabolic traits typically depend on multiple dedicated genes.*

Problem 5: Radical innovation — the need for new protein folds

The previous problem makes it clear that a realistic treatment of metabolic innovation has to explain more than a single new enzyme function. Explaining how a new enzyme function might appear is a key part of the problem, but it is not the *whole* problem for several reasons. The first, as just discussed, is that new metabolic traits typically require multiple new enzyme functions, not just one. The second is that these new functions often call for new protein folds, which adds the problem of *structural* innovation to the already mushrooming problem of functional innovation. The problem of converting an existing fold to a new function is very modest compared to the problem of generating a stable new fold with enzymatic activity from scratch.

The basis for thinking that such structural innovation is typically beyond the reach of Darwinian evolution has been described [1]. The next question is how prevalent structural innovation appears to have been in the early history of life. More specifically, how often did metabolic innovation involve structural innovation? We can get at least a rough answer to this in a couple of different ways. One is to estimate the number of distinct fold types used by a typical bacterial species and divide that by the number of metabolic pathways that these folds serve. This avoids the need to reconstruct history by giving us an average value — the average number of new folds that have to be explained per pathway explanation. A previous analysis found this average to be about four (991 distinct folds serving 263 pathways [1]), which means that the vast majority of early metabolic pathways required new folds.[7]

A complementary approach is to get a rough lower-bound estimate of the total number of distinct protein folds used in bacterial life. Analysis of the bacterial genomes that have been sequenced so far indicates that a substantial majority (>80%) of the 1,962 known protein folds are used in at least one bacterial species.[8] Although there is no reliable way to estimate the actual total number of folds, that result suggests that bacterial life uses most of them. Currently, about 40% of the proteins known to exist are known only by the sequence of their encoding gene (i.e., nothing is known of their structure or function [31]). As more genomes are sequenced, the list of these uncharacterized proteins continues to grow, and again a substantial fraction of them (about 50%) are of bacterial origin [31]. A concerted effort has been made in recent years to target these proteins for structural analysis, with interesting results. Of 248 newly determined structures described by Jaroszewski *et al.* [31], 44 are completely new folds, and another 23 have only partial similarity to known folds. Thus, the folds that have been identified so far may be only the tip of a very large 'iceberg.'

Fifth obstacle: *Accounts of metabolic innovations must recognize that they often depend on new protein folds.*

---

[7]Using the Poisson distribution with an expectation of 991/263 = 3.8 new folds per pathway gives a 98% likelihood of at least one new fold having been needed for a randomly chosen pathway.

[8]Based on analysis of *Superfamily* assignments for 1,392 bacterial genomes (version 1.75; see http://supfam.cs.bris.ac.uk/SUPERFAMILY/).

Problem 6: Causal circularity

Kun, Papp, and Szathmáry have described the problem of "kick-starting metabolic networks" [32]. Their abstract begins, "If chemical *A* is necessary for the synthesis of more chemical *A*, then *A* has the power of replication." Accordingly, they apply the term "autocatalytic" to *A*. To avoid confusion, we suggest that this term ought to be reserved for cases where *A* is *sufficient* for the production of itself (with no extraordinary preconditions). By contrast, *A* being *necessary* for making *A* does not mean that supplies of *A* are self-renewing. Rather, it means that the absence of *A* assures its continued absence. We will use the term *causal circularity* to describe this case.

Whenever a biosynthetic process exhibits causal circularity (requiring its product, *A*), selection-based accounts of the origin of this process encounter complications. In the first place, since the biosynthesis of *A* as we now see it requires not just the genes encoding the enzymes that produce *A* but also *A* itself, a satisfactory account has to go beyond gene origins. The current biosynthetic apparatus for making *A* must, in such a case, not only come into existence but also be primed with pre-existing *A* in order to begin working. But this presents another complication. If *A* was pre-existing, how would acquiring a way of making *A* provide a selective advantage? Although it is possible to construct answers to this, they all suppose circumstances beyond the simple fact that *A* is useful, which makes the final explanation only as compelling as those suppositions are.

How common is causal circularity, though? By analyzing metabolic network models for various microbial species, Kun and coworkers showed that ATP production involves causal circularity in all organisms, with other metabolites showing circularity in some organisms but not in others [32]. However, because their analysis focused on net reactions rather than on the actual physical requirements for them to occur, they may have underestimated the generality of this phenomenon.

A few examples will illustrate this. One is the biosynthesis of cysteine in bacteria. The reactant that provides the sulfur atom for incorporation into cysteine is hydrogen sulfide (H_2S),[9] which itself must be produced from sulfate (SO_4^{2-}) in a multi-step enzymatic process.[10] The final step of this process is catalyzed by sulfite reductase, an enzyme that depends upon a prosthetic group consisting of four iron atoms bridged by four inorganic sulfur atoms (an Fe_4S_4 iron-sulfur cluster [33]) and coordinated to the protein by means of four cysteine side chains (Fig. 4).

[9] http://BioCyc.org/ECOLI/NEW-IMAGE?type=PATHWAY&object=CYSTSYN-PWY.

[10] http://BioCyc.org/ECOLI/NEW-IMAGE?type=PATHWAY&object=SO4ASSIM-PWY.

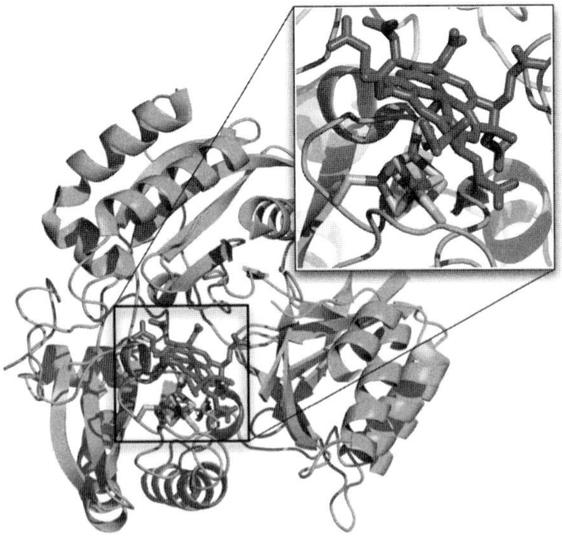

Fig. 4. The enzyme sulfite reductase. As shown in the expanded view, the active site of sulfite reductase uses two prosthetic groups. The larger of these is siroheme (dark honeycomb structure). Coupled down below the iron center of siroheme is the cube-like iron-sulfur cluster, which is held in place by four cysteine side chains.

Consequently, without those coordinating cysteine residues, sulfite reductase cannot produce H_2S, and without H_2S, cysteine synthase cannot produce cysteine. So cysteine biosynthesis is a striking example of causal circularity. Other amino acid pathways provide additional examples. The biosynthesis of arginine depends on ornithine carbamoyltransferase (ArgF),[11] which has an essential arginine residue in its active site [34], and the biosynthesis of lysine depends on diaminopimelate decarboxylase (LysA),[12] which requires a lysine residue to form a Schiff-base linkage to its PLP prosthetic group.[13]

In fact, there is a simple way to generalize the principle of causal circularity. Since life is a prerequisite for *all* biosynthesis, any biosynthetic product that is necessary for life in its present form is also necessary for its own biosynthesis in modern life. So causal circularity exists for *all* essential biosynthetic products. In some cases the loop is extremely tight. LysA, for example, embodies a causal loop in itself by both producing and requiring lysine directly. More often the causal loop involves multiple activities. Biotin production is a good example of this,

[11] http://biocyc.org/ECOLI/NEW-IMAGE?type=PATHWAY&object=ARGSYN-PWY&detail-level=2.

[12] http://biocyc.org/ECOLI/NEW-IMAGE?type=PATHWAY&object=DAPLYSINESYN-PWY.

[13] See PDB entry 1KNW and reference 35.

biotin being necessary for fatty acid biosynthesis, which is necessary for building the cell membrane, which is necessary for life, which is necessary for the biosynthesis of everything, including biotin.

So, in order to conceive of an evolutionary origin of biotin biosynthesis, we must suppose that prior to this origin either A) cells were making their membranes without biotin, or B) cells had an abiotic source of biotin. Either way, the question of how the ability to make biotin would have been beneficial is raised. To answer it, we have to contrive a selective scenario that goes well beyond plain facts, which means we end up having to justify both a contrived selection story and a seemingly unlikely supposition (either A or B) about the state of life prior to biotin biosynthesis. Of course it is *possible* to suppose any number of additional things in an attempt to do this, but each of these suppositions adds to the complication of an already complicated story.

> ***Sixth obstacle:*** *The fact that life depends on numerous components jointly means that no simple relationship exists between the functions of these components and the selective story that would be needed for them to have arisen as simple adaptations.*

Discussion

When the key shortcomings of neo-Darwinism are examined in any detail, it is hard to escape the impression that the theory is unraveling. All theories encounter unsolved problems, but for a solid theory these are challenges in the positive sense of the word — opportunities to prove itself further. With neo-Darwinism, on the other hand, things appear to be moving in the other direction. As we learn more about biological systems, we encounter apparently insoluble problems at every level. To make matters worse, as we have seen here the interdependence of these individual failures compounds them greatly, making repair of the theory seem very unlikely.

As negative as this may sound, it has a positive side: the insights we gain from identifying the obstacles facing neo-Darwinism can and should inform the construction of a new theory to take its place. That is, in pinpointing the key problems with the old theory we are identifying crucial respects in which its replacement must differ from it. We ourselves have become convinced that intelligent causation is essential as a starting point for any successful theory of biological innovation. If this is so, what is needed now is an elaboration of the general principles by which living things have been designed. Accordingly, we have attempted to identify design principles from each of the problems described above. The six principles, paired with the obstacles they address, are as follows:

> ***First obstacle:*** *Because gene expression is costly, it should not be assumed that weakly converted enzyme functions isolated by laboratory selection would provide net selective benefit in wild populations.*
>
> ***First principle:*** *Innovations are more like investments than quick cash. They must be well implemented to offset their cost, and even then the benefits tend to accrue over a long period.*
>
> ---
>
> ***Second obstacle:*** *Beneficial mutations appearing less than about once per generation in a global bacterial population may remain unfixed for a billion years or more.*
>
> ***Second principle:*** *For innovations to be established reliably they need to be carried past a critical 'tipping point' in numerical representation, beyond which they become self-establishing.*
>
> ---
>
> ***Third obstacle:*** *Adaptations requiring duplication and modification of an existing gene should not be presumed feasible if they require more than two specific base substitutions, which seems to exclude most functional conversions.*
>
> ***Third principle:*** *The substantial reworking of a homologous structure needed to give it a genuinely new function is more suggestive of reapplication of a concept than adjustment of a physical thing.*
>
> ---
>
> ***Fourth obstacle:*** *Accounts of metabolic innovation must recognize that beneficial metabolic traits typically depend on multiple dedicated genes.*
>
> ***Fourth principle:*** *Useful innovations tend to require the simultaneous solution of multiple new problems, which means they tend to be compound innovations.*
>
> ---
>
> ***Fifth obstacle:*** *Accounts of metabolic innovations must recognize that they often depend on new protein folds.*
>
> ***Fifth principle:*** *Useful innovations often involve both the reapplication of proven design concepts and the de novo invention of new ones.*
>
> ---
>
> ***Sixth obstacle:*** *The fact that life depends on numerous components jointly means that no simple relationship exists between the functions of these components and the selective story that would be needed for them to have arisen as simple adaptations.*
>
> ***Sixth principle:*** *The implementation of innovation is nearly the opposite of ordinary physical causation. It is the top-down arrangement of matter in such a way*

that the resulting bottom-up behavior of that matter serves the intended purpose of the innovator.

Even in this rough form these principles suggest some interesting things. One is that biological innovation seems similar in essence to human innovation, though certainly beyond it in degree. This realization is attracting an increasing number of engineers to biology with the aim of reapplying biological innovations in human technology [36]. Although that field of study, known as *biomimetics*, has practical ambitions, the fact that it exists (and is thriving) also implies an essential similarity between intelligent design in engineering and intelligent design in life.

Another interesting aspect of the above set of principles is that while they were drawn from observations at the molecular scale of metabolic innovation, they do not appear to be restricted to that scale. Indeed, they have the appearance of general rules that make sense irrespective of the particulars of the innovation, including its physical scale. Since universality of that kind is precisely what we expect of a useful theory, this suggests that the principles may be a starting point for framing the first successful theory of biological innovation.

Next, and perhaps most significantly, it is clear that this new theory will be of an entirely different *kind* than the one it hopes to replace. Darwinism is purely mechanistic in its approach, in that it offers a bottom-up causal explanation for the origin of all biological forms and phenomena. In this respect it is also intrinsically reductionistic — it takes physical causation to be the *fundamental* explanation of all origins events. The design-based theory hinted at in this paper will differ radically in both respects. The new theory will be fundamentally top-down in its approach and therefore fundamentally non-reductionistic. It will focus mainly on design principles rather than on mechanisms. Just as students of engineering and design focus mainly on high-level principles that leave a great deal of freedom as to their physical implementation, so too students of the new theory will focus mainly on the principles that inform biological designs [37] rather than on the processes by which these designs may be implemented.

Might this new theory transform biology beyond the topic of origins? Most who reflect on the current state of biology sense a need for understanding to catch up with the enormous flow of new data. Sydney Brenner, one of the pioneers of modern molecular biology, has concluded that "biology urgently needs a theoretical basis to unify it and it is only theory that will allow us to convert data to knowledge [38]." He continues by pointing out that the trend toward performing measurements on whole systems instead of their isolated parts (one of the emphases of *systems biology*) brings us no closer to the needed theory, but his suggestion that we should return to hard-core reductionism also misses the mark:

> Our approach directly reflects the structure of biological systems and, as we reduce each level to the level below — organisms to cells and cells to molecules — we can then confidently complete the reductionist programme because the properties of molecules can be reduced to physics [38].

The problem with this approach is that reducing a living thing to its simplest material causes does not lead to an understanding of it. By way of analogy, those who want to understand software should have some exposure to the zeros and ones of machine language, but they would do well to spend *most* of their time studying principles of software design that are nowhere to be found among the bits. More generally, one can acquire a great deal of knowledge of the *operation* of a complex innovative system without having the slightest grasp of the genius behind it. To grasp that, we need to consider how it was designed.

In the end Brenner's search for a new theory seems to be hamstrung by the old theory. He thinks "we need to remember that whereas mathematics is the art of the perfect and physics is the art of the optimal, biology, because of evolution, is only the art of the satisfactory [38]." We think it may actually be much more than that.

References

1. Axe DD (2010) The case against a Darwinian origin of protein folds. BIO-Complexity 2010(1):1–12. doi:10.5048/BIO-C.2010.1
2. Gauger AK, Axe DD (2011) The evolutionary accessibility of new enzyme functions: a case study from the biotin pathway. BIO-Complexity 2011(1): 1–17. doi:10.5048/BIO-C.2011.1
3. Ohno S (1970) Evolution by gene duplication. Springer-Verlag, New York.
4. Jensen RA (1976) Enzyme recruitment in evolution of new function. Annu Rev Microbiol 30: 409–425.
5. Dykhuizen D (1978) Selection for tryptophan auxotrophs of *Escherichia coli* in glucose-limited chemostats as a test of the energy conservation hypothesis of evolution. Evolution 32: 125–150. doi:10.2307/2407415.
6. Stoebel DM, Dean AM, Dykhuizen DF (2008) The cost of expression of *Escherichia coli* lac operon proteins is in the process, not in the products. Genetics 178: 1653–1660. doi:10.1534/genetics.107.085399.

7. Wagner A (2005) Energy constraints on the evolution of gene expression. Mol Biol Evol 22: 1365–1374. doi:10.1093/molbev/msi126.
8. Gauger AK, Ebnet S, Fahey PF, Seelke R (2010) Reductive evolution can prevent populations from taking simple adaptive paths to high fitness. BIO-Complexity 2010(2):1–9. doi:10.5048/BIO-C.2010.2
9. Mira A, Ochman H, Moran NA (2001) Deletional bias and the evolution of bacterial genomes. Trends Genet 17: 589–596. doi:10.1016/S0168-9525(01)02447-7
10. Lang GI, Murray AW, Botstein D (2009) The cost of gene expression underlies a fitness trade-off in yeast. Proc Natl Acad Sci USA 106: 5755–5760. doi:10.1073/pnas.0901620106.
11. Kuo C-H, Ochman H (2010) The extinction dynamics of bacterial pseudogenes. PLoS Genet 6(8): e1001050. doi:10.1371/journal.pgen.1001050.
12. Behe M (2010) Experimental evolution, loss-of-function mutations, and "the first rule of adaptive evolution." Q Rev Biol 85: 419–445.
13. Schmidt DM, Mundorff EC, Dojka M, Bermudez E, Ness JE, *et al.* (2003) Evolutionary potential of $(\beta/\alpha)_8$-barrels: functional promiscuity produced by single substitutions in the enolase superfamily. Biochemistry 42: 8387–8393. doi:10.1021/bi034769a.
14. Lynch M (2005) Simple evolutionary pathways to complex proteins. Protein Sci 14: 2217–2225. doi:10.1110/ps.041171805.
15. Lynch M, Conery JS (2003) The origins of genome complexity. Science 302: 1401–1404. doi:10.1126/science.1089270.
16. Milkman R, Stoltzfus A (1988) Molecular evolution of the *Escherichia coli* chromosome. II. Clonal segments. Genetics 120: 359–366.
17. Lenski RE, Rose MR, Simpson SC, and Tadler SC (1991) Long-term experimental evolution in *Escherichia coli.* I. Adaptation and divergence during 2,000 generations. The American Naturalist 138: 1315–1341. doi:10.1086/285289.
18. Levin BR (1981) Periodic selection, infectious gene exchange and the genetic structure of *E. coli* populations. Genetics 99: 1–23.
19. Alexeev D, Alexeeva M, Baxter RL, Campopiano DJ, Webster SP *et al.* (1998) The crystal structure of 8-amino-7-oxononanoate synthase: a bacterial PLP-dependent, acyl-CoA-condensing enzyme. J Mol Biol 284: 401–419. doi:10.1006/jmbi.1998.2086.
20. Schmidt A, Sivaraman J, Li Y, Larocque R, Barbosa J *et al.* (2001) Three dimensional structure of 2-amino-3-ketobutyrate CoA ligase from *Escherichia coli* complexed with a PLP-substrate intermediate: Inferred reaction mechanism. Biochem 40: 5151–5160. doi:10.1021/bi002204y.
21. Axe DD (2010) The limits of complex adaptation: an analysis based on a simple model of structured bacterial populations. BIO-Complexity 2010(4):1–10. doi:10.5048/BIO-C.2010.4.
22. Gerlt JA, Babbitt PC (2009) Enzyme (re)design: Lessons from natural evolution and computation. Curr Opin Chem Biol 13(1):10–18. doi:10.1016/j.cbpa.2009.01.014.

23. Romero and Arnold (2009) Exploring Protein Fitness Landscapes by Directed Evolution. Nature Reviews: Molecular Cell Biology 10:866–875. doi:10.1038/nrm2805.
24. Del Campillo-Campbell A, Kayajanian G, Campbell A, Adhya, S (1967) Biotin requiring mutants of Escherichia coli K-12. J Bacteriol 94: 2065–2066.
25. Rolfe B, Eisenberg MA (1968) Genetic and biochemical analysis of the biotin loci of *Escherichia coli* K-12. J Bacteriol 96: 515–524.
26. Streit WR, Entcheva P (2003) Biotin in microbes, the genes involved in its biosynthesis, its biochemical role and perspectives for biotechnological production. Appl Microbiol Biotechnol 61: 21–31.
27. Teichmann SA, Rison SCG, Thornton JM, Riley M, Gough J (2001) Small-molecule metabolism: an enzyme mosaic. Trends Biotech 19: 482–486. doi:10.1016/S0167-7799(01)01813-3.
28. Horowitz NH (1945) On the evolution of biochemical syntheses. Proc Natl Acad Sci USA 31:152–157. doi:10.1073/pnas.31.6.153.
29. Horowitz NH (1965). The evolution of biochemical syntheses - retrospect and prospect. In: Bryson V, Vogel HJ (eds) Evolving genes and proteins, pp 15–23, Academic Press, New York.
30. Rison SCG, Thornton JM, Riley M, Gough J, Chothia C (2001) The evolution and structural anatomy of the small molecule metabolic pathways in *Escherichia coli*. J Mol Biol 311: 693–708. doi:10.1006/jmbi.2001.4912.
31. Jaroszewski L, Li Z, Krishna SS, Bakolitsa C, Wooley J (2009) Exploration of uncharted regions of the protein universe. PLoS Biol 7: e1000205. doi:10.1371/journal.pbio.1000205.
32. Kun A, Papp B, Szathmáry E (2008) Computational identification of obligatorily autocatalytic replicators embedded in metabolic networks. Genome Biol 9: R51.
33. Crane BR, Siegel LM, Getzoff ED (1995) Sulfite reductase structure at 1.6 Å: Evolution and catalysis for reduction of inorganic anions. Science 270: 59–67. doi:10.1126/science.270.5233.59.
34. Kuo LC, Miller AW, Lee S, Kozuma C (1988) Site-directed mutagenesis of *Escherichia coli* ornithine transcarbamoylase: Role of arginine-57 in substrate binding and catalysis. Biochem 27: 8823–8832. doi:10.1021/bi00424a021.
35. Momany C, Levdikov V, Blagova L, Crews K (2002) Crystallization of diaminopimelate decarboxylase from *Escherichia coli*, a stereospecific D-amino-acid decarboxylase. Acta Crystallogr D Biol Crystallogr 58: 549–552. doi:10.1107/S0907444902000148
36. Bar-Cohen Y (2006) Biomimetics: biologically inspired technologies. CRC Press, Taylor and Francis Group, Boca Raton.
37. Csete ME, Doyle JC (2002) Reverse engineering of biological complexity. Science 295:1664–295. doi: 10.1126/science.1069981.
38. Brenner S (2010) Sequences and consequences. Phil Trans R Soc B 365: 207–212. doi:10.1098/rstb.2009.0221.

Section Four — Biological Information and Self-Organizational Complexity Theory: Introductory Comments

Bruce L. Gordon — Section Chairman

No discussion of new perspectives on biological information would be complete without consideration of the anti-reductionist approach of the self-organizational school of thought. The reductionist approach focuses on systematically taking apart complex systems and analyzing their individual components, seeking to explain the behavior of the whole in terms of its parts. This strategy has been very fruitful and such research undoubtedly will continue, but, like the intelligent design scientists and researchers exemplified by the editors and other contributors to this volume, self-organizational theorists believe that new theoretical approaches are necessary to understand the hierarchically integrated information networks that undergird morphogenesis in developmental biology and evolution. How do systems of genes and proteins integrate into holistic information structures? How do dynamic organelle structures form in cells? What controls cell growth, division and differentiation in organisms? How is genomic information regulated in the construction of an organism? How do selective environmental pressures integrate through time with organismal development to affect the evolution of species? How do integrated ecosystems form and evolve? Both self-organizational theorists and intelligent design (ID) theorists believe that natural selection operating on random genetic mutation is an insufficient basis on which to explain the origins of biological complexity and irrelevant to the origin of life. ID theorists also believe that the self-organizational capacities of physical systems are limited, falling far short of the order we observe, so the *ultimate* source of information for the origin of life and hierarchically integrated morphogenesis in both organismal development and speciation must be extrinsic to biological systems and their physical environments. In contrast, self-organizational researchers argue that global pattern development, including the highly complex hierarchical information structures characteristic of life, can emerge solely from the interactions of lower-level components and part-whole dynamics *without* ultimate or proximate goal-directed input. Whether biological information is somehow self-originating is thus a central point of disagreement between intelligent design theorists and self-organizational complexity theorists.

Taking their cue from non-equilibrium thermodynamics, self-organizational theorists maintain that living systems rely on a continuous flow of energy to maintain themselves far from equilibrium, and it is this constant flux of energy and

From the fact that we do not know the ever-changing phase space of biological evolution, it will follow that we do not know the "sample space" of what I call the "Adjacent Possible" of the evolution of the biosphere. From this it follows that standard notions of Information Theory, such as Shannon and Kolmogorov, cannot be applied, since both require prestatement of the sample space of the process. For example, for Shannon, prestatement of the set of possible messages — the sample space — is needed to compute the entropy of the information of the Source. If we do not know the sample space of evolution, Shannon's starting point is moot.

Moreover, if we do not know the sample space of the process of biological evolution, then probability calculations utilized by Intelligent Design scholars are also either moot, or deeply suspect.

These issues mean we need to invent a new concept of biological information. No adequate formulation now exists. I will propose the start (only) of such a formulation.

If no law entails the evolution of the biosphere, then we must ask what forms of laws, if any, we can have. One approach that I will discuss is the study of ensembles of systems [4]. For example, the study of (i) ensembles of model genetic regulatory networks controlling cell differentiation and ontogeny, (ii) ensembles of reaction networks capable of catalysis of the same reactions to form collectively autocatalytic sets for the origin of molecular reproduction and life, and (iii) ensembles of tunably rugged fitness landscapes [5]. Two major features of this ensemble approach are: (i) A search for statistical laws, despite the absence of entailing laws. As more facts are learned about the systems in question, more refined ensembles can be built for better statistical laws. (ii) Remarkable evidence for profound self organization has been found, for example, as typical, or generic, properties of ensembles of genetic regulatory networks. This self organization almost surely plays a role with selection in evolution. A generic phase transition has been found, in chemical reaction networks, to the self-organized emergence of collectively autocatalytic sets capable of molecular reproduction that are likely to play a role in the origin of life. Furthermore, a remarkable linkage has been found between species co-evolving on tunably rugged landscapes and the very structure of those landscapes also evolving, such that evolution itself can tune the structure of fitness landscapes on which evolution occurs, to lower or even perhaps minimize the rate of extinction, and hence maximize species lifetimes. In all these cases, we find both statistical laws without entailing laws for the evolution of the biosphere, and unexpected and powerful self organization that mingles with natural selection in the panorama of life's becoming.

This article is organized as follows: In section 1, I discuss work with senior French/Italian mathematician Giuseppe Longo that is the strongest case we can currently make that no law entails the detailed evolution of life. Hence my conclusion that this spells "the end of a physics worldview." In this discussion, I expand on my own work [3] and that of F. Bailly and Longo [2], both of which claim and demonstrate that the phase space of evolving life persistently changes in ways we cannot say. In section 2, I discuss the stunning fact that evolution, without selection, creates its own "adjacent possible" empty niches, which it may fill. Hence evolution, in a kind of "natural magic", builds the very possibilities it becomes. That is, I demonstrate the truly astonishing fact that, without natural selection acting at all, the evolving biosphere creates the ever new Adjacent Possible empty ecological niches that evolution may/will fill. Thus, without any selection acting to create this astonishing aspect of evolution, evolution itself is building the very possibilities that evolution becomes. Here the claim from Heraclitus that "Life Bubbles Forth" seems right and deeply new. In section 3, I lay out the claim that we do not know the sample space of the evolutionary process, so standard information theory is moot. In section 4, I relate the above results briefly to the hopes of the Intelligent Design community to demonstrate Irreducible Complexity [6], and its vast improbability by normal evolutionary processes [7]. In section 5, I describe three examples of the use of the "ensemble approach" to find statistical laws in the absence of entailing laws for the detailed becoming of the biosphere. I discuss models of ensembles of genetic regulatory networks, the emergence of collectively autocatalytic sets, and the statistical features of evolving fitness landscapes. All also exhibit the self organization alluded to above [5].

I. Evolution Is Beyond Entailing Law

At the dawn of Western philosophy and science, some 2,700 years ago, Heraclitus declared, roughly, that "the world bubbles forth". There is, in this fragment of thought, a natural magic, a creativity beyond the entailing laws of modern physics. I believe Heraclitus was right about the evolution of the biosphere and human life. We live beyond entailing law in a kind of natural magic we co-create.

Early sociologist Max Weber said that with Isaac Newton, we became disenchanted and entered Modernity. He was right. Before Newton, our tradition, from *Genesis*, saw a creator God whose divine agency, rather like the natural magic of Heraclitus, created the world also beyond entailing law.

With Newton's three laws of motion, universal gravitation, and the differential and integral calculus, our world transformed profoundly. Given the initial conditions of billiard balls' positions and momenta on the table, the boundary conditions of the shape of the table, and the motions of the balls given in differential equation form using the laws of motion, then integration, a form of deduction, yielded the entire future and past trajectories of the balls.

With Pierre-Simon Laplace, this became the bedrock of reductionism: Given the positions and momenta of all the particles in the universe, a vast intelligence could, using Newton's laws, deduce the entire future and past of the universe. For Laplace, the complete determinism of Newton's laws co-existed with a capacity for accurate prediction. With Poincaré and the three body gravitational problem, deterministic chaos was discovered. Here the system remains deterministic but unpredictable because of sensitivity to initial conditions and the fact that any measurements of initial conditions require a finite interval of space and time — a point that Bailly and Longo stress [2]. Thus, in modern classical physics, determinism does not imply predictability.

The framework of entailing laws remains in the twin pillars of twentieth-century physics — classical physics with General Relativity, and quantum mechanics — in both cases with differential equations and their entailed integration. Bailly and Longo [2], stress that in physics, it is always possible to prestate the phase space of the system, typically derived from its underlying symmetries. In classical physics, a least action principle assures that the actual behavior of the classical system in its "possible phase space" is always a unique "shortest path", or geodesic, on some manifold. In quantum physics, given the indeterminism of quantum mechanics, the analogous behavior is a geodesic of a probability distribution. In short, the framework of physics prestates its phase space in which, via laws of motion in ordinary or partial differential equation form, initial and boundary condition, and integration, yield the entailed geodesic behavior of the system.

I believe we reach a terminus of this physics worldview at the watershed of life. As we will see, it seems Heraclitus was right: Life bubbles forth in a kind of "natural magic". A purpose of this article is to spell out this natural magic, which exhibits itself as the evolving biosphere literally constructs, without selection, its own future possibilities.

First, and of truly central importance, evolution itself defies both the completeness of quantum mechanics and the completeness of classical mechanics, yet unites them both. We know this, but never say it. Mutations are often quantum random and indeterminate events yielding Darwin's heritable variations. Yet evolution itself is not random, as the phenomenon of convergent evolution demonstrates. For example, the eye has evolved independently eleven times. And the convergence of the independently evolved vertebrate and octopus camera eye to a stunning near identity, the result of powerful natural selection, is obviously not random. More examples are found in the convergent evolution of marsupials and mammals.

Thus, in blunt terms, biological evolution is neither quantum indeterminate random, nor deterministic classical mechanics. The living world really is "new". Quantum mechanics alone and classical physics alone each seem to be incomplete.

The fact of evolution, mixing quantum and classical physics for which each alone is insufficient, is clear. What might this truth mean?

One very important possibility is that, after 85 years of unsuccessful attempts to unite quantum mechanics and General Relativity, it may really not be possible to unify them into the single theory whose "Dream" is that of Weinberg [1]. We may have to live with quantum mechanics and classical physics un-united. In this case, evolution itself demonstrates that both nevertheless "mix" together: quantum indeterminate yet random mutations united with the non-random effects of natural selection acting at the level, at least in part, of classical physics, and thus the camera eye evolved in octopus and vertebrates. But this requires something that seems not to be entailed in current physics: Let a quantum indeterminate random DNA mutation occur, then natural selection act to evolve toward the tuned camera eye. As this largely classical physics evolution occurs, different alleles of mutated genomes are selected in the evolving population. Thus, when quantum random and indeterminate mutations creating yet new alleles occur, the very possibilities of what those quantum event mutations might be, i.e., in which gene sequences they may occur, has changed due to largely classical physics natural selection. In turn, the quantum random indeterminate mutations alter what natural selection will do. Taken together, evolution is both quantum indeterminate and also non-random.

Given this mixture of quantum indeterminate random, and classical physics non-random natural selection, it seems very hard on this basis alone to conceive of a single law that entails the detailed evolution of the biosphere.

In a related intellectual effort to link quantum mechanics and the mind-brain system, inspired by Sir Roger Penrose, but taking a different track, I have proposed in "Answering Descartes: Beyond Turing" [8], that even in the mind-brain system, perhaps in synapses, a similar non-determinate yet non-random mixture of quantum indeterminate and yet non-random classical behavior can occur. It may be important that there now appears to be a Poised Realm, where systems, via decoherence and recoherence, can hover back and forth between quantum coherence and classicality for all practical purposes (FAPP). This hovering may play a role in organisms and be indeterminate, yet non-random. There may also be no entailing law for this behavior.

In short, if we cannot unite quantum mechanics and General Relativity under a single law, this may not be an intellectual tragedy, but may free us, after the 350 years since Newton, from the dream of universal entailment. Then true novelty, beyond entailment, can arise. Life can "bubble forth". I now discuss further reasons to believe that no law entails the evolution of the biosphere.

Second, biological evolution concerns "Kantian wholes" [9], where the whole exists for and by means of the parts and the parts exist for and by means of the whole. An instance is a collectively autocatalytic set of peptides, as produced by

Gonen Ashkenazi of Ben Gurion University in his nine-peptide autocatalytic set. This is a clean example of a Kantian whole. No peptide catalyzes its own formation from two fragments of itself, but instead catalyzes the formation of one of the other nine peptides from two fragments of that peptide. The set of peptides as a whole catalyses the entire set of reactions by which the set of nine peptides reproduces itself in a collectively autocatalytic set. If we call catalyzing a reaction a "catalytic task", then the set achieves a "closure" in catalytic task space. All the reactions that require catalysis are catalyzed by one or more members of the set. Note that, given a Kantian whole, the "function" of a given peptide can be defined as its role in sustaining the reproduction of the whole nine peptide collectively autocatalytic set.

In his forthcoming book, *Incomplete Nature* [10], Terrence Deacon, a professor at U.C. Berkeley, points out that philosopher Jaegwon Kim has argued that even such Kantian wholes do not preclude deduction upward from particles to wholes. But, points out Kim, according to Deacon, who agrees, that argument rests on classical "materialism," i.e., the classical physics of point particles and fields. Deacon rightly notes that quantum mechanics, as in Feynman's sum over all possible pathways that, e.g., a photon might take through the two slits, obviates such a naive materialism. The position and momentum of a particle cannot be jointly measured with precision; quantum mechanics precludes point particles existing prior to measurement, and multi-particle quantum systems are, ineluctably, "wholes". Thus the collectively autocatalytic set is a Kantian "Organized Being," whose ever-changing atoms and molecules exist in the universe — when most complex things above atoms will never exist — as a united whole, an entity which is sustained existing in the universe by the linked dynamical classical and quantum processes of parts and whole enabling one another. The specific peptides may come and go, yet the Kantian whole remain as non-equilibrium, self-sustaining, partly quantum, partly classical, and perhaps partially Poised Realm, processes.

Third, a living, dividing cell is both a collectively autocatalytic set, and thus a Kantian whole. But of central importance, it achieves a task closure in a much wider set of tasks than mere catalysis. Proteins are vectored to specific cell locations, energy is transduced, pumps operate in work cycles, and chromosomes are separated in mitosis, completing a set of task closures in some wide set of tasks such that the dividing cell reproduces. The function of each such task, typically a subset of the causal consequences of the physical processes involved, is its role in sustaining the reproduction of the cell as a Kantian whole.

Fourth, and of deep importance is this: We cannot name all the causal consequences or uses of any object, say a screw driver, alone or with other objects. The set of uses appears to be both unbounded or "indefinite", and un-orderable. But that means we cannot know that we have ever "listed" all the uses of a screw driver alone or with other objects or processes.

Now consider an evolving cell in which one or more objects or processes, each with myriad causal consequences, finds a novel use that we cannot prestate, but which enhances the fitness of the cell, and so is grafted by natural selection into the evolving biosphere. This "finding of a novel use that we cannot prestate" occurs all the time. The famous flagellar motor of some bacteria made use, via Darwinian preadaptations or exaptations (discussed further below), of fragments of its flagellar proteins that were serving entirely different functions in other bacteria.

Fifth, Darwinian preadaptations are typically not prestatable. A Darwinian preadaptation is a causal consequence of a part of a process in an organism of no selective significance in the current environment that "finds a use" in a novel selective environment and is selected, typically, for a novel function. Preadaptations occur all the time in evolution. I give but one example. Some fish have swim bladders, sacs partly filled with air and water, whose ratio adjusts neutral buoyancy in the water column. Paleontologists believe that swim bladders evolved from the lungs of lung fish: water got into some lungs, and then there was a sac partly filled with air and water, poised to evolve into a swim bladder. I now raise three questions: (i) Did a new function come to exist in the biosphere? Yes. Neutral buoyancy in the water column. (ii) Did the evolution of the swim bladder alter the future evolution of the biosphere? Yes, the possibilities of new species with swim bladders, new proteins, and new ecological niches came into existence: for example, a bacterium or worm might evolve that is only able to live in swim bladders. I return to this example below. (iii) Do you think you could prestate all the possible Darwinian preadaptations just for humans in the next million years? We all say "no". Here is why: We cannot finitely prestate all the possible uses of parts, alone or together, of an organism, for they are indefinite in number and unorderable. We cannot know we have completed the list of uses. Next, we cannot say all possible selective environments for which such uses might be found to be useful. How would we know we had listed all possible selective environments?

But this means something terribly important. Let me call the set of possible next Darwinian preadaptations the Adjacent Possible of the evolution of the biosphere via preadaptations. We do not know what this set of possibilities is! Thus, and of central importance, we do not know the "sample space" of the evolution of the biosphere by Darwinian preadaptations.

But the fact that we do not know the sample space means we cannot make normal probability statements. Consider instead flipping a fair coin 10,000 times and calculating the probability of 4698 heads using the Binomial theorem. We can do this, but notice that we know ahead of time "all possible outcomes", all heads, all tails, all 2 to the 10,000 power possible outcomes of our 10,000 flips. We know the sample space, so we can erect a probability measure.

In contrast, for the evolution of the biosphere by preadaptations, we do not know the sample space and so seem entirely unable make normal probability statements. In turn, I think this inability has its roots in the indefinite set of uses of any part or set of parts or processes in a cell or organism, which set is also unorderable. We cannot know we have listed all the possible uses, nor the set of all adjacent possible selective environments. We do not know what features alone or together in, say, a dividing cell, may find a novel use in some environment and be grafted by natural selection into the Kantian whole, creating a novel function and a novel Kantian whole in the evolving biosphere.

Sixth, mathematics requires that we have the concepts beforehand of the relevant variables, say, mass and length of a pendulum, for the law of the pendulum. The older view of mathematics as mere formal manipulation of syntactic symbol strings given uninterpreted axioms, has given way to a more modern "constructivist" mathematics, as Bailly and Longo argue [2], in which the settled concepts with their semantics, not just syntax, is central to the development of mathematics. For Newton, F = MA rested on a pre-Newtonian notion of "mass".

But unlike physics, where the phase spaces are always prestated, in evolution the phase space is always changing [2, 3], and as we shall see, even more stunningly, building without selection, the very possible ways it may change its phase space. Thus, for evolution of the biosphere by ever new causal consequences, which may "find some unprestatable use" by Darwinian preadaptations in evolving Kantian wholes that constitute cells with ever changing Task closure, we do not know the relevant variables, so we cannot write down the laws of motion for the evolving biosphere.

Seventh, we do not know ahead of time the emerging novel Adjacent Possible empty niches, such as the fish swim bladder into which some worm or bacteria could evolve to live. But those niches constitute the boundary conditions on natural selection shaping the evolution of the worm or bacterium to live in the swim bladder.

Newton taught us that we need the laws of motion, which by point six above we do not have, and we need the initial and the boundary conditions, to integrate the laws of motion for the trajectory of, say, the billiard balls on the billiard table. But we do not know the boundary conditions that the swim bladder, when it may evolve, will constitute, so we cannot integrate the laws of motion, (which we do not have anyway), for the evolution of the biosphere. Lacking the boundary conditions would be like trying to integrate the motions of the billiard balls with no idea of the shape of the billiard table. We do not even have a mathematical model if we lack the boundary conditions!

In summary of these points, first through seventh, no law entails the detailed evolution of the biosphere. If this is true, it is the end of a physics worldview.

II. Life Bubbles Forth

Heraclitus was right: Life bubbles forth, beyond entailing law. Consider the evolution of the swim bladder above by a Darwinian preadaptation. Did natural selection act to craft a well-functioning swim bladder in an evolving population of fish? Of course. But did natural selection act to craft the new adjacent possible empty ecological niche that the swim bladder constituted? NO. No natural selection acted to create the new adjacent possible empty niche into which the worm or bacteria might evolve to live.

But this means that, without any selection at all, the biosphere is building its own adjacent possible pathways of evolution. The biosphere is building, without selection, its own future possibilities. By a kind of "natural magic", the biosphere creates its own future. Heraclitus was right: Life bubbles forth beyond entailing law.

If the above is true, we must give up our deep belief, at least since Newton, if not the Greeks, that without entailing law, the world cannot become in a coherent way: The biosphere has been doing just fine for 3.7 billion years of becomings as Kantian wholes make their largely self-consistent but ever-changing worlds ever anew with one another. We need to think anew how this becoming, even with extinction avalanches, can be coherent without entailing law.

More, if Max Weber is right that with Newton we became disenchanted and entered Modernity, my hope is that the "natural magic" of life bubbling forth, and, a fortiori, human life, can re-enchant us. Perhaps we can move beyond Modernity.

III. Beyond Standard Information Theory to Embodied Information

I begin with Shannon's famous information theory [11]. Shannon chose, on purpose, to ignore any semantics, and concentrate on purely syntactic symbol strings, or "messages" over some pre-chosen symbol alphabet, most simply 0 and 1. Then he considered an Information Source filled with diverse bit string "messages", say bit strings of length N. Each message might occur once or many times in the source. Let p_i be the frequency of the ith message. Then $-\sum p_i \ln(p_i)$ over the set of messages in the source is the "entropy" of the source. Given a measure of the entropy of the source and a noisy channel with a decoder at the far end, he could study information transmission down the channel from source to decoder.

It is clear that Shannon's invention requires that the ensemble of all possible messages, here the possible 2 to the Nth power bit strings length N, be stable head of time. Without this statement, the entropy of the information source cannot be defined.

Now let's turn to evolution. We saw above that we cannot prestate the adjacent possibilities of the evolution of the biosphere by Darwinian preadaptations. Thus, we cannot construct anything like Shannon's probability measure over the future evolution of the biosphere; thus, in turn, we cannot apply information theory in any obvious way to that evolution.

This blunt statement ignores further huge difficulties in applying Information Theory in biology. For Shannon, a bit is a bit, 0 or 1, hence the only "features" are the members of the alphabet of pre-chosen symbols, here 0 or 1. But in biological evolution, where we cannot finitely state the causal consequences of uses of any one or many features or processes in cells or organisms, where the set is both indefinite and unbounded, even if we prestated the "features" we could not state the alphabet of their relevant causal consequences or uses. It is precisely because of these causal consequences alone or together that "find a use" in an evolving cell or organism that these ever new features are grafted into the evolving biosphere.

More, what counts as a "feature"? Any causal consequence of "one" or many parts or processes which alone or together "find a use" that enhances fitness of the Kantian whole so enters the biosphere. We cannot even prestate what aspects of a cell may constitute a feature. In terms of Shannon, we don't even know the "alphabet".

The same concerns arise for Kolmogorov [12], who again requires a defined alphabet and symbol strings of some length distribution in that alphabet. Again, Kolmogorov uses only a syntactic approach. Life is deeply semantic with no prestated alphabet, no "Source", no definable entropy of a source, but unprestatable causal consequences which alone or together may find a use in an evolving Kantian whole of a cell or organism.

In summary, standard information theory, both purely syntactic and requiring a prestated sample space, is largely useless with respect to evolution. On the other hand, there is a persistent becoming of ever novel structures and processes that constitute specific novel and integrated functionalities in the Kantian wholes that co-create the evolving biosphere. Note that the causal consequences and uses in Kantian wholes have a deep semantic content in embodied cells and organisms living in an embodied physical world. We need a new theory of embodied functional information in a cell, ecosystem or the biosphere.

A start of such a theory is taken in Kauffman [13]. The issues include these: (i) How do we measure the diversity of functions embodied in one or a community of Kantian wholes making their worlds together at any point in their evolution?

(ii) How do we measure the "degree of organization" of the processes carried out in those embodied functions? Consider the heart. Its function is to pump blood. But it makes heart sounds and jiggles water in the pericardial sac. The function of the heart is to pump blood, not make heart sounds or jiggle water in the pericardial sac. Thus, the function of a part of an organism is typically a subset of its causal consequences.

In Kauffman [13] I propose the steps of: (i) Distinguishing the system into a set of "parts and processes". (ii) For each of these, list its set of immediate causal consequences. (iii) Find that choice, for each of the distinguished "parts", of that one of its causal consequence, such that, when taken over all the parts together, that choice of one causal consequence per part maximizes a measure of the total diversity of processes of the total system. This measure is called Set Complexity. This maximal Set Complexity measure with its identified single causal consequence, among all the causal consequences of each part, will hopefully pick out the causal consequence of each part which is the true functions of that part. Thereby this will measure the total diversity of functions in the total system. (iv) For work processes, measure the power efficiency per unit fuel consumed of that process as a macroscopic measure of the "degree of organization" of that functional work process. Power efficiency per unit fuel consumed for work processes picks out an optimal displacement from equilibrium, hence is of considerable interest as a measure of the degree of organization of a process. (v) Multiply each identified functional work process of each part by its power efficiency and sum over the parts in the system to get an overall measure of the total diversity of organized processes.

I do not know how to generalize this to functions in cells or organisms which are not work processes.

If we could invent a measure along these lines, we could measure the diversity of organized processes in an ecosystem, or even the biosphere, at any moment of time. Then this diversity is a natural measure of the "embodied information" in the Kantian wholes co-creating their worlds. With this measure, should we get it, we could measure the change, presumably an average secular increase over evolutionary time, of the embodied information of the biosphere.

IV. Implications for Intelligent Design

The underlying concept of Intelligent Design, ID, is perfectly sensible but perhaps in a restricted set of scientific contexts. For example, ID can be taken to ask: (i) given an alphabet and messages, is the set of received messages highly improbable given the entropy of the Shannon source? (ii) Alternatively, given absorption or emission signals from atoms from stars, is the observed time sequence so improbable that it suggests "design". SETI has just this legitimate problem.

At issue is whether Intelligent Design is well founded in asking this question of biology. Here there are at least two issues: First of all, Irreducible Complexity [6], exemplified by the bacterial flagellar motor, is a phenomenon said by ID advocates to be too specifically complex to have arisen by random variation and natural selection. But this approach ignores Darwinian preadaptations where old parts, selected for different purposes, are recombined for a new function — e.g., the flagellar motor itself, assembled, it is thought, from proteins serving different functions in different bacteria.

Secondly and more deeply, Intelligent Design seeks to accomplish the analogue of SETI. But if, as above, we can construct no probability measure for the emergence and evolution into the ever changing adjacent possible of the evolving biosphere, it would seem that such calculations are either moot or questionable at present.

Whether the attempt to show that evolution is, in some definable sense, more "ordered" than some new and yet to be defined measure of randomness concerning what the myriad branching pathways of evolution, with some confidence level, would allow, remains to be seen. It would seem that Intelligent Design researchers — indeed, all of us — need to begin to cope with the amazing bubbling forth of new niches without selection, allowing new directions of evolution as life itself bubbles forth.

V. The Ensemble Approach to Statistical Laws and Self Organization with No Entailing Law

The "ensemble approach" [4] may prove useful. I will give four examples where it has been applied: (1) genetic regulatory networks, (2) the origin of life, (3) statistical features of "rugged fitness landscapes, and (4) in physics, spin glasses. I discuss the first in detail.

The Ensemble Approach to Genetic Regulatory Networks

As a young man, I thought about cell differentiation. How could different cells in us, all having the same genes, be different, liver, kidney, etc? It was known that in different cells types, different genes were active making specific and different sets of proteins. In 1961 and 1963, French microbiologists, F. Jacob and J. Monod, cracked the problem when they showed that, in E. coli bacteria, one gene, say A, could make a protein, say A, that bound to a DNA region next to another gene, say the B gene, and turn on or turn off the B gene's formation of its own B protein. In a seminal 1963 paper [14], they argued that if two genes, A and B, each repressed, or turned off, the other gene, the little two gene circuit had two dynamical steady states: (1) A on and B off, or (2) A off and B on. Hence, they said, the SAME set of genes could express different proteins corresponding to two cell types [14].

All biologists recognize that Jacob and Monod set the now central question of Systems Biology: what is the genetic regulatory network among 23,000 human genes, of which about 2,200 genes coding for transcription factors, and others coding for microRNAs, regulate one another's activities and regulate the rest of the 23,000? Here we need to know which genes regulate which genes, and by what "dynamical rules". Then we need to "integrate" the equations of motion of such a network to discover its integrated behavior. Just as Newton's laws for billiard balls yield, upon integration with given initial and boundary conditions, the trajectory of the balls, so for a classical physics genetic network, the behavior of the system has a trajectory from each initial state, i.e., from each pattern of gene expression among all 23,000 genes. These flow through a sequence of patterns, or states of gene expression, and typically the flows, or "trajectories", end up on small subsets of states, called "attractors", each of which drains a "basin of attraction". Cell types probably correspond to attractors and differentiation corresponds to noise or signal induced flow among attractors [5, 15].

Here is the "ensemble approach": I wondered if natural selection had to struggle to create very specifically selected, hence "engineered", networks to achieve controlled differentiation from the fertilized egg, or zygote, called "ontogeny", or, I hoped, some broad class or "ensemble" of networks would all have "good enough" dynamical behavior to underlie ontogeny with just some tuning by natural selection.

To ask this question I idealized the behavior of a gene as an on-off device, a light bulb, and asked if there was a class of large genetic networks that yielded "orderly behavior". To ask this question is inherently to take the Ensemble Approach: it asks whether there are typical (i.e. generic) behaviors in different classes or "ensembles" of networks. In my case I imagined N genes, each with K inputs. There are vastly many networks, an entire "ensemble" of networks, with $N = 23{,}000$ genes, and $K =$ say 2 inputs per gene. To study the typical properties of this ensemble, one approach is to sample at random from this ensemble. Thus, I chose the $K = 2$ inputs to each gene at random from among the N, and for each I assigned at random one of the 16 possible logical, or "Boolean functions" prescribing the behavior of the regulated gene at the "next time moment", given the on or off states of its two inputs at the current moment. The "AND" function is such a Boolean function. It says the regulated gene will be "on" at the next moment only if both its inputs are "on" at the present moment.

To summarize many years of work by many on Random Boolean Networks, it turns out that they behave in three regimes: Ordered, Chaotic, and a "Critical" "edge of chaos" regime which is a phase transition between order and chaos. $K = 2$ networks turn out to be critical for the ensemble of networks with randomly chosen Boolean functions. Critical networks can have other values of K greater than 2 by using non-random choices of Boolean functions of K [5].

Now three essential facts: (i) Critical and Ordered networks exhibit very ordered, and also multiple, attractors, hence the generic behaviors of these networks exhibits a new form of SELF ORGANIZATION — generic order in an Ensemble of systems. These ordered attractors are so ordered that the different attractors could explain the order of the different cell types in an organism. (ii) It is becoming clear that differentiated cell types are almost certainly "attractors" [14]. (iii). More amazingly, cells appear to be "Critical", to live on the edge of chaos [16–19].

Note three essential feature of the Ensemble Approach: (i) There is a vast ensemble of *NK* Random Boolean Networks, or more realistic models of genetic networks, all of which are dynamically critical. In short, "criticality" is a feature of an entire ENSEMBLE of networks, not just of one. ii. Importantly, this means that the generic behaviors of this class of networks is independent of the physics of any specific member of the ensemble. iii. Critical networks are a subset of all Random Boolean Networks, those at the edge of chaos. If cells are critical, Natural Selection must hold networks at the edge of chaos for adaptive reasons — here is the mixture of Ensemble Self-Organization AND Natural Selection.

The Ensemble Approach Can Yield Statistical Laws Beyond Entailing Laws

As stressed at the start of this article, no law entails the detailed evolution of the biosphere, including the evolution of genetic regulatory networks. This means we cannot deduce *ab initio* what those networks are. But the Ensemble Approach allows statistical laws about the typical features and behaviors of the entire ensemble of critical networks. More profoundly, evolution does NOT follow geodesics. Thus evolution is NOT entailed. It follows myriad pathways mixing quantum random indeterminate mutations and non random natural selection. The Ensemble Approach is the natural way to seek statistical laws about the behaviors of genetic regulatory networks, without needing to know the details of any specific genetic regulatory network. As we learn more about real networks we can refine the specifications of the ensemble, hence the generic behaviors of the refined ensemble, for better statistical predictions.

In short, the Ensemble Approach marries to the lack of entailing law for evolution, to yield one viable approach to statistical laws beyond entailing law.

The Ensemble Approach to the Emergence of Collectively Autocatalytic Sets as a Generic Phase Transition in Complex Chemical Reaction Networks

Perhaps the central problem concerning the origin of life is the onset of molecular reproduction, given a "soup" of prebiotic organic molecules such as amino acids, lipids, nucleotides and other organic molecules. These molecules may have been present on the early earth due to meteorite infall, abiotic synthesis on the early earth, or both.

Such small organic molecules, say in confined spaces such as tidal pools or rocks with interconnected hollow chambers, may be a necessary condition for the onset of molecular reproduction, but not sufficient. In 1971, the received view was that life must be based on template replication of arbitrary sequences of single stranded DNA, RNA, or similar molecules. The hope was that a single, say, RNA template strand would line up free A,U,C, and G nucleotides to Watson-Crick match the arbitrary single template strand, then the free nucleotides would be bonded by 3'–5' phosphodiester bonds to make a second complementary strand, then the two strands would melt apart and cycle. This would create, without enzymes, a self replicating arbitrary RNA sequence. Forty years of intense work has so far failed.

A current approach, pinioned on the observation that RNA molecules can act as enzymes, called ribozymes, is a search for an RNA ribozyme, single stranded, able to copy a second complementary RNA strand, then copy that complementary strand back into a copy of the initial strand. Such a ribozyme would, acting as a "polymerase", be able to copy any single-stranded RNA molecule, including itself. Some progress has been made, but I have concerns: (i) Such molecules may exist but be very rare, so unlikely to arise by chance. (ii) How does such a molecule build the network of metabolism around itself? (iii) If the ribozyme is error prone, its copies will have more errors, and their copies yet more errors, and may create a runaway "error catastrophe" if the mutation rate is stronger than the selective advantage of the good ribozyme(s).

In 1971 [20, 21, 5], I took the ensemble approach based on a different conception. What was needed, I thought, was a set of molecules that were collectively autocatalytic, as is Gonen Ashkenasi's nine-peptide collectively autocatalytic set [22]. His set, by the way, conclusively demonstrates that molecular reproduction need have nothing to do with DNA, RNA, or nucleotides.

To approach my question in an ensemble sense, I asked this: Given a set of molecules, M in number, with R reactions among them, and some distribution of which, if any, of the R reactions, each of the M might catalyze, could one find conditions under which, generically, collectively autocatalytic sets would arise? The answer can be yes. Under simple assumptions in which, as a 0th order hypothesis, each molecule among M has a probability P to catalyze each of the R reactions, it is a theorem that, as the diversity of M and the greater diversity of R and hence R/M increases, a phase transition is reached at which collectively autocatalytic sets emerge with probability near 1. Importantly, the same results arise with more realistic models of *which molecules* catalyze *which reactions* by a local "matching" rule [21, 5].

This work has been confirmed and extended in a number of ways. It is the ensemble approach, for it says that independent of the detailed chemistry and physics, it is a typical or generic property of complex reaction networks — whose molecules are also candidates to catalyze the reactions — that collectively autocatalytic sets will arise. (I emphasize that this remains theory and is not confirmed experimentally yet, but is fully open to being tested using libraries of random peptides, RNA, DNA, or mixtures of the above.) Here are the important features of this ensemble approach: (i) The emergence of collectively autocatalytic sets as a phase transition in complex reaction networks is a powerful example of self organization. (ii) Since DNA, RNA, and peptide collectively autocatalytic sets have been synthesized by good chemists [23, 24, 22], such sets are our only current examples of self reproducing molecular systems and are Kantian wholes as noted above. (iii) The theory of the emergence of such autocatalytic sets is again independent of the specific underlying physics, so it cannot be reduced to any specific physics, such as the choice of a specific set of molecules that happens to be one among trillions of collectively autocatalytic sets. The routes to molecular reproduction lie in chance and number, not specific physics. (iv) It now turns out that such systems in hollow lipid vesicles called liposomes can, in silico, synchronize the division of the liposome with that of the autocatalytic set [25], and can undergo open ended evolution [26]. (v) With the inclusion of inhibition of catalysis as well as catalysis, such systems can exhibit alternative attractors and critical dynamical behavior, like model genetic regulatory networks [26]. Thus, if critical autocatalytic sets are selectively advantageous, as I suspect, there will be a vast ensemble of such possible networks among a larger set of non-critical autocatalytic sets, so selection will have interacted with self organization to yield the useful ensemble, again a marriage of self organization and selection.

The Ensemble Approach to Tunably Rugged Fitness Landscapes

The concept of fitness landscapes, introduced by Sewall Wright into biology [27], is well established. Briefly, in one concrete case, over a set of haploid genotypes, each has a fitness in some fixed environment. This fitness can be thought of as a "height" over a large-dimension space of all the gene sequences under consideration. Now, in genetics, it is known that the fitness contribution of one version, or allele, of a gene at one "locus", may depend upon the alleles and other loci. This dependence is called "epistasis". The ensemble approach I utilized was borrowed with modification from "spin glasses" in physics [28]. I presumed *N* genes, each with two alternative alleles, or versions. I assumed that each gene allele's fitness contribution depended upon the allele of that gene and the alleles of *K* other genes. The rest of this *NK* model was randomly constituted, hence the ensemble approach. I assigned the *K* epistatic inputs to each gene at random among the *N*. I assigned the fitness contribution of a given gene, *i*, for each of the 2 to the $K + 1$ alleles of that gene and the *K* other input genes, at random from the uniform interval from 0 to 1. I defined the fitness of a given vector, or list, or state of the alleles of the *N* genes, and the average of their fitness contributions. These simple ensemble assumptions yield, for each randomly built *NK* model, a fitness landscape over all 2 to the Nth power haploid genotypes. Hence any *NK* model is a random sample, having fixed *N* and *K*, of an entire ensemble of fitness landscapes with the same *N* and *K* [5].

The result is an ensemble of fitness landscapes, whose statistical properties depend upon *N* and *K*. Briefly, for $K = 0$, each allele of each gene makes a fitness contribution that is independent of all other genes. There is a Fujiyama fitness landscape with one peak and smooth sides. For $K = N - 1$, its maximum value, the fitness landscape is random, there are 2 to the *N* divided by $(N + 1)$ local fitness peaks on the landscape, and many other statistical features. These features are tuned as *N* and *K* are tuned [5].

It is clear that the *NK* model inquires into the typical or generic properties of fitness landscapes only as a function of the epistatic coupling *K*, and the size of the system *N*. *K* captures conflicting constraints, hence as *K* increases the landscapes become more rugged. This model has found use in the economics of learning curves, maturation of the immune response, molecular evolution over rugged landscapes, and even in management models [27].

Surprisingly, if species co-evolve on *NK* landscapes and can both invade one another's niches, and when they do, they carry their own landscape ruggedness parameter, *K*, which varies in the population and can itself evolve, the system evolves to a state that increases species life-time distributions, smooths landscapes to an intermediate ruggedness, and yields a power law distribution of avalanches of extinction events matching the evolutionary record. Hence, evolution can modify the landscapes upon which evolution occurs [29].

Conclusion

I have offered rather radical views. Most notably, it may well be true that there is no law which entails the evolution of the biosphere. If so, what I speak of is, in fact, the end of a physics worldview, of the dream of reductionism to find a fundamental "final theory" that entails all that occurs in the universe. This is a very large claim requiring careful investigation. But if true, it begins to appear that it is not the tragedy we may have feared for so long. In its place is a vast creativity in the living world, far beyond what we have imagined. In *Answering Descartes: Beyond Turing*, (8), I hope I have been able to articulate similar ideas that could give a new account of major problems in the philosophy of mind and neuroscience concerning how mind can act "acausally" on matter via decoherence, and how we might have a responsible free will by a similar marriage of quantum random indeterminism and classical determinism in what I call Trans-Turing systems operating in the Poised Realm that hovers reversibly between quantum-coherent and classicality-FAPP behaviors.

If no law entails the becoming of the biosphere, we do not know the sample space of evolution, for its phase space persistently changes. Hence, we need to invent a new form of Embodied Information, which is laden with the semantics of the functions of parts of Kantian wholes in sustaining the existence and co-existence of such Kantian wholes. I have proposed what may be a start of such embodied information that seeks the "diversity of organized processes" in a cell, organism, ecosystem, or the biosphere, as a measure of the embodied information in the biosphere and how it may grow over time. Such growth would be a true form of information creation, beyond entailing law, and since merely syntactic information in a prestated sample space is of no use in biological evolution, whose phase space, as stressed, keeps changing in ways we do not know.

Self organization, as in the emergence and evolution of collectively autocatalytic sets as a generic property in ensembles of complex chemical reaction networks, and in ensembles of genetic regulatory networks, must play a profound role in the emergence of functional order, beyond entailing law, in co-evolving Kantian wholes. With natural selection, the entire process, beyond entailing law, has created a functional biosphere that has persisted and flourished for 3.7 billion years. We are thus invited to new science and a new view of what is required for order and organization to emerge and flourish beyond entailing law.

Acknowledgments

This work has been partially supported by the TEKES Foundation in Finland at the Tampere University of Technology, and by the University of Vermont.

References

1. Weinberg S (1992) Dreams of a final theory. Pantheon, NewYork.
2. Bailly F, Longo G, (2011) Mathematics and the natural sciences: The physical singularity of life. Imperial College Press, London.
3. Kauffman S (2008) Reinventing the Sacred. Basic Books, New York.
4. Kauffman S A (1969). Metabolic stability and epigenesis in randomly connected nets. J Theoret Biol 22:437.
5. Kauffman S (1993) Origins of Order. Oxford University Press, New York.
6. Behe M (1996) Darwin's Black Box. Free Press, New York.
7. Dembski W, Wells J (2008) The Design of Life. Foundation for Thought and Ethics, Dallas.
8. Kauffman S (2012). Answering Descartes: Beyond Turing in The Once and Future Turing: Computing the World, (Eds. SB Cooper and A Hodges), Cambridge University Press, Cambridge, England.
9. Kant, I (1952) Critique of Judgment. Translated by James Creed Meredith. Oxford University Press, New York.
10. Deacon T (2011) Incomplete Nature. W. W. Norton & Co.
11. Shannon C (1948) A mathematical theory of communication. Bell System Technical Journal [1].
12. Kolmogorov A N (1965). Three approaches to the quantitative definition of information. Problems Inform. Transmission 1(1):1–7.
13. Kauffman S (2011) Complexity Theory: Normal Science and Frontiers Classical and Quantum in Templeton Arizona State University Conference. In Press, Cambridge University Press, Cambridge, England.
14. Jacob F, Monod J (1963). Genetic repression, allosteric inhibition, and cellular differentiation. In: Locke M (ed) Cytodifferentiation and Macromolecular Synthesis. Academic Press, New York.
15. Huang S, Eichler G et al (2005) Cell fates as high-dimensional attractor states of a complex gene regulatory network. Phys Rev Le94(12): 128701.
16. Serra R, Villani M, Graudenzi A, Kauffman SA (2007): Why a simple model of genetic regulatory networks describes the distribution of avalanches in gene expression data. J Theor Biol 249: 449–460.

17. Rämö P, Kesseli J, Yli-Harja O (2006) Perturbation avalanches and criticality in gene regulatory networks. J Theor Biol 242:164–170.
18. Balleza E, Alvares-Buylla ER, Chaos A, Kauffman S, Shmulevich I et al (2008). Critical Dynamics in Genetic Regulatory Networks: Examples from Four Kingdoms. PLoS ONE 3(6): e2456. doi:101371/journal.pone.0002456.
19. Nykter M, Price ND, Aldana M, Ramsey SA, Kauffman SA, Hood L, Yli-Harja O, Shmulevich I (2008). Gene expression dynamics in the macrophage exhibit criticality. Proc Natl Acad Sci USA 105(6): 1897–1900.
20. Kauffman S, (1971) Cellular homeostasis, epigenesis and replication in randomly aggregated macromolecular systems. J Cybernetics 1: 71–96.
21. Kauffman SA (1986) Autocatalytlic sets of proteins. J Theor Biol 119:1–24.
22. Wagner N, Ashkenasi G (2009) Symmetry and order in systems chemistry. J Chem Phys 130: 164907. Published on line 27 April 2009.
23. Seivers D, von Kiedrowski G (1994). Self-replication of complementary nucleotide-based oligomers. doi:10.1038/369221a0 Nature 369:221–224 published online 19, May 1994.
24. Lam BJ, Joyce GF (2009). Autocatalytic aptazymes: ligand-dependent exponential amplification of RNA. Nature Biotechnol 27(3): 288–292.
25. Filisetti A, Serra R, Carletti T, Poli I, Villani M (2008) Synchronization phenomena in protocell models. BRL: Biophysical Reviews and Letters 3 (1/2): 325–342.
26. Fernando M, Vasas V, Santos M, Kauffman S, Szathmary E. (2011). Spontaneous formation and evolution of autocatalytic sets within compartments. Submitted.
27. Wright S (1931) Evolution in Mendelian populations. Genetics 16: 97.
28. Edwards SF, Anderson PW (1975). Theory of Spin Glasses. J. Phys. F5:965.
29. Kauffman S (1995) At Home in the Universe. Oxford University Press, New York.

Towards a General Biology: Emergence of Life and Information from the Perspective of Complex Systems Dynamics[1]

Bruce H. Weber

Department of Chemistry & Biochemistry, California State University Fullerton and Division of Science & Natural Philosophy, Bennington College

Abstract

I argue that Darwinism is best described as a research tradition in which specific theories of how natural selection acts to produce common descent and evolutionary change are instantiated by specific dynamical assumptions. The current Darwinian research program is the genetical theory of natural selection, or the Modern Evolutionary Synthesis. Presently, however, there is ferment in the Darwinian Research Tradition as new knowledge from molecular and developmental biology, together with the deployment of complex systems dynamics, suggests that an expanded and extended evolutionary synthesis is possible, one that could be particularly robust in explaining the emergence of evolutionary novelties and even of life itself. Critics of Darwinism need to address such theoretical advances and not just respond to earlier versions of the research tradition.

Key words: complex systems dynamics; Darwinian Research Tradition; emergence; expanded/extended evolutionary synthesis; genetical theory of natural selection; Modern Evolutionary Synthesis; origin of life; self-organization

My thesis is that the Darwinian Research Tradition, defined below, is being enriched, extended and expanded by new information and concepts and that a Darwinian evolutionary synthesis deploying background assumptions of complex systems dynamics can robustly guide further research into biological phenomena and lead to the

[1]The Wistar Institute held a conference in 1966 to explore the adequacy of the neo-Darwinian interpretation of evolution, the proceedings of which were subsequently published by the Wistar Institute Press as *Mathematical Challenges to the Neo-Darwinian Interpretation of Evolution* [1]. In addition to mathematical critiques of the version of population genetics upon which the neo-Darwinian Synthesis, or more accurately the Modern Evolutionary Synthesis, was based, there were presentations, particularly by Conrad Waddington, that pointed out that the synthesis had not adequately included developmental biological phenomena and was by implication incomplete. Two of the key figures in the development and deployment of the second phase of the neo-Darwinian synthesis, Richard Lewontin and Ernst Mayr, were participants, defending the Modern Evolutionary Synthesis even as they provided some criticism of the limitations of one version of the neo-Darwinian program that reduced all biological phenomena to population genetics.

development of a theory of general biology. Such a general theory could and should address issues of the emergence of life, topics properly previously screened off in the Darwinian discourse. After reviewing the history of neo-Darwinism and the Modern Evolutionary Synthesis in the Darwinian Research Tradition,[2] and making the case for shifting background dynamical assumptions to those of complex systems, I will focus specifically on the current status of "origin of life" research and how such work may contribute to a theory of general biology. Finally, I will argue that intelligent design theory does not provide a suitable scientific alternative in that it does not provide a conceptual framework for empirical and theoretical research on the phenomena of emergent complexity.[3] However, criticisms from intelligent design theorists, among others, of on-going efforts to develop a new Darwinian evolutionary synthesis can help sharpen the deployment of such a research program.

The Modern Evolutionary Synthesis and the Darwinian Research Tradition

In *Darwinism Evolving* and subsequent publications, David Depew and I have argued that there is not a single Darwinism synonymous with evolutionary theory,

---

This paper had its origins in a 2007 conference in Boston organized by Bruce Gordon under the auspices of the Center for Science and Culture at Discovery Institute, which funded the event, In the style and spirit of the Wistar Conference, it was meant to explore, some forty years later, the robustness of the earlier neo-Darwinian mathematical population-genetics theory of evolution in light of the progress in molecular and developmental biology, as well as in ecology, in the intervening time. A number of the critics of Darwinism present at the conference articulated an alternative explanation of functional biological complexity known as 'intelligent design' or more succinctly ID. Others present, like myself, while moving beyond the specific program based upon population genetics, defended the more general concept of a Darwinian evolutionary synthesis under a 'self-organizational' rubric.

[2]Since there was a research program known as neo-Darwinism in the late nineteenth century based upon Weismannian inheritance that was taken to preclude any Lamarckian mechanisms of heredity, many historians of biology prefer to use the term 'Modern Evolutionary Synthesis' rather than neo-Darwinism, or neo-Darwinian synthesis, to characterize the genetical theory of evolution based upon population genetics (see discussion in [2]). I will use neo-Darwinism to mean the specific program based upon early Mendelian genetics and Modern Evolutionary Synthesis for a more broadly conceived synthesis that includes the version based upon population genetics. I will use the term 'Darwinian Research Tradition' to refer to an interlinked set of research programs over time that share a commitment to natural selection as a major, though not sole, source of biological adaptation, order, and innovation, even as the concept of natural selection is articulated against different sets of background assumptions about systems dynamics.

[3]This is not to say that there cannot be a productive research program based upon assumptions of intelligent design, particularly in areas studying cultural artifacts and social and cultural phenomena more generally. Also, I can imagine productive programs so based for studying atemporal aspects of biological phenomena.

nor is the Modern Evolutionary Synthesis (often called neo-Darwinism, but see footnote 2) a monolithic research program [2–6]. Rather, we see a Darwinian Research Tradition, which has itself changed over time in light of new empirical data and conceptual advances, and which has assimilated new information and resolved entailing theoretical problems through a process of modifying underlying assumptions about the nature of biological systems and the dynamics of their changes over time. For example, we see "Darwin's Darwinism" as being informed by Newtonian systems dynamics that emphasized differential survival of individual organisms in populations and saw natural selection as analogous to a Newtonian force that acted gradually, instantaneously equilibrating with other forces (such as variation), to produce adaptation. For the two to three decades following the rediscovery of Mendelian genetics in 1900 the discrete nature of mutations seemed to contradict the notion of small, continuous variation that was assumed by Darwin in his Newtonian conceptual framework. Indeed, many critics saw and/or hoped for the demise of Darwinism.

After all, Darwinism was not the only research tradition that addressed the phenomena of evolutionary biology. There were many evolutionary biologists in the nineteenth and early twentieth centuries who worked within a Lamarckian, a Geoffroyean, or a Spencerian conceptual framework and research program, in which internal factors, developmental processes, or natural laws of complexification, respectively, were taken as the driving force of evolution rather than natural selection as a Newtonian-type of force. All three of these alternatives seemed to be gaining adherents in the early twentieth century, even when such scientists called themselves Darwinians, which was for some just a label for accepting descent by modification. As Depew and I recount, the great conceptual advance brought about by Sergei Chetverikov, J.B.S. Haldane, Ronald Fisher, and Sewall Wright that produced the basis of the "genetical theory of evolution." This move, which formed the basis of the "Modern Evolutionary Synthesis," involved shifting the underlying concepts of systems and systems dynamics from Newtonian to Boltzmannian. This shift took advantage of statistical insights used by Boltzmann in his development of statistical mechanics in which macroscopic, thermodynamic properties of matter and physical processes were re-described in terms of the aggregate behavior of the microscopic atomic and molecular constituents. The analogy of the action of selection on the frequencies of genes in populations with statistical mechanics was explicitly formulated by Fisher in his seminal *The Genetical Theory of Natural Selection* [7]. What mattered in this view was that the gradual shifting of the frequencies of a number of genes within an interbreeding population of a species due to the action of adaptive natural selection, by which change the fitness of the overall population was increased.

Not only did this first phase of the Modern Evolutionary Synthesis resolve the apparent conflict of discontinuous Mendelian genetical variation and gradualistic Darwinian natural selection by changing the background systems dynamics, it was attractive since it provided biologists with a mathematical theory of population genetics that could be rigorously tested. Further it placed biology within the broader "statistical revolution" that had already occurred in the physical sciences. Finally, during the 1930s and 1940s it provided the basis for a second phase and a broader synthesis of a number of areas of biology within the rubric of population genetics. The creative work of Theodosius Dobzhansky, Julian Huxley, Ernst Mayr, George Gaylord Simpson, and G. Ledyard Stebbins produced a more general synthesis of evolutionary biology, based upon population genetics, that incorporated much of biology including botany, paleontology, systematics and population ecology [8,9]. This version of the Modern Evolutionary Synthesis, as noted above, is sometimes called neo-Darwinism or the Synthetic Theory of Evolution and continues to provide a basis for a robust program of empirical and theoretical biology [10].

Despite any misgivings about the completeness of the Modern Evolutionary Synthesis, its advocates assumed that the action of natural selection on gene frequencies over generational time ("microevolution" see [11]) could account for the phenomena of common descent over geological time ("macroevolution"). But this synthesis was not complete, as Conrad Waddington repeatedly argued, since it bracketed off developmental biological phenomena, which were assumed to be merely the readout of the genes in the conceptual framework of neo-Darwinism [12–14]. Similarly bracketed off were aspects of ecology, such as energy flow and community interactions that went beyond population ecology [15–20]. Despite expectations that knowledge of the molecular sequence structures of biological macromolecules (DNA, RNA, proteins) would fit neatly into the neo-Darwinian framework, such knowledge has raised interesting puzzles and identified new evolutionary phenomena that need to be either incorporated into an expanded version of the Modern Evolutionary Synthesis or serve as the basis for a new, yet Darwinian, Expanded and Extended Modern Evolutionary Synthesis [2, 21–23]. Paleontologists Stephen Gould and Niles Eldredge have argued that the Synthesis is unfinished and needs a hierarchical expansion with selection acting in different ways at different levels of the biological hierarchy [24–26]. Scott Gilbert has continued Waddington's efforts to call for taking developmental phenomena seriously in an expanded and extended evolutionary synthesis, especially in light of the advances in "evo-devo" [27–32]. Gilbert sees development as a complementary process working with natural selection, producing variation and novelty, rather than replacing population genetics [28]. Mary Jane West-Eberhard has shown how developmental plasticity can provide variation even when there are no changes in the genome and how such phenomena impact evolutionary theory in ways that are not anticipated in the Modern Evolutionary Synthesis even though they are consistent with a more broadly conceived Darwinism [33,34].

Toward an Expanded Darwinian Synthesis and a General Biology

More innovative approaches to catch evolutionary phenomena in a expanded synthesis have relied upon a variety of tools from the still developing sciences of complexity. One example is that of Daniel Brooks and E.O. Wiley who, along with John Collier and Jonathan Smith, have sought to expand the evolutionary synthesis by introducing concepts from information theory and non-equilibrium thermodynamics to robustly account for the appearance of new biological information and pattern as well as natural selection itself via a process of 'infodynamics' [35–42; see also 43]. Using non-equilibrium thermodynamics in a more conventional usage Jeffrey Wicken sought to "expand the Darwinian program" not only to account for the emergence of new information in biological systems but to extend a kind of Darwinian approach to the problem of the origin, or more properly the emergence, of life [44]. Stuart Kauffman applied concepts of non-linear dynamics and self-organization to both developmental genetic systems and to the problem of the origin of life, to the latter of which he also brought in non-equilibrium thermodynamic considerations as well as consideration of the emergence of 'agency' [45–47]. I will return to the issue of the origin of life below. With regard to the inclusion of developmental biology into evolutionary theory, Depew and I have argued that the shift to such systems dynamics employing insights from the behavior of complex systems can provide the conceptual context within which a synthesis both can be effected while staying within in the Darwinian Research Tradition, if not narrowly formulated versions of neo-Darwinism as espoused by Richard Dawkins, for example [48–50]. One attempt to forge such a synthesis is known as Developmental Systems Theory (see contributions in [51] as well as in [52]). It shows a range of commitment from some form of Darwinism (see [53]) all the way to embracing instead an alternative research tradition, such as the Lamarckian [54–57] or the Geoffroyean [58–60]. Jablonka and Lamb argue that since in later editions of *On the Origin of* Species Darwin's hypothetical mechanism of inheritance had a Lamarckian character their inclusion of epigenetic factors could be considered as a recovery of Darwin's original vision [56–57]. A recent review of developmental genetics and epigenetics by Robert Reid argues for an evolutionary theory that is in his own terms outside the Darwinian tradition but more at home in a Lamarckian or Geoffroyean one [61].

A current research program, which we might denote as 'emergentist' as a convenient label, has the goal of developing a theory of general biology, that is, a theory of structural and functional complexity and the emergence of novel structure/function as well as new information and phenomena [45–47,62–77]. This is a program very much in its early stages, but one that holds the promise of eventually developing a theory of biological organization that would hold not only for terrene biology but also for possible biological phenomena elsewhere in the universe. Such a general biology would be part of a more general theory of emergence (see contributions to [66]).

Cautionary Considerations and a Perspective on Emergence

When we are evaluating the sufficiency or inadequacy of the Modern Evolutionary Synthesis, or of Darwinism more generally as a research tradition in some new synthesis, or of rival naturalistic research traditions, or of theories such as intelligent design that posit sources of order and information outside of natural processes, it is important that we take care in being explicit about what we are discussing. Some evolutionary thinkers, such as Gould or Corning, see their approaches, for all the new empirical and theoretical content, as closer in conceptual stance to Darwin's original Darwinism than to a narrowly construed Modern Evolutionary Synthesis. Others, such as Deacon, Depew, Kauffman, Wicken and myself, see the deployment of the new complex systems dynamics leading to a totally new version of Darwinism, but still a research program within the Darwinian Research Tradition. Critics of Darwinism, such as Stanley Salthe, Eva Jablonka, and Robert Reid, are not rejecting evolutionary phenomena nor are they calling for sources of order outside nature. Rather, they are arguing for a different set of naturalist assumptions and dynamics that they regard to be better suited to guide future research. As a commitment to methodological naturalism does not logically entail a commitment to philosophical materialism, so we should not take any version of Darwinism as being a synonym or a placeholder for philosophical materialism, unless such a move is self avowed or can be demonstrated, as is the case in writers such as Dawkins and Dennett.

In what follows, I am going to examine current research on emergence theory as well as current work on emergence of life. Even though this issue of the origin of life historically lies outside the orbit of the Darwinian Research Tradition, I will take the cue from Wicken, as well as Kauffman, and Terrence Deacon that the processes and phenomena are rightfully the topic of a general biology and can and should be incorporated in any expanded version or new synthesis of Darwinism. I will assess the value of any theoretical approach in terms of its potential fecundity and robustness in the development of such a new synthesis and theory of general biology and of emergence. This means I am viewing science not as a body of established facts only, but rather as a process of exploring nature and deepening our understanding of natural phenomena.

Emergence of Emergence as Paradigm

The latter part of the twentieth century saw the rise of a new way of understanding nature, employing complex systems dynamics to explore and explain phenomena of self-organization and emergence (for an overview see for example [2,3,45,46, 65–68,71–73,78–88]). Self-organization, or more properly systems-organization, in which the interaction of the system and its environment under particular initial and boundary conditions leads to the emergence of novel order and structure, occurs widely in nature as well as under laboratory conditions and can be considered as a natural phenomenon [89,90]. Developing a theory of such emergent organization has as its goal providing natural explanations for such phenomena. This is very much a work still in progress but the insights gained so far provide a conceptual framework for thinking about and guiding research on the problem of the origin of life.

I define emergence as the appearance of novel properties, structures, and/or patterns in a system that are not present in the constituent components or easily predicted (weak form) or explained (strong form) from the laws and processes affecting the constituents of the system. The new level of phenomena and the lower level of constituents have mutual constraints and the arrows of causal explanation point in both directions. If we are tracking the process of the appearance of the new phenomena we are speaking of diachronic emergence in which the lower-level causality exceeds that of the upper level, but when the system has settled to a steady state we than have an instance of synchronic emergence in which the constraints fully mutual. In any event, the emergentist view is that the new, upper-level structure/properties/processes/phenomena represent real natural phenomena and not epiphenomena. In reductionism the lower level is the locus of causality and the upper-level properties are regarded as merely epiphenomenal, that is, without causality; in holism the upper level has the causality and the lower levels are epiphenomenal.

It is the strong form of emergence that will be of concern here, especially with regard to the emergence of life. In strong emergence, the emergent phenomena are novel in that they have properties not contained in the components, and are irreducible in sense that the emergent phenomena are not identical to their composition. Emergent systems exhibit a kind of holism in that the emergent phenomena cannot be analyzed into their parts without losing sight of their essential character. Further, in strong emergence the emergent phenomena obey laws that rely, in at least part, on their novel properties, that is, some of the processes and laws themselves are emergent, even as the process of their emergence itself operates under general natural laws (including for example a putative 'fourth of thermodynamics' in addition to other natural laws [45,46]). Finally, in strong emergence the emergent phenomena can impose conditions on their constituents that depend on the nature of the identity of the emergent phenomena, that is, such systems can exhibit downward causation.

Following Deacon's analysis I will further distinguish three types of emergence: first-order or supervenience, second-order or self-organization, and third order or evolution [67]. In supervenience, the higher-order properties of an *aggregate* are determined by the statistical or stochastic properties of the ensemble. For example, the liquid properties of water are said to *supervene* on the properties of individual water molecules. Second-order emergence, or self-organization, occurs on a higher hierarchical level than first-order emergence but as in all hierarchical systems the lower level continues to operate. In self-organization the configurations of individual components and the unique interactions in the system exert an organizing effect on the entire ensemble. Initial conditions and outliers can strongly affect the ensemble properties. Self-organization occurs in systems *open* to matter/energy flows that keep the systems *away from equilibrium*, resulting in *macroscopic structures* such as convection cells. Second order emergence also includes phenomena associated with nonlinearity and chaos. It is characteristic of all second-order emergent systems that they have a spatially distributed re-entrant causality that allows microstate variation to amplify and influence macrostate development, even as the macro-relationships undermine, constrain and bias micro-relationships. Snowflakes, Benard convection cells, tornados, chemical waves in the Belousov-Zhabotinskii reaction are examples of such second-order emergence. Self-organizing systems that generate and store information that is useful for system stability and survival *evolve*. Such informational memory produces *recursive*, *self-referential* self-organization that exerts a causal, cumulative (over time) influence over the future of the system. Fitness, function, and natural selection itself can be seen as examples of third-order emergence. Third-order emergence biases across iterations or generations, as in biological development or biological evolution, and can be viewed as an autopoiesis of autopoieses. "So life, even in its simplest forms, is third-order emergent. That is why its products cannot be fully understood apart from either historical or functional concerns" [67, p 300]. Both second and third order emergence exhibit a diachronic symmetry breaking not seen in first order emergence. Although higher levels in the hierarch are based upon the lower ones they can exhibit properties not seen at the lower levels because of this symmetry breaking.

The formation of Benard convection cells is an example of a self-organizing process in which the macroscopic structure of the convection flow allows for more efficient dissipation of the energy gradient, giving a thermodynamic "reward" for the production of structure. The process of formation of such convection cells involves a type of selection process working with self-organization. Rod Swenson has shown that the initial formation of convection cells produces macroscopic structures of various sizes and shapes, but that the system quickly settles down into a pattern of hexagonal cells of uniform size [91,92]. Thus there is a sorting or selection process working with self-organization. Brian Goodwin saw the shape and size selection as an instance of physical selection for the most stable [80,85]. To this Swenson added selection of the most dissipatively efficient. For complex chemical systems exhibiting self-organization there is additionally selection for the catalytically efficient, in addition to that for thermodynamic efficiency and physical stability. Thus, even before there is biological selection for the reproductively fit, emerging with the emergence of life, there exists in nature interplay of self-organization and selection at the level of physical and chemical phenomena [2–4,45,46,68,69,71,92].

Is the Origin of Life a Darwinian Problem?

Darwin himself carefully avoided the issue of the origin of life since he was concerned with explaining how living beings and their lineages changed over time and how novelties could arise through the action of natural selection upon heritable variation. For example, "How a nerve becomes sensitive to light hardly concerns us more than how life itself originated" [93, p187] was consistent with his accepting that life was "breathed into a few forms or into one" [93, p490] (Darwin [1859] 1964, 490). This position served to distinguish Darwin's theory of evolution from Lamarck's in which "active matter" spontaneously and continuously generated life [see 94–96]. Privately, Darwin was willing to speculate about the origin of life, as he did in a letter to Joseph Hooker in 1871, "But if (and oh what a big if) we could conceive in some warm little pond with all sorts of ammonia and phosphoric salts, light, heat, electricity and etc., present, that a protein compound was chemically formed, ready to undergo still more complex changes" (Cambridge University Library Manuscript Collection: DAR 94: 188–89).

Herbert Spencer argued that biological evolution is a part of a general, cosmic process of the universe becoming less homogeneous and more complex in which the origin of life was a specific instance [97]. Josiah Royce reasserted the more narrow claims of Darwinism as distinguished from those of the Spencerians [98]. With the rise of the Modern Evolutionary Synthesis, the demarcation of the problem of the origin of life from matters Darwinian was reasserted and continues today in mainstream evolutionary discourse [99,100].

However, one of the founders of the Modern Evolutionary Synthesis, J.B.S. Haldane, along with Alexander Oparin and J.D. Bernal (Marxists all), argued that advances in biochemistry and geochemistry meant that serious scientific study of the origin of life is possible, even if not required by the theories of the Darwinian Research Tradition [101–105]. They recognized that from their commitment to philosophical materialism it was necessary that the origin of life be the result of natural processes only. Opponents of Darwinism and also of philosophical materialism similarly argue that the origin of life is conflated with Darwinian theories [106–110]. Indeed, some neo-Darwinian advocates, such as Richard Dawkins, accept this conflation. In order to reduce biological phenomena to "selfish genes" Dawkins assumes that, however improbable, all that was needed for the appearance of life was to get a nucleic acid molecule that could replicate itself, although later this "naked replicator" decorated itself over time with proteins, lipids, etc. to produce better "survival machines" [49,50]. Alex Rosenberg attempts to achieve reduction of all biology to molecular genetics by a slightly different move at the origin of life [111]. He argues that natural selection has to be grounded in chemical and physical selection during the process of life's origin. During the process of life's origins, I agree; but this attempt at reduction points instead toward an emergentist account [112,113,118]. In what follows, I will consider experimental and theoretical approaches to the *emergence* of life as well as the implications of the dynamics of emergent complexity for our understanding of biological organization and how it arises.

Current Perspectives on the Emergence of Life

Whether a reductionist or emergentist approach is taken to the origin of life, the possible reactions and routes to the organized complexity of living entities is constrained by the properties of matter and the laws of chemistry and physics [43,113–118]. Not all types of bonding arrangements and compounds are possible [119]. In aqueous environments, for example, phosphate has unique properties that make it essential for life and even for proto-life. Only phosphoanhydrides had the needed mix of thermodynamic instability and kinetic stability to serve as an intermediate for capturing and providing energy. One consequence is that polypeptides can be synthesized abiotically from amino acids, polyphosphate (a phosphoanhydride) and magnesium cation [120]. Of course, life may be possible using non-aqueous chemistry, and such possibilities should be explored in a theory of general biology. Steven Benner has suggested that what is essential for the emergence of life is some sort of solvent system, the chemical elements carbon, hydrogen, nitrogen, sulfur, phosphorous, and oxygen, along with thermodynamic disequilibrium and temperatures consistent of chemical bonding [121].

However, for the present it is a sufficient challenge to address what might have happened during the emergence of life on earth. Given that, we can proceed with the understanding that the possibility space of chemical reactions in living systems is not unconstrained, nor random, but rather reflects in part structural, thermodynamic, kinetic, and combinatorial constraints. Overall, the transition to life and the subsequent evolution of living systems involves retention of reduced compounds in the presence of the resulting ever more oxidizing environment [114]. With an on-going influx of energy and matter the complexity of chemical reactions would be expected to increase as well as non-sequence specific macromolecules under pre-biotic conditions [44].

The minimal elements that need to be considered in any account of the emergence of life are:

- An energy source (gradient) and a mechanism to capture energy such that the entropy of the 'system' decreases even as the entropy of the system + environment increases
- Abiotically produced component molecules (subsequently produced by autocatalytic networks in proto-cells, and later in cellular metabolism
- Autocatalytic sets of catalysts (polypeptides, polynucleotides)
- Closure in both the sense of physical closure (an osmotic barrier) that separates the system from everything else, and chemical or catalytic closure
- Some means of reproduction and variation at the level of autocatalytic sets and thermodynamic cycles
- Templates for replication and for coding for catalysts.

It is an open question as to which of these steps must be prior to others or if some ensemble of factors is needed before the transition to life could occur. In an emergentist approach it would be expected that several steps could arise concurrently and act synergistically to give rise to more complex structures and phenomena, among which would be included natural selection [43,113,122].

Stanley Miller, working in the laboratory of Harold Urey, demonstrated that a number of amino acids could be produced via chemical processes that might have occurred on the primitive earth [123]. Although the atmosphere globally might not have been as reducing as Miller assumed, mainly due to escape of hydrogen gas, there would be local regions that were, such as near volcanoes or deep-ocean hydrothermal vents [124]. Alternative pathways to amino acids are plausible from carbon dioxide and from hydrogen cyanide [124]. Further, the presence of amino acids in the interior of meteorites indicates that they can be produced elsewhere in the universe by natural processes; indeed, extraterrestrial sources of organic compounds might have been up to three orders of magnitude greater than terrestrial ones for the primitive earth [117 p49,125]. Further, similar such putative processes involving electrical discharge and/or solar-driven photochemical reactions involving hydrogen cyanide, formic acid, hydrogen sulfide, and methane have been shown to produce sugars and purine and pyrimidine bases [for reviews see 113,124,126–129]. Chirality in such momoners could arise in a geologically short period of time due to asymmetry in cosmic radiation that was bombarding the earth [130]. Such monomers could polymerize to form polypeptides and proteins under plausible ambient temperatures [129,131]. Alternatively, hydrogen cyanide polymers form spontaneously when hydrogen cyanide is exposed to an electrical discharge; when such polymers react with water they yield polypeptides, and even polynucleotides [132–134]. Yet another alternative for generating such polymers is considered below involving chemiosmotic-type mechanisms.

Theorizing about the abiotic generation of the organic molecules that are the building blocks of living entities has given rise to a "prebiotic soup" model of increasingly complex molecules, driven by energy flows, from which macromolecules arise allowing the emergence of directed synthesis of catalysts, from which protocells would eventually be possible, followed by metabolism in true cells [44,135]. Alternative approaches follow a "metabolism first" approach, harkening back to Haldane, Oparin, and Bernal, often invoking the catalytic capacities of clays [136–138]. A third group of approaches assumes the early presence of some sort of encapsulating barrier, a "proto-cell first" model in which chemical processes occur in high and sequestered concentrations, within which emerge the catalytic polymers and ultimately directed synthesis [77,139,140]. In this scenario the mutual interaction of catalytic macromolecules and the reactions of a protometabolism within an osmotic barrier provides the "theatre" within which specified information can emerge.

Regardless of the approach, at some point catalytic polymers would be expected to emerge and open new chemical possibilities. Polypeptides and proteins produced abiotically would initially have a random sequence [44]. But such sequences have a high probability (at least 25%) of assuming a compact, globular tertiary structure and can exhibit some weak catalytic activity [117,141]. Given that many different sequences of amino acids fold up into the same or similar three dimensional structure, the number of such possible folds is a relatively rather small number [142]. Further, completely different and unrelated sequences can produce the same active-site geometry and catalytic function, that is they overlap in the map of catalytic task space [143]. Thus a highly specified informational content is not necessary for a polypeptide to serve as a catalyst. However, when such a specification process became available via nucleic acid templates, there would be an enormous advantage to such specified information, selected on the basis of catalytic and thermodynamic efficiency.

The "hard problem" in origin-of-life research is not so much how the monomers and even polymers might have arisen by physical and chemical processes, but rather how it came to be that a digital-type code in nucleic acids came to specify the analogical information in the thousands of proteins that catalyze metabolism and are involved in signally and information processing [43,45–47,69,108,109,113,118,144–148]. It is here that the new sciences of complexity can have their greatest impact.

The Complex Systems View of the Emergence of Life

As Kauffman has analyzed in his simulations, "protein sequence space" can cover what he terms the "catalytic task space" of all possible chemical reactions that can be catalyzed by polypeptides [45]. Thus, even an ensemble of random peptides would be able to provide such coverage. Such an ensemble can be self-sustaining when it can catalyze the formation of more such catalytic polymers in what is called an autocatalytic cycle. When such a set of autocatalytic cycles can produce their components such that they are self-sustaining, a condition termed catalytic closure is said to obtain. Such catalytically closed, autocatalytic cycles can be maintained, grow, and complexify if they also have some mechanism by which they can tap available energy gradients so as to drive the ensemble away from chemical equilibrium [44,46]. In such emergent systems there would be physical selection of clusters of amino acid sequences that are soluble in water and more stable in an aqueous environment since the less stable structures would tend to degrade and less soluble to precipitate. There would also be a chemical selection of those sequences that were more efficient catalysts or which more efficiently contributed to the autocatalytic cycles and/or more efficiently extracted energy from ambient gradients as the ensembles to which they occur would tend to persist longer. Kauffman, who suspects that such an emergence of organization and complexity, an emergence of life, would be an expected consequence of natural law, possibly a fourth law of thermodynamics, writes: "We can think of the origin of life as an *expected emergent collective property* of a modestly complex mixture of catalytic polymers" [45, xvi, emphasis in original]. Such ensembles of catalytic polymers would be expected to show weak inheritance due to the action of physical and chemical selection. Such systems as those modeled by Kauffman currently are being experimentally studied by Reza Ghardiri to document their dynamics as

exploration of only a small subset of the possible patterns of organization [46]. When a sufficiently complex organization emerges, not only does natural selection arise, but also the autonomous agency exhibited by living entities. He seeks a possible fourth law of thermodynamics that would account for the emergence of life and new organization. Deacon seeks to develop a broader theory of general biology through expanding our conception of organism [69]. His *autaea* are the chemical systems that exhibit autonomous self-maintenance, in contrast to all other configurations of matter, and include autocells. Autocells have coherent and integrated organization as well as self-reproduction in that they can reproduce by direct morphological means. Such *morphota* would include not only autocells, but also bilayer vesicles capable of reproduction or reproduction of autocatalytic sets. The transition to life comes when it is possible to transmit information of representation via genetic coding, so living things as we know them are also examples of *semeota*. The criteria Deacon develops for these categories and the specific example he explores can give us insight as to how to frame questions as to whether some entity encountered elsewhere in the universe is living or to delineate the logical requirements for the emergence of life. In Deacon's view, as in that of Weber and Depew, natural selection emerges as a phenomenon along with the phenomenon of the emergence of life, which in turn is a specific instance of the interaction of self-organizational principles with each other and with general selectional principles [3,43,67,69,71,113,118,163,164].

Implications of an Emerging Emergence Paradigm

We are in the very early stages of the development of the emergentist research program. If successful and if widely adopted such theories of emergent organization and general biology may in time become a new paradigm. Even in these early years it is generating new theoretical and experimental approaches that are particularly relevant to the problem of the emergence of life. When a more complete picture of how life might have emerged is available and we see how it fits into a broader theory of general biology, it will be time to assess whether the Darwinian Research Tradition, if not the Modern Evolutionary Synthesis, can encompass such insights, or if some new conceptual synthesis will be required. At this point we can acknowledge that Conrad Waddington's intuitions were fecund but needed the developments in biochemistry, molecular biology, developmental genetics, computer simulations, and complexity theory to be cashed out.

The complexification of abiotic chemical reactions is driven primarily by non-equilibrium thermodynamics, exploring state space in an ergodic fashion. When the transition occurs to living systems, a much larger state space of combinatorial possibilities, provided by catalytic (and templating) polymers, is explored by a combination of self-organizing and selecting processes via what Kauffman terms the "adjacent possible" [45,46]. Though thermodynamics provides the driving force for self-organization, it is the kinetic mechanisms that afford the pathways of emergence. With the emergence of life there is a shift to an extreme expression of kinetic control in which thermodynamic requirements play a supporting rather than directing role. Replication is an instance of this kinetic control. From this emerges the teleonomic and semiotic character of living entities.

In the emergentist perspective, organisms are begotten not made, that is they are the result of developmental processes individually and of evolving lineages. In both cases these phenomena are viewed the result of an on-going interplay of selection and self-organization. What organisms, or their constituent parts, are not, are artifacts. Although emergentist and reductionist approaches to biology share a commitment to methodological naturalism, they view organisms differently in this sense of the importance of epigenetic processes. What the reductionist version of the Modern Evolutionary Synthesis and proponents of intelligent design theory share is a view of biological traits and molecules as artifacts, something made by a designer or by the process of random variation and selection. Emergentists argue that natural and artifactual systems should not be conflated; by anchoring the emergence of life and natural selection in natural laws and processes of thermodynamics and kinetics, a conceptual wedge is driven between natural organization and design.

Elsewhere I address my more general philosophical problems with design arguments [165–167]. Here I am attempting only to argue that whereas the emerging theory of general biology is generating novel theoretical insights, predictions, and experimental approaches by which we can deepen our understanding of the emergence of life, ID theory does not suggest how to proceed theoretically or experimentally as to how life originated, other than to place the causes outside of scientific scrutiny. ID seems to me to provide only a negative capability by criticizing proposed naturalistic and emergentist explanations for the origin of life. Good critics are always helpful in the process of scientific research, but any research program worth its salt also has to guide in the generation of new experiments and theories. The latter is being achieved by those, such as Deacon, Deamer, Ghardiri, Kauffman, Luisi, Morowitz, and Wicken among others, seeking to understand the emergence of life, but not yet substantially by those advocating design arguments.[4]

[4]ID advocates would, of course, dispute these assessments, arguing that intelligent causes can reliably be distinguished from unintelligent (undirected natural) causes, and that intelligent causation therefore forms a significant part of our understanding of the cause-and-effect structure of the world under uniformitarian assumptions and constraints. As noted above, emergentists would argue against

Through processes of emergence, life itself may be viewed as begotten, not made, from underlying natural laws and a dialectic of self-organization and selection.

What Might We Expect from a Theory of General Biology About the Origin of Life?

We not only have to acknowledge the difficulty of the problem of how life might have emerged here on earth, let alone how it might emerge and instantiate elsewhere in the universe, but we need to accept that we should not expect a single narrative trajectory for life's emergence. Not only would the earliest true living beings destroy the traces of earlier transitional forms, but the action of living systems alters in fundamental ways the chemistry of their environments. Thus, we can only hope to elucidate plausible pathways of emergence, tested by simulations, experiments, and what geological data is available. This is not unlike the point Keith Miller makes about the paleontological record, in which we do not have all the details but do have some general patterns to explain [168]. Thus, we need to explore all possible routes of chemistry and proto-biochemistry to develop a range of plausible scenarios for life's emergence on earth and to eliminate those that are unlikely, through theoretical analysis, computer simulations, and experimentation.

In complex systems not only is the whole defined by closure conditions (physical and catalytic) but there is redundancy and parallelism. Thus even weakly insipient functional patterns of structure and interaction can persist due to greater stability and/or efficiency. With functionality comes pressure for improved structures/stability/efficiency, through an on-going process of selection and self-organization. Thus in the origin of life, we should not expect one function to be perfected, say replication, before others appear, but that there would be an inherent holism in the process by which cellular life arose [43,45,46,113,118,140,147].

If there is not grandeur in this view of the emergence of life at least there is a reasonable hope for progress, through application of the tools of complex systems dynamics, towards developing a theory of emergence and of general biology.

---

this conflation of natural and artifactual systems. To be fair to ID advocates, however, a more substantial ID research program seems to be brewing as of late, as evidenced in the research being done through the Evolutionary Informatics Lab (http://www.evoinfo.org) and in the work of Biologic Institute (http://www.biologicinstitute.org) and its journal *BIO-Complexity* (http://bio-complexity.org/ojs/index.php/main/index). Indeed, this present volume is part of that general trend. The only thing that can be said is that we must wait and see whether these efforts will go anywhere. For a broader discussion of these issues from a variety of perspectives, both supportive of ID and critical, see Gordon and Dembski [169].

References

1. Moorehead PS, Kaplan MM (eds) (1967) The Mathematical challenges to the Neo-Darwinian interpretation of evolution. Wisar Institute Press, Philadelphia, PA.
2. Depew DJ, Weber BH (1995) Darwinism Evolving: Systems Dynamics and the Genealogy of Natural Selection. MIT Press, Cambridge, MA.
3. Weber BH, Depew DJ (1996) Natural selection and self-organization: Dynamical models as clues to a new evolutionary synthesis. Biol and Phil 11: 33–65.
4. Weber BH, Depew DJ (1999a), Does the second law of thermodynamics refute the neo-Darwinian synthesis? In: Koslowski P (ed) Sociobiology and Bioeconomics: The Theory of Evolution in Biological and Economic Theory, pp. 50–75. Springer Verlag, Berlin.
5. Weber BH, Depew DJ (1999b) The modern evolutionary synthesis and complex systems dynamics: Prospects for a new synthesis. In: Taborsky E (ed), Semiosis, Evolution, Energy: Towards a Reconceptualization of the Sign, pp. 263–281. Saker Verlag, Aachen.
6. Weber BH (2007c) Fact, phenomenon and theory in the Darwinian research tradition. Biological Theory 2: 168–178.
7. Fisher RA (1930) The Genetical Theory of Natural Selection. Oxford University Press, Oxford.
8. Jepsen GL, Mayr E, Simpson GG (eds) (1949) Genetics, Paleontology, and Evolution. Princeton University Press, Princeton, NJ.
9. Mayr E, Provine WB (1980) The Evolutionary Synthesis: Perspective on the Unification of Biology. Harvard University Press, Cambridge, MA.
10. Nowak MA (2006) Evolutionary Dynamics: Exploring the Equations of Life. Harvard University Press, Cambridge, MA.
11. Mayr E (1988) Toward a New Philosophy of Biology: Observations of an Evolutionist. Cambridge MA: Harvard University Press.
12. Waddington CH (1940) Organizers and Genes. Cambridge University Press, Cambridge.
13. Waddington CH (ed) (1968–72) Towards a Theoretical Biology. Aldine, Chicago.
14. Waddington CH (1975) The Evolution of an Evolutionist. Cornell University Press, Ithaca NY.
15. Lotka AJ (1924) Elements of Physical Biology. Williams and Wilkins, Baltimore, MD.
16. Odam E (1969) The strategy of ecosystem development. Science 164: 262–270.
17. Odam HT (1988) Self-organization, transformity, and information. Science 242: 1132–1139.
18. Ulanowicz RE (1986) Growth and Development: Ecosystems Phenomenology. Springer Verlag, New York.

19. Ulanowicz RE (1997) Ecology, The Ascendent Perspective. Columbia University Press, New York.
20. Ulanowicz RE (2007) Emergence, naturally! Zygon 42: 945–960.
21. Nei M (1987) Molecular Evolutionary Genetics. Columbia University Press, New York.
22. Gillespie JH (1991) The Causes of Molecular Evolution. Oxford University Press, New York.
23. Lewin R (1999) Patterns in Evolution: The New Molecular View. Scientific American Library, New York.
24. Gould SJ (1982) Darwinism and the expansion of evolutionary theory. Science 216: 380–387.
25. Gould SJ (2002) The Structure of Evolutionary Theory. Harvard University Press, Cambridge, MA.
26. Eldredge N (1985) Unfinished Synthesis: Biological Hierarchies and Modern Evolutionary Thought. Oxford University Press, New York.
27. Gilbert SF (2006a) Developmental Biology, 8th edn. Sinauer, Sunderland, MA.
28. Gilbert SF (2006b) The generation of novelty: The province of developmental biology. Biological Theory 1: 209–212.
29. Gilbert SF, Opitz JM, Raff RA (1996), Resynthesizing evolutionary theory and developmental biology. Devel Biol 173: 357–372.
30. Carroll SB, Grenier JK, Weatherbee SD (2001), From DNA to Diversity: Molecular Genetics and the Evolution of Animal Design. Blackwell Science, Malden MA.
31. Carroll SB (2005) Endless Forms Most Beautiful: The New Science of Evo-Devo. Norton, New York.
32. Forgas G, Newman SA (2005) Biological Physics of the Developing Embryo. Cambridge University Press, Cambridge.
33. West-Eberhard MJ (2003) Developmental Plasticity and Evolution. Oxford University Press, New York.
34. West-Eberhard MJ (2007) Dancing with DNA and flirting with the ghost of Lamarck. Biol Phil 22: 439–451.
35. Brooks DR, Wiley EO (1986) Evolution as Entropy: Toward a Unified Theory of Biology, 2nd edn. University of Chicago Press, Chicago.
36. Brooks DR (1998) The unified theory and selection processes. In: Van De Vijver G, Sathe SN, Delpos M (eds) Evolutionary Systems: Biological and Epistemological Perspectives on Selection and Self-Organization, pp 113–128. Kluwer, Dordrecht.
37. Brooks DR, McClennan, DA (1999) The nature of the organisms and the emergence of selection processes and biological signals. In: Taborsky E (ed) Semiosis, Evolution, Energy: Towards a Reconceptualization of the Sign, pp. 185–218. Aachen: Saker Verlag.
38. Collier J (1986) Entropy in evolution. Biol Phil 1: 5–24.
39. Collier J (1998) Information increase in biological systems: How does adaptation fit? In: Van De Vijver G, Sathe SN, Delpos M (eds) Evolutionary Systems: Biological

and Epistemological Perspectives on Selection and Self-Organization, pp 129–139. Kluwer, Dordrecht.

40. Collier J (1999) The dynamical basis of information and the origins of semiosis. In: Taborsky E (ed) Semiosis, Evolution, Energy: Towards a Reconceptualization of the Sign, pp. 111–138. Saker Verlag, Aachen.
41. Smith JDH (1998) Canonical ensembles, evolution of competing species, and the arrow of time. In: Van De Vijver G, Sathe SN, Delpos M (eds) Evolutionary Systems: Biological and Epistemological Perspectives on Selection and Self-Organization, pp 141–153. Kluwer, Dordrecht.
42. Smith JDH (1999) On the evolution of semiotic capacity. In: Taborsky E (ed) Semiosis, Evolution, Energy: Towards a Reconceptualization of the Sign, pp. 283–309. Saker Verlag, Aachen.
43. Weber BH (1998) Emergence of life and biological selection from the perspective of complex systems dynamics. In: Van De Vijver G, Sathe SN, Delpos M (eds) Evolutionary Systems: Biological and Epistemological Perspectives on Selection and Self-Organization, pp 59–66. Kluwer, Dordrecht.
44. Wicken JS (1987) Evolution, Information and Thermodynamics: Extending the Darwinian program. Oxford University Press, New York.
45. Kauffman SA (1993) The Origins of Order: Self-Organization and Selection in Evolution. Oxford University Press, New York.
46. Kauffman SA (2000) Investigations. Oxford University Press, New York.
47. Kauffman SA (2004), Autonomous agents. In: Barrow JD, Davies, PCW, Harper Jr. CL (eds) Science and Ultimate Reality: Quantum Theory, Cosmology, and Complexity, pp. 654–666. Cambridge University Press, Cambridge.
48. Weber BH, Depew DJ (2001) Developmental systems, Darwinian evolution, and the unity of science. In: Oyama S, Griffiths PE, Gray RD (eds) Cycles of Contingency: Developmental Systems and Evolution. MIT Press, Cambridge, MA.
49. Dawkins R (1976) The Selfish Gene. Oxford University Press, Oxford.
50. Dawkins R (1989) The Selfish Gene. Oxford University Press, Oxford.
51. Oyama S, Griffiths PE, Gray RD (eds) (2001) Cycles of Contingency: Developmental Systems and Evolution. MIT Press, Cambridge, MA.
52. Weber. BH, Depew DJ (2003) Evolution and Learning: The Baldwin Effect Reconsidered. MIT Press, Cambridge, MA.
53. Griffiths PE (2007) The Phenomena of Homology. Biol Phil 22: 643–658.
54. Jablonka E, Lamb MJ (1995) Epigenetic Inheritance and Evolution: The Lamarckian Dimension. Oxford University Press, Oxford.
55. Jablonka E, Lamb MJ (1998) Bridges between development and evolution. Biol Phil 13: 119–145.
56. Jablonka E, Lamb MJ (2005), Evolution in Four Dimensions: Genetic, Epigenetic, Behavioral ad Symbolic Variation in the History of Life. MIT Press, Cambridge, MA.

57. Jablonka E, Lamb MJ (2007), The expanded evolutionary synthesis — a reply to Godfrey-Smith, Haig, and West-Eberhard. Biol Phil 22: 453–472.
58. Salthe SN (1993) Development in Evolution: Complexity and Change in Biology. MIT Press, Cambridge, MA.
59. Salthe SN (1998) The role of natural selection theory in understanding evolutionary systems. In: Van De Vijver G, Sathe SN, Delpos M (eds) Evolutionary Systems: Biological and Epistemological Perspectives on Selection and Self-Organization, pp. 13–20. Kluwer, Dordrecht.
60. Salthe SN (1999) Energy, Development and Semiosis. In: Taborsky E (ed.) Semiosis, Evolution, Energy: Towards a Reconceptualization of the Sign, pp. 245–261. Saker Verlag, Aachen.
61. Reid R (2007) Biological Emergences: Evolution by Natural Experiment. MIT Press, Cambridge, MA.
62. Kauffman SA (2007) Beyond reductionism: Reinventing the sacred. Zygon 42: 903–914.
63. Kauffman SA, Clayton P (2006) On emergence, agency, and organization. Biol Phil 21: 501- 521.
64. Kauffman SA, Logan RK, Este R, Goebel R, Hobill D, Shmulevich I (2008) Propagating organization: an enquiry. Biol Phil 23: 27–45.
65. Clayton, P. (2004) Mind and Emergence: From Quantum to Consciousness. Oxford University Press, Oxford.
66. Clayton, P. and P. Davies (2006) The Re-emergence of Emergence: The Emergentist Hypothesis from Science to Religion. Oxford University Press, New York.
67. Deacon TW (2003) The hierarchic logic of emergence: Untangling the interdependence of evolution and self-organization. In: Weber BH, Depew DJ (eds) Evolution and Learning: The Baldwin Effect Reconsidered, pp. 273–308. MIT Press, Cambridge, MA.
68. Deacon TW (2006) Emergence: The hole at the wheel's hub. In: Clayton P, Davies P (eds) The Re-emergence of Emergence: The Emergentist Hypothesis from Science to Religion, pp. 111–150. Oxford University Press, New York.
69. Deacon TW (2006) Reciprocal linkage between self-organizing processes is sufficient for self-reproduction and evolvability. Biological Theory 1: 136–149.
70. Sherman J, Deacon TW (2007) Teleology for the perplexed: How matter began to matter. Zygon 42: 873–901.
71. Weber BH, Deacon TW (2000) Thermodynamic cycles, developmental systems, and emergence. Cybernetics and Human Knowing 7: 21–43.
72. Johnson S (2001) Emergence: The Connected Lives of Ants, Brains, Cities, and Software. Scribner, New York.
73. Salthe SN (2004) The spontaneous origin of new levels in a scalar hierarchy. Entropy 6: 327–343.

74. Schneider ED, Sagan D (2005) Into the Cool: Energy Flow Thermodynamics and Life. University of Chicago Press, Chicago.
75. Corning P (2003) Nature's Magic: Synergy in Evolution and the Fate of Humankind. Cambridge University Press, Cambridge.
76. Corning P (2005) Holistic Darwinism: Synergy, Cybernctics, and the Bioeconomics of Evolution. University of Chicago Press, Chicago.
77. Morowitz H, Smith E (2007) Energy flow and the organization of life. Complexity 13: 51–59.
78. Casti JL (1994) Complexification: Explaining a Paradoxical World through the Science of Surprise. HarperCollins, New York.
79. Crofts AR (2007) Life, information, entropy, and time. Complexity 13: 14–50.
80. Goodwin, BC (1994) How the Leopard Changed Its Spots: The Evolution of Complexity. London: Weidenfeld and Nicolson.
81. Juarrero, A. (1999), Dynamics in Action: Intentional Behavior as a Complex System. MIT Press, Cambridge, MA.
82. Morowitz, H. (2002), The Emergence of Everything: How the World Became Complex. Oxford University Press, New York.
83. Schneider, ED, Sagan, D (2005), Into the Cool: Energy Flow Thermodynamics and Life. University of Chicago Press, Chicago.
84. Silberstein M (2002) Reduction, emergence and explanation. In: Machamer P, Silberstein M (eds). The Blackwell Guide to the Philosophy of Science, pp. 80–107. Blackwell, Malden, MA.
85. Solé R, Goodwin BC (2000) Signs of Life: How Complexity Pervades Biology. Basic Books, New York.
86. Taylor MC (2003) The Moment of Complexity: Emerging Network Cultures. University of Chicago Press, Chicago.
87. Ulanowicz RE (2002) Ecology, a dialog between the quick and the dead. Emergence 4: 34–52.
88. Wheeler TJ (2007) Analysis, modeling, emergence & integration in complex systems: a modeling and integration framework & system biology. Complexity 13: 60–75.
89. Camazine S, Deneubourg J-L, Franks NR, Sneyd J, Theraulaz G, Bonabeau E (2001) Self-Organization in Biological Systems. Princeton University Press, Princeton NJ.
90. Milsteli T (2001) The concept of self-organization in cellular architecture. Journal of Cell Biology 155: 181–185.
91. Swenson R (1989) Emergent attractors and the law of maximum entropy production: Foundations to a theory of general evolution. Systems Research 6: 187–197.
92. Swenson, R. (1998) Spontaneous order, evolution, and autocatakinetics: The nomoloical basis for the emergence of meaning. In: Van De Vijver G, Sathe SN, Delpos M (eds) Evolutionary Systems: Biological and Epistemological Perspectives on Selection and Self-Organization, pp. 155–180. Kluwer, Dordrecht.

93. Darwin CR ([1859] 1964) On the Origin of Species by Means of Natural Selection, or the Preservation of Favoured Races in the Struggle for Life. Facsimile of the English first edition published by John Murray; with introduction by Ernst Mayr, Harvard University Press, Cambridge, MA.
94. Desmond A (1989) The Politics of Evolution. University of Chicago Press, Chicago.
95. Fry I. (2000) The Emergence of Life on Earth: A Historical and Scientific Overview. Rutgers University Press, New Brunswick, NJ.
96. Strick JE (2000) Sparks of Life: Darwinism and the Victorian Debates over Spontaneous Generation. Harvard University Press, Cambridge, MA.
97. Spencer H (1864), The Principles of Biology. Williams and Norgate, London.
98. Royce J (1904) Herbert Spencer: An Estimate and Review. Collier, New York.
99. Miller KR (1999) Finding Darwin's God: A Scientist's Search for Common Ground between God and Evolution. Harper Collins, New York.
100. Scott EC (2004) Evolution vs. Creationism: An Introduction. Greenwood, Westport CT.
101. Haldane JBS ([1929] 1967) The origin of life. Rationalist Animal. Reprinted in: The Origin of Life, Bernal JD (ed), pp. 242–249. World, Cleveland OH.
102. Oparin AI (1924) Proiskhozhdenie zhizy. Moskovski Rahochii, Moscow. (English translation reprinted as: The origin of life, in Bernal JD (ed) The Origin of Life, pp. 199–241. World, Cleveland, OH.
103. Oparin AI (1938) The Origin of Life. Macmillan, London.
104. Bernal JD (1951) The Physical Basis of Life. Routledge and Kegan Paul, London.
105. Bernal JD (ed) (1967) The Origin of Life. World, Cleveland, OH.
106. Behe MJ (1996) Darwin's Black Box: The Biochemical Challenge to Evolution. Free Press, New York.
107. Behe MJ (2007) The Edge of Evolution: The Search for the Limits of Darwinism. Free Press, New York.
108. Meyer SC (2003) DNA and the origin of life, information, specification, and explanation. In: Campbell JH and S.C. Meyer SC (eds) Darwinism, Design, and Public Education, pp. 223–285. Michigan State University Press, Lansing, MI.
109. Meyer SC (2009) Signature in the Cell: DNA and the Evidence for Intelligent Design. HarperOne, San Francisco.
110. Bradley WL (2004) Information, entropy, and the origin of life. In: Dembski WA, Ruse M (eds) Debating Design: From Darwin to DNA, pp. 331–351. Cambridge University Press, Cambridge.
111. Rosenberg A (2006) Darwinian Reductionism: Or, How to Stop Worrying and Love Molecular Biology. University of Chicago Press, Chicago.
112. Weber BH (2007) Back to basics. Nature 445: 601.
113. Weber BH (2007) Emergence of life. Zygon 42: 837–856.
114. Williams RJP, Fraústo Da Silva JJR (2003) Evolution was chemically constrained. J Theor Biol 220: 323–343.

115. Williams RJP, Fraústo Da Silva JJR (2006) The Chemistry of Evolution. Elsevier, Amsterdam.
116. Lurquin PF (2003) The Origins of Life and the Universe. Columbia University Press, New York.
117. Luisi PL (2006) The Emergence of Life: From Chemical Origins to Synthetic Biology. Cambridge University Press, Cambridge.
118. Weber BH (2009) On the emergence of living systems. Biosemiotics 2: 343–359.
119. Deamer DW (2011) First Life: Discovering the Connections between Stars, Cells, and How Life Began. University of California Press, Berkeley.
120. Yamagata Y, Inomata K (1997) Condensation of glycylclycine to oligoglycines with trimetaphosphate in aqueous solution II: Catalytic effect of magnesium ion. Origins of Life and Evolution of the Biosphere 27:339–344.
121. Benner SA, Ricardo A, Carrigan MA (2004) Is there a common chemical model for life in the universe? Curr Opin Chem Biol 8: 672–689.
122. Shapiro R (2007) A simpler origin of life. Scientific American 296(6): 46–53.
123. Miller SL (1953) A production of amino acids under possible primitive earth conditions. Science 117: 528–529.
124. Bada JL (2004) How life began on earth: A status report. Earth and Planetary Science Letters 226: 1–15.
125. Ehrenfreund P, Irvine W, Becker L (2002) Astrophysical and astrochemical insights into the origin of life. Reports of Progress in Physics 65: 1427–1487.
126. Lahav N (1999) Biogenesis: Theories of Life's Origin. Oxford University Press, New York.
127. Wills C, Bada J (2000) The Spark of Life: Darwin and the Primeval Soup. Perseus, Cambridge, MA.
128. Bada JL, Lazcano A (2003) Prebiotic soup — revisiting the Miller experiment. Science 300: 745–746.
129. Leman L, Orgel L, Ghadiri MR (2004) Carbonylsulfide-mediated prebiotic formation of peptides. Science 306: 283–286.
130. Kondepudi D (1988) Parity violation and the origin of biomolecular chirality. In: Weber BH, Depew DJ, Smith JD (eds) Entropy, Information, and Evolution: New Perspectives on Physical and Biological Evolution, pp. 41–50. MIT Press, Cambridge, MA.
131. Kumar JK, Oliver JS (2002) Proximity effects in monolayer films: Kinetic analysis of amide bond formation in the air-water interface using proton NMR spectroscopy. J Am Chem Soc 124: 11307–11314.
132. Matthews CN, Moser RE (1967) Peptide synthesis from hydrogen cyanide and water. Nature 215: 1230–1234.
133. Liebman SA, Pesce-Rodriguez RA, Matthews CN (1995) Organic analysis of hydrogen cyanide polymers: Prebiotic and extratrerrestrial chemistry. Advances in Space Research 15(3): 71–80.

134. Minard RD, Hatcher PG, Gourley RC, Matthews CN (1998) Structural investigations of hydrogen cyanide polymers: New insights using TMAH Thermochemolysis/ GC-MS. Origin of Life and Evolution of the Biosphere 28: 461–473.
135. Fox SW (1988) The Emergence of Life. Academic Press, New York.
136. Wächterhäuser G (1988) Before enzymes and templates: Theory of surface metabolism. Microbiological Reviews 52: 452–484.
137. Wächterhäuser G (1992) Groundworks for an evolutionary biochemistry: The iron-sulfur world. Progress in Biophysics and Moleculr Biology 58: 85–201.
138. Wächterhäuser G (1997) The origin of life and its methodological challenge. J Theor Biol 187: 483–494.
139. Morowitz H, Deamer DW, Smith T (1991) Biogenesis as an evolutionary process. J Mol Evol 33: 207–208.
140. Morowitz H (1992) Beginnings of Cellular Life: Metabolism Recapitulates Biogenesis. Yale University Press, New Haven.
141. Shakhnovich EI, Gutin AM (1990) Implications of Thermodynamics of Protein Folding for Evolution of Primary Sequences. Nature 346: 773–775.
142. Bowie JU (1990) Deciphering the message in protein sequences: tolerance of amino acid substitutions. Science 247: 1306–1310.
143. Hazen RM, Griffin PL, Carothers JM, Szostak JW (2007) Functional information and the emergence of biocomplexity. Proc Natl Acad Sci USA 104: 8574–8531.
144. Axe DD (2010) The case against a Darwinian origin of protein folds. BIO-Complexity 2010(1):1–12. **doi:**10.5048/BIO-C.2010.1
145. Fox RF (1997) The origins of life: What one needs to know. Zygon 32: 393–406.
146. Meyer SC, Nelson PA (2011) Can the origin of the genetic code be explained by direct RNA templating? BIO-Complexity 2011(2): 1–10.
147. Weber BH (2010) What is Life? Defining life in the context of emergent complexity. Origins of Life and Evolution of Biospheres 40: 221–229.
148. Yockey HP (2005) Information Theory, Evolution, and the Origin of Life. Cambridge University Press, Cambridge.
149. Joyce GF, Orgel, LE (1993), Prospects for understanding the origin of the RNA world. In: R.F. Gesteland RF, Atkins JF (eds) The RNA World, pp. 1–25. Cold Spring Harbor Press, Cold Spring Harbor MA.
150. Schwartz AW (1997) Speculation on the RNA precursor problem. J Mol Biol 187: 523–527.
151. Deamer DW, Pashley RM (1989) Amphiphilic components of the Murchison carbonaceous chondrite: Surface properties and membrane formation. Origin of Life and Evolution of the Biosphere 19: 21–38.
152. Deamer DW, Dworkin JP, Sandford SA, Bernstein MP, Allamandola LJ (2002), The first cell membranes. Astrobiology 2: 371–381.

153. Ourisson G, Nakatani Y (1994) The terpenoid theory of the origin of cellular life: The evolution of terpenoids to cholesterol. Curr Biol 1: 11–23.
154. Bachmann PA, Luisi PL, Lang J (1992) Autocatalytic self-replication micelles as modes for prebiotic structures. Science 357: 57–58.
155. Mitchell PD (1961) Coupling of phosphorylation to electron and hydrogen transfer by a chemiosmotic type of mechanism. Nature 191: 144–148.
156. Williams RJP (1961) Possible functions of chains of catalysts. J Theoret Biol 1: 1–17.
157. Weber BH, Prebble JN (2006) An issue of originality and priority: The correspondence and theories of oxidative phosphorylation of Peter Mitchell and Robert J. P. Williams, 1961–1980. Journal of the History of Biology 39: 125–163.
158. Prebble J, Weber BH (2003) Wandering in the Gardens of the Mind: Peter Mitchell and the Making of Glynn. Oxford University Press, New York.
159. Lane N (2006) Power, Sex, Suicide: Mitochondria and the Meaning of Life. Oxford University Press, Oxford.
160. Deamer DW (1992) Polycyclic aromatic hydrocarbons: Primitive pigment systems in the prebiotic environment. Advances in Space Research 12: 1–4.
161. Martin W, Russell MT (2003) On the origins of cells: A hypothesis for the evolutionary transitions from abiotic geochemistry to chemoautotrophic prokaryotes, and from prokaryotes to nucleated cells. Phil Trans Royal Society of London (Series B) 358: 59–85.
162. Carter CW, Kraut J (1974) A proposed model for interaction of polypeptides with RNA. Proc Natl Acad Sci USA 71: 283–287.
163. Weber BH (2000) Closure in the Emergence and Evolution of Life. NY Acad Sci 901: 132–138.
164. Weber BH (2003a), Emergence of mind and the Baldwin effect. In: B.H. Weber BH, Depew DJ (eds) Evolution and Learning: The Baldwin Effect Reconsidered, pp. 309–326.
165. Weber BH (2003b) Biochemical complexity: Emergence of design? In: J.A. Campbell JA, Meyer SC (eds) Darwinism, Design, and Public Education, pp. 455–462. Michigan State University Press, East Lansing, MI.
166. Weber BH, Depew DJ (2004) Darwinism, Design, and Complex Systems Dynamics. In: Dembski WA, Ruse M (eds) The Appearance of Design in Nature, pp. 173–190. Cambridge University Press, Cambridge UK.
167. Weber BH (2011) Design and its discontents. Synthese 178 (2): 271–289.
168. Miller KB (2003) Common descent, transitional forms, and the fossil record. In: Miller KB (ed) Perspectives on an Evolving Creation, pp. 152–181. Eerdmans, Grand Rapids MI.
169. Gordon BL, Dembski WA (eds) (2011) The Nature of Nature: Examining the Role of Naturalism in Science. ISI Books, Wilmington, DE.

Index